MARE Publication Series

Volume 25

Series Editors
Maarten Bavinck, University of Amsterdam, Amsterdam, Noord-Holland,
The Netherlands
Svein Jentoft, Norwegian College of Fishery Science, UiT-The Arctic University
of Norway, Tromsø, Norway

The MARE Publication Series is an initiative of the Centre for Maritime Research (MARE). MARE is an interdisciplinary social-science network devoted to studying the use and management of marine resources. It is based jointly at the University of Amsterdam and Wageningen University (www.marecentre.nl). The MARE Publication Series addresses topics of contemporary relevance in the wide field of 'people and the sea'. It has a global scope and includes contributions from a wide range of social science dis-ciplines as well as from applied sciences. Topics range from fisheries, to integrated management, coastal tourism, and environmental conservation. The series was previously hosted by Amsterdam University Press and joined Springer in 2011. The MARE Publication Series is complemented by the Journal of Maritime Studies (MAST) and the biennial People and the Sea Conferences in Amsterdam.

Editors:
J. Maarten Bavinck, University of Amsterdam, The Netherlands
j.m.bavinck@uva.nl
Editors Svein Jentoft, UiT - The Arctic University of Norway, Norway
svein.jentoft@uit.no

Stefan Partelow • Maria Hadjimichael
Anna-Katharina Hornidge
Editors

Ocean Governance

Knowledge Systems, Policy Foundations
and Thematic Analyses

 Springer

Editors
Stefan Partelow
Leibniz Center for Tropical Marine
Research (ZMT)
Bremen, Germany

Center for Life Ethics
University of Bonn
Bonn, Germany

Anna-Katharina Hornidge
German Institute of Development
and Sustainability (IDOS) &
University of Bonn
Bonn, Nordrhein-Westfalen, Germany

Maria Hadjimichael
Cyprus Marine and Maritime Institute
Larnaca, Cyprus

ISSN 2212-6260   ISSN 2212-6279   (electronic)
MARE Publication Series
ISBN 978-3-031-20739-6   ISBN 978-3-031-20740-2   (eBook)
https://doi.org/10.1007/978-3-031-20740-2

This work was supported by Germany Institute of Development and Sustainability (IDOS)

Cover illustration: Tadeu Jnr on Unsplash.

This Springer imprint is published by the registered company Springer Nature Switzerland AG
The registered company address is: Gewerbestrasse 11, 6330 Cham, Switzerland

Foreword

May 4, 2022, Vancouver, Shores of the Salish Sea, Planet Ocean

Yesterday, I went for a walk and had sushi for dinner on the shores of the Salish Sea. As I do most days living in this place, I pondered both the wonders and the complexity of the ocean. From my vantage point in the coastal city of Vancouver, I could see clearly how human society interfaces with the ocean – people seeking solace on the shore, a jumble of port infrastructure, fishing boats heading out to and back from sea, barges and ships moored in the bay.

And, underlying it all, but invisible to the human eye, was a patchwork of governance institutions, processes, and decisions that structures what activities can happen and where, whether and how the ocean is managed sustainably, and who has access to space and resources.

Yet, it has only been a drop in the bucket of human time since the seas were viewed as a common resource that was free for all. As human activities in and pressures on the oceans have increased, so have efforts to sustainably govern the oceans. Thus, in the few short decades since the United Nations Convention on the Law of the Seas has come into force, we have seen exponential growth in the layering and complexity of different governance arrangements and actions within the oceans.

This timely book makes a critical and constructive contribution to our knowledge of and the scholarship on ocean governance by examining the past, present, and future. In particular, the book details the historical developments that have led to current issues across a variety of problem and policy contexts – including chapters touching on governance topics related to fisheries, aquaculture, food systems, shipping, marine plastics, seabed mining, and the blue economy. The authors of the chapters include recognized experts from around the world, who apply interdisciplinary perspectives to analyze issues at various scales from local to national to global. But, it is more than a collection of chapters – the editors bookend the volume with an introduction to the field and conclude with a summary of insights and lessons learned that are pertinent for the pursuit of sustainable ocean governance.

The volume is destined to become a critical resource book for students, senior scholars, and practitioners who have an interest in oceans, sustainability, and/or governance. A broad readership will enjoy and benefit from this book. It will be best enjoyed while sitting near or pondering the ocean that we all depend on.

Principal, The Peopled Seas Initiative &   Nathan J. Bennett, PhD
Chair, People and the Ocean Specialist Group,
International Union for the Conservation of Nature
Gland, Switzerland

Preface

Ensuring that ocean governance approaches work constructively towards achieving transformative change towards sustainable outcomes across differentiated contexts is entering a phase of critical societal urgency. Our oceans are being rapidly developed, faster and at a broader scale than any other time in history. Rare minerals are being mined, new shipping routes are being established, energy installations are being built, ports are expanding, and water is being desalinized but also re-entering with sedimentation and pollution from human use. All the while, hundreds of millions of mostly politically silent and largely unseen small-scale livelihoods remain dependent on healthy oceans for fishing, tourism and aquaculture, where the need for inclusive conservation approaches to resolve the 'paper parks' and non-compliance problems is paramount. At the same time, waves of coastal urban migration continue to grow, putting pressure on local coastal ecosystems for food, recreation and infrastructure needs, while also increasing demand for rare metals and minerals mined in the deep sea for electronics and goods shipped worldwide across the ocean surface. Coastal areas, while being steadily built up, are also threatened by rising seas and increased storm intensity and frequency from climate change. This is coincided by major political and business organizations such as the OECD, World Bank, World Economic Forum and the United Nations shaping a Blue Growth development agenda with Blue Economy strategies largely proclaiming the ocean as the next frontier of development to achieve human prosperity.

While scholars and practitioners are largely aware of the problems facing our oceans, we are arguably now at a critical turning point in recognizing that process, plurality, participation and social-ecological differentiation are key ingredients for achieving any claims to prosperity or sustainability in our ocean governance and politics. This is not to ignore the challenges in reversing the trend that pushes for further intensification of ocean uses within the current political and economic environment. It is nevertheless important that we as a society can learn to cook with these ingredients in order to influence the politics and facilitate in the (re)creation of ocean institutions where needed. There is a need to be both honest with the current state and hopeful with current efforts to track pragmatic paths forward. Nonetheless, among the tides of often disheartening news and tragic events, there are reasons for growing optimism. The twenty-first century will enable us to 'see the ocean' like we have never seen it before, both the physical activities and features that happen on and below the surface enabled by technology, but also in the calls and movements to foster transparency and justice in governing institutions. Inspiration can find many paths, and there is no shortage of catalyzing individuals leading vocal movements for positive social, cultural, economic and political change seen around the world. However, seeing the ocean will only lead to ocean transformations to sustainability if persistent actions of engagement and empowerment are actively pursued. Conscious efforts are needed at all levels of our ocean societies and politics, alongside societal shifts in norms and behaviors among consumers, users and voters.

The manifestation of this book has been an effort to mobilize sets of existing knowledge to foster continued ocean engagement, scholarship and stewardship. We have brought together a diverse range of ocean governance scholars to engage in discussions and analysis of the current topics and critical perspectives facing our oceans today. Importantly, this includes the role prior events, institutions and governing activities have played in shaping our current issues and future ocean trajectories. The book hopes to inform and inspire students and early career scholars to emerge and continue engaging in the research, policy and practice needed to enable sustainable ocean-based development. If we take the tagline of the United Nations Decade of Ocean Science for Sustainable Development seriously – 'The Science We Need for the Ocean We Want' – it embodies a call for both continued engagements into a diversity of sciences that help us know the ocean, while also recognizing that what we want the ocean to be is a choice, a normative one that raises issues of how those choices are made and who gets to choose. Imagining the ocean we waht guides and facilitates these discussions.

The ocean offers opportunities to reconcile persistent political challenges within a new global context. On the one hand, for example, the 2022 Intergovernmental Panel on Climate Change (IPCC) report has never been more clear about the role of humans in global environmental change as well as the impacts on and role of our ocean's in mitigation and adaptation. On the other hand, the uptake and use of this knowledge for sustainability transformation remains contested and divisive. How knowledge is (co-)produced, communicated and utilized to spark action is part of the ocean governance puzzle that requires sustained attention. On top of this, cooperation, coordination and deliberative processes will be needed to resolve the collective action problems in both resource use and institutional development and change.

Bremen, Germany   Stefan Partelow
Larnaca, Cyprus   Maria Hadjimichael
Bonn, Nordrhein-Westfalen, Germany   Anna-Katharina Hornidge

Acknowledgments

The editors would like to thank and acknowledge the support from the European Cooperation in Science and Technology (COST) organization for funding the COST Action 'Ocean Governance for Sustainability – Challenges, options and the role of science, CA15217', where the content and network of contributing researchers for this book originated. The COST Action would not have been possible without the substantial support from numerous individuals, institutes and universities throughout Europe. The process of preparing this edited volume has been hindered by various incidents, with the COVID-19 pandemic being an important one delaying the finalization of the book. We are grateful to all authors for patiently working with us to see this work completed. We would like to particularly extend gratitude to the Leibniz Centre for Tropical Marine Research (ZMT) in Bremen, Germany, for initially hosting the COST Action with institutional support. Similarly, we would like to substantially thank the German Institute of Development and Sustainability (IDOS)/Deutsches Institut für Entwicklung und Nachhaltigkeit, Bonn, Germany (formerly known as German Development Institute/Deutsches Institut für Entwicklungspolitik (DIE)), for hosting the COST Action in the latter stages and for making the open access publication possible with financial support from the Federal Ministry for Economic Cooperation and Development (BMZ) and the state of Northrhine-Westphalia (NRW). Without the support from Frauke Domgoergen, managing and coordinating the development of the book would have been a far greater struggle, and we are grateful for her efforts.

Contents

About the Editors

Stefan Partelow is a researcher at the Leibniz Centre for Tropical Marine Research (ZMT) in Germany, where he jointly completed his Ph.D. in Political Science (2018) with Jacobs University. Starting in 2023, he will be a Senior Research Fellow at the Center for Life Ethics at the University of Bonn, Germany. His research focuses on the governance of commons and the environment, including institutional development and change, and social-ecological systems analysis. His work advances interdisciplinary and transformative change research towards sustainability at the science-society interface, where he has focused extensively on coastal systems such as fisheries, aquaculture and tourism. He is widely published in leading international journals, with conceptual, review and empirical research across numerous levels and scales.

Maria Hadjimichael is a Senior Associate Scientist at the Cyprus Marine and Maritime Institute. She conducts research on the fields of political ecology, environmental politics and governance of the Commons, with a focus on the sea and the coastline, how is the understanding of the sea and the coastal space as a 'Common' or 'Common Heritage' affected by international agreements or national law, and how are such institutional arrangements being instrumentalized to expand the privatization of the marine and coastal space. She has published widely in this area, including in the journals *Sustainability Science*, *Marine Policy*, *Ocean and Coastal Management* and *Political Geography*. She is co- editor of the *Island Studies Journal*.

Anna-Katharina Hornidge is director of the German Institute of Development and Sustainability (IDOS)/Deutsches Institut für Entwicklung und Nachhaltigkeit, Bonn, Germany (formerly known as German Development Institute / Deutsches Institut für Entwicklungspolitik (DIE)), and Professor of Global Sustainable Development at the University of Bonn. In her research, Ms. Hornidge works on knowledges and innovation development for development, as well as questions of natural resource governance in agriculture and fisheries in Asia and Africa. Ms. Hornidge serves as expert advisor at national, EU and UN levels: as a member of the German Advisory Council on Global Change of the German Government (WBGU), co-chair (with Gesine Schwan) of SDSN Germany and as part of the executive council of the German UNESCO-Commission.

Contributors

Pinar Ertör-Akyazi Institute of Environmental Sciences, Boğaziçi University, Istanbul, Turkey

Diva Amon SpeSeas, D'Abadie, Trinidad and Tobago & Marine Science Institute, University of California, Santa Barbara, CA, USA

Fernanda C. B. Araujo Universidade de Brasilia, Brasilia, Brazil

Milena Arias-Schreiber University of Gothenburg, Gothenburg, Sweden

Maria Baker Ocean and Earth Sciences, University of Southampton, Southampton, UK

María José Barragán-Paladines Charles Research Station, Charles Darwin Foundation, Santa Cruz, Galapagos, Ecuador

Nina Bednaršek Marine Biology Station Piran, National Institute of Biology, Ljubljana, Slovenia
Cooperative Institute for Marine Resources Studies, Oregon State University, Hatfield, Oregon

Helena Calado UAc/FCT/MARE - University of the Azores/Faculty of Science and Technology and Marine Environmental Science, Ponta Delgada, Portugal

Donata Melaku Canu Istituto Nazionale di Oceanografia e di Geofisica Sperimentale, Sgonico, Italy

Wenting Chen Norwegian Institute for Water Research, Oslo, Norway

John Childs Lancaster Environment Centre, Lancaster University, Lancaster, UK

Ratana Chuenpagdee Geography Department, Memorial University of Newfoundland, St. John's, NL, Canada

Marta Conde Centre for Social Responsibility in Mining (CSRM), University of Queensland, Brisbane, Australia
ICTA, Autonomous University of Barcelona, Barcelona, Spain

Agnese Cretella Trinity Centre for Environmental Humanities, Trinity College Dublin, Dublin, Ireland
Department of Philosophy and Communication, University of Bologna, Bologna, Italy

Daniel Depellegrin Landscape Analysis and Management Laboratory, Department of Geography, University of Girona, Girona, Spain

Winny Collot d'Escury Independent Researcher, Leiden, The Netherland

Irmak Ertör The Ataturk Institute for Modern Turkish History, Boğaziçi University, Istanbul, Turkey

Richard A. Feely NOAA Pacific Marine Environmental Laboratory, Seattle, WA, USA

Carmel Finley Oregon State University, Corvallis, OR, USA

Wesley Flannery School of Natural and Built Environment, Queen's University Belfast, Belfast, UK

Charles Galdies Institute of Earth Systems, University of Malta, Msida, Malta

Blaženka Gašparović Division for Marine and Environmental Research, Ruđer Bošković Institute, Zagreb, Croatia

Jelena Godrijan Division for Marine and Environmental Research, Ruđer Bošković Institute, Zagreb, Croatia

Sabine Gollner Royal Netherlands Institute for Sea Research (NIOZ), Texel, The Netherlands

Henriette Grimmel Independent Researcher, Zurich, Switzerland

Roberta Guerra Department of Physics and Astronomy, University of Bologna, Bologna, Italy
Centro Interdipartimentale di Ricerca per le Scienze Ambientali (CIRSA-UNIBO), University of Bologna, Bologna, Italy

Bleuenn Guilloux European Institute for Marine Studies, Laboratory for Law and Economics of the Sea, Plouzane, Brittany, France

Maria Hadjimichael Cyprus Marine and Maritime Institute, Larnaca, Cyprus

Johannes Herbeck Sustainability Research Center (artec), University of Bremen, Bremen, Germany

Anna-Katharina Hornidge German Institute of Development and Sustainability (IDOS) & University of Bonn, Bonn, Nordrhein-Westfalen, Germany

Kerstin Knopf University of Bremen, Bremen, Germany

Paul Lawlor School of Architecture, Building & Environment, Technological University Dublin, Dublin, Ireland

Kristin Magnussen Menon Economics, Oslo, Norway

Alenka Malej Marine Biology Station Piran, National Institute of Biology, Piran, Slovenia

Aletta Mondre Institute of Political Science, Kiel University, Kiel, Germany

Fabiana Moniz FGF/UAc/FCT, Fundação Gaspar Frutuoso, University of the Azores/Faculty of Sciences and Technology, Ponta Delgada, Portugal

Md. Mostafa Monwar Institute of Marine Sciences, University of Chittagong, Chittagong, Bangladesh
Australian National Centre for Ocean Resources and Security (ANCORS), University of Wollongong, Wollongong, Australia

Julia Nakamura Law School, University of Strathclyde, Glasgow, UK

Ståle Navrud School of Economics and Business, Norwegian University of Life Scienes (NMBU), Ås, Norway

Eva A. Papaioannou Independent Researcher, Athens, Greece
Present Affiliation: GEOMAR – Helmholtz Centre for Ocean Research Kiel, Kiel, Germany

Stefan Partelow Leibniz Centre for Tropical Marine Research (ZMT), Bremen, Germany
Center for Life Ethics, University of Bonn, Bonn, Germany

Greg Pelletier Washington State Department of Ecology, United States (Independent Researcher), Bellingham, WA, USA

Jerneja Penca Euro-Mediterranean University, Piran, Slovenia

Kimberley Peters Helmholtz Institute for Functional Marine Biodiversity, Oldenburg, Germany

Irina Rafliana University of Bonn and the German Development Institute (DIE), Bonn, Germany

Alicia Said Department of Fisheries and Aquaculture, Luqa, Malta

Pekka Salmi Natural Resources Institute Finland, Turku, Finland

Cordula Scherer Trinity Centre for Environmental Humanities, Trinity College Dublin, Dublin, Ireland

Achim Schlüter Leibniz Centre for Tropical Marine Research, Bremen, Germany
Jacobs University, Bremen, Germany

Michael Schoon School of Sustainability, Arizona State University, Tempe, AZ, USA

Hendricus A. Simarmata Universitas Indonesia (UI), Indonesian Association of Urban and Regional Planners (IAP), Jakarta, Indonesia

Simona Simoncelli Istituto Nazionale di Geofisica e Vulcanologia, Sezione di Bologna, Bologna, Italy

Pradeep A. Singh Institute for Advanced Sustainability Studies (IASS), Potsdam, Germany
Research Centre for European Environmental Law (FEU), University of Bremen, Bremen, Germany

Rapti Siriwardane-de Zoysa Leibniz Center for Tropical Marine Research, Bremen, Germany

Cosimo Solidoro National Institute of Oceanography and Applied Geophysics (OGS), Trieste, Italy
International Centre for Theoretical Physic (ICTP), Trieste, Italy

Roger Spranz Making Ocean Plastic Free e.V., Bali, Indonesia
Making Ocean Plastic Free e.V., Freiburg, Germany

Philip Steinberg Department of Geography, Durham University, Durham, UK

Kristina Svels Natural Resources Institute Finland, Turku, Finland

Valentina Turk Marine Biology Station Piran, National Institute of Biology, Piran, Slovenia

Jan P. M. van Tatenhove Centre for Blue Governance, Aalborg University, Aalborg, Denmark
Van Hall Larenstein University of Applied Sciences, Leeuwarden, The Netherlands

Marta Vergílio Trisolaris Advanced Technologies, Lda., Ponta Delgada, Portugal

Klaas Willaert Faculty of Law and Criminology, Maritime Institute, Ghent University, Ghent, Belgium

Serena Zunino Istituto Nazionale di Oceanografia e di Geofisica Sperimentale, Sgonico, Italy

Chapter 1
Ocean Governance for Sustainability Transformation

Stefan Partelow, Maria Hadjimichael, and Anna-Katharina Hornidge

Abstract This introductory chapter focuses on selected key events, features and policies of ocean governance that have had, or are likely to be needed in transforming how and why we govern the ocean sustainably. In doing so we outline examples of prominent historical events, important thematic areas of global development, policy instruments and the principles of governance processes that can transform the way society engages with the ocean. However, we acknowledge that such an overview cannot fully capture all issues, particularly how each is differentiated at regional and local levels. Accordingly, we introduce globally relevant issues and general principles, which will require further inquiry to fully unpack at the relevant levels and scales for engaged students, researchers, policy-makers and practitioners. Thus, we provide an overview of these topics from a multi- and inter-disciplinary perspective, supported by up-to-date literature. This is followed by a brief explanation of how the chapters in the book are organized into three parts, and how each chapter contributes to the book's content, including a final chapter that outlines the takeaway points for students, researchers and policy-makers in pursuing ocean governance for sustainability transformation.

S. Partelow (✉)
Leibniz Centre for Tropical Marine Research (ZMT), Bremen, Germany

Center for Life Ethics, University of Bonn, Bonn, Germany
e-mail: stefan.partelow@leibniz-zmt.de; sbpartelow@gmail.com

M. Hadjimichael
Cyprus Marine and Maritime Institute, Larnaca, Cyprus

A.-K. Hornidge
German Institute of Development and Sustainability (IDOS) & University of Bonn, Bonn, Nordrhein-Westfalen, Germany

1.1   Focal Areas, Policies and Processes for Sustainable Ocean Governance

Human relationships with our oceans date back millennia. They have shaped the rise of civilizations, provided food and story, and seeded a diversity of coastal cultures and engagement practices around the world. However, they have also been a source of conflict, oppression and turmoil. Human-ocean stories are not new, but the magnitude of changes now incurred from these relationships are. Historical human interactions were once limited to near shore areas, however, technological advances now enable remote access and previously unimaginable exploitation opportunities for minerals, energy, shipping, food and political power (Jouffray et al. 2020). Looking back on our human-ocean past, we can see a plurality of governance narratives that have emerged, yet most remain relevant in the ocean governance debates of today. Some societies approached stewardship and use as synonymous activities, forming an embedded cultural ethic and respect for both the bounty and mystery of oceans. Others saw oceans as a source of social and economic power. If the oceans could be controlled, navigated and utilized, gains could be made and power over others could be leveraged. Such symbolic power has been tightly coupled with the promise of material gains, whether by facilitating transport to new territories or by harnessing resources deep below. Oceans have further offered opportunity of undiscovered potential. Often they signify hope, such as embedded in the Agenda 2030 of the UN or the Blue Economy discourses in Europe or parts of Africa. Like no other ecosystem on earth, the oceans have consistently fueled narratives of endless potential for human flourishing – a new life across them, adventure, power, discovery, food, spirituality and wealth.

Viewing governance as a system of systems, with connectivity across multiple levels and scales, is critical for understanding how transformative changes in governing manifest. Ocean governance is no different. Governance comprises not only the policies and politics of state-level decision making, but the processes, coordination and collaboration with and throughout civil society. Knowledge sharing, learning, deliberation and communication are increasingly put forth as important features of modern processes of governing that include equality, justice and sustainability as desired outcomes. Ultimately, governance aims to consciously transform our human-ocean interactions toward sustainability, however, transformation is also an emergent property of current social, economic and political systems. There is no single lever, key actor, politician or policy that will cause cascading effects toward desired goals. Rather transformation emerges in response to the amalgamation of incentives, tradeoffs, aggregate actions and largely unforeseeable current events in everyday life.

Governance is always situated in a context, where the material and non-material nature of what is being governed, by whom and for whom, dictates how governance activities will function and what they can achieve. From this perspective, ocean governance faces challenges of being seen, often far from shore or below the surface, negotiated out of sight in the spaces where the activities and actors doing the direct interactions occur. Ocean governance is challenged by the need to embrace and acknowledge its often invisibility, to foster transformative change processes as an opportunity for building constructive collaborations and pursuing moral actions. More broadly, peripheral domestic and international politics undoubtedly shape ocean issues, positioning them in a matrix of agendas, motivations and challenges for achieving change towards sustainable practices that are not necessarily tied to environmental realties or local social and economic needs. Thus, rethinking and reshaping ocean governance towards a governance of the ocean and its resources in a more sustainable manner than before indeed requires trans-regional and cross-scalar 'transformational alliances', coined by Dirk Messner (2015), and actor networks.

The ocean provides a unique context to explore how human-nature narratives are being constructed and discourses shaped, guiding actors in their decision-making, in forming cognitive, policy-making and –implementing structures. We physically see the ocean as an endless surface, which leaves no traces of past events in its ever-shifting and elastic fluidity. We know boats have crossed, animals have splashed and food has been harvested, yet on its surface we see little evidence. We are forced to remember and imagine, until we can rediscover, interpret and (re-)govern. The ocean is constant in its fluidity, similar to our discourses about it, changing and evolving to shape our experiences with it. Importantly, discourses of the ocean that portray them as vast expanses with limitless resources have been some of the most powerful in history. Yet, this discourse is being steadily reformed and retold. Perhaps most importantly, ocean governance discourses are shifting towards sustainability transformation.

Sustainability transformation is understood as the urgent and intentional change in the composition, structure and/or condition of human-environmental relationships with our oceans, to ensure human well-being, social justice and environmental stewardship (Patterson et al. 2017; Bennett et al. 2019; UN 2019). Intentional and concerted governance engagement is needed to achieve such transformations, importantly, the setting of goals and agendas for action. The Sustainable Development Goals (SDGs) have incorporated 'Life Below Water' (SDG 14), which has provided multilateral momentum for mobilizing ocean stewardship awareness and activities. More broadly, the Global Sustainable Development Report (2019), produced by an independent group of scientists appointed by the United Nations, suggests six transformational fields for sustainable development and four transformational levers to actualize them. These can be envisioned to frame ocean sustainability transformations, linked to specific themes and activities (Table 1.1).

Furthermore, the United Nations has initiated the UN Decade of Ocean Science for Sustainable Development (https://www.oceandecade.org/), taking place between 2021 and 2030. The Ocean Decade is aimed at achieving seven broadly defined outcomes (Box 1.1), and provides a global platform for networking, cooperation and other actions on related to ocean science and practice. The puzzle of governing often disparate activities is nonetheless an interconnected system of systems, both multi-level and multi-scale, where partnerships linking public and private goals and activities around all of the SDGs, through knowledge co-creation processes, will

Table 1.1 The Global Sustainable Development Report (2019) produced by an independent group of scientist appointed by the United Nations suggests six transformational fields to focus sustainable development on, and four transformational levers to actualize them (left). A non-exhaustive list of fields and levers specific to ocean and coastal governance are highlighted for each (right)

Global Sustainable Development Report		Examples within ocean governance
Transformational fields	Human well-being and capabilities	Supporting small-scale & traditional blue livelihoods
	Sustainable and just economies	Inclusive property rights and tenure recognition
	Food systems and nutrition patterns	Enabling fisheries and aquaculture transformation
	Energy decarbonization & universal access	Offshore renewables while ending fossil fuel extraction
	Urban and peri-urban development	Just access to coastal spaces while adapting to sea level rise
	Global environmental commons	Conserving high seas and seafloor ecosystems
Transformational levers	Governance	Transparency, inclusion & deliberation in multi-use spaces
	Economy and finance	Ending fisheries subsidies and ocean resource grabbing
	Individual and collective action	Changing plastic use norms and mobilizing political action
	Science and technology	Satellite vessel tracking for monitoring and enforcement

Box 1.1: The Seven Desired Outcomes from the UN Decade of Ocean Science for Sustainable Development (https://www.oceandecade.org/vision-mission/)

1. **A clean ocean** where sources of pollution are identified and reduced or removed.
2. **A healthy and resilient ocean** where marine ecosystems are understood, protected, restored and managed.
3. **A productive ocean** supporting sustainable food supply and a sustainable ocean economy.
4. **A predicted ocean** where society understands and can respond to changing ocean conditions.
5. **A safe ocean** where life and livelihoods are protected from ocean-related hazards.
6. **An accessible ocean** with open and equitable access to data, information and technology and innovation.
7. **An inspiring and engaging ocean** where society understands and values the ocean in relation to human wellbeing and sustainable development.

play a key role in solving challenges and finding joint solutions. Such solutions cannot leave out local actors, smallholders, least developed groups, indigenous communities or historical stewards. Inclusion, participation and incorporating diversity needs to be better prioritized in deliberation and decision-making processes to deliver outcomes that better serve humanities wide range of people and interests, rather than an elite few. This includes the science community in rethinking who creates knowledge, how it is created (e.g., through which processes, and with what purpose and interests) and how knowledge from scientific communities is used as a tool with power for decision-making and practical change.

Today, human-ocean interactions are indeed rapidly transforming. Some as conscious efforts for sustainable change, others as self-emergent responses to the incentives of markets, capitalization and politics. In turn, societies are tasked with balancing new ocean-based development opportunities with environmental stewardship and social sustainability goals, and thus engaging with governance in a pluralistic and place-based manor (Allison et al. 2020). Engaging with a diverse range of governance activities – research, practice, policy – can provide the tools societies need to transformation our interactions with the oceans towards desired sustainability goals. This is no easy challenge. Social, economic and environmental issues are complexly intertwined, and the amalgamation of institutions, people, places that encompass ocean governance are co-shaped and often contested processes that require focused attention and societal investment to make successful.

Governing the ocean is arguably the collective responsibility of humanity (Allison et al. 2020). Who governs, who participates in governing, who is allowed to have a stake in the process and for what purpose, is where the contention, tradeoffs and political interests interact to make governing a complex and pluralistic pursuit. Ocean governance practices that adopt principles of sustainability are no different (Gissi et al. 2022). Governance broadly refers to the social processes that guide human behavior, inclusive of all stakeholders, and is thus a composite societal process of laws, norms, rule systems, institutions, discourses, power dynamics and organizational hierarchies that intermix to shape our behavior, decision making and practical actions (Davidson and Frickel 2004; Lemos and Agrawal 2006; Partelow et al. 2020a).

However, ocean governance has not evolved independently, as noted by Steinberg (1996), "ocean governance systems are influenced by three elements that, in turn, influence each other: the organization of land-based society, the dominant uses of the sea by land-based society, and the physical characteristics of the sea as experienced by users." Models and approaches to land-based environmental governance have historically shaped aquatic ones, although they often do not fit biophysical characteristics of ocean fluidity or the types of social-economic interactions that characterize ocean-based human activities. For example, in Chile, aquaculture property rights models that have mirrored the success of terrestrial farming and small-scale capture fisheries tenure rights face challenges of being immovable and fixed under constantly changing environmental and economic conditions which require adaptation for aquaculture (Tecklin 2016).

In parallel, many international organizations including the World Bank, OECD and FAO are advocating for and driving Blue Economy agendas, framing ocean-based development activities as the new horizon for twenty-first century social-economic prosperity. The term 'Blue Economy' emerged from discussions on the 'Green Economy' during the 2012 UN Conference on Sustainable Development (Rio+20). Since then, major international organizations have launched sustained Blue Economy efforts such as the World Bank's PROBLUE Blue Economy program, the FAO's Blue Growth Initiative, the OECD's 'Ocean Economy in 2030' report, the Global Ocean Alliance's 30-by-30 campaign, and the World Economic Forum's Sustainable Blue Economy theme supporting the Virtual Ocean Dialogues. Both critiques and praise have been raised in response to Blue Economy framings. Critics have raised concerns that such agendas aim to extend capital intensive investments with growth based economic framings into the sea without learning the lessons from the decades of similar approaches applied on land which have led to environmental degradation and the erosion of culturally rich and small-scale livelihood practices under the promise of technological solutions, scalability and efficiency within the political economy discourse of globalism (Golden et al. 2017; Voyer et al. 2018; Farmery et al. 2021). Further neoliberalizing the oceans risks prioritizing the decision-making and interests of those with power in it, often over the silent or silenced ocean-dependent majority whose livelihoods and wellbeing are more directly linked to ocean health (Bennett et al. 2021). On the other side, Blue Economy agendas bring light to the long ignored sustainability issues of oceans and coasts, and can be seen as an opportunity to more appropriately steward ocean-based economic development activities for advancing societies, while recognizing small-holder dependencies and vulnerability, in line with contextually rooted but globally recognized sustainability ambitions. Across this spectrum of critique and optimism are many nuanced positions and arguments, such as which governance strategies at the national level and below can most effectively adapt economic development strategies to local challenges within existing institutional frameworks (Voyer et al. 2021).

Societal organization remains a key practical and scholarly question for governance. How should we organize our activities in a joint way, to ensure goal development and implementation in a timely matter, while also including the necessary diversity of stakeholders and effective deliberation on key issues? Procedural justice, equality and developing capacities for co-production and participation will be central to successful ocean governance efforts, as they are elsewhere in sustainable development processes. This is easier said than done, and the right approach is likely to differ across contexts. Investments into capacity building for representation and self-organization is needed at all levels and in all sectors, particularly for vulnerable small-holder groups. Thematic specialists, facilitators, technical experts and group representatives of resources users, resource stewards, governments, civil society groups, industry and academia need to be incentivized to pursue constructive engagement opportunities and be supported in doing so.

Beyond procedural and capacity issues, specific governing models and institutions require nuanced attention. Many ocean governance issues involve property rights, such as the rights to access, use, manage and exclude others from activities in specific areas. Ocean rights are three dimensional, where rights in the vertical water column, or on the sea floor, are equally important and as differentiated as two dimensional surface space. However, the ocean is humanity's least privatized environmental entity (Schlüter et al. 2020), and the allocation of further property rights need to consider sustainability issues such as the distributive and procedural justice dynamics as well as spillover or path dependency implications (Partelow et al. 2019). Much of the ocean is a commons, for humanities shared use, where no jurisdiction of any single government applies, and only voluntary international conventions have acted as a guide for use and stewardship. The United Nations Convention on the Law of the Sea (UNCLOS), implemented in 1982, provided the first international legal framework establishing ocean property rights for individual countries in their offshore waters. The UNCLOS Exclusive Economic Zones enable countries to manage and exploit resources up to 200 nautical miles off their shore, or until another EEZ is met, Beyond these Exclusive Economic Zones for individual states, the ocean remains common property upheld by voluntary agreements of use and stewardship. In many instances, rights are synonymous with power. Common property arrangements on our shores and seas involve power sharing, but also require collective action to organize sharing in fair and responsible ways. Private property concentrates rights, and thus concentrates power, but also internalizes costs and can motivate quick action for either use or protection. Focused efforts are needed to ensure that if and when rights are allocated, they are done so in recognitional, distributional and procedurally just ways.

One of the major challenges with pursuing transformative governance and sustainability agendas is acknowledging the potential risks. Blythe et al. (2018) examine how the discourse supporting transformation as apolitical or inevitable has potential to generate significant and counterproductive risks. In other words, fostering social, political and economic change can be very difficult and come with unforeseen costs (Table 1.2). Although the outlined risks are not specific to ocean governance, they can be easily applied. Transformations in ocean governance can risk shifting the burden of change to vulnerable groups, despite the origins or problems coming from more powerful actors in wealthier politically and economically dominant countries. For example, due to historically high carbon emissions in the United States and Europe leading to increased ocean acidification, local low-income fishers may be forced or crowded out of coastal spaces where conservation areas are established with Global North support to protect resilient varieties of coral or seagrass to increased acidification and warming sea surface temperatures, without offering fishers an alternative livelihood opportunity or compensation. Transformation can also be used to justify business as usual, often expressed in critiques of Blue Economy agendas that seem to extend unsustainable growth-based neoliberal logic into the oceans masked in sustainability terminology. Furthermore, social science has shown for decades the need for differentiating social context in economic and political decision-making to avoid implementing initiatives and policies that don't consider local practices, culture and history. This has been supported in natural resource governance literature, that panacea solutions fail to deliver

Table 1.2 Five latent risks associated with the shift from descriptive to prescriptive engagements with the concept of transformations to sustainability, taken from Blythe et al. (2018)

Sustainability transformation risk	Examples within ocean governance
Risk 1: Transformation Discourse Risks Shifting the Burden of Response onto Vulnerable Parties	Resettling informal coastal settlements for elite real-estate developments. Aquaculture increases seafood prices, reducing access to essential nutrients for poor.
Risk 2: Transformation Discourse May Be Used to Justify Business-As-Usual	Blue Economy framings draw investments that require growth and returns for elites, reinforcing capitalistic market incentives that crowd-out just and equitable resource use and development ambitions.
Risk 3: Transformation Discourse Pays Insufficient Attention to Social Differentiation	Governance uses generic policies to solve context specific problems such as coastal protected area spatial planning, use rules and rights. What works for diverse people and cultures is likely to substantially vary.
Risk 4: Transformation Discourse Can Exclude the Possibility of Non-Transformation or Resistance	Risks emerge when transformation is framed as inevitable, positive or singular in its directionality. Such as establishing more conservation areas which may fail to recognize that coupling stewardship and use may be optimal or that more time may be needed to shift society in just ways.
Risk 5: Insufficient Treatment of Power and Politics Threatens the Legitimacy of Transformation Discourse	Efforts to shift local plastic use and pollution behavioral norms fail to consider structural economic incentives and industry lobbying. In contrast, policies for reduction through legislation fail to consider equally harmful alternatives available to producers, or consumer preferences shaped by marketing and contrasting political views.

sustainable outcomes when they do not allow for tailored approaches and local implementation, often by failing to include local stakeholder inputs who have useful and practical non-scientific knowledge (Brock and Carpenter 2007; Ostrom et al. 2007). Transformation can also crowd-out possibilities of non-fundamentally transformative changes as valid solutions, or the emergence of resistance for unforeseen reasons in different stakeholder groups, perhaps due to historical mistrust or lack of inclusion. Finally, the role of power in politics can threaten legitimacy and acceptability at all levels of governance.

1.2   Key Events in the History of Ocean Governance

For millennia, countless events have shaped the human relationship with our oceans. There is a rich history of triumph, societal expansion and cultural development, but also of oppression and struggle. Here we focus on some of the key events dating back to the early twentieth century, to highlight a limited but influential set of key government actions and policies, scientific advancements, and society and environment activities that have influenced current perspectives and trajectories (Table 1.3).

Table 1.3 Selective ocean governance related events in (1) governance and policy, and (2) science and society

Years	Governance and policy	Science and society
1900–1950	German naval blockade (1939–1945) United Nations (1945) International Whaling Commission (1946)	Northwest Passage (1906) Titanic sinks (1912) Panama Canal (1914) Acoustic sea floor exploration (1914) Meteor maps seafloor[a] (1925) Bathysphere invented (1934) Aqua-Lung SCUBA diving (1943) WWII Naval advances (1939–1945)
1950s	UNCLOS I[b] (1956) Antarctic Treaty by 12 nations (1959)	The Sea Around Us (Carson, 1951)
1960s	UNCLOS II (1960) Intergovernmental Oceanographic Commission of UNESCO (IOC) (1960)	Silent Spring (Carson, 1962) Santa Barbara oil spill (1969)
1970s	UNEP Regional Seas Program[c] (1974) OSPAR: Oslo & Paris Conventions[d] (1972) HELCOM: The Baltic Marine Environment Protection Commission founded (1974)	First Earth Day (1970) NOAA established[e] (1970) Blue Marble photo from Apollo 17 (1972) International Decade of Ocean Exploration (IDOE) (1971–1980) Greenpeace first anti-whaling campaign (1975)
1980s	Abidjan Convention[f] (1981) UNEP COBSEA (1981)[g] UNCLOS III adopted along with International Seabed Authority (1982) Nairobi Convention[h] (1985) Moratorium on whaling (1986) Basel Convention[i] (1989)	Our Common Future[j] (1987) Exxon Valdez oil spill – Alaska (1989)
1990s	Rio Earth Summit[k] (1992) UNCLOS comes into force (1994) Marine Stewardship Council (1996)	Argo project[l] (1990) Atlantic cod fishery collapse (1992) First UN State of World Fisheries and Aquaculture report (1994) Fishing Down Marine Food Webs (Pauly et al. 1998)[m] Oceana founded[n] (1999)
2000s	EU Marine Strategy Framework Directive (2008) USA Ocean Policy Task Force (2009) UK Marine and Coastal Access Act (2009)	The Blue Planet series (2001) Indian Ocean earthquake & tsunami[o] (2004) Hurricane Katrina, USA[p] (2005) 5 Gyres Institute[q] (2009)

(continued)

Table 1.3 (continued)

Years	Governance and policy	Science and society
2010s–present	Aquaculture Stewardship Council (2010) Blue Economy from Rio+20 (2012) Global Partnership on Marine Litter[r] (2012) FAO Small scale fisheries guidelines (2014) UN SDG 14 'Life below Water' (2015) COBSEA Strategic Directions (2018–2022)[s] African Union Blue Economy report[t] (2019) ASEAN Blue Economy declaration[u] (2021) EU Blue Economy strategy report[v] (2021) UN Decade of Ocean Science for Sustainable Development (2021–2030) International Seabed Authority has issued 31 deep sea mining contracts[w] (2022)	Census of Marine Life (2010) Fukushima nuclear disaster (2011) Solo Dive in Mariana Trench (2012) Blackfish documentary (2013) Global Fishing Watch[x] (2016) Seabed 2030 project[y] (2017) Global coral bleaching (2016–2017) UN State of the World Fisheries and Aquaculture[z] (2020)

[a]https://en.wikipedia.org/wiki/German_survey_ship_Meteor
[b]First United Nations Conference on the Law of the Sea
[c]https://www.unep.org/explore-topics/oceans-seas/what-we-do/regional-seas-programme
[d]https://www.ospar.org/convention
[e]https://www.noaa.gov/
[f]Cooperation for the Marine and Coastal Environment of the Atlantic Coast of West, Central and Southern Africa
[g]https://www.unep.org/cobsea/
[h]https://www.nairobiconvention.org/
[i]http://www.basel.int/TheConvention/Overview/tabid/1271/Default.aspx
[j]https://sustainabledevelopment.un.org/content/documents/5987our-common-future.pdf
[k]https://www.un.org/en/conferences/environment/rio1992
[l]https://argo.ucsd.edu/
[m]https://www.science.org/doi/10.1126/science.279.5352.860
[n]https://oceana.org/
[o]https://en.wikipedia.org/wiki/2004_Indian_Ocean_earthquake_and_tsunami
[p]https://en.wikipedia.org/wiki/Hurricane_Katrina
[q]https://www.5gyres.org/
[r]https://www.gpmarinelitter.org/
[s]Satellite tracking of human activity at sea (https://globalfishingwatch.org/)
[t]https://www.unep.org/cobsea/resources/policy-and-strategy/cobsea-strategic-directions-2018-2022
[u]https://www.au-ibar.org/sites/default/files/2020-10/sd_20200313_africa_blue_economy_strategy_en.pdf
[v]https://asean.org/wp-content/uploads/2021/10/4.-ASEAN-Leaders-Declaration-on-the-Blue-Economy-Final.pdf
[w]https://eur-lex.europa.eu/legal-content/EN/TXT/PDF/?uri=CELEX:52021DC0240&from=EN
[x]https://www.isa.org.jm/deep-seabed-minerals-contractors
[y]100% of the ocean floor mapped by 2030 (https://seabed2030.org/)
[z]https://www.fao.org/documents/card/en/c/ca9229en/

Early twentieth century exploration included the first navigation of the northwest passage, an arctic sea route shortening the distance from the Atlantic to the Pacific Ocean with access to Asia. Today, Arctic sea routes remain contested spaces with receding summer sea ice due to climate change easing access. The ability to establish rights and norms for navigating the Arctic and dealing with the competition and resource exploitation remain a contested multi-lateral issue. Early scientific achievements include acoustic seafloor exploration and bathymetry science, which allowed early expeditions to map large areas of the ocean with more accuracy. Entering a phase of global turmoil, World War II showed the power that control over the sea can have on politics and the economy, largely shaping outcomes with substantial naval technology advances displayed in both the North Atlantic and Pacific. Following the war period, the newly formed United Nations established various conventions, including the Convention on the Law of the Sea (UNCLOS) which first met in Geneva in 1956. Subsequent UNCLOS conventions lasted until consensus was reached in 1982, coming into force in 1994. The UNCLOS convention enabled various state level provisions shaping our current ocean governance landscape including the 12 nautical mile territorial zone and 200 nautical mile Exclusive Economic Zone (EEZ).

Starting in the 1950s and 60s, public awareness of environmental issues began to grow, catalyzed by influential events and books such as the The Sea Around Us (1951) and Silent Spring (1962) by Rachel Carson. The 1972 'Blue Marble' photo taken from the space ship Apollo 17 provided one of the first public and simple pieces of evidence that the oceans both dominate life on our planet, but also have limits, and that our political borders dissolve at the level of planetary stewardship. The 1969 Santa Barbara and 1989 Exxon Valdez oil spills awakened public awareness to the risks of carelessly exploiting our oceans, risking the public goods oceans provide for human health, recreation and food. Greenpeace, one of the most well-known environmental NGOs, was founded in the early 1970s in a first attempt to raise awareness and stop US nuclear weapon tests off the coast of Alaska, an area considered at the time to be out of sight and out of mind. The 1992 collapse of the northwest Atlantic cod fishery showed us the ocean has material limits, leading to recognition that social, economic and political turmoil are coupled to environmental health. The fishery's collapse sparked changes in how scientists, fishers and politicians interact to govern fisheries today.

In the 1980s and 1990s, awareness and public policy increased on specific topical and regional issues. HELCOM spurred Nordic cooperation in the Baltic Sea, while the Abidjan (1981) and Nairobi (1985) Conventions mobilized management activities among countries along the Eastern and Western African coastlines respectively. The United Nations Conference on Environment and Development, also known as the Rio Earth Summit, took place in 1992 and catalyzed international actions and the formation of many conventions for environmental protection and action today such as the Convention on Biological Diversity (CBD) and Framework Convention on Climate Change (UNFCCC). The summit further spurred the formation of non-governmental organizations (NGO) focused on environmental issues (Partelow et al. 2020b). One the key global data collection and monitoring efforts in our oceans, the United Nations State of World Fisheries and Aquaculture report (FAO 2020), was first published in 1994. The report series and its data continue to provide much of national, regional and global seafood production and development data for scientists and policymakers despite challenges with maintaining accuracy and consistency in reporting across highly diverse political and economic contexts.

The 2000s saw many societal events that further catalyzed societies dynamic relationship with the ocean, coastline and the need for disaster risk reduction investments and planning. The Indian Ocean earthquake and tsunami in December 2004 devastated parts of low lying coastal Indonesia, Thailand, Sri Lanka, India and the Maldives, among other areas. The event triggered substantial humanitarian efforts in the immediate aftermath, spurred ongoing debates on coastal security and warning systems, and raised critique on the role of foreign aid in enabling long-term recovery and resilience. Hurricane Katrina in 2005 flooded substantial sections of the city of New Orleans, USA and surrounding areas, raising awareness to coastal hazards, government response and the impacts of climate change. Furthermore, the large earthquake off the coast of Japan in March 2011, and subsequent tsunami, led to the meltdown of the Fukushima Daiichi nuclear power plant, contaminating the surrounding coastal area, raising debates regarding nuclear security and coastal protection worldwide. Later, in 2016 and 2017, subsequent ocean warming periods led to widespread global coral bleaching events, raising awareness of the impacts climate change is having on marine biodiversity and its dependent economy.

More recent events indicate the rising political awareness, along with regional and international efforts to mobilize action for ocean management, protection and science. The Food and Agricultural Organization of the United Nations (FAO) Small Scale Fisheries Guidelines were released in 2014, recognizing the importance of small-scale livelihoods in protection and management. The United Nations Agenda 2030, announced in 2015, included the 17 Sustainable Development Goals (SDG), with SDG 14 focused on 'Life Below Water' with the aim to conserve and sustainably use the oceans, seas and marine resources for sustainable development. In economic and political spheres, declarations and strategic reports for the Blue Economy were released by the African Union (2019), ASEAN (2021) and the European Commission (2021). Looking forward, the UN Decade of Ocean Science for Sustainable Development started in 2021, with the intent to mobilize and coordinate global action and activities surrounding our oceans over the next decade and beyond.

1.3   Key Themes of Ocean Governance

Many themes and topics are emerging as critically important for our oceans, for engagement at all levels, and for achieving the ambitions outlined in SDG 14. Below we highlight a select few that have been, remain or have emerged as influential in ongoing ocean governance arenas. Most notably, fisheries have been a central focus of ocean governance efforts over the last half century. Nonetheless, many fisheries globally remain overexploited and under-recognized in their contributions to food and livelihood security (Pauly and Zeller 2016). This is not the sole responsibility of fishers, but often of politics on the multilateral and regional levels regarding state subsidies and industry interests. It is not unusual that fishery contracts have been bundled into development aid and economic trade agreements that put fishing rights in negotiation with multilateral financial reform and the privatization of public service provision, for example, in countries in West Africa (Gagern and Bergh 2013; Gegout 2016; Hornidge and Keijzer 2021). Numerous governance strategies have been suggested and advocated to reform the policies and practices of industrial fishing, such as those suggested in Box 1.2. Importantly, Hornidge and Keijzer make the necessary distinction between small and large scale fisheries. Small scale fisheries account for roughly 50% of the global catch, but roughly 90% of the sectors employment, and tend to be rooted in community-based practices that support local culture, food security and livelihoods (FAO 2020). However, this doesn't mean small-scale fisheries do not face substantial sustainability issues and governance challenges themselves, although they are often overlooked in policy making and economic development arenas (Smith and Basurto 2019).

Private sector supported initiatives are leading numerous ocean governance activities. Global Fishing Watch, an international nonprofit organization founded by Oceana, Skytruth and Google, is revolutionizing the potential for ocean governance through data driven analytics that utilize automatic identification system (AIS) technology to track the movement of boats with satellites worldwide (https://globalfishingwatch.org/). This global data has revealed previously unobservable observations and patterns on transshipment (Boerder et al. 2018), distant water fishing (Tickler et al. 2018b), vessel identification strategies and regional movement patterns (Taconet et al. 2019), forced labor issues (McDonald et al. 2021), and the outsized role of wealthy nations in global industrial fishing (Mccauley et al. 2018). Furthermore, science and industry partnerships are now emerging to tackle the practices and incentives for ocean stewardship through cooperative open-dialogue and transdisciplinary scientific engagement, such as the Seafood Business for Ocean Stewardship (SeaBOS) initiative (Österblom et al. 2017), bringing together some of the largest seafood producing companies to develop sustainability commitments (https://seabos.org/). However, these activities need further adoption and scaling, as the industrial fishing industry remains plagued by its environmental impacts and human-rights abuses in the form of modern day slavery (Tickler et al. 2018a) and human trafficking (Mileski et al. 2020).

Box 1.2: Action Items for Fisheries Reform in International Cooperation and Development (Hornidge/Keijzer 2021)

1. Eliminate subsidies for industrial fisheries.
2. A ban on all high-sea fishing activities.
3. Institutional strengthening and capacity development of regional fisheries management.
4. Special support for small-scale and coastal fisheries in developing and middle-income countries.
5. Targeted development of local fish-processing industries and (trans-) regional marketing, including gender-sensitive job creation measures, social and environmental standards, capacity development and training.
6. Promoting cross-sector cooperation and coordination in ocean-based branches of the economy.

Following rapidly behind capture fisheries is aquaculture, where South and Southeast Asian countries, led by China, India, Indonesia and Vietnam, have undergone blue food agricultural revolutions, demonstrating that the world can farm seafood at scale. This has not been done with overly advanced technology and high capital investments in the ocean, but rather through low tech rural development in inland and coastal brackish ponds, quietly demonstrating that the often utopic visions of Blue Economy aquaculture expansion for high value and high trophic level species in the open sea overlook the need for small-scale livelihood and food security in shaping agriculture transformation rather than technology (Edwards et al. 2019). However, aquaculture is expanding in many forms globally, and has been the fastest growing food production sector globally for the last two decades, now producing more tonnage of farmed products than capture fisheries (FAO 2020). Similar to capture fisheries, much of aquaculture is small-scale, and its emergence as a sustainable means of seafood production will require specific policy attention and regulation to curb environmental impacts while bolstering livelihood opportunities, food access and safety through supply chain innovations and transformation (Belton et al. 2020). Aquaculture is a newly emerging sector, and although it is highly reliant on environmental commons such as water quality, water quantity, feed sourcing and nutrients, it is likely that a regulatory landscape already exists to govern those commons in other competing sectors, where institution building will likely require cross-sector collaboration and adaptation (Partelow et al. 2021).

Open marine space is increasingly viewed as a "commodity frontier", something necessary to procure rights over (Campling 2012; Schlüter et al. 2020), but there have been parallel voices calling for a reconsideration of the intensification of humanity's relation with the ocean (Hadjimichael 2018; Ertör and Hadjimichael 2020). Enclosure and territoriality is not a new feature of the ocean commons and still continue today (Constantinou and Hadjimichael 2021). For example, in the South China Sea, with implications for capture fisheries, fossil fuel and mineral extraction coupled with strategic political and economic interests in securing navigation, use and management rights (Manlosa et al. 2021a, b). The South China Sea example showcases how international legal frameworks are used and disputed to expand maritime claims for different geopolitical interests, and for retaining or acquiring fishing rights, or access to seabed resources. Governing oceanic commons has been approached through international cooperation in the Antarctic, where the Antarctic Treaty was signed in 1959 stipulating peaceful use of the region in the interest of fostering publically available science, with 54 parties in agreement to the treaty today. However, in the Arctic, the decreasing presence of summer sea ice due to climate change is making shipping passage through Arctic routes a realistic option for tourism and large container ships, but also for previously inaccessible natural resource exploitation interests that remain open to negotiation and are still contested.

Only what is known and cognitively grasped can be governed, leaving what is happening offshore and underwater less seen and at risk. We can now find examples of our ungoverned and hidden ocean past, leading to reinterpretations and the reframing of our human-ocean narratives (Table 1.3). Installations of wind farms in the European North Sea are regularly challenged by the presence of thousands of illegally dumped barrels of explosive and corrosive World War II ammunitions. Off the coast of southern California, thousands of barrels of the agricultural pesticide DDT (Dichloro-diphenyl-trichloroethane) were illegally dumped in the 1950s and 1960s. DDT was banned in California in the 1970s in part due to the observation that nesting seabird eggs became inviable due to shell thinning, influenced by Carson's 1962 book Silent Spring. Making the out-of-sight ocean visible to the public and policy makers is challenging, for example, to govern seabed mining. Seabed mining is of increasing interest for the extraction of minerals and metals due to terrestrial depletion, and is occurring in both areas beyond national jurisdiction and on near-shore continental shelves (Wedding et al. 2015; Levin et al. 2020). Minerals such as copper, cobalt, nickel, zinc and lithium are needed for many electronic devices including electric vehicles and transportation as well as renewable energy generating devices desired for transitioning to low carbon economies (Levin et al. 2020). The International Seabed Authority established in tangent under UNCLOS, is in charge of regulating human activities on the seabed beyond the continental shelf, and has issued 31 contracts for mining. However, many questions and uncertainties exist regarding environmental impacts, scale of operations and legal ambiguities (Miller et al. 2018).

As seen above through aquaculture and seabed mining, ocean systems and ocean governance are not isolated, they interact strongly with land-based coastal systems and climate. Governing climate change mitigation and adaptation is synonymous with governing our oceans. The oceans not only absorb carbon, but also show the direct implications climate change with sea level rise and increasing storm intensity and frequency, threatening hundreds of millions of people globally. Entire countries such as Bangladesh, the Maldives and the Marshall Islands face existential threats in the loss of territory with future sea level rise projections. Climate justice is an ever-present issue, as those countries have been among the lowest contributors to global greenhouse emissions. The oceans are also a climate buffer because they absorb carbon dioxide from the atmosphere, most effectively when they have intact ecosystems. However, the side effect is increased ocean acidification through higher amounts of carbonic acid that reduce carbonate availability for calcifying organisms such as coral. The oceans also promise renewed efforts into oil and gas exploration, with billions of dollars invested yearly by the largest fossil fuel corporations to find new reserves under the sea floor. Many of these corporations still receive substantial financial subsidies and regulatory support from state governments (Rentschler and Bazilian 2017), while also making pledges for climate action.

The ocean can't be governed in isolation. Many of the negative impacts on our oceans originate with governance challenges on land. Fertilizers, pesticides, plastics and other hazardous materials, when mismanaged on land, end up in our waterways and eventually our oceans. Socially, there has been steady increases in the percentage of the global population living in coastal areas. Other economic, cultural and political issues such as drought, conflict, housing speculation or health trends can drive interest in coastal development or change demand for coastal resource use, for example in the demand for specific types of seafood. Nearly the entire global fishery for sea cucumbers is driven by cultural interests and markets in China (Eriksson et al. 2015). In real estate, islands such as Cyprus and Malta, have extensively developed their coastline in recent years, in an attempt to increase real estate prices on picturesque coastlines to attract foreign investment, with criticized citizenship for sale schemes that ultimately crowd out coastal access and use for local residents.

1.4   Organization of the Book

The chapters in this book are organized into three parts. Chapters in each of the parts address a range of specific focal topics. As the book is an edited volume, the specific topics, analyses and insights are written and derived by a diverse group of scholars who specialize in each subject area. The catalyst for the book originates from the European Cooperation in Science and Technology (COST) Action on 'Ocean Governance for Sustainability – challenges, options and the role of science'. The focus of the Ocean Governance COST Action was focused around six working groups, each with specific thematic topics: (1) Land-Sea Interactions, (2) Area-Based Management, (3) Seabed Resource Management, (4) Nutrition Security and Food Systems, (5) Ocean, Climate Change, and Acidification, and (6) Fisheries Governance. The focus of the chapters loosely represents these six thematic areas, but also link to topics beyond them with a global scope. Overall, while the book can certainly not address the full spectrum of ocean governance topics and issues, it provides a baseline of up-to-date multi- and inter-disciplinary literature that intends to foster pluralistic understanding and capacity to think about and engage with ocean governance in a way that enables critical thinking, systems thinking and sustainability analytical capacity about past, present and future ocean challenges and opportunities.

1.4.1   Part I – Knowledge Systems for Ocean Governance

How we as a society – as researchers, policy-makers, students, practitioners and citizens – know the ocean is essential for understanding our actions, perceptions and framings around it. Chapter 2 by Hornidge et al., (2022) examines how we 'Know the ocean', exploring patterns of science collaboration through a lens of epistemic inequalities. The synthetic overview brings together prior reviews and critical perspectives to examine differences in knowledge production trends across disciplines, genders and transregional networks in the context of the UN Agenda 2030 and the Decade of Ocean Science for Sustainable Development. Chapter 3, provided by Barragán Paladines et al., (2022), focuses on the history of fisheries governance in Latin America, with a specific focus on Ecuador, and to what extent politics, power and knowledge have deeply influenced policies and practices in the use and management of marine and terrestrial resources and at managing fish and seafood. Chapter 4, by Finley (2022), provides a detailed historical narrative of Japanese contributions to ocean science and the construction of recruitment fisheries ocean-ography, the study of the effects of climate and ocean variability on fish abundance.

1.4.2 Part II – Policy Foundations of Ocean Governance

Many policies at the international, transregional and regional levels have shaped human interaction with the sea. In Chap. 5, Flannery (2022), examines how Marine Spatial Planning (MSP) has become one of the key components of marine gover-nance, and outlines the scholarly debates critiquing the ability of MSP to transform unsustainable marine governance and management practices within the context of emerging Blue Economy and Green Deal policy ambitions. Chapter 6, from authors Singh and Araujo (2022), aim to reflect on the past, present and future of ocean governance within fisheries at sea, marine area-based management tools and inter-national seabed mineral resources. The three case studies demonstrate how the law of the sea has evolved, particularly with respect to the challenge of protecting and preserving the marine environment through the sustainable use of marine resources. In Chap. 7 written by Calado et al., (2022), the authors review the diverse legal and regulatory frameworks for the marine environment in the North Atlantic and assess where differences between countries exist and at which governance level they are being created. In Chap. 8, Nakamura (2022) examines the past and future of inter-national fisheries law, providing examples and analyses of how legal developments have been shaped and can adapt to new challenges such as climate change going forward. Chapter 9, from Lawlor and Depellegrin (2022), review the marine and coastal management systems in Ireland, Romania, Spain and France under the Marine Strategy Framework Directive committed to delivering Good Environmental Status. They assess their capacity to manage land sea interactions, and provide con-crete recommendations to assist EU member states going forward.

1.4.3 Part III – Thematic Analyses of Ocean Governance

Ocean governance span a wide range of topics and contexts. In this part, numerous topics are explored in specific detail highlighting context specific problems, chal-lenges and directions forward for good governance and sustainability transforma-tion. Chapter 10, from Cretella and Scherer (2022), unpack the issues connected to seafood consumption in Ireland's coastal capital Dublin examining behavioral shifts in consumption towards more sustainable local seafood by rediscovering historical recipes and cultural heritage. In Chap. 11, van Tatenhove (2022) gives insight into marine governance challenges in the context of Arctic shipping. Drawing on theory of reflexive institutionalization, governance interactions related to three Arctic shipping routes are examined including the Northwest Passage (NWP), the Northeast Passage and Northern Sea Route (NEP/NSR), and the Transpolar Sea Route (TSR). Chapter 12, by Wenting et al., (2022), draw on assemblage theory to examine ecological, legal and practical insights into seabed mining, drawing on interdisciplinary perspectives to connect the debates surrounding seabed mining issues. In Chap. 13, Salmi et al., (2022) draw on interactive governance theory to compare Finnish and Swedish small-scale fisheries governance challenges, concluding that the present governance system is incompatible and that new co-governance arrangements are needed to include small-scale fishers' interests, values and local knowledge. Chapter 14, by Spranz and Schlüter (2022), explores the behavioural and cultural reasons for the high consumption and pollution by plastic bags on Bali, Indonesia, identifying promising approaches that can effectively support local initiatives and awareness campaigns. In Chap. 15, from Simarmata et al., (2022), Indonesia is again examined exploring the two distinct and interrelated concepts supporting archipelagic thinking – 'Nusantara' and 'Tanah Air'. The role of each in shaping the island nation's development trajectories are critically explored under ambitions for continued Blue Economy expansion. Chapter 16 from Penca and Said (2022) explores the multiscale contributions of small-scale fisheries by focusing on recently developments across the Mediterranean with impacts on the supply chain and the marketing of their products, concluding that such market interventions challenge the conception of small-scale fisheries as a non-innovative sector. In Chap. 17, Ertör and Ertör Akyazi (2022) examine small-scale fisher movements and food sovereignty issues, by exploring their local and global initiatives and role in food justice movements. To conclude the part, Chap. 18 by Bednaršek et al., (2022) analyze ocean acidification as a governance challenge for fisheries and aquaculture in the Mediterranean Sea, and produce depth-related pH and aragonite saturation state exposure maps overlaid with the existing aquaculture industry to demonstrate potential risk for farming fish in the future.

To conclude and in part summarize the book's key messages, we provide Afterword, a brief synthetic overview of the main lessons learned and practical take-away messages for each of the book's target audience groups: students, researchers, and policy-makers. This chapter, acting as an Afterword, aims to provide explicit points for each group to guide further study, research or policy-making agendas across ocean governance topics.

References

Allison EH, Kurien J, Ota Y, Al E (2020) The human relationship with our ocean planet. Washington, DC

Belton B, Reardon T, Zilberman D (2020) Sustainable commoditization of seafood. Nat Sustain 3(9):677–684

Bennett NJ, Blythe J, Cisneros-Montemayor AM, Singh GG, Sumaila UR (2019) Just transformations to sustainability. Sustainability (Switzerland) 11(14):1–18

Bennett NJ, Blythe J, White CS, Campero C (2021) Blue growth and blue justice: ten risks and solutions for the ocean economy. Mar Policy 125(January):104387

Blythe J, Silver J, Evans L, Armitage D, Bennett NJ, Moore ML, Morrison TH, Brown K (2018) The dark side of transformation: latent risks in contemporary sustainability discourse. Antipode 50(5):1206–1223

Boerder K, Miller NA, Worm B (2018) Global hot spots of transshipment of fish catch at sea. Sci Adv 4(7):1–11

Brock WA, Carpenter SR (2007) Panaceas and diversification of environmental policy. Proc Natl Acad Sci U S A 104(39):15206–15211

Campling L (2012) The tuna "commodity frontier": business strategies and environment in the industrial tuna fisheries of the western Indian Ocean. J Agrar Chang 12(2–3):252–278

Constantinou CM, Hadjimichael M (2021) Liquid entitlement: sea, terra, law, commons. Glob Soc 35(3):351–372

Davidson DJ, Frickel S (2004) Understanding environmental governance: a critical review. Organ Environ 17(4):471–492

Edwards P, Zhang W, Belton B, Little DC (2019) Misunderstandings, myths and mantras in aquaculture: its contribution to world food supplies has been systematically over reported. Mar Policy 106(January):103547

Eriksson H, Österblom H, Crona B, Troell M, Andrew N, Wilen J, Folke C (2015) Contagious exploitation of marine resources. Front Ecol Environ 13(October):435–440

Ertör I, Hadjimichael M (2020) Editorial: blue degrowth and the politics of the sea: rethinking the blue economy. Sustain Sci 15(1):1–10

FAO (2020) The state of world fisheries and aquaculture: sustainability in action. The State of the World, Rome

Farmery AK, Allison EH, Andrew NL, Troell M, Voyer M, Campbell B, Eriksson H, Fabinyi M, Song AM, Steenbergen D (2021) Blind spots in visions of a "blue economy" could undermine the ocean's contribution to eliminating hunger and malnutrition. One Earth 4(1):28–38

Gagern A, van den Bergh J (2013) A critical review of fishing agreements with tropical developing countries. Mar Policy 38:375–386

Gegout C (2016) Unethical power Europe? Something fishy about EU trade and development policies. Third Word Q 37(12)

Gissi E, Maes F, Kyriazi Z, Ruiz-Frau A, Santos CF, Neumann B, Quintela A, Alves FL, Borg S, Chen W, Fernandes MDL, Hadjimichael M, Manea E, Marques M, Platjouw FM, Portman ME, Sousa LP, Bolognini L, Flannery W, Grati F, Pita C, Văidianu N, Stojanov R, van Tatenhove J, Micheli F, Hornidge AK, Unger S (2022) Contributions of marine area-based management tools to the UN sustainable development goals. J Clean Prod 40(July 2021)

Golden JS, Virdin J, Nowacek D, Halpin P, Bennear L, Patil PG (2017) Making sure the blue economy is green. Nat Ecol Evol 1(January):1–3

Hadjimichael M (2018) A call for a blue degrowth: unravelling the European Union's fisheries and maritime policies. Mar Policy 94:158–164

Hornidge A, Keijzer N (2021) Global fisheries – still a blind spot in international cooperation. Rural 21(4):6–9

Jouffray J-B, Blasiak R, Norström AV, Österblom H, Nyström M (2020) The blue acceleration: the trajectory of human expansion into the ocean. One Earth 2(1):43–54

Lemos MC, Agrawal A (2006) Environmental governance. Annu Rev Environ Resour 31:297–325

Levin LA, Amon DJ, Lily H (2020) Challenges to the sustainability of deep-seabed mining. Nature Sustain 3(10):784–794

Manlosa AO, Hornidge A-K, Schlüter A (2021a) Aquaculture-capture fisheries nexus under Covid-19: impacts, diversity, and social-ecological resilience. Maritime Stud 2014

Manlosa AO, Hornidge AK, Schlüter A (2021b) Institutions and institutional changes: aquatic food production in Central Luzon, Philippines. Reg Environ Chang 21(4)

Mccauley DJ, Jablonicky C, Allison EH, Golden CD, Joyce FH, Mayorga J, Kroodsma D (2018) Wealthy countries dominate industrial fishing. Sci Adv 4

McDonald GG, Costello C, Bone J, Cabral RB, Farabee V, Hochberg T, Kroodsma D, Mangin T, Meng KC, Zahn O (2021) Satellites can reveal global extent of forced labor in the world's fishing fleet. Proc Natl Acad Sci 118(3):9

Messner D (2015) Deutschland als Gestaltungsmacht in der globalen Nachhaltigkeitspolitik – Chancen und Herausforderungen unter den Bedingungen „umfassender Globalisierung". Zeitschrift für Außen- und Sicherheitspolitik 8(S1):379–394

Mileski JP, Galvao CB, Forester ZD (2020) Human trafficking in the commercial fishing industry: a multiple case study analysis. Mar Policy 116(103616)

Miller KA, Thompson KF, Johnston P, Santillo D (2018) An overview of seabed mining including the current state of development, environmental impacts, and knowledge gaps. Front Mar Sci 4(Jan)

Österblom H, Jouffray JB, Folke C, Rockström J (2017) Emergence of a global science–business initiative for ocean stewardship. Proc Natl Acad Sci U S A 114(34):9038–9043

Ostrom E, Janssen MA, Anderies JM (2007) Going beyond panaceas. Proc Natl Acad Sci U S A 104(39):15176–15178

Partelow S, Abson DJ, Schlüter A, Fernández-Giménez M, von Wehrden H, Collier N (2019) Privatizing the commons: new approaches need broader evaluative criteria for sustainability. Int J Commons 13(1):706–747

Partelow S, Schlüter A, Armitage D, Bavinck M, Carlisle K, Gruby RL, Hornidge A-K, Le Tissier M, Pittman JB, Song AM, Sousa LP, Văidianu N, Van Assche K (2020a) Environmental governance theories: a review and application to coastal systems. Ecol Soc 25(4)

Partelow S, Winkler KJ, Thaler GM (2020b) Environmental non-governmental organizations and global environmental discourse. PLoS ONE:12–15

Partelow S, Schlüter A, Manlosa AO, Nagel B, Paramita AO (2021) Governing aquaculture commons. Rev Aquac

Patterson J, Schulz K, Vervoort J, van der Hel S, Widerberg O, Adler C, Hurlbert M, Anderton K, Sethi M, Barau A (2017) Exploring the governance and politics of transformations towards sustainability. Environ Innov Soc Trans 24:1–16

Pauly D, Zeller D (2016) Catch reconstructions reveal that global marine fisheries catches are higher than reported and declining. Nat Commun 7:1–9

Rentschler J, Bazilian M (2017) Reforming fossil fuel subsidies: drivers, barriers and the state of progress. Clim Pol 17(7):891–914

Schlüter A, Bavinck M, Hadjimichael M, Partelow S, Said A, Ertör I (2020) Broadening the perspective on ocean privatizations: an interdisciplinary social. Ecol Soc 25(3):20

Smith H, Basurto X (2019) Defining small-scale fisheries and examining the role of science in shaping perceptions of who and what counts: a systematic review. Front Mar Sci 6(May)

Steinberg PE (1996) Three historical systems of ocean governance: a framework for analyzing the law of the sea. World Bull 12(5–6):1–19

Taconet M, Kroodsma D, Fernandes JA (2019) Global Atlas of AIS-based fishing activity

Tecklin D (2016) Sensing the limits of fixed marine property rights in changing coastal ecosystems: salmon aquaculture concessions, crises, and governance challenges in Southern Chile. J Int Wildl Law Policy 19(4):284–300

Tickler D, Meeuwig JJ, Bryant K, David F, Forrest JAH, Gordon E, Larsen JJ, Oh B, Pauly D, Sumaila UR, Zeller D (2018a) Modern slavery and the race to fish. Nat Commun 9

Tickler D, Meeuwig JJ, Palomares ML, Pauly D, Zeller D (2018b) Far from home: distance patterns of global fishing fleets. Sci Adv 4(8):4–10

UN (2019) Global Sustainable Development Report 2019: the future is now – science for acheiving sustainable development. Page Independent Group of Scientists appointed by the Secretary-General, New York

Voyer M, Quirk G, McIlgorm A, Azmi K (2018) Shades of blue: what do competing interpretations of the Blue Economy mean for oceans governance? J Environ Policy Plan 20(5):595–616

Voyer M, Quirk G, Farmery AK, Kajlich L, Warner R (2021) Launching a Blue Economy: crucial first steps in designing a contextually sensitive and coherent approach. J Environ Policy Plan 23(3):345–362

Wedding LM, Reiter SM, Smith CR, Gjerde KM, Kittinger JN, Friedlander AM, Gaines SD, Clark MR (2015) Managing mining of the deep seabed. Science 349(6244):144–145

Part I
Knowledge Systems

Chapter 2
Knowing the Ocean: Epistemic Inequalities in Patterns of Science Collaboration

Anna-Katharina Hornidge, Stefan Partelow, and Kerstin Knopf

Abstract Ocean governance requires us to know the ocean. However, the knowledge systems that have shaped how and why we know the current ocean have been historically limited. In the present, they often subdue other knowledge systems that, if and when recognized and included into governing processes, not only move towards social justice and inclusion but can also improve decision-making and practical outcomes. The concept of epistemic inequalities encapsulates the disparities between different ways of knowing and their influence in ocean governance. For example, since the rise of colonial Europe, European-centric white male ideologies have long dominated global development practices. Within science, some disciplines have substantially more power than others, represented by funding and policy influence. In turn, local and indigenous knowledge systems, feminist ideologies and a broader range of highly valuable ways of knowing and doing in the sciences are far from equally participating in shaping ocean development discourses, decision-making and governance processes affecting the future of ocean sustainability. This chapter provides a theoretical basis for unpacking such epistemic inequalities in ocean governance, and thus setting a foundation for critically reflecting on the context and knowledge within the chapters of this book.

A.-K. Hornidge (✉)
German Institute of Development and Sustainability (IDOS) & University of Bonn, Bonn, Nordrhein-Westfalen, Germany
e-mail: Anna-Katharina.Hornidge@idos-research.de

S. Partelow
Leibniz Centre for Tropical Marine Research (ZMT), Bremen, Germany

Center for Life Ethics, University of Bonn, Bonn, Germany

K. Knopf
University of Bremen, Bremen, Germany

2.1   Introduction to Knowing & Governing Our Ocean

Governing our ocean requires us to know them: their structures, functions, internal processes, the resources and services they provide, as well as their carrying capacities, stressors and triggers of change. In-depth research forms the basis for our use, management and governance of the ocean, as well as how those actions shape sustainability outcomes (Campbell et al. 2016; Partelow et al. 2020b; Rudolph et al. 2020). However, these are not the only influences. Millennia of experiential knowledge of our marine and terrestrial ecosystems are embedded in our cultural practices, stories and ethics across coastal societies in the form of local and Indigenous knowledge (Drew 2005; Martin et al. 2007). Numerous studies have now shown the benefits of marine and coastal governance and management outcomes when knowledge integration can be achieved between different scientific, local, traditional and Indigenous knowledge systems and integrated in decision-making (Alexander et al. 2019; Porten et al. 2021; Poto et al. 2021). Nonetheless, epistemic inequalities remain widespread in ocean governance in terms of what types of knowledges are recognized, valued, supported and utilized as a form of power to inform decision-making.

How we know the ocean varies substantially around the world with regard to the respective ecosystems at hand, level and scale, disciplinary perspective, geographic area, method of data collection and analysis as well as with which thematic foci we approach the ocean. What individuals, communities and societies regard as knowledge or 'non-knowledge', and by that, what is worth knowing, protecting, sharing and further developing, represents different forms of past, present and future realities. Thus, how people see and read their realities and environments is determined not only by hypothesis testing and empirical positivism, but also by processes of meaning-construction and sense-making. These processes in turn shape societal norms, rules, and institutions. However, the sequence of effects also works in reverse through institutional structures – and the materialities those have resulted in – influencing processes of sense-making. While this ensures global diversity in engaging with earth systems, and in knowing and governing them, substantial global imbalances prevail in the systematic scientific assessment of local and regional ecosystems, with respective effects on how we globally know and can locally govern our earth systems.

The United Nations (UN) 2030 Agenda for Sustainable Development, and especially the Sustainable Development Goal (SDG) 14 'Life below Water – Conserve and sustainably use the worlds ocean, seas and marine resources for sustainable development' – marked a paradigmatic shift in the ways in which life on earth, whether terrestrial or aquatic, is to be globally valued and sustained. There is increasing awareness of the relevance of ocean-related science in the context of sustainable development, framing the biosphere as the base for all other SDGs in the ocean-climate-biodiversity nexus. Furthermore, the overall production of global ocean science is increasing (IOC-UNESCO 2017, 28). However, the ocean is not yet sufficiently   included   in   concepts   of   sustainable   development,   particularly concerning interlinkages, synergies, circular processes and trade-offs. This lack in mainstreaming ocean-related issues leads to underestimating given opportunities of ocean science in terms of narratives, models, theories of change and monitoring. The UN has declared 2021–2030 as the UN Decade of Ocean Science for Sustainable Development, with the tagline "The Science We Need for the Ocean We Want", addressing the many off-track indicators under SDG14 and challenges of ocean-related science. There are seven envisioned outcomes of the Decade, with the last entitled as 'An Inspiring and Engaging Ocean'. This explicitly supports the development of transformative ocean science as a means for globally fostering ocean literacy, meaning a thorough understanding of the ocean and its needs, in society. In doing so, the UN Ocean Science Decade refers to the Agenda 2030 as a guiding framework. Celebrated at the UN "Our Ocean" Conference in New York in June 2017, SDG 14 offers a global (while exclusive) platform for (re-)negotiating, overcoming and (re-)affirming hierarchies within and between different marine knowledge systems. Yet, what are marine knowledge systems? Furthermore, how are they characterized across different cultural and marine-environmental science contexts? In sum, what are these ocean knowledge systems that are being addressed by the UN Ocean Science Decade 2021–2030, and in particular by its aim to foster transformative ocean science and contribute to societal ocean literacy around the globe?

This chapter – in an overview manner – assesses these questions with regard to the ocean. How do we know the ocean? What characterizes the (largely) scientific and (less) non-scientific knowledge systems that engage with and study the ocean? Which infrastructures are in place, financed by whom? Which disciplinary organization do we find? Which thematic foci guide agenda setting processes and how basic versus applied are the questions asked and the answers given?

We reflect on these questions (1) by bringing together insights from the Global Ocean Science Report by the Intergovernmental Oceanographic Commission of the United Nations Educational, Scientific and Cultural Organization (IOC-UNESCO 2017), (2) by providing a synthesis of a series of review publications focused on analyzing the current state of marine science knowledge in published literature in specific fields (Barboza and Gimenez 2015; Aksnes and Browman 2016; Kim et al. 2016; Costa and Caldeira 2018; Mazaris et al. 2018; Partelow et al. 2018, 2020a; Pauna et al. 2019; Syed et al. 2019; Tolochko and Vadrot 2021; Cesarano et al. 2021), and (3) through a discussion linking ocean governance theory and practice.

Based on these, we argue that substantial 'epistemic inequalities' (Wellmon and Piper 2017) exist with regard to globally knowing the ocean and immensely hamper any regional and global attempts of coordinated or collaborative ocean governance. A globally comparable knowledge base, required for the implementation of, for example, a 'Common Heritage of Mankind' principle for the seafloor, is not given – as confirmed in the United Nations Convention of the Law of the Sea for the Area Beyond National Jurisdiction. Alice Vadrot and colleagues even go as far as arguing that the international world order is being contested through the principle in the field of marine biodiversity (Vadrot et al. 2021, 2022). As the challenges of our earths' ecosystems nevertheless require coordinated and collaborative global responses in the twenty-first century, the UN Ocean Science Decade thus sees itself challenged to overcome some of these immense inequalities in how we know the ocean and to create platforms for (a) substantially strengthening local and regional ocean knowledge systems, and (b) putting them in dialogue with each other on transregional and global levels. As we argue below, a solid and transregionally nurtured and anchored knowledge base with regard to the ocean is absolutely necessary for Ocean Governance in the coming years.

2.1.1 Knowledge System Diversity

Substantial scholarly work exists, assessing the manifold nature of different epistemic cultures and knowledge systems in subsistence and larger-scale agriculture in developing contexts (Wall 2008; Sanginga et al. 2009; Hornidge and Antweiler 2012; Hornidge et al. 2016). These works empirically document and analyze the interrelationships between high nature dependency in situations often characterized by rural peasant lifestyles, high social inequalities, and local ecology-related knowledge systems. However, there is substantially less knowledge assessing marine ecosystems and fisheries-related knowledge systems in comparably rural, subsistence-level lifestyles in developing contexts (Bavinck and Verrips 2020). We know surprisingly little about the unique characteristics, internal logics, negotiation powers, and peculiarities of marine knowledge systems of marine ecosystem-dependent communities, and how they may differ contextually, which may not allow us to make assumptions about those knowledge systems based on what we know from terrestrial systems.

'Knowledge systems' is a term we understand with reference to Karin Knorr Cetina's concept of 'epistemic cultures' as "those amalgams of arrangements and mechanisms – bonded through affinity, necessity, and historical coincidence – which, in a given field, make up how we know what we know" (Knorr-Cetina 1999, 1). Knorr Cetina illustrates in her own work that these epistemic cultures include small, clearly defined environments of knowledge production, as well as larger and less clearly defined environments of these environments, their preconditions, and their characterising elements. Processes of meaning-construction and sense-making determine how we see and interpret our environments while ourselves being influenced by the environments that surround us. Based on these constructions, we then establish norms, rules, and a wide range of different types of institutions for regulating our everyday lives. With respect to what is regarded as meaningful and how, the processes of sense-making themselves are influenced by former inter-subjectively shared interpretations of reality, by the institutional structures and materialities they have resulted in, and by guiding actors in their everyday practices towards the realisation of future imagined realities. These insights into the social and communicative construction of reality from the sociology of knowledge perspective provide a foundation for research into particular knowledge systems (Schütz 1932; Berger and Luckmann 1966; Schutz and Luckmann 1974). However, they say little about the qualitative nature of these epistemic realities specific to particular environmental contexts, or about the power structures shaping and shaped by them. This chapter thus aims to – in an overview manner – bring together insights on marine and ocean related knowledge systems as basis for ongoing discussions regarding transformative ocean science and the nurturing of ocean literacy in societies as part of the UN Ocean Science Decade 2021–2030. Below, we therefor seek to assess existing hierarchies and the contestation thereof of different stocks of marine resource-related knowledge in order to understand the underlying rationales, logics, and power interests in different subjective and objective interpretations of marine resource realities.

2.2   Synthesis of Ocean Science Knowledge and Capacities

2.2.1   Ocean Science Infrastructures

The Global Ocean Science Report by the Intergovernmental Oceanographic Commission of the United Nations Educational, Scientific and Cultural Organization (IOC-UNESCO 2017) globally assesses – for the first time ever – the status quo and current trends in ocean science capacity. By taking stock of who, how and where ocean science is conducted, the report states that "[t]he USA has the highest number of research institutions varying in size (p. 315) – roughly equal to the total number of research institutions in Europe combined and greatly exceeding the number of institutions operated in Asia and Africa". Assessing the type of researchers working in the field, the report interestingly states that the participation rate of female scientists in ocean research was 10% higher than the global share of female researchers across all natural scientific disciplines, and that they comprised on average 38% of all researchers across the marine sciences (p. 8). Underlining the importance of ocean science institutions, marine laboratories and field stations in more detail, the report identifies amongst the five largest Ocean science budgets in terms of percentage of national research and development funding those by the USA, Australia, Germany, France and the Republic of Korea (p. 27). The overall 784 marine field stations counted by the report are located in Asia (23%), Africa (8%), South America (10%) and Oceania (5%), as well as Europe (22%), North America (21%), and Antarctica (11%) (UNESCO-IOC 2017). Furthermore, the report counts 325 research vessels globally that were – at the time of writing the report – in operation and of which more than 60% belong to the Russian Federation, USA and Japan together. These range from 10 m to more than 65 m in length, with some built more than 60 years ago, while others have been in operation for less than 5 years. The average age of national fleets varies between <25 years (Norway, Bahamas, Japan and Spain) and >45 years (Canada, Australia and Mexico). As well, the report states that more than 40% of all research vessels focus on coastal research, while 20% engage in open ocean research (p. 26) (see Fig. 2.1).

The data collected for the report show differences in national stocktaking of the infrastructures and personnel in the sector. Despite these shortcomings they nevertheless indicate substantial differences in technological equipment and scientific

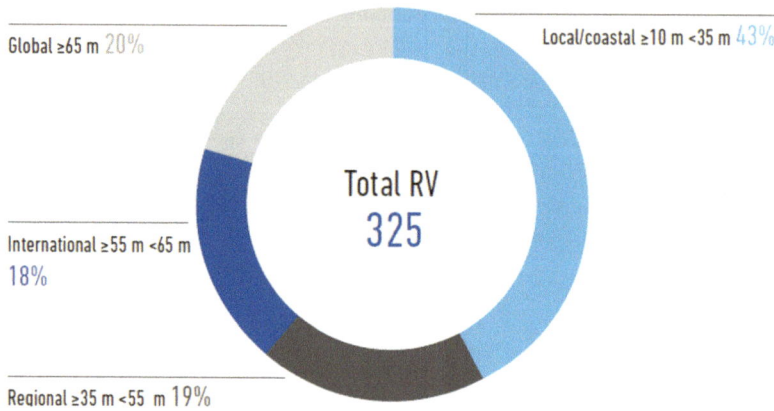

Fig. 2.1 Relative proportion of the different ship sizes summarizing all research vessels. (IOC-UNESCO 2017, 26)

capacity for studying the ocean. These differences in resources determine the knowledge production in the marine context due to varying capacities to actually conduct research on marine topics as well as differing access to specific research areas and equipment (e.g., research vessels, instruments for deep-sea activities and resource extraction). In addition, marine sciences are not bound to specific disciplines, but instead span the disciplinary range from natural to social sciences (Glaser et al. 2012; Markus et al. 2017; Partelow et al. 2018) with the common research objective of understanding coastal ecosystems, their functioning, use, management and governance, acting as a defining and uniting frame. Thus, specific knowledge systems and traditions shape ocean sciences and its research priorities. Due to existing hierarchies in knowledge production and sharing in the marine context, many actors worldwide are dependent on the research, which is conducted by the knowledge systems financed, organised and fostered by the above-mentioned nation states. These dependencies lead to international asymmetries, a limited range of databases and analyses, restricted access as well as gaps in our understanding of what the ocean is. This is not to say that the advancement and funding of research by the few dominant actors does not contribute substantially to global knowledge advancement, but rather that the interests and agendas of those states have taken precedent in shaping what we know, how we know, and what is done with that knowledge in a way that lacks global intellectual and cultural diversity. Furthermore, a few actors substantially influence the contextual insights that shape and fund what is valued, and thus pursued in practice, as a knowledge creation activity, as well as have control over who benefits from that knowledge and for what reason. In addition, we further know that prior knowledge shapes interest in what future knowledge creation pursuits should be. This is a form of path dependency, where past players largely control what we think is interesting scientifically, such as the research questions, methods and geographies, largely steering globally limited scientific capital. This has, historically, been limited to a select group of states that has largely missed the knowledge needs and values of more diverse world regions, as synthesized below.

2.3 Ocean Science in Publishing: Collaboration Patterns Across Countries and Regions

Global knowledge about the ocean is not equal across space, time, thematic areas or disciplinary lenses. Nor is it even in who, how or where it is produced. In practice, ocean knowledge production exists within, and is reinforced by, interdependent networks of science collaboration (Barboza and Gimenez 2015; Aksnes and Browman 2016; Kim et al. 2016; Costa and Caldeira 2018; Mazaris et al. 2018; Pauna et al. 2019; Syed et al. 2019; Partelow et al. 2020a; Tessnow-von Wysocki and Vadrot 2020; Tolochko and Vadrot 2021). Transregional network patterns and the actors within them are iteratively co-shaping each of their roles (or lack thereof) in those networks, leaving a science system with substantial path dependencies (likely future trajectories guided by historic patterns) and epistemic imbalances (what is worth knowing, why and who benefits) in terms of who is able to produce and access knowledge (and on which topics). It can be argued that this creates and reinforces scientific partnerships largely driven by access to material and immaterial infrastructures such as finance, language, thematic expertise and networks (Partelow et al. 2020a, b). As shown below, the challenge of deconstructing those path dependencies to foster eye-level science systems with valued contributions built on robust cooperative networks within and between Global North and Global South science systems is a distant reality, but one with steady progress.

As a necessary step towards fostering more comprehensive ocean literacy (Marrero et al. 2019), and to move towards a more equal and just version of that literacy, a bibliometric understanding of current scientific literature is a necessary starting point. In a systematic review of peer-reviewed publications in the field of tropical marine science, Partelow et al. (2018) highlight the dominance of natural science publication output compared to the social sciences in nearly every world region, with Southeast Asia being relatively balanced (Fig. 2.2b). Similarly, the spatial distribution of knowledge about tropical marine regions is unequal. Far more knowledge exists on Southeast Asia and northern Australia (classified separately), followed by the Pacific Islands, Central America and the Caribbean. East African knowledge has a comparatively little share, but is far ahead of West African and Sub-Saharan African research which represents a substantial gap in global ocean science. Similarly, Liquete et al. (2013) review patterns of global marine and coastal ecosystem service research. They importantly highlight a large number of case studies in Northern Europe and North America, which are primarily being done by researchers from those countries. In contrast, they show that there are indeed case studies in Central and South America, Africa, the Pacific as well as South and

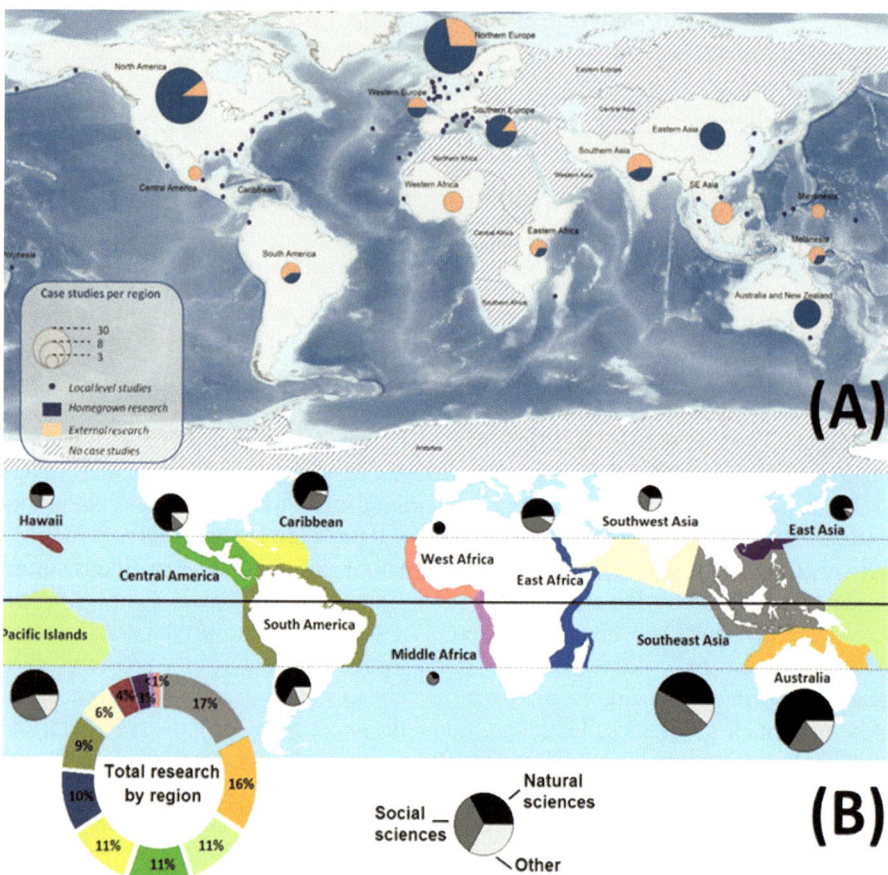

Fig. 2.2 (a) Spatial distribution of marine and coastal ecosystem service research taken from Liquete et al. (2013). Pie charts split by origin of research authors, domestic (blue) or external (orange). Most tropical research done by the UK and US. (b) Spatial distribution of tropical marine research taken from Partelow et al. (2018). Large knowledge gaps exist in West and Middle Africa as well as Southwest Asia

Southeast Asia, but the majority, if not all cases in those regions, are done by authors from outside those regions, predominantly the UK and US (Fig. 2.2a). Similar disparities have been shown in other global sustainability research areas, such as urbanization, where knowledge on the Global South is primarily produced by researchers in the Global North, although Global South sustainability challenges are fundamentally different (Nagendra et al. 2018).

The paradigmatic shift towards orienting both fundamental and applied science towards solving real world problems is an important driver for understanding patterns of emergent ocean literacy and discursive framing. This thematic area knowledge, or problem orientations, within the tropical marine sciences are also skewed. As a percentage of the literature, dominant social science problem framings are conservation (30.9%), commercial resource use (19.7%), tourism (9.7%), pollution/ degradation (9.0), subsistence resource use (7.9%) and none (5.6%). Dominant natural science problem frames are firstly, none (37.0%), followed by pollution/degradation (23%), conservation (10.9%) and commercial resource use (9.1%) (Partelow et al. 2018). Coral reefs dominate the ecosystem focus in the marine tropics, followed far behind by mangroves, estuaries/lagoons, intertidal ones, deep sea and others. In total, ~57% of tropical marine research is locally focused, compared to regional (36%) and global focused (7%). When split into specific scales, focus on ecosystem, spatial, management and temporal scale research far exceeds research on knowledge, institutional, jurisdictional or network scale research (Partelow et al. 2018). In addition, the majority of all research across both scale and discipline is skewed towards producing system knowledge (i.e., descriptive system functionality) with only a smaller subset of social science producing target knowledge (perspectives, values, goals) and transformative knowledge (actionable pathways for change). The more specific social and ecological system processes that tropical marine science has focused on are shown in Fig. 2.3.

Scientific collaboration networks can be measured using bibliometric data on co-authorship patterns as a broadly representative indicator of other formal and informal transregional cooperation. Drawing on data from Partelow et al. (2020a, b), co-authorship patterns in the field of tropical marine science are moving towards more international collaboration nearing 40%–50% of all peer-reviewed journal articles in 2016, 2017 and 2018, with domestic collaborations (all authors have the same country affiliation) increasing proportionally with the publication inflation rate over time. Single author papers have drastically decreased as a percentage of total output in tropical marine science research. Similarly, in the global fisheries

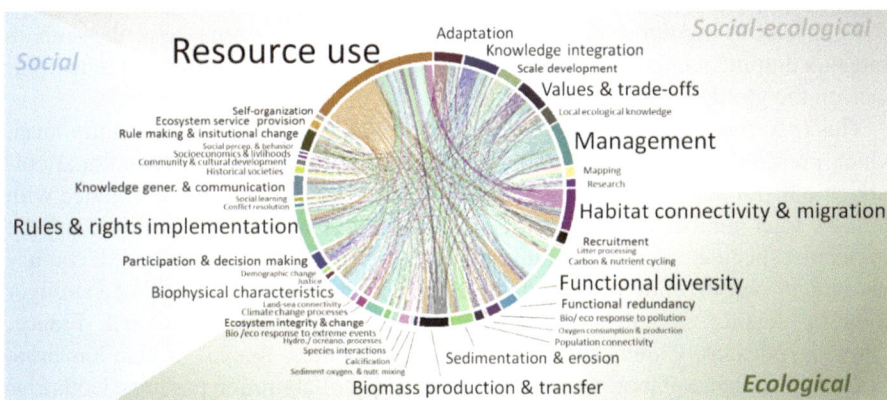

Fig. 2.3 Circle plot of the frequency and combined focus areas of publications that examine at least two system functions or processes, taken from Partelow et al. (2018). The proportion of the research focus that each process receives within multi- or interdisciplinary research is shown. This is visualized by the font size and the size of the colored segment of the circle. Also, process connectivity is shown. A connection between processes means that both processes were examined in the same publication

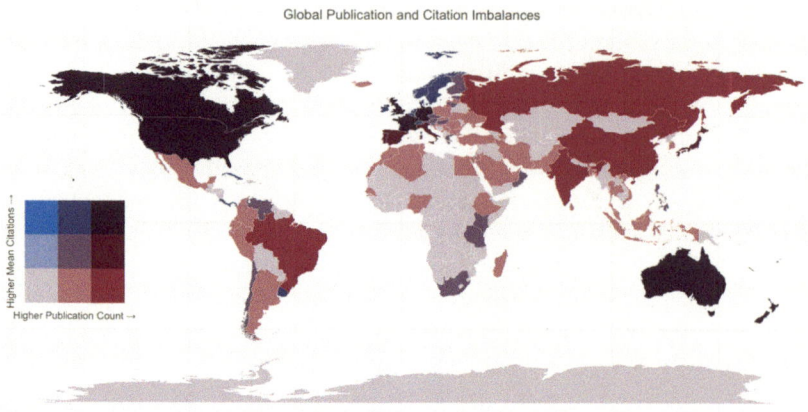

Fig. 2.4 Taken directly from Tolochko and Vadrot (2021), showing the geographic distribution of the total amount of articles and average citation count by country in English language peer-reviewed marine biodiversity literature between 1990 and 2018

science literature, Syed et al. (2019) provide a comprehensive analysis indicating that international collaboration outputs are increasing and single author outputs are decreasing. In the field of marine biodiversity research, Tolochko and Vadrot (2021) examine global collaboration networks, which show the dominance of the United States, European Union member states (namely Germany, France, UK), and Australia. They also provide data on the relationship between high output and high citations (Fig. 2.4), and while countries such as Brazil, India, China and Russia have high publication outputs, they have comparatively less citations. The Tolochko and Vadrot (2021) study considers only English language publications, and while the findings lead to numerous speculations as to why such patterns exist, the authors note that dominant countries have the highest 'collaboration capital' and thus influence on the global science system.

This is further more supported by Partelow et al. (2020a, b) at the country level, which presents findings indicating that the ratio of domestic to international collaborations (all publications classified as one or the other), is highly correlated with both the total number of collaborations a country has with other countries, and the number of specific countries a country collaborates with. More simply, if a country has a larger portion of domestic collaboration outputs (broadly indicating a stronger domestic science system such as in the UK, USA, Australia, Brazil, France, Germany, Mexico, China, India, Indonesia, Philippines, Kenya), it also has more total international collaborations and more specific collaboration partners. Countries with a larger portion of international collaboration outputs than domestic (perhaps indicating stronger dependence on external science systems), also have less total collaborations and less total specific countries with which they collaborate (e.g., small European countries, Chile, Cambodia, Argentina, Ghana, Pakistan). In tropical regions, the largest number of in-coming international collaborations are in Southeast and Southwest Asia as well as East Africa, with the fewest in West Africa,

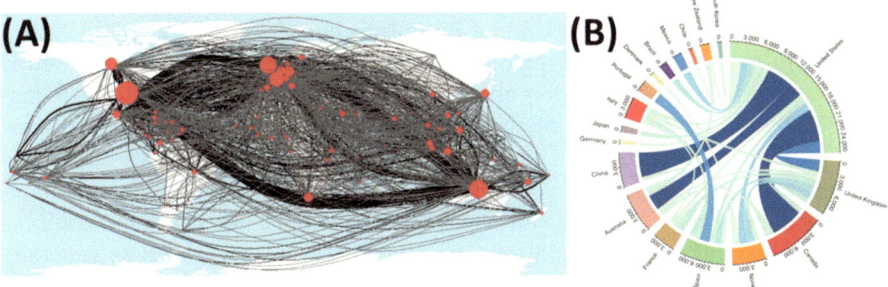

Fig. 2.5 (**a**) International co-authorship patterns between countries in the tropical marine science literature taken from Partelow et al. (2020a, b), with a dominant nexus between North America, European countries and Australia. (**b**) International co-authorship patterns between countries in global fisheries research taken from Syed et al. (2019), dominated by the US, Canada, European countries, Australia and China

indicating where international research partnerships exist (Fig. 2.5a). Globally, within tropical marine research, there is a Western-dominated nexus of science cooperation between Australia, North America and Europe (Partelow et al. 2020a, b). Syed et al. (2019), focused on global fisheries science networks, also show that the science powerhouses of USA, Canada, Japan, Australia, UK and Norway are now being joined by China, India, Mexico and Brazil. However, they also state that "as the field has become increasingly collaborative, historical links between European and North American countries have intensified" (p. 7), suggesting similar historical science cooperation dependencies (Fig. 2.5b).

Partelow et al. (2020a, b) also observe that the emergence of thematic areas or science agendas, indicated by clusters of terminology use over regions and time, are being driven by Australia, North America and Europe (as terminological anomalies i.e., new sets of words and phrases, emerge there first), later spreading to other world regions as part of a more mainstream discourse driven by Global North countries. This trend is supported in more specific fields such as within the 'ocean literacy' literature. Costa and Caldeira (2018) show that the concept of ocean literacy was started in the United States, and is currently dominated by publications from the United States, with other countries only beginning to adopt the term and publish on it years later. Back in the tropical marine science literature, Australian, North American and European countries lead the number of citations per publication per year with 5.8, 4.0 and 3.6 respectively, with all other regions below 3. Furthermore, Syed et al. (2019), in their global fisheries science analysis, find that North American and European countries publish in journals with higher impact factors and have higher rates of citations per paper. These findings are largely supported in a similar bibliometric analysis of global fisheries science literature, showing that there are no countries who have higher citation rates than the world average in the regions of South America, Africa or Asia except for China and South Korea (Aksnes and Browman 2016). Pauna et al. (2019) additionally show the dominance of the US, UK, Germany, France and Australia in marine microplastics research, with more diverse groupings of transregional cooperation although all cooperation clusters of countries are dominated by the US or a European country.

In more specific studies on regions and thematic areas, disparities in scientific collaboration patterns and outputs are broadly similar with variations in each context. Kim et al. (2016), analyzing the marine biodiversity research literature, show that European countries, USA, Canada and Australia are the dominant co-authorship partners for China, Japan and South Korea. Mazaris et al. (2018) show the dominance of the UK and USA in sea turtle research, both in the number of international co-authorship collaborations and total outputs. However, they also note generally increasing collaboration globally, with the increased role of some countries in maintaining regional networks such as Croatia, Tunisia and Costa Rica. In contrast, although a rapidly growing collaboration hub, they highlight Southeast Asia as a sea turtle research cooperation gap. In the field of marine microplastics pollution research, Barboza and Gimenez (2015) provide findings showing an increase in domestic and international collaborative outputs globally, although dominated by Europe and the US, but also in Japan and numerous Southeast Asian countries.

In sum, current ocean literacy is primarily dominated by the values, leadership and outputs of Global North science systems, namely North America, Europe and Australia, although other large economies are starting to play a larger role such as Japan, China, Brazil and Mexico. Despite the exponentially increasing amount of published science on the ocean, what we know is not based on a complete empirical picture. Many spatial, disciplinary and thematic area gaps exist, and many domestic science systems are not yet developed to the extent to which they can become mutually beneficial eye-level cooperation partners within global and regional science cooperation networks.

2.4 Discussion of Theory and Ocean Governance Practice

2.4.1 Epistemic Inequalities Between Knowledge Systems

In order to discuss the above trends and implications on ocean science systems, we begin with an overview of how to frame the epistemic inequalities between knowledge systems. When we speak of 'epistemic inequalities', we mean focusing on those between knowledge systems, and the different types of knowledge systems or ways of knowing such as those between different world regions, between scientific disciplines as well as between genders and sexual orientations, ethnicities, and other possibly defining lines. These 'epistemic inequalities' (Wellmon and Piper 2017) rest on structural path dependencies related to the science systems in different countries (Morgan et al. 2018; Partelow et al. 2020a) and determine the possibilities and limitations available for governance in a globally coordinated, jointly devised manner. It is important to stress that none of these 'knowledge systems', whether commonly regarded as originating in or connected to a particular world region, discipline, sex, age group or ethnicity, can be or is here regarded as a closed entity. Neither is any of them characterized by perceived homogeneity on the inside, or defined by clear-cut borders (thus representing container spaces). Instead, these knowledge systems are dynamic with porous borders, continuously (co-)evolving in and through the interaction, the exchange of ideas, ontological, and epistemological building blocks, and manifold forms of social, geographic, and epistemic mobilities (Mielke and Hornidge 2017a; Hornidge et al. 2020). Thus, rather than perceiving there to be variations and heterogeneity within one global knowledge system, these dynamics speak of different knowledge systems, which is further confirmed by existing hierarchical differences. Thus, knowledge systems are important to be assessed as units in their own right.

Not all knowledge systems are equally valued or even recognized, and thus a limited set of knowledge systems is more influential in shaping how and why scientific knowledge is created, and is utilized in decision-making, politics and governance. More simply, knowledge and power are closely intertwined. In the Foucauldian tradition, power and knowledge are understood to be inextricably related (Foucault 1980; Burchell et al. 1991). The nexus of power and knowledge can be productive as well as constraining: it can limit but also open new ways of acting and thinking. For example, the dominance of male Eurocentric understandings and practices of knowledge still affects patterns of knowledge systems such as which countries adopt and prioritize certain scientific disciplines, topics or governance approaches. In the ocean context, high nature dependencies, social inequalities and traditional/local knowledges have to be taken into account to analyse marine knowledge systems and power structures (Drew 2005; Martin et al. 2007). But many less adopted knowledge systems of traditional or Indigenous origin lack validation as useful and thus lack integration into decision-making forums that impact them directly. In the sense of everyday knowledge systems constructed in public-discourses at the interface of scientific, non-scientific, every day and traditional/local knowledges, analyses also need to consider political implications of marine knowledge systems including non-regarded and marginalized readings of the ocean (Cash et al. 2003; Ommer et al. 2012; Weichselgartner and Marandino 2012; Bennett 2016). They are shaped by given power structures and result in context-specific politics of knowledge.

In order to overcome existing asymmetries between knowledge systems that originated in unequal power structures and in turn constantly strengthen these power relations, marine knowledge systems need to be contextualized (Ommer et al. 2012; Weichselgartner and Marandino 2012). Still more research has to be conducted to further understand the unique characteristics, international logics, negotiation powers, and peculiarities of marine knowledge systems (Campbell et al. 2016; Blythe et al. 2021). Against this backdrop, a particular focus on marine ecosystem-dependent communities supports the assessment of existing hierarchies, and contestation thereof, of marine knowledge. Consequently, questions can be addressed of what the underlying rationales, logics, and power interests in different interpretation of marine realities are.

2.4.2 Epistemic Inequalities Between Scientific Disciplines

Within science, hierarchies between different types of scientific knowledge and structural processes of knowledge production are the result of the constant struggle for credibility and scientific authority via the search for the best argument or scientific findings. Outlining this struggle over epistemic authority, Gieryn assesses: "What science becomes, the borders and territories it assumes, the landmarks that give it meaning depend upon exigencies of the moment – who is struggling for credibility, what stakes are at risk, in front of which audiences, at what institutional arena?" (Gieryn 1999, x–xi). These struggles determine the defining boundaries of and hierarchies between basic versus applied sciences, between disciplines, but also, as empirically developed by Kohler (2002), between field and lab research. Based on a historical account of biological research, he argues: "Since the mid-nineteenth century, field biologists have lived in a world where lab disciplines have the greater credibility and authority, and they do still" (Kohler 2002, 307). Similar distinctions and structurally nurtured hierarchical differences can also be observed with regard to different disciplines. Especially scholarly work on the organisation of interdisciplinary research endeavours, bringing together natural and social sciences, empirically illustrates the need to overcome these hierarchies as precondition for cooperation at eye-level and interdisciplinary forms of knowledge production in its own right. Peter Mollinga (2008, 2010) for instance argues for the 'rational organisation of dissent' in interdisciplinary research settings as a crucial determinant for academic excellence without being apoliticised.

2.4.3 Epistemic Inequalities in Gendered Ocean Science

The patriarchal organization of the vast majority of societies practiced globally over centuries has resulted in gendered epistemes, in all aspects of social organization in which strong gender divisions in terms of exercising tasks prevailed. Gendered lenses in defining what is regarded as knowledge in and by society were the consequence (Doucet and Mauthner 2006). In connection with women's very late admittance to universities, also the breadth of women's academic achievements was largely truncated and only a selective list of women pioneers in their disciplines heralded. And while these forms of historically generated appropriations of women's knowledge are increasingly challenged, substantial shifts in male-dominated hierarchies in academia are statistically seen still outstanding (Fatnowna and Pickett 2002). Kristie Dotson, drawing on Miranda Fricker (1999), for instance, speaks of 'epistemic oppression' and points to "the persistent epistemic exclusion that hinders one's contribution to knowledge production" (Dotson 2014, 115). For developing her argument, she refers to postcolonial and gender-related contexts of exclusion, illustrating the interplay, but also succinct differences, between social, political, and epistemic oppression as well as 'privilege'. Ian James Kidd, José Medina, and Gaile Pohlhaus take this further and have developed 'epistemic injustice' as a research category that integrates a variety of research topics and areas across major social and intellectual movements and fields, such as philosophy, feminism, hermeneutics, critical race theory, disability studies, and decolonising and queer epistemology studies (Fricker 2007; Kidd et al. 2017).

In response to this, in the 1980s, feminist interventions started developing feminist epistemologies and methodologies (e.g., Code 1981; Harding 1987; Haraway 1988; Lennon and Whitford 1994; Longino 1997; Fawcett and Hearn 2004; Doucet and Mauthner 2006). The authors built on the premise that women due to being socialized into particular gender-specific role patterns and social identities regard the world in many ways differently from their male counterparts. It was argued that through the development of feminist methodologies, female epistemologies could be empirically assessed and advanced in public, official, and academic discourses, while at the same time grappling with basic questions such as the nature of knowledge, epistemic agency, justification, and objectivity in general (Alcoff and Potter 1993; Doucet and Mauthner 2006).

Within the marine field, the gendered life worlds of marine-based societies, whether in the context of industrial and small-scale fisheries, or within the multifarious realities aboard ships and vessels, have been amply documented, particularly in terms of how sailing, surfing, maritime navigation, and other forms of seafaring have historically been perceived as distinctly "masculinized" practices (Mack 2011, 30; Laderman 2014). Yet these (interpretative) gendered essentialisms have also been critiqued across anthropological and transcultural scholarship spanning Oceania and the Mekong borderlands to Madagascar (cf. Astuti 1995; Probyn 2014; Gissi et al. 2018), which in turn illustrate the (internally diverse) livelihood practices, ontologies, and epistemologies of distinct sub-groups such as female pearl divers or Indigenous fisherwomen. However, in the context of scientific knowledge production, the gendered inequalities in marine epistemes come to be revealed in the relative (in)visibility of diverse stocks of knowledge about how marine life is perceived, experienced, and differently studied. Moreover, nascent scholarship in interdisciplinary fields such as Science and Technology Studies (STS) that explore epistemic cultures of knowledge production, particularly in the marine realm, often barely address the gendered nuances in science-oriented meaning-making (cf. Helmreich 2009), while conceptual strands such as feminist and postcolonial STS have conventionally dealt with questions that have largely been driven by terrestrially-oriented disciplines (e.g., botany, forensic science, clinical research), often produced in firmly 'grounded' spaces such as chemical laboratories, engineering, and medical institutes (cf. Harding 2011; Subramaniam 2014). Thus the gendered epistemic dynamics inherent in the liminal floating worlds of knowledge production (for example on submarines and research vessels) are only but beginning to be explored across the marine humanities and the social sciences.

2.4.4   Transregional Networks of Knowing & Governing

Knowledge production with regard to the world ocean is – as all knowledge production on global commons – shaped by the transregional networks driving it and thus by the interests, values, logics and (legal, financial) structures shaping these. As such, the above outlined scientometric analysis on peer-reviewed journal articles in the field of tropical marine sciences identified a set of material and immaterial path dependencies co-shaping how we know the systems of tropical coastal waters (Partelow et al. 2020a, b). Material path dependencies include equipment, labs, and access to research vessels and marine research stations. Immaterial path dependencies include access to funding and donor landscapes, language of research and teaching, science networks and discipline. These link with larger discussions by postcolonial scholars on historically grown knowledge hierarchies (emerging out of the Enlightenment period of Europe) between normatized standard European and neglected non-European knowledge systems. As such, Gloria Emeagwali (2006), Dipesh Chakrabarty (2000), and other scholars have pointed to the 'intellectual dominance' of the West as being legitimized by way of colonial histories, which have resulted in presumably 'destined' trajectories that re-ordered the world and 'naturalized' cultural hierarchies, and of thus 'grown' all-encompassing epistemologies rooted in the Greco-Roman worlds. Gayatri Spivak criticizes that Enlightenment humanism did not include non-European cultures in its understanding of 'man', who was rather understood as the "settler-colonial white man" (Spivak 1999, 26). Chakrabarty, with his concept of 'provincializing Europe' (Chakrabarty 2000), seeks to unveil the constructed nature of universalist assumptions and to engage Western and non-Western histories and knowledges in equilibrious negotiation in order to "displace a hyperreal Europe from the center" (p. 45). Walter Mignolo connects the "coloniality of power (economic and political)" with the "coloniality of knowledge and of being (gender, sexuality, subjectivity and knowledge)" as entangled characteristics of modern society that constantly reproduce "coloniality" and calls for a 'pluriversality' of knowledge production (2007, 450–53, 2012, 49, 51–60). These ideas are further taken up by scholars such as Linda Tuhiwai Smith (1999), Margaret Kovach (2010), and Gurminder K. Bhambra (2010), who make clear that prioritized Western-based research practices and policies reproduce colonial relationships in the academy and that the epistemological challenge is to achieve a "systemic shift in the ideology of knowledge production" (Kovach 2010, 28; cf. Knopf 2018a, b).

More recent debates in Area Studies (Mielke and Hornidge 2017a, b; Derichs 2017; Middell 2013; Jackson 2017) bring these postcolonial assessments together with increasing geographic, social, and epistemic mobilities, and thus with questions on how the travel of goods, people, ideas, capital, lifestyles, and symbols render perspectives on the world as divided into particular world regions, each defined by a set of cultural characteristics and languages on the inside and different ones on the outside (i.e., defining regions as 'container spaces'). Instead, de-territorial perspectives on how social realities are being negotiated are being discussed

(Hornidge et al. 2020). Here, the ocean is also gaining attention as a transregional water body and global common that challenges and offers substantial opportunity for joint understanding and governing (Mielke and Hornidge 2017b; Alff and Hornidge 2019; Tessnow-von Wysocki and Vadrot 2020).

2.5 Final Remarks: Regimes of Knowing for a World Beyond 2030

Knowing the world ocean is necessary for living with global challenges. Yet, knowing it requires pluriversality, and thus transregional dialogue processes that are structured by reduced hierarchies that allow mutually understanding and learning. Recognizing the many different ways of knowing, and valuing the contributions and different epistemic views and knowledges is essential for social justice and inclusion, and also for the progression of science and governance. As the focus on ocean matures into twenty-first century development discourses, policies and governance practices, enabling the transition towards more equal epistemic ways of knowing and doing will require re-shaping the structures of knowledge production. This will entail large, self-reflective and proactive efforts to materialize, where the processes of recognition and actions towards change themselves will play a large role in manifesting new integrated knowledge landscapes premised on pluralism. Building on the introduction to this book, and as we will see in the forthcoming chapters, ocean governance for sustainability requires knowing the ocean, and how and why we know the ocean in part to be a reflection of the science-policy interfaces and knowledge governance practices that enable and constrain its diversity, integration and uptake. This chapter has provided an overview of some of the theoretical foundations with which the chapters in this book can be reflected upon. In many ways, the book is about the nexus of knowledge and governance – a nexus shaping society's path towards sustainability.

References

Aksnes DW, Browman HI (2016) An overview of global research effort in fisheries science. ICES J Mar Sci 73(4):1004–1011
Alcoff L, Potter E (1993) Feminist epistemologies. Routledge, New York
Alexander SM, Provencher JF, Henri DA, Taylor JJ, Lloren JI, Nanayakkara L, Johnson JT, Cooke SJ (2019) Bridging indigenous and science-based knowledge in coastal and marine research, monitoring, and management in Canada. Environ Evid 8(1):1–24
Alff H, Hornidge A-K (2019) 'Transformation' in the study of international development: across disciplines, knowledge hierarchies and oceanic spaces. In: Baud ISA, Basile E, Kontinen T, Itter SV (eds) Building development studies for the new millennium. Palgrave Macmillan, Cham, pp 141–162

Astuti R (1995) People of the sea: identity and descent among the Vezo of Madagascar. Cambridge University Press, Melbourne

Barboza LGA, Gimenez BCG (2015) Microplastics in the marine environment: current trends and future perspectives. Mar Pollut Bull 97:5–12

Bavinck M, Verrips J (2020) Manifesto for the marine social sciences. Marit Stud 19(2):121–123

Bennett NJ (2016) Using perceptions as evidence to improve conservation and environmental management. Conserv Biol 30(3):582–592

Berger PL, Luckmann T (1966) The social construction of reality: a treatise in the sociology of knowledge. Anchor Books, New York

Bhambra GK (2010) Sociology after postcolonialism: provincialized cosmopolitanisms and connected sociologies. In: Rodríguez EG, Boatcă M, Costa S (eds) Decolonizing European sociology: transdisciplinary approaches. Routledge, pp 33–47

Blythe JL, Armitage D, Bennett NJ, Silver JJ, Song AM (2021) The politics of ocean governance transformations. Front Mar Sci 8(July):1–12

Burchell G, Gordon C, Miller P (1991) Governmental rationality: an introduction. In: Burchell G, Gordon C, Mille P (eds) The Foucault effect: studies in governmentality. The University of Chicago Press, Chicago, pp 1–52

Campbell LM, Gray NJ, Fairbanks L, Silver JJ, Gruby RL, Dubik BA, Basurto X (2016) Global oceans governance: new and emerging issues. Annu Rev Environ Resour 41(1):517–543

Cash DW, Clark WC, Alcock F, Dickson NM, Eckley N, Guston DH, Jäger J, Mitchell RB (2003) Knowledge systems for sustainable development. Proc Natl Acad Sci U S A 100(14):8086–8091

Cesarano C, Aulicino G, Cerrano C, Ponti M, Puce S (2021) Scientific knowledge on marine beach litter: a bibliometric analysis. Mar Pollut Bull 173(PB):113102

Chakrabarty D (2000) Provincializing Europe: postcolonial thought and historical difference. Princeton University Press, Princeton

Code L (1981) Is the sex of the knower epistemologically significant? Metaphilosophy 12:267–276

Costa S, Caldeira R (2018) Bibliometric analysis of ocean literacy: an underrated term in the scientific literature. Mar Policy 87(October 2017):149–157

Derichs C (2017) Knowledge production, area studies and global cooperation. Routledge, Abingdon/Oxon

Dotson K (2014) Conceptualizing epistemic oppression. Soc Epistemol 28(2):115–138

Doucet A, Mauthner D (2006) Feminist methodologies and epistemology. In: Peck DL, Bryant CD (eds) Handbook of 21st century sociology, vol 2. Sage, Thousand Oaks, pp 36–43

Drew JA (2005) Use of traditional ecological knowledge in marine conservation. Conserv Biol 19(4):1286–1293

Emeagwali G (ed) (2006) Africa and the academy: challenging hegemonic discourses on Africa. Africa World Press, Trenton

Fatnowna S, Pickett H (2002) The place of indigenous knowledge systems in the post-postmodern integrative paradigm shift. In: Hoppers CAO (ed) Indigenous knowledge and the integration of knowledge systems: towards a philosophy of articulation. New Africa Education, Claremont, pp 257–285

Fawcett B, Hearn J (2004) Researching others: epistemology, experience, standpoints and participation. Int J Soc Res Methodol 7(3):201–218

Foucault M (1980) Power/knowledge: selected interviews and other writings 1972–1977. Gordon C. (eds). Pantheon Books, New York

Fricker M (1999) Epistemic oppression and epistemic privilege. Can J Philos 25:191–210

Fricker M (2007) Epistemic injustice: power and the ethics of knowing. Oxford University Press, Oxford

Gieryn T (1999) Cultural boundaries of science: credibility on the line. University of Chicago Press, Chicago

Gissi E, Portman ME, and Hornidge A (2018) "Un-Gendering the Ocean: Why Women Matter in Ocean Governance for Sustainability." Marine Policy 94 (November 2017): 215–19. https://doi.org/10.1016/j.marpol.2018.05.020.

Glaser M, Christie P, Diele K, Dsikowitzky L, Ferse S, Nordhaus I, Schlüter A, Schwerdtner Mañez K, Wild C (2012) Measuring and understanding sustainability-enhancing processes in tropical coastal and marine social–ecological systems. Curr Opin Environ Sustain 4(3):300–308

Haraway D (1988) Situated knowledges: the science question in feminism and the privilege of partial perspective. Fem Stud 14:575–599

Harding S (1987) Conclusion: epistemological questions. In: Harding S (ed) Feminism and methodology. Indiana University Press, Bloomington, pp 181–190

Harding S (2011) Beyond postcolonial theory: two undertheorized perspectives in science and technology. In: Harding S (ed) The postcolonial science and technology studies reader. Duke University Press, Durham, pp 1–31

Helmreich S (2009) Alien Ocean: anthropological voyages in microbial seas. University of California Press, Berkeley

Hornidge A-K, Antweiler C (2012) Environmental uncertainty and local knowledge: Southeast Asia as a laboratory of global ecological change. Hornidge A-K, Antweiler C (eds). Transcript Verlag, Bielefeld

Hornidge A-K, Shtaltovna A, Schetter C (2016) Agricultural knowledge and knowledge systems in post-soviet societies. Peter Lang, Bern

Hornidge A, Herbeck J, Zoysa RS, Flitner M (2020) Epistemic mobilities: following sea-level change adaptation practices in southeast Asian cities. Am Behav Sci 64(10):1497–1511

Jackson P (2017) The neoliberal university and global immobilities of theory. In: Mielke K, Hornidge A-K (eds) Area studies at the crossroads: knowledge production after the mobility turn. Palgrave Macmillan, New York, pp 27–44

Kidd IJ, Medina J, Pohlhaus G (2017) The Routledge handbook of epistemic injustice. Routledge, London

Kim J, Lee S, Shim W, Kang J (2016) A mapping of marine biodiversity research trends and collaboration in the East Asia region from 1996-2015. Sustainability (Switzerland) 8(10):1075

Knopf K (2018a) Introduction. In: Indigenous knowledges and academic discourses; Les savoirs autochtones et les discours scientifiques; Indigenes Wissen und Akademische Diskurse. Special Edition of Zeitschrift für Kanada-Studien (ZKS) 67(1): 10–23. http://www.kanada-studien.org/wp-content/uploads/2020/03/ZKS_2018-67_2Knopf.pdf

Knopf K (2018b) Indigenous knowledges, ecology, and living heritage in North America. In: Arregui A, Mackenthun G, Wodianka S (eds) DEcolonial heritage: natures, cultures, and the asymmetries of memory. Waxmann, Münster, pp 175–202

Knorr-Cetina KD (1999) Epistemic cultures: how the sciences make knowledge. Harvard University Press, Cambridge

Kohler RE (2002) Landscapes and labscapes exploring the lab-field border in biology. University of Chicago Press, Chicago

Kovach M (2010) Indigenous methodologies: characteristics, conversations, and contexts. University of Toronto Press, Toronto

Laderman S (2014) Empire in waves: a political history of surfing. University of California Press, Berkeley

Lennon K, Whitford M (1994) Introduction. In: Lennon K, Whitford M (eds) Knowing the difference: feminist perspectives in epistemology. Routledge, London, pp 1–14

Liquete C, Piroddi C, Drakou EG, Gurney L, Katsanevakis S, Charef A, Egoh B (2013) Current status and future prospects for the assessment of marine and coastal ecosystem services: a systematic review. PLoS One 8(7):e67737

Longino HE (1997) Feminist epistemology as a local epistemology. Aristot Soc 71:19–35

Mack J (2011) The sea: a cultural history. Reaktion Books, London

Markus T, Hillebrand H, Hornidge A-K, Krause G, Schlüter A (2017) Disciplinary diversity in marine sciences: the urgent case for an integration of research. ICES J Mar Sci 75(2):502–509

Marrero ME, Payne DL, Breidahl H (2019) The case for collaboration to foster global ocean literacy. Front Mar Sci 6(February 2007):1–2

Martin KS, McCay BJ, Murray GD, Johnson TR, Oles B (2007) Communities, knowledge and fisheries of the future. Int J Glob Environ Issues 7(2/3):221

Mazaris AD, Gkazinou C, Almpanidou V, Balazs G (2018) The sociology of sea turtle research: evidence on a global expansion of co-authorship networks. Biodivers Conserv 27(6):1503–1516

Middell M (ed) (2013) Self-reflexive area studies, Global history and international studies 5. Leipziger Universitätsverlag, Leipzig

Mielke K, Hornidge A-K (2017a) Area studies at the crossroads: Knowledge Production after the Mobility Turn. Palgrave MacMillan, New York

Mielke K, Hornidge AK (2017b) Area studies at the crossroads: knowledge production after the mobility turn. In: Area studies at the crossroads: knowledge production after the mobility turn. Palgrave Macmillan, New York

Mignolo WD (2007) Delinking: the rhetoric of modernity, the logic of coloniality and the grammar of de-coloniality. Cult Stud 21(March/May 2/3):449–514

Mignolo WD (2012) Local histories/global designs: coloniality, subaltern knowledges, and border thinking. Princeton University Press, Princeton

Mollinga PP (2008) The rational organisation of dissent: boundary concepts, boundary objects and boundary settings in the interdisciplinary study of natural resources management

Mollinga PP (2010) Boundary work and the complexity of natural resources management. Crop Sci 50(Supplement 1):S-1-S-9

Morgan AC, Economou DJ, Way SF, Clauset A (2018) Prestige drives epistemic inequality in the diffusion of scientific ideas. EPJ Data Sci 7:1

Nagendra H, Bai X, Brondizio ES, Lwasa S (2018) The urban south and the predicament of global sustainability. Nat Sustain 1(7):341–349

Ommer RE, Ian Perry R, Murray G, Neis B (2012) Social–ecological dynamism, knowledge, and sustainable coastal marine fisheries. Curr Opin Environ Sustain 4(3):316–322

Partelow S, Schlüter A, von Wehrden H, Jänig M, Senff P (2018) A sustainability agenda for tropical marine science. Conserv Lett 11(1):1–14. http://doi.wiley.com/10.1111/conl.12351

Partelow S, Hornidge A, Senff P, Stäbler M, Schlüter A (2020a) Tropical marine sciences: knowledge production in a web of path dependencies. PLoS One 15(2): https://doi.org/10.1371/journal.pone.0228613

Partelow S, Schlüter A, Armitage D, Bavinck M, Carlisle K, Gruby RL, Hornidge A-K, Le Tissier M, Pittman JB, Song AM, Sousa LP, Văidianu N, Van Assche K (2020b) Environmental governance theories: a review and application to coastal systems. Ecol Soc 25(4):1–21. https://doi.org/10.5751/ES-12067-250419

Pauna VH, Buonocore E, Renzi M, Russo GF, Franzese PP (2019) The issue of microplastics in marine ecosystems: a bibliometric network analysis. Mar Pollut Bull 149(September):110612

Poto MP, Kuhn A, Tsiouvalas A, Hodgson KK, Treffenfeldt MV, Beitl CM (2021) Knowledge integration and good marine governance: a multidisciplinary analysis and critical synopsis. Hum Ecol 50:125–139

Probyn E (2014) Women following fish in a more-than-human world. Gender, Place & Culture 21(5):589–603. https://doi.org/10.1080/0966369X.2013.810597

Rudolph TB, Ruckelshaus M, Swilling M, Allison EH, Österblom H, Gelcich S, Mbatha P (2020) A transition to sustainable ocean governance. Nat Commun 11(1):1–14

Sanginga P, Waters-Bayer A, Kaari S, Njuki J, Wettasinha C (2009) Innovation Africa. Enriching farmers' livelihoods. Earthscan, London

Schütz A (1932) Grundzüge einer Theorie des Fremdverstehens. In: Der Sinnhafte Aufbau der Sozialen Welt. Springer, Vienna

Schutz A, Luckmann T (1974) The structures of the life-world. Heinemann, London

Spivak GC (1999) A critique of postcolonial reason: toward a history of the vanishing present. Harvard University Press, Cambridge

Subramaniam B (2014) Ghost stories for Darwin: the science of variation and the politics of diversity. University of Illinois Press, Champaign, IL, USA

Syed S, ní Aodha L, Scougal C, Spruit M (2019) Mapping the global network of fisheries science collaboration. Fish Fish 20:830–856. https://doi.org/10.1111/faf.12379

Tessnow-von Wysocki I, Vadrot ABM (2020) The voice of science on marine biodiversity negotiations: a systematic literature review. Front Mar Sci 7(December):20201223

Tolochko P, Vadrot ABM (2021) The usual suspects? Distribution of collaboration capital in marine biodiversity research. Mar Policy 124:104318

Tuhiwai Smith L (1999) Decolonizing methodologies: research and indigenous peoples. Zed Books, London

UNESCO-IOC (2017) Global ocean science report: the current status of ocean science around the world. UNESCO, Paris

Vadrot ABM, Langlet A, Tessnow-Von Wysocki I, Tolochko P, Brogat E, Ruiz-Rodríguez SC (2021) Marine biodiversity negotiations during covid-19: a new role for digital diplomacy? Glob Environ Polit 21(3):169–186

Vadrot ABM, Langlet A, Tessnow-von Wysocki I (2022) Who owns marine biodiversity? Contesting the world order through the 'common heritage of humankind' principle. Environ Polit 31(2):226–250

von der Porten S, Ota Y, Mucina D (2021) Indigenous knowledge, knowledge-holders and marine environmental governance. In: Thornton TF, Bhagwat SA (eds) The Routledge handbook of indigenous environmental knowledge, 1st edn. Taylor & Francis, London

Wall C (2008) Argrods of Western Uzbekistan. Knowledge control and agriculture in Khorem. LIT Verlag, Münster

Weichselgartner J, Marandino C a (2012) Priority knowledge for marine environments: challenges at the science–society nexus. Curr Opin Environ Sustain 4(3):323–330

Wellmon C, Piper A (2017) Publication, power, and patronage: on inequality and academic publishing. Crit Inq 21:1–20

Chapter 3
Managing Fish or Governing Fisheries? An Historical Recount of Marine Resources Governance in the Context of Latin America – The Ecuadorian Case

María José Barragán-Paladines, Michael Schoon, Winny Collot D'Escury, and Ratana Chuenpagdee

Abstract The narratives and images about ocean and its resources governance, their use and value have deep roots in human history. Traditionally, the contemporary images of fish and fisheries have been shaped under the cultural construction of power, wealth and exclusion, and also as one of poverty and marginalization. This perception was formed on early notions of natural (marine) resources access and use that were born within the colonial machinery that ruled the world from the Middle Ages until late XVII. This research explores the historical overview of marine resources usage and governance in Latin America, from a 'critical approach to development' perspective, by following a narrative description based on a 'three-acts' format. It illustrates how and to what extent politics, power and knowledge have deeply influenced policies and practices at exploring the marine and terrestrial resources and at managing fish and seafood, historically, and how the fisheries resources' management practices are influenced by principles of appropriation, regulation and usage, put in place already in the XV century that were imposed at the conquering and colonization of the Americas, disregarded previous governance practices. This article argues that fisheries governance cannot be improved without some appreciation for the social, historical, geopolitical, and cultural significance of the fishing resources themselves, of the perceptions of them by humans, and of the interactions Global North-Global South. The analysis also opens the dialogue about

M. J. Barragán-Paladines (✉)
Charles Research Station, Charles Darwin Foundation, Santa Cruz, Galapagos, Ecuador
e-mail: mariajose.barragan@fcdarwin.org.ec

M. Schoon
School of Sustainability, Arizona State University, Tempe, AZ, USA

W. C. D'Escury
Independent Researcher, Leiden, The Netherlands

R. Chuenpagdee
Geography Department, Memorial University of Newfoundland, St. John's, NL, Canada

© German Institute of Development and Sustainability (IDOS) 2023   47
S. Partelow et al. (eds.), *Ocean Governance*, MARE Publication Series 25,
https://doi.org/10.1007/978-3-031-20740-2_3

what kind of ocean and governance "science" we want, to support decisions, policies and practices regarding fisheries governance. Final thoughts highlight a reflection about whose knowledge is created and used to support decision and policy making in Ecuador.

3.1 Early Images of Fisheries – The Notions of "Fish"

From very early moments in human history, the sea called the fascination and attention of humans, and provided them with food, livelihoods and resources. In fact, the long-lasting relationship between humans and the sea critically defined how humans evolved, where they blossomed and how marine systems became vital for us. Since the emergence of ancestors of *Homo sapiens*, modern humans were already engaged in exploration, discovery, usage, management and governance of the ocean goods and services, at different scales and with diverse formats. Under that line, the sea, the fish and other marine resources that humans have used and taken advantage of since early moments in human history have been associated with the varied meanings and images of the unknown dimension of the 'salty world'. These images, created and recreated along with human existence, were not formed in a vacuum. In fact, they have been shaped by the values, culture, political and social dimensions, circulating around the fish and fisheries, in specific moments of history and under specific circumstances.

The following sections describe two historical moments where the relationship between the humans and the marine resources were shaped under formats driven by politics, geopolitics, economy and varied interests, some of which still are functional. The examples featured fall within two key periods in the human history, and illustrate how the fish and marine resources were imagined and governed along millennia, until our times with deep implications within the historical, geopolitical, social and cultural dimensions.

This chapter starts by presenting an overview of a couple of key narratives that dominated past and recent debates on fish, fisheries and their governance. The analysis reveals how different images (past and current), views and interpretations prevail amongst 'rulers', 'users' and 'experts' about what the governing issues of ocean and marine resources are, and consequently what the solutions addressing these difficulties should be. We then explore how the sustainability idea developed within those narratives and how this notion relates to the question of improved ocean governance. We focus on a description that essentially illustrates the history of marine resources governance and discuss how these formats of appropriation relate to unsustainable fishing and other marine related activities and practices. We conclude with a discussion of the future implications for fish and marine resource governance, of the current governing formats, policies and practices.

For the purpose of this chapter, the story is presented following a 'three-act' structure, based on the historical facts of marine resources usage, management and governance in the past, that fundamentally influenced the way we currently look at and imagine 'fish' and 'fisheries' and other marine resources. Act I – The '*Discovery*' of the commons; Act II – The '*Appropriation*' of the commons; and Act III – The '*Benefits*' from the ocean commons governance.

3.1.1 Roman Times

One basic idea that has deeply influenced the first notion of the ocean by humans and all that it fosters is the closeness with 'fish' and all the dimensions associated to it. In fact, making a living by the sea, implicitly connected members of those societies to the growing development poles around the known world. Already in 200 d.c., the factories of fermented fish (i.e., "salazón" in Spanish or *garum* in Latin) became an important economic activity, producing a highly valuable commodity, which was traded around the entire Mediterranean (Cayo Plinio II, 23–79, 1999). In those years, the production of *garum*, first mentioned by Ateneo de Naucratis and Dífilo de Sinope (in IV a.C), was described as the "*well known salazón from Sexi, of Hispania*". Already the roman naturalist Cayo Plinio the Second, in his "*Naturales Historine*", mentions the existence of a "*abundant fish in the coast of Sexi*", and related it to the "industry of the Iberian *salazón*", a Hispanic-roman gourmet tradition that used an authentic and expensive fermented fish-based sauce, also called '*garum*' (Bernal et al. 2018), that was produced, exported and traded by the romans within the borders of their imperium (Bernal-Casasola et al. 2016). Other references (Portillo Sotelo et al. 2020) have mentioned *garum* and highlighted the importance of this product in the Mediterranean diet. Within the Roman Empire's economy and industry, *garum* was an important good that, being originated as a fish-based-processed commodity, its production, logistics, and social prestige associated to its usage at a gourmet scale, had great implications within the geopolitics of the largest empire at that time and with the marine resources governance (Asingh and Damm 2020).

3.1.2 Middle Age – Colonial Mindset – Fifteenth Century

During the late Middle Ages, the notions of fish, fisheries, the ocean, and its creatures were imagined as mysterious, ghostly, and even devilish shapes that threatened the spiritual and corporal human wellbeing (Mc Dughann 2002). Along those years, the consumption of fish and fish-related produce was sustained, and some societies privileged some species against others. During those times, the ownership of ships, factories, slaves, or having the 'know-how' to produce *garum* were already representations of diverse formants of marine resource governance, within the entire value chain levels. Additionally, some management tools, that could be seen as 'modern instruments' were already in place on early moments of marine resource governance, such as resource ownership, and areas of exclusion. In fact, at that time, their design and implementation were not based on scientific knowledge, and neither were they unbiased, objective and impartial technical-driven tools, but rather value laden and interest-based instruments. One illustration of 'marine resource ownership allocation' was presented by Cristobal Colón, on 15th August 1498, when he took possession of the islands Margarita and Cuabagua (in front of the current Venezuela). That strategic movement was driven by the large quantity of pearls in the surrounding waters, which triggered the greed of the conquerors (Arveras 2021).

It has been recognized that the age of great 'discoveries' in the fifteenth Century, was in essence, the age of the discovery of the sea, where the control of the world trade, and thus the political control, was placed in the hands of a reduced number of states who were able to build enough ships to operate around all the world, simultaneously (Parry 1989). In that sense, Parry notes that the exploration and discovery not only had political and economic interest. It also had other advantages, like access to unexploited fisheries and fertile islands with productive lands, both of which were available for those who wanted to take them (*ibid*). And the travels and explorations of the seas conducted at that time, were not intended to discover the 'unknown'. Instead, they were used as usable maritime routes to link isolated regions that were separated from the inhabited and known world (Parry 1989). Later on, one example of images utilized to appropriate and govern marine resources may well be linked to the 'managerial ecology' notion which could be seen as a modern utilitarian approach to nature, having philosophical roots in the Enlightenment and the revolutionary economic, political and scientific order (Merchant 1980). This approach, he claims, began to emerge in the sixteenth and seventeenth century Europe and later on with the industrial capitalism that dramatically changed human attitudes toward and interaction with nature (Bavington 2005) and its resources.

It is recognized that human values very often focus on the efficiency and returns obtained by the sustained use of nature, for human benefits, which relate and are organized around the dictates and dynamics of the markets (Merchant 1980:238). Since around 1500, the cod represented one of the most important fisheries for the Bretons, Normans, Basques and English fleets, that seasonally harvested this resource in the Western Atlantic between 1500s–1990s (Lowitt 2011). The early fifteenth Century became the time for the recently encountered American territory, with the ambition to fish all the cod fisheries found in the Northeastern Atlantic region of Canada (i.e., Newfoundland). This fishery and the cod abundance in the waters of Newfoundland and Labrador supported the largest ground fishery in the world (Bavington 2005), harvesting approximately 100 million tons of cod between 1500 and 1992 (Rose 2003), an activity that determined the depletion of the great "northern" cod (*Gadus morhua*) and the subsequent fishery collapse. Yet, the cod fishery that had been pursued for over 500 years came to a sudden end when in July second 1992, Canada's Fisheries Minister John Crosbie (a Newfoundlander citizen) announced an immediate cod fishery moratorium (Bavington 2005). With this action, centuries of intensive fishing activities, mostly by Basque and Spanish fleets, were terminated. That moment illustrated the largest industrial layoff the modern fisheries discourse knows (Rose 2015; Thornhill 2020). Yet, after twenty-eight years, this extreme and radical fisheries management action still has large scale consequences in fishing communities around the entire Newfoundland. Still today, it is possible to witness entire fishing towns that were abandoned (Britain 1979; Haggan 2000), and with them, all the fishing practices, history, and cultures associated with fishing near this island (Butler 1983). This case shows the best example of managerial failure of a marine resource (Finlayson and McCay 1998) under the lenses of the evolution of the cod fishery in Newfoundland, along centuries. This activity changed from a migratory format to a resident (settler) fishery when the moving Spanish fleets stopped coming and after the local or migrant fishers established themselves as the beneficiaries from the cod and from whaling (Cervera 2021). Additional key aspects were the ups and downs of the ever-changing fishing industry; the fiscal and political context when Newfoundland joined the Canadian Confederation, the implications of the responsibility allocated to Canada for the fishery; and when the earliest warning signs of a collapse of the east coast cod fishery were on the horizon (Rose 2015).

Differently to those early images of fish and fisheries, and despite traditional consumption of fish and other seafood sources, some Europeans continued looking at fish and fisheries as an occupation that rhymes with 'poverty' (Bené 2003) and one of last resort (Bené 2004; Jentoft et al. 2011). Since the Renaissance, Western science and political and economic development in the so-called developed world have been closely related and connected. In fact, even sciences considered "unbiased and objective" such as evolutionary biology, have had significant influence from the social and political context (Fichman 1997) to the present. In the context of protected areas, Western science and the generation of scientific knowledge has dominated, since the 1980s, the various management approaches promoted as an alternative to conserve the biological diversity of the so-called "hot sites" of biodiversity and endemism.

3.2   Methodological and Theoretical Approach

This chapter is theoretical in nature and follows a narrative format that seeks to communicate the stories and the meanings associated with the notion of 'fish', 'fisheries' and 'marine resource governance'. It addresses the governance systems of marine resources, following a reflective approach on the history and geopolitics that originated the current governing forms of marine resource usage, management and governance, going from the global to the Latin American and the Caribbean (LAC) scale. The chapter circulates around the idea that the early encounter of those meanings represented a collision of worldviews and a shift in the comprehension of the two ideas we address: "fish" and "fisheries".

This reflective piece is inspired by the 'Critical approach to development' (Escobar 1994, 2008; Gudynas and Acosta 2011; Gudynas and Acosta 2011; Gudynas 2011), that contests the recent and current trend of development that the LAC region has followed in the last decades and the notion of development, linked to the fisheries sector, imposed in LAC by states of the Global North, on early 60s.

The approach also includes, as a key element, the construction of an 'alternative' ideal for development (*"Alternatives-to-development"*) that is nicely illustrated by Escobar (2007) in his book '*The Invention of the Third World*'. This conceptual governing format was fostered in some Latin-American countries (e.g., Bolivia and Ecuador) between 2007–2017 and provoked a paradigm shift in the notion of 'development' worldwide. Combined with this approach, we also look at other key aspects in the narrative related to natural (marine) resources usage and governance, especially reflecting in the *"Alternative (South) epistemologies"*, the scholarly work developed by De Souza Santos (2010). Within this perspective, we propose to explore other ways to look at the 'Southern-born world-views' and at the '*Buen Vivir*' paradigm, as strategic approaches to put the 'conservation and sustainability' of marine resources usage and governance, within a broader, more comprehensive and historical-sensitive perspective. We thus, argue that one cannot study 'fisheries' without some appreciation for the historical and cultural significance of the human communities who make a living from fisheries and of the social construction about fisheries that have been shaped during the last century in the LAC region, where the geographical focus of the paper is.

3.2.1 Transdisciplinarity and Knowledge

In the seminal paper written by Rittel and Webber "Dilemmas in a General Theory of Planning" (1973), it is said that the search for scientific bases for confronting problems of social policy is bound to fail, because of the nature of these problems. These problems, these authors argue, are "wicked" problems, whereas science has developed to deal with "tame" problems. Therefore, we claim that the many complex problems of and about fisheries cannot be definitively described nor solved. Within this comprehensive approach to fisheries issues, and within pluralistic societies, we could argue that there are no recipes nor "optimal solutions" for success, nor are there only technical solutions for societal problems in the sense of definitive and objective answers. In fact, our view recognizes that fisheries policies that respond to social problems cannot be meaningfully correct or false, unless severe qualifications are imposed first. Along the lines of this chapter, we tackle complexity and heterogeneity in/of science, problems and organizations, from an historical perspective. This lens looks first at '*knowledge*' as one key attribute in the creation of the 'fish' and 'fisheries' and its quality of hybridity, non-linearity and reflexivity, that transcend any academic disciplinary structure notion, and second at '*transdisciplinarity*' as a way to deal with real-world topics, where those involved have a major stake in the issue, where there is societal interest in improving the situation, and when the issue is under dispute (e.g., mangroves deforestation for shrimp farming, mangrove water pollution, biodiversity loss). Within the small-scale fisheries thus, it becomes especially relevant to articulate between these two dimensions, given the coincidence at generating knowledge and at addressing societal problems that aims to *"bridge the gap between knowledge derived from research and decision-making processes in society*" (Mollinga 2008). The 'transdisciplinary approach (Bergmann and Keil 2012; Pohl and Hirsch-Hadorn 2007, 2008) that has been used in this study fits with the scale and complexity of the matter under the scope. This case in particular uses the inter- transdisciplinary perspective to illustrate the natural sciences in general and the fisheries research being conducted in Ecuador and Galapagos Islands (History, Anthropology, Sociology, Economy, Biology, and Ecology), as windows through which we can look at such diverse, dynamic, and complex systems.

3.3   Act I: 'The Discovery' – The Wild and "Empty" Space Exploration Under the Colonial Machinery

The arrival of Christopher Columbus to the so-called 'New World', represented a breaking point in the entire cosmovision of the world, until then. The 'landing' of the Santa María, La Niña and La Pinta vessels to a Caribbean island, was set as the illustrating image of the '*encounter of two worlds*'. That image, however, also represented the '*encounter of the images about those two worlds*' which could be counted as the initial mismatch on how the story about marine resources was interpreted by the actors, given the colonial and imperial power-based scope of their enterprises put in place in the late fifteenth Century. At that time, said Bernabéu Albert et al. (2015), "[...] *the geographic colonization, was also an economic and a spiritual colonization, with the domain of the Catholic Church who was concerned about the competing faith, brought by the Anglican Church*". Additionally, these authors mention "[...] *the fragmentation of the Pacific lands, increased the reconnaissance and occupation of neighboring inhabitants and also of European explorers. All this, as a consequence of the colonization of the Pacific territories, by the large imperial powers from XVI onwards*". With that, it started the appropriation and exploitation of people, land, sea, and their resources in LAC, by expanding their geostrategic mechanisms that were successfully applied in other regions of the world, already under their control (e.g., Philippines).

Colonial Machinery and the meanings of small-scale fisheries during the Middle Age was the dominating governing format in the globe. In 1492, the Castille and Aragón Kingdoms, unified under a dominant Christian discourse, and thanks to their support, the 'discovery' enterprise took Cristopher Columbus on a journey that transformed the world. What at the time was called the "*Descubrimiento*" became the strategic movement for one imperial power to dominate the usage, benefits, and profit of the known world at that time, and their resources?

In later centuries, the colonial power executed by Spain, Portugal, England and Germany, paid more attention to the strategic position of LAC and thus promoted, motivated, and financed overseas travels, which significantly contributed to the development of the scientific agendas of countries and regions, to the creation of new livelihoods at a 'globalized mode', and to control the political structures of the

European Colonies in the Americas. Examples of the development of the colonial 'scientific agenda' in LAC, are the expeditions of scientists (mostly white European men), to Ecuador, following certain disciplinary interests that are illustrated by Table 3.1.

During that period, some specific cases were highlighted by the prestige that the scientists got due to their enterprises, or due to the nature of the interests that triggered their expeditions, benefiting the power of elites. In later centuries, colonial powers, this time integrating Spain, Portugal, England and Germany, paid more attention to the strategic position Latin America had and promoted, motivated and financed overseas travels, which produced significant contributions to the development of the scientific agendas, at European scale. Here are some examples:

Samuel Fritz (Czech Jesuit, Missionary, and Geographer) – Period in South America: 1703–1707. This priest was the first to describe, using cartographic-inspired notions, the Amazonas River basin. His maps are the first graphic and officially recognized records of the Amazonas. The original title of his work: "*Cours De Fleuve Maragnon autre dit des Amazones Par le Samuel Fritz Missionaire de la Compagnie de Jesus*" (1707), later published in German-language *Der Neue Welt-Bott*. Augsburg, 1726, I.

Charles Marie de la Condamine – (French Geographer, Astronomist, and Mathematician, Member of the French Academy of Sciences and of the French Geodesic Mission). Period in Ecuador: 1733–1743. Measured the length of a degree latitude at the equator and created the first map of the Amazon region, based on astronomical observations. The other French Scientists accompanying La Condamine were: Godin, Bouguer, and Joseph de Jussieu.

Alexander von Humboldt (German Geographer, Botanist, Volcanologist, Zoologist). Period in South America: 1799–1804. Naturalist who first explored the wilderness of this region with scientific interest. His multidisciplinary training and interests triggered his expedition to South America. He described the first Ecological Altitudinal-based Map, describing Andean ecosystems on the mountain of Chimborazo still serves as a reference for current researchers. Von Humboldt could be considered the first recognized scientist who started making connections ('*cause-effect*') of biological and ecological phenomena, including Climate Change. Even more, his scientific contributions were also argued to have been instrumental in the awakening of the 'freedom' spirit of the Spanish colonies, as stated by Wulf (2015) ("*[…] it was Humboldt, with his pen, who awakened Latin America*", Simón Bolívar"), which takes his scientific work to another dimension, with political and economic implications.

Charles Darwin (English Naturalist, Geologist). Period in Galapagos: five weeks in 1835. His contribution to scientific knowledge made a breaking point in what science was at that time. After his trip around the globe on board the MS Beagle and inspired by his observations in Galapagos Islands, Darwin published his "Evolution Theory based on Natural Selection'' (1859) which became the seminal work for any biological and ecological research until present. At the same time, it was not free of controversy, since it contradicted the postulates of nature, creation and God of religious faiths. Interestingly enough, it is said that Darwin benefited from less known

Table 3.1 Scientists and explorers going to the New World (eighteenth, nineteenth and early twentieth centuries)

Year	Name	Country	Topic
1735–1743	Joseph de Jussieu	France	Geodesic Mission
1711–1786	**Pedro Franco Dávila**	**Ecuador (Paris)**	**Natural history**
1761–1816	Thaddäus Peregrinus Xaverius Hae	Czech (Austria–Hungary)	Botany
1734–1807	Luis Née	France	Botany
1755–1811	Juan José Tafalla	Spain	Botany
1793	Juan Agustín Manzanilla	Spain	Botany
1775–1813	**José Mejía del Valle y Lequerica**	**Ecuador**	**Botany**
?–1807	Anastasio Guzmán	Spain	Chemistry, pharmacy
1769–1859	Friedrich Wilhelm Heinrich Alexander von Humboldt	Germany	Botany, geology
1773–1858	Aimé Jacques Bonpland	France	Botany
1815–1825	Karl (Carl) Sigismund Kunth	Germany	Botany
1771–1816	Francisco José de Caldas	Colombia	Botany
1796–1873	William Jameson	England	Botany
1831–1832	Francis hall	England	Botany
1831–1832	Jameson and Bousingault	France	Geology
?–1861	Richard Brinsley hinds, George W. Barclay, Andrew Sinclair	England	Botany
1825	David Douglas, John Scouler	England	Botany – Galapagos
???	James Macrae	England	Botany
1829	Hugh Cumming	England	Botany
1835	Charles Darwin	England	Natural history
1838	Abel Aubert du petit-Thouars, Adolphe-Simon Neboux	France	Botany
1846	Thomas Edmonton, John Goodridge	England	Botany
1845–1851	Berthold Carl Seemann	England	Botany
1809–1863	William Lobb	England	Natural history
1841–1842	Karl Theodor Hartweg	England	Botany

(continued)

Table 3.1 (continued)

Year	Name	Country	Topic
1844–1856	Gustav Karl Wilhelm Herman Karsten	Germany	Botany
1845–1853	Joseph Warscewicz Ritter von Rawicz	Lituania	Botany (life orchids)
1852	Nils Johan Andersson	Sweden	Botany
1855	Joseph Pitty Couthouy	USA	Malacologist
1856	Jules Ezechiel Rémy	France	Botany
1857–1863	Richard spruce	England	Bryologist
1858–1859	Moritz Friedrich Wagner	Germany	Botanist and zoologist
1862–1865	Juan Isern y Batlló	Spain	Botany
1865–1868	Gustav Wallis	Germany	Botany
1879–1909	Luis (Luigi-Aloisius) Sodiro	Italy	Botany
1870–1874	Alphons Stübel	Germany	Volcanology, botany
1873	Benedikt Roezl	Czech	Botany
1876	Édouard François André	France	Botany
1876–1881	Karl Friedrich Lehmann	Germany (German consul in Popayan)	Botany
1879–1880	Edward Whymper	England	Geography, alpinism, botany
1889–1892	Niels Gustaf Lagerheim	Sweden	Botany
1890–1943	August Rimbach	German	Botany and zoology
1891–1897	Henrik [Heinrich] Franz Alexander baron von eggers	Denmark	Soldier and botanist
1868–1869	A. Habel	Austria	Botany in Galapagos
1872	Franz Steindacher	Austria	Ichthyology
1875	Franz Theodor wolf	Germany	Geology, mineralogy, botany (Galapagos)
1884	Gaetano Chierchia and Cesare Marcacci	Italy	Botany
1898–1899	R. E. Snodgrass (dates/nation. Unknown) and Edmund Heller	USA	Botany
1891	Alexander Emmanuel Rodolphe Agassiz	Switzerland	Botany
???	George Baur	???	Botany
???	Luis mille	Belgium	Botany

(continued)

Table 3.1 (continued)

Year	Name	Country	Topic
1911	**Luis Cordero**	**Ecuador**	**Botany**

In bold: Ecuadorians (3)
Woman: none

explorers, like William Dampier, (Preston and Preston 2010) who authored cartography or batimetry charts that were later used by Darwin. Additionally, it is claimed that Darwin also benefited from the immense volume of work that was produced by Alexander von Humboldt) ("[…] *Darwin would not have been Darwin without Humboldt* […]." Andrea Wulf in the "Invention of Nature – Alexander von Humboldt's New World (2017).

Franz Theodor Wolf (German Theologist, Geographer, Botanist, Geobotanist, Volcanologist, and Mineralogist). Period in Ecuador: 1876–1890. His contributions to the discipline of Volcanology, after his explorations of the volcanoes in mainland Ecuador and in the Galapagos, substantially increased the knowledge and interest of this less-known discipline. His legacy still serves as the basis for much current scientific knowledge about volcanology.

After these few examples of scientific-driven expeditions, we can identify varied interests that were influential at exploring the Americas. It could be claimed that these interests were mostly moved by the intention to collect, describe new species, name them (taxonomically or with other scientifically-accepted format), map, store them in European museums, publish new findings in academic journals, create and be part of scientific associations, academies and disciplines. Interestingly enough, it was not until 1699, when a woman, Maria Sibylla Merian, a seventeenth century entomologist and scientific adventurer, embarked on a purely scientific expedition in history, going to Suriname to illustrate new species of insects (Latty 2019). However, in Ecuador it was not the case. No evidence was found of their involvement, although it could be argued that they were part of such expeditions, but no mention has been made of their presence.

These references and expeditions show to us that during those years, the Western scientific imaginary was inspired by a positivist tradition – which surprisingly, remains to this day, within the most orthodox branches of the scientific agendas. Those images of science promoted and conceived biological and numerical science as the only and/or most important answers of the scientific endeavor. In fact, during this first moment in the exploration of Galapagos by Darwin, both ignorance and curiosity sowed the basis for a science focused on natural objects that did not necessarily recognize, in their development, the influence of the global geopolitical agenda. Yet, despite the role humans played in the dynamics of those early explored systems and the recognized importance of humans effects on the environment, there was an intentional avoidance to address it due to the implicit complexities associated to it ("*You ask whether I shall discuss "man"; I think I shall avoid whole subject, as so surrounded with prejudices, though I fully admit that it is the highest and most interesting problem for the naturalist*" Charles Darwin's letter to Alfred Wallace, 22nd December 1857 – Source: Tapia et al., 2009).

The Galapagos Islands are often called the *'natural laboratory of evolution'*, a phrase that became a powerful metaphor that has shaped the Galapagos territory and its inhabitant's mindset, since the last century. This image, it is said, communicates a way of understanding space through scientific research, conservation and tourism (Hennesey 2018). In this sense, science in the Galapagos has played and continues playing a preponderant role in the communication of the meanings associated with the terrestrial and marine systems, their resource, usages, and governance, which are linked to each sector of Galapagos society and its visitors.

All these examples show that contemporary images of fish and fisheries, and other natural resources, have generally been shaped under the rationalities associated with the knowledge (mostly western-minded) being produced and reproduced by those with access to that, in this case, positivist natural scientific knowledge.

3.4   Act II – The 'Appropriation' of the Commons – By Regulating the Usage and Governance of Marine Resources

Once the 'New World' was 'discovered', the Spanish Crown started to spatialize their new properties, in varied formats, and following geographical distribution patterns that coincided with more-or-less, equitable spaces around the Americas. By conducting this 'spatial' distribution of their 'new properties', the right of usage and profit of those new spaces, automatically became the exclusive right of the Spanish Queen, and, proportionally, of those subjects on site, who were in charge of their control, administration, and trade. The most important administrative unit was the '*Virreynato*', which represented the power and authority of the Spanish Queen in the Americas. The *Virreynatos* were considered the actual representation of the interests of Spain in the colonies, executed by the *Virrey*. Based on those interests, the conquerors, who mainly were Spanish, low socially ranked men, explored the Americas searching for gold and prosperity, knowing that by 'discovering' and 'claiming' the so-called 'new-territories' for the greatness of Her Royal Highness of Spain, they also were granted rights to exploit and benefit from the exploitation of those resources. However, on their way to the 'El Dorado', they also encountered hazards, when the new 'discovered' territory was not the environment they were used to, like the following example illustrates when referring to the mangroves encountered.

> *[....] some knights decided to continue discovering by boat, since the terrestrial route was so arduous, due to the thickness of the mangroves, and also due to the many rivers full of ferocious alligators and mosquitos that tormented them [....]*
>
> Chapter IX:164. Chronicles of Spaniards soldiers accompanying Pizarro in the conquest journey of Peru.

[....] meanwhile, Francisco Pizarro and his pals crossed between those rivers and mangroves, being tormented by the mosquitos, experiencing unbearable efforts and misfortunes, and were tired of being forced to walk in that hell, and they all wanted to go back to Panama [....]
    Chapter XII:170. Chronicles of Spaniards soldiers accompanying Pizarro in the conquest journey of Peru.
    Ref: Pedro Cieza de León – La Crónica del Perú (1541–1550) the `discovery´ of the Pacific Coast.

Those sorts of difficulties prevented other conquerors from exploring those areas, yet their greed and pursuit of riches moved them to continue exploring even those places considered as 'hellish', which later on, actually represented important harvesting areas for marine resources that produced large revenues for the Spanish Queen. The case illustrated here, tells us the story of the exploitation of pearls, whose harvesting became an attractive business in the Caribbean at that time.

*[....] the interest of the Royal Highness to fish "margaritas" (*sic. for pearls*) and her proposal to colonize the San Diego Harbor* (currently Venezuela) *to harvest them.* (ES.41091. AGI, Archivo General de Indias, 1719-2-18 Madrid).

But the exploitation of marine resources was only possible by granting rights of usage and benefits to specific representatives of the Spanish crown, under the assumption that tributes and taxes would be paid to the Queen based on the marine resources' profitability. One of the strategies to proceed with those 'private rights' allocation, was what could be considered as the first 'fishing rights allocation', to a private user (i.e., by establishing a Fishing Maritime Company) in the early eighteenth Century, for the exploitation of marine mammals.

Request to establish along the coast of the Pacific Ocean, in our Meridional America, a whale fishery and a factory of oil and sperm candles. (Expedient about the establishment of a fishing company, Archivo General de Indias, 1789-7-18).

These references illustrate how and to what extent the conquering and colonization of the Americas and the spatialization of their power were used to impose new legal instruments. These, put in place by the most powerful sector of the societies at that time, also served as ways to disregard previously existing governance practices that dealt with harvest of fish and seafood. We will reflect further on this finding, which has substantial implications in the way marine resources are still managed and governed. The strategies to regulate the usage and harvest of the already 'appropriated' marine resources were, in this case, the design and imposition of a 'management tool' for marine resources, which could be named as the first 'fishing-gear ban' and by its establishment, a sort of a 'protected area' creation. At that time, the Royal Decree issued by the Spanish Crown read *'the prohibition of the use of purse seines'*, in response to his demand to restrict the use of marine resources exploitation in the East Indies (i.e., Caribbean islands). This shows that strategies to 'ensure' certain practices for the 'ocean resources governance', were already in place in the sixteenth Century. This instrument is thought to have been created first, to protect highly valued resources (i.e., pearls), when conducting other fishing practices, second, to gain benefits from any other fishing-related practice conducted in the harvesting areas of pearls, and third, to make profitable for the Spanish Queen, any business-related activity, conducted by her subjects, in the colonies.

> *Cédula Real to Fray Tomás de Berlanga, Terra Firme Bishop (1538), in response to his request, as protector of the recently discovered West Indies, to prohibit the usage of purse seiner to fish close to the Pearl Island* (sic. Today's Isla Margarita, Venezuela). Registro de Oficio y Partes Tierra Firme, Archivo General de Indias, 1538-6-26, 1542-3-10.
>
> Cédula Real de Oidores de la Audiencia de Tierra Firme: "[....] *for any Spaniard to fish with purse seiner in the Pearl Island, a right which has been granted to the Marquez Don Francisco Pizarro, and if some person would like to do so without him, a license could be conferred, for which a fee (one fifth) should be paid to her Royal Majesty.* Audiencia de Panamá, 235, L.7., F.74 V – 75 R, Archivo General de Indias, 1539.103.
>
> *It is then stated the necessity to pay the fee (one tenth) for fish, salt, fruits, vegetables, chickens and other stuff.* Archivo General de Indias, Indiferente, 1538-6-26. Registro de Oficio y Partes Tierra Firme, Archivo General de Indias, 1538-6-26, 1542-3-10.

3.5 Act III – The Blessing from Ocean Commons Governance: The Development Ideal

After the Independence wars freed the Spanish colonies in the Americas, between 1809 and 1832, a series of discontinuous, but related events took place. Once the newborn Republics in the Americas became sovereign states, the rules of the game shifted from being led by Spanish rulers to being led by the new *Criollo*-elites, linked to the previous Spanish power, but still, being perceived as the 'new local legitimate' new authorities. From the start of the Ecuadorian Republic in 1832, it was not until the early twentieth Century, when other strata of Ecuadorian society started to become part of the discussion of marine resources usage following a global-driven policy, facilitated by the Food and Agriculture Organization (FAO) in order to start the exploitation of the fishing resources, in the entire LAC.

This global 'development' agenda was mainly influenced by the interest in the economic development of the coastal communities and the development of a promising industry that was increasingly demanded by a global rising economy, awakening after the dark years of WWII. At this point western societies, politicians, businesses, and their scientists were eager to leave behind a season of limitations and scarcity and were motivated to start and/or resume their postponed agendas. This became especially relevant within all the aid initiatives, promoted and carried out by the United States of America, through their global US/AID program, along the varied fronts their geopolitical interests looked at: the post-war Europe, Asia, and LAC. This trend determined that the dominant 'Development Discourses', promoted and fostered between 1950s and 1970s, greatly contributed to the three main phases of fisheries resources harvesting at an industrial level in Ecuador: (1) the installment of political, institutional and technical capacities; (2) training, education and technological advancement of those involved with the fisheries activities; and (3) the application of the fishing and fisheries' 'know-how' from abroad. More or less at the same time when the fishing activities were started regionally, the agriculture sector also received substantial support, by governments in the Global North, who looked at the Global South as the breadbasket of the world. These large-scale development agendas for both the fisheries and agriculture sectors were critical and influenced how they evolved up to today.

Later, during the 1980s, the second 'boom' of the primary economy in Ecuador began. This followed the initial oil industry success in late 60s-early 70s and the start of banana-export businesses. In 1980–1981, the world's desire to expand their access to global markets, started to demand more and more exotic produce (e.g., bananas, fish, timber) from the Global South. That triggering of market forces pushed the recently launched oil-based economies to expand their offerings and begin exploring other resources that could result in attractive and profitable initiatives for the Ecuadorian state. Additionally, the already developed aquaculture business in Asian countries, that also followed the FAO initiatives for the large-scale production of fishing and other marine resources, showed that the shrimp harvesting industry was a promising one and the idea was imported to Ecuador. With that, a new business landed in a country with 30% of its coastline covered by mangroves and with an economy circulating around oil exports.

The first movement to help the new industry prosper was to enable the existing policy on land property rights and rights of use and allocation to expand into the mangroves areas. Previously, usage and ownership were exclusive rights of the Ecuadorian state. With the arrival of this new business model, the entire image of the mangroves shifted from being the '*hell where humans suffer due to the unbearable conditions*' and was transformed to be '*the frontier to be conquered and the promising wealth coming from its taming*'. At that moment, the technology needed for and associated with the taming of the mangroves was utilized as a means to spatially distribute the shrimp-farming areas in the mangroves, to plan and map the shrimp farms, and to allocate the right of use in those areas under the 'concession' format for one hundred years to the elite, powerful and wealthy, who were already part of the government and business groups. Yet, the image of mangroves, associated with the 'illegal' notion that started when African slaves survived a shipwreck along the Esmeraldas' coast, escaped and hidden on the mangroves of that region, remained. This negative image association was enhanced by the later arrival of guerrillas groups to the mangroves areas in the border region between Ecuador and Colombia.

In the following years (1970s–1990s) and with the arrival to Ecuador of more naturalists and explorers following the 'Darwin legacy', new 'images' about the marine and terrestrial resources and systems were created and introduced. These newly shared 'natural' images coincided with a flourishing industry demanding more and more diverse products to offer their markets: nature-based tourism. It was then, when the idea of nature conservation, in the form we know it now, was developed and promoted as a twofold strategy. On the one hand, it aligned with the recently launched environmental movement that asked for bans in the usage of aerosols, as measures to diminish their negative effects over the ozone layer, and on the other hand, it also highlighted the nature as the product to be sold, by tourism industry operators, for an always growing tourism sector.

By the 1980s and 1990s, scientific interest was mainly focused on the recently described 'hot-spots' of biological diversity, as priority areas for conservation (Olson and Dinerstein 1998; Olson et al. 2002), with a view of conservation that is achieved through the implementation of protected areas and closed areas (Nederveen Pieterse 2010), all under an approach of restriction of use and limitation of access to users. It was during this period that the paradigm of 'sustainability' started to gain currency, and a new notion of conservation began to propagate at a global level, generating novel research approaches towards biodiversity conservation 'with communities' as a strategy to achieve successful conservation initiatives.

Those conservation strategies at that time, were ones influenced by the ideal of 'wilderness' that fascinated the tourism-centered discourse, and at the same time, one of conservation of endangered and iconic species, which originated lines of thought and images still in currency: whales and dolphins in the marine realm, and panda bears and other charismatic megafauna in the terrestrial world. Those species and the images associated with them are still very powerful and influential political and economic discourse on finding the best mechanisms and strategies to improve or change marine and ocean governance policies and practices, worldwide. In parallel with those images, the establishment of protected areas (marine and terrestrial) was also an illustration of ways to enable the conservation initiatives, whereas at the same time, guaranteeing that the natural capital, over which local and non-local actors make a living, would be protected.

Yet despite the 'sustainable development' notion coined in 1997, it took at least ten more years (2007–2010) until this 'sustainability' discourse, actively promoted by grass-roots movements and some academic debate, actually reached the economic and political agendas of states, multilateral agencies, and even business, to finally foster a more economically profitable, socially just, and environmentally friendly global economy. This new approach in sustainable marine resources usage and governance, for instance, could be considered the 'keystone' of the way the 'improved ocean governance' is envisioned. This approach was also reflected in the way 'sustainable development' was promoted in Ecuador, both on the mainland and in the Galapagos Islands. In fact, between the 2000s–2010s, research formats integrated, for the first time, a more holistic approach, contemplating natural and social dimensions, and giving them an equitable value, in the best understanding of the problems that affect the Galapagos (González et al. 2008; Tapia et al. 2009). These new proposals, therefore, highlight the need for integration of multiple disciplinary and methodological traditions, all this, as a more realistic mechanism to understand and mitigate the challenges that the Archipelago is facing. And this also involves the 'fisheries' sector, which until three years ago, was still being looked at only through a 'hard science' lens.

In recent years, after a decade of political turns towards the leftish-minded social agendas in some South American countries, in 2018 the development paradigm in several countries in the region changed back to a conservative position. The new government of Ecuador, representing the traditional bank-and-business oriented policy and economy, allocates to the governance of marine resources (especially the tuna-fish large scale fisheries industry and the shrimp industry of Ecuador, which significantly contribute to the national economy), high relevance and political attention. Thus, deep and responsible discussions and reflections are needed regarding which of the current development formats (i.e., more socially-oriented policies or business-minded schemes) are desired, all this, for an equitable, fair and sustainable economic recovery after the so-called "lost decade" of the 2010s.

3.6 Discussion

3.6.1 Ocean Governance – Inter & Transdisciplinary Approach

Governance of marine and ocean resources demands a substantial amount and quality of knowledge. We have been told that decision and policy making should be done based on 'sound scientific evidence'. It may, however, be useful to open the dialogue to what kind of "science" we want and what we need. By opening the door to other ways of knowing (traditional knowledge of local fish users, for instance), a transdisciplinary approach would enable the integration of diverse epistemological formats that can provide us with lights to understand "local" realities and "local" views. One critical part in fisheries research, for instance, has been and still is the 'technologies'. In fisheries research, currently going on in the Galapagos, for instance, for the first time, a reconciliation of diverse formats of knowledge has been achieved, by integrating an interdisciplinary perspective. One example is the use of high-tech lab equipment, to study otoliths and larvae, as part of the fisheries biology section, whereas, at the same time, focus groups and participant observation are also used to describe and better understand the fisheries sector dynamics and complexities. In that light, the search for classical scientific bases for confronting problems of social policy is bound to fail because of the nature of these problems. They are "wicked" problems, whereas science has developed to deal with "tame" problems. This is a challenge with the reductionist approach to science, which works well in trying to advance foundational scientific explorations as in particle physics. It works poorly, however, in understanding wicked problems, which fundamentally require a holistic approach that classical science is unequipped to perform. Policy problems cannot be definitively described nor solved. Moreover, in a pluralistic society there is nothing like the indisputable "truth" and there is no objective definition of "equity".

3.6.2 Policy for Governing the Marine Resources

Policies respond to societal problems yet, they cannot be fully nor meaningfully correct or false since the so-called "optimal solutions" still are value, power and knowledge influenced. In fact, the potential "solutions" for these 'wicked' scenarios within ocean and marine resource governance, are not definitive and objective answers to those problems. Rather, they are instruments, created within a specific context, that need to be prioritized, discussed and negotiated. And this iterative negotiation is the space where the values, principles, interests and power behind the policies are made evident. It is here when the meanings the actors involved allocate to critical notions like "improved governance", "wellbeing", "development", "sustainable", "prosperous", "growth", are known. This article has shown that oceans and marine resources governance has historically been implemented by biased strategies and practices. Thus, we argue that improving the governability of oceans and seas requires the encounter of common grounds about those dimensions, and that should be placed over the "differences" that block the negotiation process. Within this scenario, aspects like relevance, urgency, priority, equity and equality, the rule of law, legitimacy of actions, transparency, accountability, social responsibility, holistic interactive governance, economic sustainability and social viability, should be placed as conditions *sine qua non*, actions, policies and practices are conducted. This, certainly will enable the promotion of a 'new era' for fisheries and marine resources governance, since it would foster the negotiation, going along mandatory guiding principles: human rights and dignity, respect of cultures, non-discrimination, gender equality and equity (Jentoft et al. 2017).

3.6.3   From Society-Driven to Enterprise-Focused Marine Resources Governance

The plurality of the consumer society, and of consumers, customers, citizens are fundamental in achieving and creating a new (improved) governance for fisheries and other marine resources. The desired change, could, for instance be triggered by a *"smart governance approach"* (G. Krause, comm. Pers., 18.09.2017) also in LAC context. Hence, under the *Buen Vivir* approach, the existence of societal institutions–fostering reciprocity, cooperation, and solidarity–is envisioned, also in the form of responsible markets, as key means to promote the good way of living that this concept involves. Seen as an ancient ontological notion that has been recently recovered (Viveiros de Castro 2004; Haidar and Berros 2015b), the *Buen Vivir* constitutes *"an alternative approach to development, and as such represents a potential response to the post development"* need (Gudynas and Acosta 2011; Acosta and Martínez 2009).

This notion involving fair, responsible, and sustainable markets, are driving forces for a new trade format that balances the dominant role that the western science-based and technocratic perspective of marine resources governance, has put in place together with political and managerial agenda, as has been illustrated by different cases of natural resources management (e.g., oil palm, shrimp, and soy monocultures) (Escobar 2016) in the Latin American context. Findings in this chapter evidence the relationship between conquering and colonizing the Americas, the spatialization of the ruling power, the value-laden scientific endeavors conducted in LAC and Ecuador, and their consequences for marine resources. Additionally, we have seen that imposed regulations for ocean and marine resources were unsuccessfully applied since they disregarded previous governance practices linked to fish and to seafood, locally. This imposition, we claim, has had large implications on how resources were and still are perceived, imagined, used, and governed, given the broken relationship between former and current users and the resources. This artificial and violent suppression of traditional policies and practices and their replacement by some new strategies, brought from abroad, has greatly influenced the perception of governing rules as 'impositions' from abroad, which for practical purposes, facilitates either the active opposition or the passive ignorance of regulations that blocks and even boycotts successful ocean and marine governing practices.

3.6.4 The New Ecuadorian Constitution – Still Useful?

In 2008, for the first time ever, the Ecuadorian Constitution (2008) granted inalienable rights to nature and recognized nature as a subject that enjoys juridical protection, at both Constitutional and legal levels (Berros 2015). In the preamble of this normative instrument, the Ecuadorian Nation State is defined as "*constitutional, rights and justice-based, social, democratic, sovereign, independent, unitary, intercultural, plurinational and secular*". Additionally, it is said that this constitutional law operates as an integrative and conciliation strategy to "[i]*ntegrate the diversity of peoples, cultures, notions* (i.e., Mother Earth or *Pachamama* and the *Sumak Kawsay*") *at all dimensions of National interest (e.g., economic, politic, financial, cultural and environmental*)." As a social bonding instrument, the 2008 Constitution successfully recovered and integrated multiple constituents of Ecuadorian society which greatly enhanced Ecuadorians' national pride, identity, and self-esteem. Since its approval, this Constitution proclaims high levels of symbolism illustrated by practices that have been recovered after their replacement by western-based habits over the previous centuries (e.g., traditional food and garment). This instrument has embraced in a rather tacit manner the nation's ancient heritage and has pleaded against discrimination of traditionally marginalized groups. In the end, and at least in theory, this constitution can be seen as a successful example of a participatory process useful to "redeem the past" of this nation state (Acosta 2009; Acosta and Martínez 2009).

3.6.5 The Buen Vivir Principle

The *Sumak Kawsay* (in Quechua language) paradigm, translated as "Good way of living" or "*Buen Vivir* "in Spanish is not a new notion. It has been present in ancient Amerindian discourses and indigenous Andean cosmovisions (or non-dualistic philosophies, Escobar 2016) that illustrate a comprehensive way of understanding life. This idea retrieves and articulates broader ontologies and epistemologies about humans, animals, and environment, and operates as an alternative construct of life. These visions, as have been said by Berros (2015), nicely align with newly-grounded ideas that currently belong to environmental- and animal-ethics fields and which are present in the juridical field (Haidar and Berros 2015a). The *Sumak Kawsay* discourses circulate around the equilibrium and the harmonic coexistence of beings – from social and natural realms – privileging the collective over the individual and solidarity over competition. *Buen Vivir* is a category in the Andean life philosophy of the indigenous societies that has lost ground due to the effects of Western rationality's practices and messages (Viveiros de Castro 2004; Duarte and Belarde-Lewis 2015) mainly due to the discredit given to this way of thinking in front of most dominant currents (e.g., the church and religious ways of thought) (Haidar and Berros 2015a).

Since the 2008 Constitution approval, the *Buen Vivir* principle has ruled the National Ecuadorian Plan for Development (or *Plan Nacional del Buen Vivir* – *PNBV* in Spanish), which questions the traditional notion of development focused on economic growth (Lind 2012). In contrast, it proposes sustainable development only as an interim goal on the way toward a paradigmatic shift in the development notion that encompasses dimensions like happiness, freedom, and equal rights, as well as sustainability (Gudynas 2011; Escobar 1996; Acosta 2008; among others). Acosta and Martínez (2009) propose to promote "alternatives for development" instead of models for "alternative development". Operating under this perspective, between 2007–2017 the state has played a critical role as a driving force for achieving social well-being in Ecuador. In that regard, policies, programs and practices constituting the full public agenda, have given the *Buen Vivir* principle an influential role within the national strategic development plan.

3.6.6   Governing Marine Resources – From Past to Now

Historically, fishing has been an important cultural, social and, only recently, economic role in Ecuador. There is evidence of pre-Hispanic communities fishing, consuming, and trading fish products at a low to mid-scale, locally and regionally (Baumann 1978; Norton 1985; McEwan and Silva 1998) even by practicing very complex fishing strategies[1] and by using diverse gears (e.g., nets, lines, and hooks) (De Madariaga 1969). In most recent times, small-scale fisheries in Ecuador used only subsistence practices until the early 1950s, when fishing started to develop as a commercial sector, mainly aided by international bodies (e.g., FAO) (Allsopp 1985; Williams 1998). Since then, small-scale fisheries have been identified as

---

[1] The Spanish conquerors recorded fishing scenes of South American indigenous tribes who used a "hunter fish" to catch bigger prey, even sharks. For a detailed description of these practices, see De Madariaga 1969:116–130).

critical for the economic growth of fishing communities in the Ecuadorian coasts, besides construction and tourism. On the contrary, the relevance of fish, first as a food source, and second as a cultural and identity-linked asset in Ecuador, has only been referred to by scattered research conducted by scholars who described very early stages of Ecuadorian's history (Norton 1985; McEwan and Silva 1998) and recounted their use of marine resources (Baumann 1978; Rotsworowski 2005).

Bolstered by the cultural construction of fisheries (Finley 2009) and due to the prevailing doctrine of free trade and markets which look at fish only as goods to trade with, the fish produced by small-scale fisheries in Ecuador has remained unnoticed under different lenses. In fact, historical, cultural, and spiritual dimensions of fisheries within fishing communities have remained unnoticed under the current governing practices developed to achieve the *Buen Vivir* paradigm in Ecuador. We claim that fisheries have not been treated according to the new constitution perspective mainly due to the incongruities and dissonant approaches in governing fisheries and other resources, at national scale (Barragán-Paladines 2015, 2017).

3.7 Science for Marine and Ocean Governance in the Future

We argue that the boundaries of interdisciplinary research are shown to be under constant negotiation, and are still far from mutual understanding or consensus, which in fact explains the often uneasy negotiations. We posit that the increasing prominence of the difficulties encountered in achieving the so-called 'sustainability' partly relies on the inability (or unwillingness) to deal with boundaries (of many sorts) and how to overpass them. Furthermore, it is here shown, how the circuit of knowledge production about fisheries in Latin America is deeply influenced and informed by history, by power, by academic research, by publishing, and by the imposition of one dominant ontological agenda for fisheries' and other marine resources appropriation, exploitation, and usage, and even by putting fish and fisheries as object of conservation practices and economic wealth. In order for us to reverse this circle of inadequate and unsuccessful ocean and marine resources governance, that has historical origins, we rather look at the positive outputs of governing interactions that strengthen and facilitate the negotiation of the principles that mobilize and enable improved governance policies and practices.

We have seen the account of the long history of the relation between human practices and marine resources, which in the last centuries has been shaped under a dominionist and colonial principle of appropriation and control. The 'modern' development notion, thus, also obeys to a contested spirit of continuous growth based on accumulation of goods, which, in the case of the marine resources, has been illustrated by fisheries. Fish and fisheries are described as marine resources that started to be governed through management instruments already in the sixteenth Century. Therefore, the conversion of fisheries, from being a subsistence-based activity until becoming a highly valued good, nicely illustrates the transformation of the meaning of fish within different contexts and moments in human history. We claim that alternatives to this notion of development are needed. It also is desirable to have a more comprehensive approach to look at fisheries, as well as coherent and fair policies and practices. A "one-fits-all' approach to deal with governability matters in fisheries needs to be revisited, and more diverse lenses to look at it need to be found. One strategic move could be, for instance, to reconcile the discussion and negotiation of competing claims of knowledge and power and to install continuous reflection processes not only about "what" development, but even more important, "whose" development do we want.

References

Acosta A (2008) El Buen Vivir, una oportunidad por construir. Ecuador Debate 75:33–47

Acosta AM (2009) Informal coalitions and policymaking in Latin America: Ecuador in comparative perspective. Routledge

Acosta A, Martínez E (2009) El buen vivir. Una vía para el desarrollo. AbyaYala, Quito

Allsopp WHL (1985) Fishery development experiences. Fishing New Books Ltd, Farnham

Arveras D (2021) De mucho más honor merecedora. Doña Aldonza Manrique, la gobernadora de la isla de las perlas. SND Editores. 256 páginas

Asingh PA, Damm J (eds) (2020) Pompeii and the fermented fish. Bound for Disaster. Pompeii & Herculaneum. Catalogue of the Exhibition Moesgaard Museum 6

Barragán-Paladines MJ (2015) Two rules for the same fish: small-scale fisheries governance in mainland Ecuador and Galapagos Islands. In: Jentoft S, Chuenpagdee R (eds) Interactive governance for small-scale fisheries. Springer, Cham, pp 157–178

Barragán-Paladines MJ (2017) The Buen Vivir and the small-scale fisheries guidelines in ecuador: a comparison (Chapter 33). In: Jentot S et al (eds) . Springer, Heidelberg, pp 695–713

Baumann P (1978) Valdivia. El descubrimiento de la más antigua cultura de América. Hoffmann und Campe Verlag, Hamburg

Bavington D (2005) Of fish and people: managerial ecology in Newfoundland and Labrador cod fisheries. Dissertation. Submitted to the Department of Geography and Environmental Studies in partial fulfillment of the requirements for Doctorate of Philosophy in Geography and Environmental Studies. Wilfrid Laurier University

Béné C (2003) When fishery rhymes with poverty: a first step beyond the old paradigm on poverty in small-scale fisheries. World Dev 31(6):949–975

Béné C (2004) Contribution of small-scale fisheries to rural livelihoods in a water multi-use context (with a particular emphasis on the role of fishing as "last resort activity" for the poor). Advisory Committee on Fisheries Research, 20

Bergmann M, Keil F (2012) Transdisciplinarity: between mainstreaming and marginalization. Ecol Econ 79(0):1–10

Bernabéu Albert S, Mena García C, Luque Azcona EJ (Coord) (2015) Conocer el Pacífico – Exploraciones, imágenes y formación de sociedades oceánicas. Editorial Universidad de Sevilla, Sevilla

Bernal D, Díaz JJ, Expósito JA, Palacios V, Vargas JM, Lara M, Pascual MÁ, Retamosa JA, Eid A, Blanco E, Portillo JL (2018) Atunes y Garum en Baelo Claudia: nuevas investigaciones (2017)/ Tuna fish & Garum at Baelo Claudia: Recent Research (2017). Al Qantir 21:73–86

Bernal-Casasola D, Marlasca R, Rodríguez-Santana CG, Ruiz-Zapata B, Gil-García MJ, Alba M (2016) Garum de Sardinas en Augusta Emerita. Caracterización arqueológica, epigráfica, ictiológica y palinológica del contenido de un ánfora. REI CRETARIÆ ROMANÆ FAVTORVM ACTA 44

Berros V (2015) Ética animal en diálogo con recientes reformas en la legislación de países latino-americanos. Revista de Bioética y Derecho 33:82–93

Britan G (1979) "Modernization" on the North Atlantic Coast: the transformation of a traditional Newfoundland fishing village. In: Anderson R (ed) North Atlantic maritime cultures. Mouton, the Hague.

Butler G (1983) Culture, cognition, and communication: fishermen's Location-finding in L'anse-a-Canards, Newfoundland. Canadian Folklore Canadien 5:7–21

Cayo Plinio Segundo (23–79) (1999) Historia Natural. Edición Facsímil de la versión de F. Hernández & J. de Huertas. Visor Libros. Universidad Nacional de México. Madrid. In: Montero A, Díaz MA, Gutiérrez MM. The knowledge of nature in the Late Middle Ages: classifications of don Juan Manuel (1282–1348) in the Libro del cavallero et del escudero (1326–1328). Bol. R. Soc. Esp. Hist. Nat. Sec. Biol., 111, 2017

Cervera C (2021) Los fuertes lazos históricos de España con Terranova: ¿llegaron los vascos a América antes que Colón? ABC Historia. 17/2/2022. https://www.abc.es/historia/abci-fuertes-lazos-historicos-espana-terranova-llegaron-vascos-america-antes-colon-202202180301_noticia.html

De Madariaga JJ (1969) La caza y la pesca al descubrirse América. Editorial Prensa española, Madrid

De Souza Santos B (2010) Refundación del Estado en América Latina – Perspectivas desde una epistemología del Sur. Instituto Internacional de Derecho y Sociedad/International Institute On Law And Society – Programa Democracia Y Transformación Global. Lima

Duarte ME, Belarde-Lewis M (2015) Imagining: creating spaces for indigenous ontologies. Cat Classif Q 53(5–6):677–702

Ecuadorian National Constitution (2008) Supra note 5, at Art. 71, 72, 73, 74, 75

Escobar A (1994) Encountering development: the making and unmaking of the third world. Princeton University Press, Ewing

Escobar A (1996) Constructing nature. Routledge, New York, pp 46–68

Escobar A (2007) La invención del Tercer Mundo. Construcción y deconstrucción del desarrollo. Serie colonialidad/modernidad/descolonialidad. Fundación Editorial El Perro y la Rana, Caracas

Escobar A (2008) Territories of difference. Place, movements, life, redes. Duke University Press, Durham/London

Escobar A (2016) Difference and conflict in the struggle over natural resources. A political ecology framework. Haenn, N./Wilk, R./Harnish, A. The environment in anthropology: a reader in ecology, culture, and sustainable living. New York University Press, New York, 362–368

Fichman M (1997) Biology and politics: defining the boundaries. In: Victorian science in context, pp 94–118

Finlayson AC, McCay B (1998) Crossing the threshold of ecosystem resilience: the commercial extinction of northern cod. In: Bavington D (2005) Of fish and people: Managerial ecology in Newfoundland and Labrador Cod Fisheries. Dissertation. Submitted to the Department of Geography and Environmental Studies in partial fulfillment of the requirements for Doctorate of Philosophy in Geography and Environmental Studies. Wilfrid Laurier University

Finley C (2009) The social construction of fishing, 1949. Ecol Soc 14(1):6. http://www.ecologyandsociety.org/vol14/iss1/art6/

González JA, Montes C, Rodríguez J, Tapia W (2008) Rethinking the Galapagos Islands as a complex social-ecological system: implications for conservation and management. Ecol Soc 13(2):13. http://www.ecologyandsociety.org/vol13/iss2/art13/

Gudynas E (2011) Buen Vivir: germinando alternativas al desarrollo. América Latina en Movimiento 462:1–20

Gudynas E, Acosta A (2011) La renovación de la crítica al desarrollo y el buen vivir como alternativa: Utopía y praxis latinoamericana. Revista Internacional de Filosofía Iberoamericana y Teoría Social 16(53):71–83

Haggan N (2000) Back to the future and creative justice. In: Coward H, Ommer R, Pitcher T (eds) Just fish: ethics and Canadian marine fisheries. ISER Books, St. John's, pp 83–99

Haidar V, Berros V (2015a) Entre el Sumak Kawsay y la "vida en armonía con la naturaleza": Disputas en la circulación y traducción de perspectivas respecto de la regulación de la cuestión ecológica en el espacio global. Revista Theomai Estudios Críticos sobre Sociedad y Desarrollo 15(32):128

Haidar V, Berros MV (2015b) Hacia un abordaje multidimensional y multiescalar de la cuestión ecológica: la perspectiva del buen vivir. Revista Crítica de Ciências Sociais 108:111–134

Hennesey E (2018) The politics of a natural laboratory: claiming territory and governing life in the Galápagos Islands. Soc Stud Sci 48(4):483–506. https://doi.org/10.1177/0306312718788179

Jentoft S, Eide A, Bavinck M, Chuenpagdee R, Raakjær J (2011) A better future: Prospects for small-scale fishing people. Chap. 20. In: Jentoft S, Eide A (eds) Poverty mosaics: realities and prospects in small-scale fisheries. Springer, London. https://doi.org/10.1007/978-94-007-1582-0_20

Jentoft S, Chuenpagdee R, Barragan-Paladines MJ, Franz N (eds) (2017) The small-scale fisheries guidelines, Global implementation. Springer, Amsterdam

Latty T (2019) Hidden women of history: Maria Sibylla Merian, seventeenth-century entomologist and scientific adventurer. School of Life and Environmental Sciences, University of Sydney. https://theconversation.com/hidden-women-of-history-maria-sibylla-merian-17th-century-entomologist-and-scientific-adventurer-112057

Lind A (2012) "Revolution with a Woman's face"? Family norms, constitutional reform, and the politics of redistribution in post-neoliberal Ecuador. Rethink Marx 24(4):536–555

Lowitt K (2011) Fish and fisheries in the evolution of Newfoundland foodways. Chap. 7. In: Chuenpagdee R (ed) World small scale fisheries – contemporary visions. Eburon, The Hague, pp 117–131

Mc Dughann S (2002) Mitos y Leyendas del Mar. El Azul Infinito. Océano Ámbar, España

McEwan C, Silva MI (1998) Arqueología y comunidad en el Parque Nacional Machalilla. In: Josse C, Iturralde M (eds) Compendio de Investigaciones en el Parque Nacional Machalilla. Nuevo Arte, Quito

Merchant c (1980) The death of nature: women ecology and the scientific revolution. Harper & Row, New York

Mollinga P (2008) The rational organisation of dissent. ZEF Working Paper Series, Center for Development Research, University of Bonn

Norton P (1985) Boletín de los museos del Banco Central N° 6. Simposio 45 del Congreso Internacional de Americanistas. Universidad de los Andes, Bogotá

Olson DM, Dinerstein E (1998) The global 200: a representation approach to conserving the Earth's Most biologically valuable ecoregions. Conserv Biol 12:502–515. https://doi.org/10.1046/j.1523-1739.1998.012003502.x

Olson DM, Dinerstein E, Wikramanayake GVN, Powell ED (2002) Conservation biology for the biodiversity crisis. Conserv Biol 16(5):1435–1437. https://doi.org/10.1046/j.1523-1739.2002.01532.x

Parry JH (1989) El Descubrimiento del Mar. Traducción castellana de J. Beltrán. Editorial Crítica, Barcelona. 362 pp

Pieterse JN (2010) Development theory: deconstructions/reconstructions, 2nd edn. Sage, London

Pohl C, Hirsch-Hadorn G (2007) Principles for designing transdisciplinary research. Proposed by the Swiss Academies of Arts and Sciences, oekom Verlag, München, 124 pp

Pohl C, Hirsch-Hadorn G (2008) Methodological challenges of transdisciplinary research. Nat Sci Soc 16:111–121. https://doi.org/10.1051/nss:2008035

Portillo Sotelo JL, Bernal-Casasola D, Eïd A (2020) Arqueología del garum baelonense: reflexiones metodológicas y excepcionales hallazgos. https://www.academia.edu/signup?a_id=65511095

Preston D, Preston M (2010) A pirate of exquisite mind: the life of William Dampier. Random House, New York

Rittel HW, Webber MM (1973) Dilemmas in a general theory of planning. Polit Sci 4(2):155–169

Rose GA (2003) Fisheries resources and science in Newfoundland and Labrador: an independent assessment. Research paper for the royal commission on renewing and strengthening our place in Canada. Government of Newfoundland and Labrador, St. John's

Rose G (2015) Northern cod comeback. Can J Fish Aquat Sci. 27 October 2015. https://doi.org/10.1139/cjfas-2015-0346

Rostworowski M (2005) Recursos naturales renovables y pesca, siglos VXI-XVII: Curacas y Sucesiones, costa norte. Instituto de Estudios Peruanos, Lima

Tapia W, Ospina P, Quiroga D, González JA, Montes C (eds) (2009) Ciencia para la Sostenibilidad en Galápagos: el papel de la investigación científica y tecnológica en el pasado, presente y futuro del archipiélago. Parque Nacional Galápagos. Universidad Andina Simón Bolívar, Universidad Autónoma de Madrid y Universidad San Francisco de Quito. Quito

Thornhill VJ (2020) Cod Collapse – lessons, legacy in Cod Collapse. Blog. February 2020. http://nqonline.ca/article/cod-collapse/

Viveiros de Castro E (2004) Perspectival anthropology and the method of controlled equivocation. Tipití: J Soc Anthropol lowland S Am 2(1):1

Williams M (1998) Aquatic resources. Education for the development of world needs. In: Symes D (ed) Fisheries dependent regions. Fishing New Books Blackwell Science, Oxford, pp 164–174

Wulf A (2015) The invention of nature: the adventures of Alexander von Humboldt, the lost hero of science: Costa & Royal Society Prize Winner. Hachette UK

Chapter 4
Post-War Reconnaissance of Japanese Fishery and Ocean Science and Its Contribution to the Development of U.S. Scientific Programs: 1947–1954

Carmel Finley

Abstract This chapter examines the over-looked contribution of Japanese scientists to ocean science and the construction of recruitment fisheries oceanography, the study of the effects of climate and ocean variability on fish abundance. After World War II, the U.S. Fish and Wildlife Service worked with the Supreme Commander Allied Powers staff in Tokyo to find and translate scientific documents about tuna and oceanography, for use by Americans trying to start fisheries in former Japanese waters. Determining the migration patterns of the fish was essential to catching them, and the Japanese translations greatly influenced *"Progress in Pacific Oceanic Fishery Investigations, 1950-53."* The document pioneers the integration of fisheries, oceanography, and meteorology to better understand the dynamic structure of the equatorial Pacific Ocean, and the importance of upwelling and frontal structures as they relate to distribution and abundance of Pacific tunas. The science of finding the fish was a critical step in the global expansion of tuna fishing throughout the subsequent decades. While the paper acknowledged the Japanese contribution to the construction of the science, the publication also masked the importance of the contribution.

4.1 Expanding the Foundation Stories about Fisheries Science

> In the last half of the 19th-Century American economy was largely based upon the development of the Great Plains. The Pacific Ocean is the Great Plains of the last half of the 20th century. (Chapman 1949)

C. Finley (✉)
Oregon State University, Corvallis, OR, USA
e-mail: finleyma@oregonstate.edu

© German Institute of Development and Sustainability (IDOS) 2023
S. Partelow et al. (eds.), *Ocean Governance*, MARE Publication Series 25,
https://doi.org/10.1007/978-3-031-20740-2_4

The short version of the foundation story of the development of fisheries science is that it built on natural history and zoological studies begun in Northern Europe and formally organized in 1902 under the direction of the International Society for the Exploration of the Seas (ICES), headquartered in Copenhagen. Its first theoretical paradigm was developed by Johan Hjort (1869–1948) in 1914, with an explanation of the natural variations in year-classes of fish (Hjort 1914). Hjort brought his ideas with him to Nova Scotia in 1914, where he met and influenced American oceanographer Henry Bryant Bigelow (1879–1967), the Harvard zoology professor and later the first director of Marine Biological Laboratory at Woods Hole (Schwach and Hubbard 2009). But how did Hjort's ideas spread to the Pacific Ocean?

A 1998 paper by two fishery scientists offered an idea: that Bigelow's two graduate students at Harvard were responsible for bringing his ideas to the Pacific. The two students, Oscar Elton Sette (1900–1972) and Lionel Walford (1905–1979), worked for the U.S. Fish and Wildlife Service while they were completing advanced degrees at Harvard under Bigelow. The federal agency transferred them to Stanford University in 1937 to lead an investigation into the collapse of the California sardine (*Sardina caerulea*) fishery. Sette wrote the first coordinated research plan for sardines in 1943, and his ideas were implemented with the creation of the California Cooperative Oceanic Fisheries Investigation (CalCOFI) after 1949. Arthur W. Kendall, Jr. and Gary J. Duker contend that the sardine plan was written to test Hjort's theories on recruitment (Kendall and Duker 1998).

Sette would not end his career with his work on sardines. In 1949 he was named director of the Pacific Oceanic Fisheries Investigation (POFI), headquartered in a new laboratory in Honolulu, with a mandate to find enough information about tuna to start an American fishery in the waters of the Mandated Islands, the former Japanese possessions now under American control. In addition to three research ships, POFI included a reconnaissance mission between the U.S. Fish and Wildlife and the Supreme Commander Allied Powers (SCAP) to find and translate Japanese documents about tunas and oceanography.

Sette published *"Progress in Pacific Oceanic Fishery Investigations, 1950-53,"* pioneering the integration of fisheries, oceanography, and meteorology to better understand the structure of the equatorial Pacific Ocean, its weather, and most importantly, the behavior of its tuna stocks (Sette 1954). This paper argues that Sette's contribution to ocean science has been systematically overlooked, as has the contribution of Japanese scientists after World War II, to the development of what is known as recruitment fisheries oceanography. Most simply, that is the study of the "effects of climate and ocean variability on fish abundance," (Wooster 1987). "Fisheries science" in this paper is used very broadly, to refer to scientists who are involved in studying fish and the catching of fish, and to the process of managing both fish and people.

Oceanography is by no means a unified science. There are four (or five) main divisions, with physical oceanography (waves, tides and energy); geological oceanography (sediments); chemical oceanography (the components of seawater): and biological oceanography (marine life). Actions by the Japanese and American governments led to the development of a new sub-field, integrating weather, currents, and fish survival.

While there has been much attention paid to the impact of the military on the development of oceanography more broadly, there is little attention to the impact of the military on the development of fisheries science. I have argued that after World War II, science became a tool of government; in particular, fisheries science became a tool of the State Department, used to structure post-war relations in terms beneficial to the U.S. But the military, with the assistance of federal scientists, was also used immediately after the war, to help create an American fishery far from the home waters, (Finley 2011).

The central conundrum for fisheries scientists is why fish populations fluctuate so much. The great seasonal herring (*Clupea harengus*) and cod (*Gadus morhua*) migrations in Northern Europe fluctuated wildly and a poor year threatened national economies. Naturalists in the 1880s at first thought that the stocks fluctuated when they took different migration routes. Johan Hjort, the Norwegian director of fisheries, was one of the first to move away from migration thinking to looking at fish as populations, then trying to understand the factors that influenced their behavior. The "critical period" for survival was during the egg and larvae stages; both life stages needed plentiful plankton as the eggs hatched and the larvae learned to swim. The key to understanding fish migration was to understand ocean currents, and what is more broadly called dynamic oceanography, the study of the ocean forces.

For generations, oceanographers had measured and mapped the oceans, such as in the volumes of the *Challenger* Expedition of 1872 to 1876. Baselines were established and changes were measured over time and interpreted. But with the turn of the twentieth century, this descriptive oceanography was being replaced by dynamic oceanography, grounded in mathematics, and trying to understand the large-scale interactions between the ocean and the atmospheric systems. The scientists who gathered in Copenhagen at the first ICES meetings increasingly were interested in a new strategy- repeated cruises, in the same area, at the same time of the year. Called intensive area studies, the objective was to create a web of hydrographic, biological and geologic data, which scientists hoped to integrate into a comprehensive analysis of fisheries problems (Brosco 1989). Such large-scale research projects needed interdisciplinary teams to delineate the patterns the data revealed (Hamblin 2014). While Hjort is credited with the theory, the research was a joint undertaking of a small group at the Directorate of Fisheries in Bergen, named the Bergen group, and in co-operation with the ICES scientists in Copenhagen, as well as state and university scientists from a variety of disciplines and member countries (Schwach 2013).

Such government-funded science was expensive, and it was paid for with the expectation that scientists would find new schools of fish for exploitation. As Norwegian historian of science Vera Schwach has noted, "the establishment of marine science as a multidisciplinary field occurred globally and was to a large extent materialized and financed within the framework of the economic utilization of fishes and fisheries management," (Schwach 2013).

Historians are now looking at how fisheries expanded globally, especially after World War II. Fishing has always been a strategy of empire, and it assumed new importance as military technologies were increasingly used by fishing boats, as were larger and more powerful engines that could fish bigger nets in deeper water. Governments played a central role in industrializing the fisheries, with the adoption of policies that encouraged investment in the development of fleets and processing facilities, as well as research into how to store and ship fish. Fishing was increasingly woven into government policies as the 1950s went on (Finley 2017).

There is an increasing body of scholarship exploring the development of marine resources in the Pacific. The patterns of development were more rapid than development in the Atlantic, where fisheries changed over centuries. Development in the Pacific was much faster and more international, with many nations using their fisheries to achieve economic and social objectives. While most of the scholarship on development in the Pacific deals with terrestrial matters, there is growing scholarship about the development of fisheries and whaling in the Pacific (Tsutsui 2013; Hee 2019; Arch 2018; Ogawa 2015).

It was not until the twentieth century that fishermen developed the skills and technologies to follow tuna throughout the oceans. Maritime countries had always taken some of the great fish as they migrated past, but they did not have the power to pursue the fish that never stop swimming, until the early 1900s (Joseph et al. 1988). Steam engines gave boats the power to chase the fish, and then to slow them down by throwing live bait into the water, attached to long slender bamboo poles; three men could work together to catch one of the giant fish; yellowfin could reach 400 pounds. The technique soon spread from the waters of Japan across the Pacific Ocean to Southern California, early in the 1900s. It was only a start for the fishers of the two nations to learn from each other and to transfer technologies. They also transferred science, sometimes involuntarily. And it was the start to a rivalry, over which nation would dominate the catch of the Pacific's great tuna runs.

There are approximately 58 species of tuna and related fish in the family, which also includes billfish, bonitos, swordfish, and mackerel. The largest species are marlins and bluefin tuna. Tuna are found in the tropical and temperate waters of the Pacific, the Atlantic, and the Indian oceans. They are unique among fish; while they are related to salmon, the two species are separated by approximately 100 million years of evolution (Dewar and Korsmeyer 2001). Biologists call tuna energy speculators, because they can invest large amounts of energy based on a payoff when they capture food. When they need it, tuna have the capacity for increased levels of oxygen uptake, delivery, utilization, and, consequently, work, allowing them to carry out many metabolic functions faster than other fish. Their circulatory system is designed to dissipate excess heat and they usually maintain a body temperature that is higher than the temperature of the water in which they swim. Tagging studies on tuna show they migrate thousands of miles across the open ocean. "These fish are alert and very difficult to catch," wrote the world's premier tuna biologist, Kamakichi Kishinouye (1923). The most important commercial species were skipjack (*Katsuwonus pelamis*), yellowfin tuna (*Neothunnus macropterus*) and albacore (*Thunnus alalunga*).

It was well known by the 1930s that the Japanese were the world's best fishermen. The sea has always been of central importance to Japan, and fishing, whaling, and shipbuilding have played prominent roles in the development of the world's largest and most sophisticated fishing fleet. A series of subsidies began in 1923, encouraging the construction of refrigerators, refrigerated boats, and ice-making systems, allowing Japanese boats to carry their fish to other countries. During 1931–1938, when fishing was at its peak, Japan's aggregate annual production ranged from 3.5 million metric tons to 4.5 million metric tons. The U.S. catch, combined with Alaskan salmon, was less than 2.5 million metric tons a year (Espenshade 1949).

But while they were the best fishermen, the quality and depth of their scientific scholarship is only recently receiving attention. They were also skilled scientists, with a rich research tradition that had been well-funded by successive governments. The *Fukuoka Gyogyoshi*, or "Description of Fukuoka's Fisheries," identifying about 100 species of fish, was compiled in the 1870s. The Hydrographic Department of the Imperial Navy was established in 1871 to make charts of ocean currents, tides, and depths in the coastal regions (Kalland 1995). The government also set up an extensive series of fisheries experimental stations and meteorological observatories. The fisheries experiment stations studied sea conditions and broadcast weather reports to the fishing industry. The marine meteorological observatories were engaged in ocean meteorology. The Central Meteorological Observatory conducted surveys of sea currents using a series of instruments placed along the Japanese coast.

The Fisheries Society of Japan was created 1882 to give direction to the general fishery activity in the country. In 1885, the Fisheries Bureau was inaugurated within the Ministry of Agriculture and Commerce. In 1890, the Fisheries Bureau established the Fisheries School for the training of technicians, while the government created the Committee of Investigation for Fisheries and the Investigation Station of Fisheries (*Japan Times and Mail* 1939a, b). The Fisheries School was reorganized into the Imperial Institute of Fisheries, located outside Tokyo. The curriculum was divided into three general areas, fishing, fisheries technology, and pisciculture. Study in each area took 3 years, and included all aspects of fishing, from navigation to gear development, canning and salting technology, and a wide range of aquaculture efforts aimed at increasing the cultivation of fishes and seaweeds. It was a uniquely comprehensive education.

By 1937, Japan was the world's leading fishing nation. Its network of fisheries was spread throughout the Pacific, and into the Indian and Atlantic oceans. The objectives of the "aquatic products industry" were to guarantee fishermen a stable livelihood and to improve the health of the nation by providing a supply of fresh protein. The development of overseas fishing and the export of fisheries products were considered extremely important to the health of the Empire. The Japanese were proud of their fisheries development, and the research that furthered the country's accomplishments. "The perfect cooperation among the aquatic industrial experimental stations…is unheard of in other countries," wrote the *Japan Times & Mail* in 1939. While fishery institutes in other countries only concentrated on the deep-sea, Japan had a far more extensive and expansive scale of fishery education, drawing requests for information from scholars in other countries. The initial structure of the School of Fisheries at the University of Washington in 1919 was modeled on the Japanese model (Stickney 1989).

After World War I, Japan had acquired control over the Micronesian islands, the Marshall, Mariana, and Caroline Islands, also known as the Mandated Islands. By the 1930s they had developed a lucrative tuna fishery. With the end of World War II, the islands and their waters, were under the control of the Americans. The Japanese fishing industry, which had dominated fishing in the Pacific during the 1930s, was now strictly confined to its home waters, opening an opportunity for the U.S. to begin developing fisheries the Japanese had discovered.

The Americans starting planning for the occupation of Japan in 1942, with a research division in the State Department (Martin 1948). The Supreme Commander Allied Powers (SCAP) arrived in Japan with a series of policies designed to completely transform Japanese life. Nine sectors were organized to carry out the Occupation. Japan was to be demilitarized and disarmed. The economy was to be transformed, the large industrial and banking combines dissolved, and the educational system modernized. Society was to be transformed from feudal and authoritarian to democratic, labor unions encouraged, and women given the right to vote, hold property, enter higher education, and run for public office. Four million acres of land was bought and sold cheaply to farmers (Le Feber 1997).

Fisheries was managed by the Natural Resources Section, along with agriculture, forestry, and mining.[1] It was headed by Col. Hubert Schenck, a paleontologist from Stanford University. SCAP's initial fisheries policy, laid out on Feb. 18, 1946, included the goal of "ensuring the maximum production of seafood products consistent with security requirements," (Yamamoto 2000). At the same time, Japanese boats were greatly restricted to their home waters, in the interests of security.

The Americans turned out to be extremely interested in reforming Japanese fisheries and giving rights to poor fishermen through the Fisheries Rights Reform bill. An undated SCAP document records a long series of meetings and correspondence over the American legislation; it covers 17 pages, with SCAP continuously urging the government to move forward with the American plans.[2] The core of the plan was to establish a fishery coordination committee to make democratic and optimum use of fishery resources. Local and regional fishermen would control the sea off their prefecture, conserving their resources for themselves and their communities. It was an attempt to break the power of the Japanese fishing companies and the government ministries.

The fisheries division staff included John L. Kask, an Army captain and a fishery graduate from the University of Washington. He published an intensive study of the ownership of the four largest Japanese fishing companies in 1949, including the names of their shareholders (Kask 1947). He wrote two other leaflets, about the

---

[1] National Archives and Research Administration (NARA), RG 331, Box 8867. Supreme Commander for the Allied Powers, "Summation of Non-Military Activities in Japan and Korea, No. 1," (Tokyo: Supreme Commander for the Allied Powers 1945) 3.

[2] NARA RG 331, Box 8867, Japanese Reconnaissance Team, Pacific Oceanic Fisheries Survey, Nov. 22, 1948.

fishing gear used in Japan, and the Japanese system of education. He found there were fisheries schools in all of the prefectures, turning out expert fishermen, cannery operators, and technicians. There were two universities doing advanced work in fisheries and oceanography, in Tokyo and Hokkaido.

A further report, in October of 1948, detailed the history of oceanography in Japan, starting in 1902, when the first cross-line observation, measurements on a wide scale, was attempted. The report contained a summary of published research for 1946, including what scientists were working on selected projects in various prefectures. The fisheries literature was "extremely voluminous," Kask wrote, and would need to be translated (Kask 1947). Japan supported 32 provincial fishery schools in 24 provinces, teaching everything from "how to row a boat and how to fish to meteorology and navigation." There are also two fisheries colleges and 70 research and training vessels (this is before the war). There were 112 provincial research stations and a large Central Fisheries Research Station in Tokyo with five strategically situated branch stations throughout the country. Even school children learned about fish.

By contrast, the American funding for ocean science had been scant and intermittent. The U.S. Fish Commission was created in 1871, after the British demanded landing taxes for American mackerel sold in Nova Scotia. The British had landing bills; the Americans no catch numbers, and Congress was unhappy about the size of the British tax bill. The first director of the new institution was Spencer Fullerton Baird (1823–1887). Baird argued that in order to understand the fluctuations in the supply of commercially valuable fish, it was necessary to understand the ocean food chain. This justified the construction of the first American oceanographic fishing vessel, the *Albatross*, a 200-foot-long steamer launched in 1882, and the construction of the Woods Hole laboratory, to process the material collected at sea and to do more intense work on marine organisms (Allard 1978).

The Depression had led to steep cuts in the budget for the U.S. Fish and Wildlife Service, and the last research ships had been mothballed early in the 1930s. There were no federal and state funds for ocean research. One of the reasons the Scripps Institution of Oceanography hired Norwegian Harold Sverdrup in 1937 was the hope that he would lead a resurgence of American research ships back to the ocean (Rainger 2000).

The fluctuations in the California sardine fishery, and its eventual collapse, created the crisis that sent American scientists back to sea. Sardines had gone from a $60 million industry down to $15 million. Despite its slim budget and small staff, the agency sent its two top Atlantic scientists to its laboratory at Stanford, to head an investigation into why the fishery was collapsing. For both Elton Oscar Sette (he preferred to go by Elton) and Lionel Walford, who were both born in California, it was chance to take Hjort's and Bigelow's ideas, and the techniques of intensive area studies, to the Pacific Ocean and the sardine problem. Sverdrup was introducing the theories of dynamic oceanography, and the need to study all of the life stages of marine life, as well as the environment in which they lived. It was an exciting time for the development of ocean science (Powell 1972).

Sette was born in California in 1900. He was 18 when he was hired to survey albacore landings at San Pedro. He would do his undergraduate work at Stanford under noted educator and ichthyologist, David Starr Jordan (1831–1951). His first academic publication, about why sardines fluctuated, is marked by its use of statistical methods to conclude that samples may not be representative of the population as a whole. Hired by the old Bureau of Commercial Fisheries, Sette was promoted to the Chief of the North Atlantic Fishery Investigations in 1928. His office was at the Museum of Comparative Zoology at Harvard, and he spent the summers acting as director of the Bureau's station at Woods Hole.

For the sardine research, the California legislature appropriated $800,000 for the Scripps Intuition of Oceanography and levied a $200,000 special tax on sardine processors. Sette's sardine plan, published in 1943, became the blueprint for the California Cooperative Sardine Research Program, re-named the California Cooperative Oceanic Fisheries Investigations, or CalCOFI. It was necessary to study all of the life stages of the sardines, as well as to study the impact of fishing on the stocks.

California state biologists were originally uneasy about the additional federal presence, but Sette soon established good relationships with state biologists (Powell 1982). With the spread of the fishery into Oregon and Washington waters in the 1930s, research into sardines also expanded to other agencies, including federal and provincial scientists in British Columbia. Sette organized annual meetings to share data and information, calling it a "cooperative research program, in the best sense," (Sette 1943).

The creation of CalCOFI, and the prospect of pushing the American tuna fishery deeper into the Pacific, generated a lot of state and federal support. Congress in 1944 passed a resolution to expand American fisheries, to develop king crab in Alaska and a high-seas tuna fishery. American boats had fished their way south to the Galapagos in the 1930s, and as far east as Hawaii. But to develop a new fishery, there would have to be substantial federal support.

As early as 1943, the U.S. military had decided on a Pacific strategy that depended on the building of military bases, some of them in the Mandated Islands, the former Japanese territories which came under U.S. control in 1946. As the fighting in the Pacific intensified, military officials were interested in finding new food sources, especially fish that could be served fresh. The Office of Economic Warfare was responsible for procurement and production of all imported materials necessary to sustain the war effort and the civilian economy. One of the board's many goals was to use local foods to supplement canned rations in war zones. For a war zone in the middle of the Pacific Ocean, that meant finding fish to feed service men.

The food situation was critical; in November of 1943, the upper Solomon Islands were so recently secured from the Japanese there were no lines of supply. Rations were dry and in short supply. There were growing numbers of troops in the Pacific. Could fish be caught to feed them? Four scientists, including Wilbert McLeod Chapman, were hired to find out. Chapman had graduated from the School of Fisheries at the University of Washington with doctorate in ichthyology in 1937. When war broke out, he had been hired as Curator of Fisheries at the California

Academy of Sciences in San Francisco. His close friend, Milner Baily Schaefer, had also graduated from the University of Washington School of Fisheries, with a Bachelor of Science in 1935. Chapman had Schaefer seconded to the fisheries investigation, but Schaefer contracted rheumatic fever in New Caledonia and would spend most of the war in military hospitals.

Chapman's initial scouting trip stretched from a few days to 3 months and 20,000 miles of air travel. He would eventually spend 14 months in all working to start fisheries in the Gilbert, Ellice, and New Caledonian islands, and then to the Solomon Islands. He started fisheries at roughly 20 different military bases, primarily in the New Caledonia, the New Hebrides, and the Solomon Islands.[3] But while the projects could catch fish to feed soldiers, it did not find a home. It was originally a Navy project, but it was transferred to the Army, and Chapman's plan to establish fisheries "in the whole South Pacific area," disappeared "and I was never again able to find the slightest trace of it," according to his account of his wartime service.[4]

Chapman's wartime plan for the military might have disappeared but he certainly retained his own plan to establish American fisheries in the South Pacific. After his return to San Francisco, he immediately started an extensive letter-writing campaign to expand American tuna fisheries deeper into the Pacific. In letter after letter, to politicians and other academics, Chapman urged for the expansion of the American tuna industry into the Pacific and insisted that federal funding was essential to the expansion.[5] Throughout Chapman's extensive writing during this time, he frequently referred to the effort the Japanese put into research and science on oceanography and tuna, far more than the Americans were funding.

In December of 1946, he asked Schaefer, who had finally been released from a military hospital, to pull together some information about the potential for an American fishery in the islands. Americans could reap a "considerable harvest," from the adjacent seas, and there were possibilities "that lie in the exploitation of other parts of Oceania by American fishermen based on scientific study of the tunas and their habitats." Schaefer went on to say the Japanese are building "new large tuna vessels and motherships. They may be expected to expand their fisheries as rapidly as the occupation forces permit."[6]

Hawaii's delegate to Congress, Joseph R. Farrington, introduced a bill in January of 1946, seeking funds to provide for the exploration and development of high seas

---

[3] University of Washington Special Collections (UWSC), Wilbert M. Chapman papers, Box 4, folder A, undated report.

[4] UWSC, Chapman papers, Box 4, Folder 1.

[5] The most complete account of Chapman's activities during this period comes from Harry Scheiber, "Origins of the Abstention Doctrine in Ocean Law: Japanese-U.S. Relations and the Pacific Fisheries, 1937–1958." *Ecology Law Quarterly* 16 (1989): 23–101; "Pacific Ocean Resources, Science, and Law of the Sea: Wilbert M. Chapman and the Pacific Fisheries, 1945–1970," *Ecology Law Quarterly* 13, no. 38 (1986), Arthur F. McEvoy and Harry N. Scheiber, "Scientists, Entrepreneurs, and the Policy Process: A Study of the Post-1945 California Sardine Depletion" *Journal of Economic History* 44, no. 2 (1984).

[6] NARA RG 331, Box 8867, Japanese Reconnaissance Team, Pacific Oceanic Fisheries Survey, Nov. 22, 1948.

fishing in the Territorial waters of the sub-tropical Pacific. The bill called for $350,000 to build the research lab in Honolulu, $700,000 for three vessels, and $350,000 as an operating budget. For a country that has stopped going to sea in the 1930s because at sea research was too expensive, it was a big step forward. Too big; critics protested that surely the fish resources of Hawai'i could never be big enough to warrant such an expenditure.

Chapman became one of the most enthusiastic proponents of Farrington's bill, speaking with the authority that came having spent 14 months in the Eastern Pacific. This was the start of his rise to a national political figure, one of the most influential scientists of his generation, appointed to a position at the State Department and deeply involved in the negotiations over several fisheries treaties, including the peace treaty with Japan.

Chapman was explicit that the objective of the bill was to provide the information needed "by American industry to risk capital in establishing fisheries in the area, particularly in the Mandated Islands."[7] The Japanese harvested more tuna from the waters of the Mandated Island than what Americans had caught in the entire Eastern Pacific, Chapman wrote, "and their fisheries there were new and still rapidly developing." The Americans developed a high-seas tuna fishery that was dependent on being able to harvest bait from near-shore waters, increasingly the waters off Mexico and Latin America. The Latin American countries were increasing the fees they charged to American boats to fish in their waters.

In his frequent publications, Chapman argued that while crops are produced from the top few inches of soil, the sea had resources throughout its water column. With the victory in the war, Chapman wrote that the nation had won "an empire of great riches, where the land is as nothing and the sea is everything—an empire in which the native people are small in numbers and restricted to small points in its vastness; an empire which no other nation save the Japanese covets and which no other nation save theirs and ours can cultivate and make produce," (Chapman 1949).

With Chapman's support, the Farrington Bill passed on a second attempt in 1949, inaugurating a new period in the development of federal fisheries science, the exploratory fishing programs. Four programs were established, the Gulf Exploratory Fishery Investigations, the Northwest Pacific Exploratory Investigations, and the North Atlantic Fishery Investigations. The lead program was POFI, and Sette was the logical scientist to direct the new laboratory and its large-scale research operation. He hired Schaefer to head the section on biology and oceanography. Schaefer was the chief scientist onboard the first POFI cruise, on a vessel called the *Oregon*, out of Honolulu. Assigned to run surveys on systematic legs, they found the ocean was so rough they sometimes could not cast live bait. Bait was scarce. Finding tuna was going to be more difficult than they thought.

While Sette was in charge of the POFI operation, Chapman was deeply involved in the reconnaissance mission. He had left the California Academy of Sciences in 1947 to take over as director of fisheries at the University of Washington. Three of

---

[7] UWSC, Papers of Edward Allen, Box 18, Folder "United Nations fisheries conference."

the scientists hired for the reconnaissance mission came from the University of Washington. The leader was Frederick "Fred" Cleaver, and included a chemistry student, David T. Miyauchi.

The most important component of the renaissance mission was a 26-year old Japanese American scientist, Bell M. Shimada (1922–1958). He was born in Seattle to immigrant parents. He showed an early aptitude for mathematics and entered the School of Fisheries at the University of Washington in 1939. With the declaration of war against Japan, he was one of thousands of Japanese people rounded up and sent to internment camps; he was sent to Minidoka in Idaho in 1941. He volunteered as an infantryman, then was selected for intelligence and Japanese language training. He was assigned to the Military Intelligence Service and embedded in the US Army Air Forces.

For the next 2 years, Shimada "hopscotched behind the Pacific frontline," as his official federal biography states. After the surrender of Japan, he moved to U.S. Army Air Forces headquarters in Tokyo, as part of the Occupation of Japan. His job was to collect and synthesize economic and infrastructure data on the effects of the strategic bombing of Japan. He was discharged from the military in February of 1946, but he stayed in Tokyo, in a civilian position as a fisheries biologist in the Natural Resources Section. He remained in Tokyo for another 9 months before returning to Seattle where he enrolled for the fall quarter at the School of Fisheries in 1947. He left Tokyo with two highly complementary letters, including one from the SCAP natural resources director, Schenck. Shimada did "superior work," Schenck wrote, completing several detailed studies of fisheries and helping the Occupation run more smoothly. His loss would be "keenly felt." A second letter, from Major John F. Janssen, wrote that Shimada's "innate ability, pleasing personality, loyalty and conscientiousness make you a valuable asset to any fisheries research."[8]

Despite the disruptions to his schoolwork, he was seventh his senior class the fall of 1947. He would graduate in December, cum laude, and stayed in on to work on a graduate degree.[9] By December of 1948, he had his Master of Science in Fisheries, and had been hired by Sette as part of the new POFI investigation. In November of 1948, he was back in Tokyo, "to gather information on the methods of fishing, methods of fish processing, methods of research, distribution, ecology, life history and other information relating to tuna."[10]

He would certainly have been welcomed back at SCAP. He kept a detailed journal of his activities in Tokyo, dealing with scientists he was meeting and copies of papers that he has acquired. He was busy from the start, finding out who to talk with, and making appointments, acquiring copies of papers that were microfilmed by an assistant. It was to be a 3-month assignment, but it stretched until June of 1949. His

---

[8] Papers of Bell Shimada, courtesy of the Shimada family.

[9] UWSC, Chapman papers, 1852-1,2,3, Box 11, Folder 26.

[10] NARA RG 331, Box 8867, Japanese Reconnaissance Team, Pacific Oceanic Fisheries Survey, Nov. 22, 1948.

journal was typed on loose-leaf lined paper and kept in a three-ring binder. Over the 9 months, he would list the documents he was seeking, and those he was able to find. In his 1951 publication of tuna, Shimada thanked the Natural Resource Section for its help, including William C. Herrington, Drs. K. Kuronuma and Y. Hiyami, as well as additional scientists (Shimada 1954).[11] It is the first publication of some Japanese scientific works in English.

Shimada kept notes of all conversations in his journal. A typical example is of his conversation with Dr. Kinosuke Kimura of the Central Experiment Station. He wrote that Kimura tagged 1700 skipjack in 3 years, of which three were recovered offshore and six were taken in the inshore fishery. Details of the tagging and the recovery were included, as was Kimura's belief that the hook tags adhered best to the fish. His recording to conversations indicates how little was known about tuna, and how all scraps of information had potential value to be passed on. Everywhere he went, he asked for copies of papers. One of the most significant that he acquired was a copy, written in English, by Kishinouye Kamakichi's 1923 publication, "Contributions to the comparative study of the so-called Scombroid Fishes."

Over the next months, he continued to visit science stations, recording details of fish landed in various ports. He was especially interested in talking with fishermen, such as the fleet at Omaezaki, in the Shizuoka Prefecture, said to be the best skip-jack fishermen in Japan. They told him that some skipjack migrated through their waters, but others were resident, said to live along the underseas ridges. "Fishermen believe that skipjack which are too weak to continue their journey drop out of the schools and remain near these ridges to feed…" He also packed up specimens for shipment to the POFI office in Honolulu.

He also found and was involved in translating the minutes of a meeting Japanese scientists held in 1940, to discuss what they knew about the spawning grounds of tuna and skipjack. Published as a Special Scientific Report, Fisheries 18, it was edited by Shimada and W.G. Van Campen, another of the SCAP translators, in April of 1950. Ten scientists and industry representatives met to pool their knowledge about tuna and to craft a research response. Shinkishi Natai, director of the Palou Tropical Biological Station and an emeritus professor from Tokyo Imperial University, was recorded as saying that almost nothing was known about the spawn-ing grounds of most fish, but especially skipjack, the species most important to the Japanese industry. Despite a decade of considering the problem with conferences every 3 or 4 years, they were no closer to a solution. "No new facts have yet been ascertained," Natai said. He hoped the group would come up with a "definite plan" of study (Shimada and Van Campen 1951).[12]

Back in Los Angeles, POFI held a conference in October of 1949, laying out the work that needed to be done to expand the fishery. Expectations were high. "The

[11] B. M. Shimada, "An annotated bibliography on the biology of Pacific tunas," U.S. Fish and Wildlife, Fishery Bulletin 56.

[12] U.S. Fish and Wildlife, Special Scientific Report, Fisheries No. 18, "Spawning grounds of tuna and skipjack," translated by B. M. Shimada and W. G. Van Campen, Pacific Oceanic Fisheries Investigations, April, 1950.

expedition is expected to locate new tuna banks that should produce from $80,000,000 to $100,000,000 worth of tuna each year," enthused *Tuna Fisherman* magazine, a new publication from San Diego, (*Tuna Fishing Magazine* 1948a, b).

The first task would be to finish the translations that had come in from Shimada and the rest of the SCAP staff in Japan. The material was of "great value," both for its information about the fish, but also about successful Japanese fisheries. POFI cruises would begin with basic studies of salinity, oxygen, and nutrients. One of the first objectives was to look at how to catch bait, the fishing system used by most American tuna boats. The area of operation was to be the Central Pacific Ocean, between the Hawaiian archipelago and the equator, where the Japanese had established a growing fishery for skipjack tuna. The fishery expanded to include larger boats to catch yellowfin and marlin.[13] But bait proved hard to find. "It may well be necessary to test and devise techniques new to American fishermen."[14]

Three exploratory vessels were assigned to the new laboratory, all named after early federal fisheries scientists The *R/V Hugh M. Smith* was a 128-foot ex-Navy auxiliary vessel, outfitted "to conduct oceanographic studies of all sorts as well as semi-commercial-scale tuna fishing by means of live bait, trolling, and long-line fishing," Sette and Schaefer wrote in a statement about the program. The *Henry O'Malley* was a sister ship to the *Hugh M. Smith* and was set up for live bait fishing and trolling on a commercial scale. The third vessel was the *John R. Manning*, a newly built 85-foot purse seiner, designed for experimental and exploratory fishing. Finding tuna in the Pacific was a tall order, even for three new research ships. As a fishing industry contribution to the conference put it, while the industry was interested in new opportunities, it was hard to find a great fish "about which we know less than we do about tunas."[15]

As Shimada continued with his research in Tokyo, the new laboratory opened in Honolulu. Sette transferred there, along with his secretary, Rae Shimojima, originally from Portland.[16] The data was beginning to come in from the first research cruises. Some of the first came from POFI's flagship, the *Hugh Smith,* and its young oceanographic officer from the University of California, Townsend Cromwell. He was setting longline gear while fishing for tuna at the equator, south of Hawaii in December of 1951. The gear drifted to the east, while the surface current drifted the ship to the west. None of the current theories about ocean circulation could account for the phenomenon. During the next five longline cruises, Cromwell found further evidence for an eastward subsurface current. The following August, he headed an investigation that made 12 direct current measurements near the equator. He had

---

[13] University of Washington Special Collections, Pacific Oceanic Fisheries Investigations, tuna industry conference, Oct. 7, 1949, Richard Van Cleve papers, 168-3-71-10, box 4, Folder, "Tuna meeting, 1949."

[14] *Commercial Fisheries Review*, May Progress Report, 27.

[15] UWSC, Papers of Richard Van Cleve, Pacific Oceanic Fisheries Investigations, tuna industry conference, Oct. 7, 1949, Box 4, Folder, "Tuna meeting, 1949."

[16] https://fish.uw.edu/2019/02/centennial-story-69-bell-masayuki-shimada-bs-1947-ms-1948-phd-1956-ba-2008-honoris-causa/ Accessed 05/06/2018

discovered what he suggested calling the Pacific Equatorial Undercurrent for this east-flowing subsurface current, (Knauss 1960).

Shimada left Tokyo in June of 1949 and began work for POFI out of Honolulu. Some of the first translations began to appear in the U.S. Fish and Wildlife literature, and in the trade press. *Pacific Fisherman* in June of 1948 heralded "SOME of the SECRETS of Japanese tuna fishing dug from archives."[17]

In June of 1948 Chapman was appointed as an assistant to the State Department, to deal with fisheries issues. He was extremely successful, overseeing the signing of the treaties to establish the International Conference of the North Atlantic Fisheries (ICNAF) and the Inter-American Tropical Tuna Commission (IATTC), both active today. He was also heavily involved the negotiations of the peace treaty with Japan, as well as the signing of the first fishery treaty among Japan, Canada, and the U.S.

The Inter-American Tropical Tuna Commission was established in La Jolla; its first director was Schaefer. Among his first acts was the hiring of several scientists from the POFI laboratory in Honolulu, including Cromwell and Shimada. The two were on their way to another expedition in Mexico when their plane plunged into a mountain in 1958, killing everyone onboard. The Pacific current Cromwell had described was re-named the Cromwell Current. The Shimada Sea Mount is located southwest of Baja, California. Both men have had research vessels named after them, as has, Sette; Wilbert Chapman was also honored by the naming of a research vessel.

The 1954 report lays out the integration of fisheries, oceanography, and meteorology to better understand the dynamic structure of the equatorial Pacific Ocean, and the importance of upwelling and frontal structures as they relate to distribution and abundance of Pacific tunas. The 80-page document contains 25 pages of footnotes, with a substantial number of entries by Japanese scholars and the scientists who helped with the translations. Sette, aided by the translations (not just from the Japanese but from German, British and Italian scholars), had been able to apply the theories of dynamic oceanography to find order in the data that had poured in from so many sources. It was a triumph of the dynamic oceanography approach (Hamblin 2014). As Sette wrote, the results of the 3 years of sea work "appear to have immediate practical fishery significance," (Sette 1954).

Sette's research showed why equatorial waters were more productive than waters to the north and south: the presence of a powerful equatorial circulation. The steady southeast trade winds brought nutrient-rich waters from ocean floor to the surface, where sunlight stimulated production of planktons, benefitting the entire food chain, and where tunas, "the final step in oceanic production line, concentrate here where there is good feeding much more of the time than elsewhere," (Sette 1954).

With the development of hydraulics after 1957, purse seining for tuna expanded rapidly, worldwide. There had been seining in the ocean during the 1920s and 1930s, but nets were made of cotton painted with tar; they were heavy and difficult to bring back onboard, requiring a tuna boat to have a large crew. Along with

---

[17] *Pacific Fisherman*, June, 1948, 37–8.

hydraulics came nylon nets, lighter, stronger, and requiring a far smaller crew. Another powerful innovation was rapid freezing technology. The surface and the inside of the tuna are frozen simultaneously, allowing ice crystals to freeze before they can clump with other ice crystals, damaging the cell walls of the fish. The technology allowed tuna to be caught, frozen at sea, and delivered anywhere in the world.

While the Americans were busy copying any papers on tuna, salmon, hatcheries, and ocean conditions, at the same time, SCAP disparaged Japanese science as being woefully behind American science. Fisheries research was not based on population studies. Too many of the research stations did technical research into how to catch fish, not biological studies. SCAP recommended "a carefully planned and coordinated research program in the natural resources field."[18]

SCAP brought three prominent American fishery scientists to Tokyo, to help Japan develop a "sound, modern fisheries research plan," according to the report, written by Willis H. Rich of Stanford University.[19] He found that research before the war was largely devoted to technology and biological studies, aimed at improving catch rates. The effort was on exploitation, with little focus on conservation and the methods of research and regulation that were "sound and effective." It was an article of faith that American fishery management was the best in the world, based on sound science. In fact, sardines and salmon were both being over-harvested, and studies at sea, which the Japanese had being doing for decades, were just getting started on the West Coast.

Yet the Americans touted their modern, science-based research. Chapman was certainly aware of how far ahead the Japanese were, and that the Soviets were rapidly escalating their fisheries and research in both the Atlantic and Pacific. "The old method of straight political regulation of fisheries in international waters is passé; the new method of regulation on straight biological grounds is not yet applicable because of our ignorance," he stated in one of his letters campaigning for the Farrington Bill.[20]

The first significant scholarship on these events comes from Berkeley law professor Harry Scheiber, who has written extensively about the development of ocean law, especially in the Pacific. Scheiber places Chapman at the center of his analysis, with the central political role he played in events between 1945 and 1952. He called Chapman "a brilliant scientific entrepreneur," who was at the center of the development of ocean law between 1945 and 1951.

Scheiber also identifies several other scientists that were catalysts of change within the science. Milner Schaefer "exemplified the possibilities that Chapman and the other heralded when they embarked on their campaign for the new oceanography in 1945," Scheiber wrote. He identifies other scientists, including Sette, but he gives more credit to Schaefer. As Scheiber tells his story, the quest was to mobilize

---

[18] UWSC, Papers of Miller Freeman, Box 11, Folders 4, SCAP, Natural Resources Section, Preliminary Study of No. 42, Fisheries Research Program of Japan, Willis H. Rich.

[19] Ibid.

[20] UWSC, Richard Van Cleve papers, Box 2, Folder "Chapman, W. M., 1940–48."

the "intellectual resources of American scientists, the fishing industry, as well as the government, to develop American ocean fishing interests," and also "developing marine fisheries management on a global scale." Missing from Scheiber's account is the influence of the military in these efforts, and the science developed by the Japanese.

The short story of the development of fisheries science needs to be amended, to include the Japanese contributions to the construction of the science.

References

Allard DC Jr (1978) Spencer Fullerton Baird and the U.S. Fish Commission. Arno Press, New York, p 182

Arch JK (2018) Bringing whales ashore: oceans and the environment of early modern Japan. University of Washington Press

Brosco JP (1989) Henry Bryant Bigelow, the US Bureau of Fisheries, and Intensive Area Study. Soc Stud Sci 19(2):239–264. Stable URL: https://www.jstor.org/stable/285142. Accessed 1 June 2019 19:23 UTC

Chapman WM (1949) The wealth of the ocean. Sci Mon 65(3):192–197

Commercial Fisheries Review, May Progress Report, 27

Dewar KE, Korsmeyer H (2001) Tuna metabolism and energetics. In: Stevens ED, Block BA (eds) Tuna: physiology, ecology, and evolution. Academic, New York, pp 35–78

Espenshade A (1949) A program for Japanese fisheries. Geogr Rev 39(1):76

Finley C (2011) All the fish in the sea: maximum sustained yield and the failure of fisheries management. University of Chicago Press, Chicago

Finley C (2017) All the boats on the ocean: how government subsidies led to global overfishing. University of Chicago Press, Chicago

Hamblin JD (2014) Seeing the oceans in the shadow of Bergen values. Isis 105(2):352–363. Stable URL: https://www.jstor.org/stable/10.1086/676573

Heé N (2019) Tuna as an economic resource and symbolic capital in Japan's "Imperialism of the Sea". In: Animals and Human Society in Asia, Palgrave Macmillan, Cham, pp 213–238

Hjort J (1914) Fluctuations in the Great Fisheries of Northern Europe viewed in the light of biological research. Conseil Permanent pour V exploration de la mer, Rapport et Procès-Verbaux des Réunions XX

Japan Times and Mail, Japan's Fisheries Industry 1939 (1939a), 34

Japan Times and Mail, Japan's Fisheries Industry 1939 (1939b), p 15

Joseph J, Witold K, Murphy P (1988) Tuna and billfish: fish without a country. Inter-American Tropical Tuna Commission, La Jolla, p 3

Kalland A (1995) Fishing villages in Tokugawa Japan. University of Hawaii Press, Honolulu, p 99

Kendall AW, Duker GJ (1998) The development of recruitment oceanography in the United States. Fish Oceanogr 7(2):69–88

Kask JL (1947) Japan's big fishing companies, U.S. Department of the Interior, Fish and Wildlife Service

Kishinouye K (1923) Contributions to the comparative study of the so-called Scombroid fishes. J Coll Agric 8(3):293–475. Imperial University of Tokyo

Knauss JA (1960) Measurements of the Cromwell current. Deep-Sea Res 6(4):265–286

Le Feber W (1997) Clash: U.S.-Japanese relations throughout history. W.W. Norton & Company, New York, pp 263–266

Martin EM (1948) The allied occupation of Japan. Stanford University Press, Stanford, pp 5–7

Ogawa M (2015) Sea of opportunity: the Japanese pioneers of the fishing industry in Hawaii. University of Hawaii Press

Powell P (1972) Oscar Elton Sette: fishery biologist. Fish Bull 70(3):525–535

Powell P (1982) Personalities in California fishery research. CalCOFI Rep XX111:43

Rainger R (2000) Patronage and science: Roger Revelle, the U.S. navy, and oceanography at the Scripps institution. Earth Sci Hist 19:58

Schwach V (2013) The sea around Norway: science, resource management and environmental concerns, 1960-1970. Environ Hist 18(1):101–110

Schwach V, Hubbard J (2009) Johan Hjort and the birth of fisheries biology: the construction and transfer of knowledge, approaches and attitudes, Norway and Canada 1890–1920. Studia Atlantica 13:20–39

Sette OE (1943) Studies on the Pacific pilchard or sardine (Sardinops caerulea). Structure of a research program to determine how fishing affects the resource. US Fish Wild Serv Spec Sci Rep 19:27

Sette OE (1954) Progress in Pacific Oceanic fishery investigations, 1950-53. U.S. Fish and Wildlife Service, Special Scientific Report: Fisheries No. 116, Washington, DC

Shimada BM (1954) An annotated bibliography on the biology of Pacific tunas. U.S. Fish and Wildlife, Fishery Bulletin 56

Shimada BM, Van Campen WG (1951) Tuna fishing in palau waters. U.S. Fish and Wildlife Service

Stickney R (1989) Flagship: a history of the fisheries at the University of Washington. University of Washington Press, Seattle

Supreme Commander for the Allied Powers (1945) Summation of non-military activities in Japan and Korea, no. 1. Supreme Commander for the Allied Powers, Tokyo, p 3

Tuna Fisherman Magazine, Vol 1, No 2 (1948a)

Tuna Fisherman Magazine, Vol 2, No 4 (1948b)

U.S. Fish and Wildlife, Special Scientific Report, Fisheries No. 18, Spawning grounds of tuna and skipjack. Translated by BM Shimada, WG Van Campen, Pacific Oceanic Fisheries Investigations (1950)

William TM (2013) 1. The pelagic empire: Reconsidering Japanese expansion. In: Miller IJ, Thomas JA, Walker BL (ed) Japan at nature's edge: The environmental context of a global power, Honolulu, University of Hawaii Press, pp 21–38. https://doi.org/10.1515/9780824838775-004

Wooster W (1987) Immiscible investigators: oceanographers, meteorologists, and fishery scientists. Bio Science 37(10) URL: https://www.jstor.org/stable/1310470. Accessed 1 May 2021 19:47 UTC

Yamamoto T (2000) Collective fishery management developed in Japan – why community-based fishery management has been well developed in Japan. ILFIT 2000 proceedings

Part II
Policy Foundations

Chapter 5
Making Marine Spatial Planning Matter

Wesley Flannery

Abstract Over the last decade, Marine Spatial Planning (MSP) has become one of the key components of marine governance. In the European Union, member states are working towards the development of their first plans under the Maritime Spatial Planning Directive. Internationally, UNESCO and the European Commission have launched their MSP Global initiative to speed up the implementation of MSP around the world. MSP is also framed as being a key mechanism for sustainably realising the benefits of the Blue Economy and emerging Green Deals. During this same period, however, a substantial body of critical academic work has emerged that questions whether the implementation of MSP will transform unsustainable marine governance and management practices. This scholarship illustrates that the current trajectory of many MSP initiatives is to preserve the *status quo* and that they fail to adequately address longstanding marine governance issues. Drawing on Flyvbjerg's vital treatise on phronetic social science, this chapter will explore: where is MSP going; who gains and loses, and how they do so; is this desirable, and if not, what can be done to make MSP matter? I particularly focus on mechanisms of winning and losing, characterising them as key tensions in MSP processes that can be unsettled to make MSP more transformative.

5.1 Introduction

Demand for marine space has significantly increased over the last two decades. The increased pressure on marine space has been particularly driven by the expansion of spatially-fixed activities such as wind farms and aquaculture development (Schütz and Slater 2019). The average size and number of offshore wind farms have increased substantially, with, for example, a 22% annual growth rate in the number of offshore farms in the North Sea between 2008 and 2018 (Xu et al. 2020). Animal aquaculture production increased on average by 5.3% annually between 2001 and

W. Flannery (✉)
School of Natural and Built Environment, Queen's University Belfast, Belfast, UK
e-mail: w.flannery@qub.ac.uk

© German Institute of Development and Sustainability (IDOS) 2023
S. Partelow et al. (eds.), *Ocean Governance*, MARE Publication Series 25,
https://doi.org/10.1007/978-3-031-20740-2_5

2018 (FAO 2020). Demand for marine space will intensify in the coming years as new energy and aquaculture technologies are scaled up. This will include the adoption of floating wind farm technology, which will enable arrays to be located further offshore, and greater deployment of tidal and wave energy devices. Furthermore, technologies such as floating solar are progressing at speed and will create additional demand for marine space. Offshore aquaculture will also become more common.

The rapid growth in spatially fixed activities has obvious socio-spatial consequences. There is concern that the growth of these activities may displace others such as fishing (Lester et al. 2018; Young et al. 2019), placing considerable pressure on ocean biodiversity. Marine Spatial Planning (MSP) has been developed as a way of tackling possible conflict among stakeholders and reducing negative environmental impacts that may emerge from the intensification of marine space usage. MSP has rapidly achieved a dominant position within discourses about improving marine governance (Toonen and van Tatenhove 2013). These discourses tend to position MSP as fundamentally different to existing sectoral and fragmented management approaches (Douvere 2008). In contrast to the top-down, piecemeal, reactive, and issue-driven approaches that preceded it, MSP is envisaged as holistic, participatory, and proactive, with the potential capacity to address a multitude of issues simultaneously across sectors and marine spaces.

Although MSP has the potential to reform existing marine management regimes, assessments of MSP in practice illustrate that it is failing to radically transform marine governance (Fairbanks et al. 2019). There is evidence that MSP initiatives have neglected to: address issues such as the continuation of uncoordinated sectoral and fragmented management (Alexander and Haward 2019; Piwowarczyk et al. 2019a); adequately resolve sectoral conflicts, address the dominance of powerful sectors or fully understand trade-offs between sectoral objectives (Flannery et al. 2018; Sander 2018; Tafon 2018; Aschenbrenner and Winder 2019; Cohen et al. 2019; Flannery et al. 2019; Schutter and Hicks 2019; Tafon et al. 2021); fail to include non-economic and/or non-spatial uses, such as diverse stakeholder values (Strickland-Munro et al. 2016) and traditional and cultural uses of the sea (McKinley et al. 2019); or foster meaningful social and governance changes (Gissi et al. 2019; Kelly et al. 2019; Saunders et al. 2020). This indicates that the implementation of MSP may do little more than preserve the *status quo* and frustrate rather than facilitate the urgent reform of unsustainable marine management processes.

Given the rapid rollout of MSP initiatives across the world (Ehler 2020), including, potentially, to the high seas (Wright et al. 2019; Toonen and van Tatenhove 2020), and the fact that it will feature in SDG, Ocean Decade, and climate change strategies (Ntona and Morgera 2018; Noble et al. 2019; Frazão Santos et al. 2020; Calado et al. 2021; Gilek et al. 2021; Reimer et al. 2021), it is critically important to develop actions that can reclaim MSP's transformative potential (Clarke and Flannery 2020). There is, therefore, an urgent need to understand both how the transformative capacity of MSP has become blunted as it moves from concept to practice, and how this can be corrected. This is not to suggest that all MSP initiatives are failing or that there has been no reformation of unsuitable practices. Rather,

I argue there is a need to reflect on the emerging body of literature that raises issues of MSP in practice and to think strategically about how we insert transformative differences into ongoing and emerging MSP initiatives (Boucquey et al. 2019).

Drawing on the central questions for phronetic social science as developed by Flyvbjerg (2001), I review recent academic literature to identify key issues with the implementation of MSP. For Flyvbjerg, phronetic social science "relates to the practical wisdom that comes from familiarity with the contingencies and uncertainties of various forms of social practice embedded in complex social settings" (Schram 2004 p.442). Phronetic social science aims to help publics question the relationships of knowledge and power in specific settings and to produce practical solutions that can implement change. The adoption of Flyvbjerg's (Flyvbjerg 2001) approach is appropriate for the task of understanding how MSP may have failed to achieve the transformation of marine management and for developing ameliorating actions. Adapting Flyvbjerg's (Flyvbjerg 2001) approach, I review recent academic literature to ask: where is MSP going; who wins and loses, and through which mechanism; is this desirable, and if not, what can be done to make MSP better? I particularly focus on the mechanisms of winning and losing and argue that five issues create an illusion of progressive change within MSP. Like Scraff et al. (Scarff et al. 2015) I characterise these issues as being key tensions (Flyvbjerg et al. 2016) in MSP processes that may provide avenues to instigate more transformative forms of MSP. "In phronetic research, tension points are power relations that are particularly susceptible to problematization and thus to change, because they are fraught with dubious practices, contestable knowledge, and potential conflict" (Flyvbjerg et al. 2012, p. 288). The five tensions I identify include the tensions between participation and legitimisation; rationality and partiality; socio-political issues and technological solutions; future orientation and path dependency; and conflict management and silencing. I describe these issues as tensions as they illustrate a strain between the promise of MSP and what it has become in practice. Focusing on tensions can reveal how governing processes serve particular interests, and where and how differences can be inserted to address unjust processes and undesirable outcomes. While recognising that there will always be a gap between concept and practice, focusing on these key tensions can instigate actions that can move MSP back towards what it originally promised.

5.2 Where Is MSP Going?

To understand where MSP is going, we must consider its origins, the issues it was conceptualised as addressing, why its uptake has been relatively quick, and how it has been translated into practice. Until relatively recently, marine governance and management were very disaggregated. Marine governance predominately adopted a sectoral approach, with distinct marine activities being governed and managed separately. This approach made it difficult to evaluate the synergistic, antagonistic and/or cumulative impacts of decisions made in one sector on other sectors. This issue was sometimes compounded by spatially and temporally fragmented marine governance, with the governance of contiguous marine areas (e.g., territorial sea and Exclusive Economic Zone) being partitioned across different governance entities, levels, and timeframes (O'Hagan et al. 2020). Such a sectoral and fragmented approach was ill-suited to sustainably addressing key management issues that were being exacerbated due to the expansion of human activities in the marine environment. Addressing both the immense environmental challenges emanating from growing human use of the marine environment, while facilitating an increased demand for marine space and avoiding user conflicts, necessitated the development of integrated marine governance approaches.

Although integrated approaches to marine management have a long history (Eger et al. 2021), MSP has risen to become the dominant marine management paradigm. As a concept, MSP is framed as a rational, place-based response to the issues that have arisen from sectoral and fragmented management (Ehler and Douvere 2009). MSP has been defined as "a public process of analyzing and allocating the spatial and temporal distribution of human activities in marine areas to achieve ecological, economic, and social objectives that are usually specified through a political process" (Ehler and Douvere 2009, p. 18). It is viewed as a way of addressing long-standing marine issues and achieving a range of objectives, including reducing cumulative negative impacts from marine activities (Kirkfeldt and Andersen 2021); implementing ecosystem-based management (Douvere 2008; Lombard et al. 2019); achieving sustainable Blue Growth (Gustavsson and Morrissey 2019; Hassan et al. 2019; Gerhardinger et al. 2020; Guerreiro et al. 2021; Luhtala et al. 2021; Surís-Regueiro et al. 2021); managing stakeholder conflict and enhancing participation (Ritchie and Ellis 2010; Yates et al. 2015; Smythe and McCann 2019; Morzaria-Luna et al. 2020); and facilitating a transition to a local carbon society (Wright 2015; Hoegh-Guldberg et al. 2019; Dundas et al. 2020; Stelzenmüller et al. 2021b).

The broad appeal of MSP is partly due to it being so fundamentally different from the sectoral and fragmented regime. But this does not fully explain its rapid uptake globally. Other integrative and transformative alternatives had been developed, including, for example, integrated coastal zone management, but these have not been supported as enthusiastically in policy and stakeholder discourses. For some, MSP's dominant status is simply due to it being a logical idea whose time has come (Ehler 2018). Adopting this view, the global embracement of MSP is seen as being appropriate at this moment; the rapid adoption of MSP is simply the outworking of increasing demands for marine space and the recognition that this demand could not be sustainably managed through existing regimes. As I have argued elsewhere Flannery and McAteer (2020), I believe that this reasoning only partly explains the current popularity of MSP and that its conceptual simplicity and purported rationality also contribute to its broad appeal.

The enthusiastic uptake of MSP may also be a result of it being more accessible and acceptable than other solutions, such as ecosystem-based management. Spatial planning is a relatively intuitive and familiar concept that can be communicated easily through policy discourses. Drawing on this familiarity, dominant discourses often portray MSP as an uncomplicated, inherently rational, and unbiased process that will simplify governance. Though MSP is regularly referred to as an ecosystem-focused approach (Foley et al. 2010), in practice it tends to be less concerned with environmental issues than other ecosystem management concepts (Macpherson et al. 2020). MSP may, therefore, be perceived as being a comparatively value-neutral concept when compared to these other approaches (Flannery and McAteer 2020). MSP is also more accessible to non-specialists than ecosystem-focused approaches, which have been critiqued for being exclusionary and privileging specific forms of knowledge (Díaz et al. 2018; Stefansson et al. 2019). Furthermore, prevailing policy discourses have adopted asocial and apolitical framings to advance MSP as an inherently "rational" means of achieving balanced management in the future (Tafon 2018). As will be outlined below, spatial planning processes are not rational and should be understood as power-laden processes wherein actors compete to shape the future of specific spaces (Tafon 2018, 2019). I argue, therefore, that we should view the dominance of MSP as a result of it being both a concept whose time has come (Ehler 2018) and due it the oversimplification of the socio-political nature of spatial planning and the problems it will address (Slater and Claydon 2020). This view is supported by recent studies that illustrate the considerable gap between how MSP has been conceptualised and how it has been implemented (Jones et al. 2016; Santos et al. 2018; Zuercher et al. 2022a).

MSP is now underway in about 50% of the nation states that have maritime waters (see Ehler (2020) for a review of MSP initiatives worldwide). While this illustrates its rapid and wide adoption, a significant and expanding body of research raises questions about its effectiveness in practice (Ritchie and Ellis 2010; Jones et al. 2016; Smith and Jentoft 2017; Smith 2018; Tafon et al. 2018; Boucquey et al. 2019; Fairbanks et al. 2019; Tafon 2019; Campbell et al., 2020). Although assessments of the effectiveness of MSP processes are dependent on local contextual factors and the selection of specific evaluative frameworks (Stojanovic and Gee 2020), a set of similar issues have been reported across different initiatives. For example, several MSP processes have been implemented in ways that are less than holistic, excluding key sectors, such as small-scale fisheries (Janßen et al. 2018; Piwowarczyk et al. 2019b; Said and Trouillet 2020) or issues, such as climate change (Rilov et al. 2020) or failing to incorporate conservation measures (Katsanevakis et al. 2020; Trouillet 2020; Kirkfeldt and Andersen 2021). Rather than being a forward-orientated process, MSP initiatives have been critiqued for merely giving spatial effect to past decisions, such as energy licenses (Jones et al. 2016; Clarke and Flannery 2020) or for being top-down processes focused on key economic sectors (Guerreiro et al. 2021). MSP initiatives have also been critiqued for reflecting existing power relations (Aschenbrenner and Winder 2019; Flannery and McAteer 2020; Páez et al. 2020; Ramírez-Monsalve and van Tatenhove 2020), and for being ambiguous both in terms of future objectives (Sander 2018; Clarke and Flannery 2020; Zuercher et al. 2022b), and monitoring processes (Stelzenmüller et al. 2015; O'Leary et al. 2019; Flannery and McAteer 2020; Stelzenmüller et al. 2021a). Although the uptake of MSP has been impressive, how it has been implemented raises questions about its effectiveness to move marine governance into a different paradigm.

To return to the questions posed at the start of this section, I argue that MSP emerged as a genuine, yet socially naïve and oversimplified answer to the inadequacies of the existing management system. As it moves towards implementation, MSP has been further simplified, erasing, or ignoring the complex socio-political context of marine spaces (Flannery et al. 2016) and the ontological assumptions that underpin prevailing approaches to ocean management (Peters 2020). There is broad acceptance that the sectoral and fragmented management regime was ill-suited to managing the increasing demand from marine space and associated pressures and conflicts. However, the popular framing of MSP as neutral, rational, and capable of producing win-win solutions, means that the form of MSP that has emerged, and that will likely be implemented more broadly in the future, is reductive, asocial, and apolitical. Continuing in this vein will mean that MSP will lose credibility as a transformative governance approach (Flannery and McAteer 2020). This retrograde direction of travel is not an inherent failure of the concept of MSP, but rather, reflects inattention to issues of power within the original literature, and an approach to implementation that fails to address the socio-political complexity of marine spaces. The broad adoption of MSP does, however, offer opportunities for doing marine governance differently (Boucquey et al. 2019; Karnad and St. Martin 2020). For example, spatialising marine governance can empower marginalised stakeholders and communities. It is crucial, therefore, to identify key tension points in existing and emerging MSP processes, and to develop actions that can unsettle their suppression of more radical and progressive forms of MSP.

5.3   Who Wins and Loses, and Through Which Mechanisms?

It is difficult to evaluate who, exactly, is winning and losing in MSP processes as they are so new and the impacts of plans are yet to be fully evaluated. However, as outlined above, academic evaluations do seem to indicate that MSP has not transformed marine governance or delivered significant social or governance changes. The winners can, therefore, be thought of as those who are resistant to radical change and who believe their interests are best served through MSP implementation that falls short of its transformative potential. On the other hand, the losers can be considered those who would benefit from a fundamental transformation of the governance regime. From a review of the literature, MSP appears to repackage the *status quo* by failing to address five interrelated tensions: 1. participation – legitimisation; 2. rationality – partiality; 3. socio-political issues – technological solutions; 4. future-orientated – path-dependent; and 5. conflict management – silencing.

5.3.1 Participation – Legitimisation

The adoption of MSP is advocated as a way to enhance participation in marine governance and to produce win-win outcomes for stakeholders (Pomeroy and Douvere 2008; Carneiro 2013). Participation is framed as being central to effective MSP as it will give local communities a voice in planning processes, objective setting, and planning decisions. Participation in MSP will also: reduce user conflict; enhance participants' knowledge of the environment and their impacts; allow for different forms of knowledge to be included in planning processes; enhance trust in planning processes; and increase the legitimacy and acceptance of planning decisions (Pomeroy and Douvere 2008; Douvere and Ehler 2009; Ehler and Douvere 2009). In theory, by spatialising marine governance, MSP should broaden the constituency of stakeholders who participate in marine governance, moving participation beyond narrow sectoral silos and towards more shared mechanisms of planning and decision-making, which includes processes of space- or place- making.

While advocates are correct to highlight the potential positive impacts of participation, how governments have implemented MSP appears, in many cases to fall short of core participatory planning principles. MSP initiatives have been evaluated as being top-down, centralised processes (Scarff et al. 2015; Jones et al. 2016), that reassert rather than address longstanding community power dynamics (Flannery et al. 2018). Broad-scale and tokenistic participatory processes are common within existing MSP initiatives. Local and less powerful actors are reported as being engaged in tokenistic ways (Jones et al. 2016; Smith and Jentoft 2017). Within these MSP processes, power can be mobilised to marginalise particular groups of marine actors and "herd their participation and ways of knowing toward achieving limited policy outcomes" (Tafon 2018, p. 258). Furthermore, several participatory approaches that governments have used in MSP initiatives, such as townhall-style meetings, tend to take place during the latter stages of planning processes and seldom have a real impact on plan objectives (Flannery et al. 2018; Quesada-Silva et al. 2019). These processes are highly tokenistic, focusing on providing the appearance of inclusion and allowing governments to fulfil participatory obligations without meaningfully engaging with the public. This may mean "that MSP is not facilitating a paradigm shift towards publicly engaged marine management, and that it may simply repackage power dynamics in the rhetoric of participation to legitimise the agendas of dominant actors" (Flannery et al. 2018, p. 32).

5.3.2 Rationality – Partiality

Dominant policy discourses have framed MSP as being inherently rational. The adoption of space as a governance mechanism is a way of making rational decisions about how and where development should occur (Douvere 2008). This reasoning reinforces the perception that there is an unproblematic spatial configuration that can be formulated to organise the many actors who compete for locations. This is, however, a highly asocial and apolitical conceptualisation of spatial planning. Comprehensive and rational planning is framed in a way that is distant from power and as having the capacity to produce broadly accepted outcomes. As Smith and Jentoft (2017, p. 34) assert, "as the theoretical foundation of Marine Spatial Planning was being laid, the issue of power was arguably not sufficiently problematized". MSP is neither neutral nor inherently rational, and like many other procedures it can, without due attention being given to power dynamics, produce unjust management outcomes that benefit some to the detriment of others (Jentoft 2017). The naïve framing of MSP as rational is founded on an uncritical understanding of the power dynamics with spatial planning. This does not mean that MSP processes cannot be made more equitable, just that greater attention needs to be paid in practice to different forms and mechanisms of power (Tafon et al. 2019; Ramírez-Monsalve and van Tatenhove 2020) and how they shape MSP processes and outcomes.

5.3.3   Socio-Political Issues – Technological Solutions

MSP has been advanced as a way of resolving a wide range of socio-political issues in the marine environment. For example, MSP is seen as a way of addressing the democratic deficit in marine governance and as a way of addressing issues such as coastal poverty. Although MSP may be able to address these topics, in practice they have tended to be pushed aside in favour of less complex issues. This may be because the spatial turn in marine governance has been accompanied by a rise in the use of geotechnologies. These geotechnologies seek to make marine space more understandable and governable but have been misapplied in ways that overgeneralise complex issues (Trouillet 2019).

The development of a Geographic Information System (GIS) database is a key part of MSP (Gimpel et al. 2018). These databases can help planners and stakeholders conceptualise marine areas and the issues within them (Shucksmith and Kelly 2014). A suite of decision-making tools has also been developed (Pınarbaşı et al. 2019). These tools can, for example, help diagnose the spatial interaction between activities, focus on cumulative effect assessments, or be part of decision support systems (Stelzenmüller et al. 2013). These databases and tools can contribute to evidence-based decision-making in MSP. Although the development of these databases and tools can benefit MSP and contribute to the development of more progressive and sustainable futures, in practice, many of them have come to be an end in themselves or are employed in ways that obscure, rather than resolve, complex socio-political marine issues (Smith and Brennan 2012; Trouillet 2019). For example, the complexity of social-ecological relations in the marine environment is increasingly simplified through the use of mapping technologies (Smith and Brennan 2012) and captured in geospatial databases (Boucquey et al. 2019), creating problematic conceptualisations of relationships as being fixed and two-dimensional (Steinberg and Peters 2015). These GIS databases are analysed by technical experts to make 'rational' decisions about marine issues that have been disembodied from their social contexts. In this manner, MSP has been reduced to a mere technocratic exercise of allocating space efficiently, dulling its potential for envisaging alternative marine futures.

5.3.4 Future-Orientated – Path-Dependent

In contrast with the reactive management regime that preceded it, MSP is considered to be a future-oriented process that allows the public and stakeholders to shape actions that could lead to a more desirable future (Ehler 2018). To achieve this, MSP processes should focus on envisioning sustainable future socio-political and environmental scenarios and develop plans to realise them. This means that management regimes must move beyond a narrow focus on the present. What the future is to be for a particular marine area is likely to be highly contested and must also acknowledge the historical tension between traditional marine uses and new and emerging activities and how they may be resolved or exacerbated in the future. MSP must consider issues beyond sectoral trends and potential trade-offs. This should include issues such as climate change and how it may impact specific social-ecological systems and the diverse adaptive capacities of different communities (Santos et al. 2020, 2022).

Evaluations of MSP in practice illustrate, however, that many are adopting path-dependent rather than future-orientated approaches to plan development (Jones et al. 2016; Kelly et al. 2019; Clarke and Flannery 2020). For example, the fragmented licensing and management regimes, the complexity of which gave rise to MSP, will remain in place even as nation states begin to implement MSP. By entrenching historic practices while claiming to be future-orientated, many MSP processes create an artifice of progressive change while doing very little to address urgent marine issues (Jones et al. 2016). These issues have arisen as many MSP initiatives have been grafted onto existing governance structures and policy frameworks without due consideration being given to their capacity to deliver transformative change. This approach fails to address institutional and policy issues that undermine efforts at transformative change (Kelly et al. 2018), meaning that MSP is often implemented in a path-dependent manner, resulting in it becoming merely the spatialisation of the existing regime or in very incremental changes being implemented. Therefore, broader consideration needs to be given to how marine futures are imagined (Merrie et al. 2018; Spijkers et al. 2021) and realised in MSP processes (Gissi et al. 2019; Kelly et al. 2019).

5.3.5 Conflict Management – Silencing

One of the key things that MSP is celebrated for is its capacity to address conflict among competing activities. Growth in marine activities brings with it an increased possibility of conflict amongst and within sectors. The holistic, integrated, and participatory nature of MSP is seen as a way to avoid or minimize conflicts and maximize synergies across interests (Douvere and Ehler 2009). MSP initiatives can do this by examining potential future scenarios to identify who benefits and who loses from planning potential decisions (von Thenen et al. 2021) and develop actions to resolve potential conflicts (de Koning et al. 2021; Steins et al. 2021).

The approach to understanding conflict in both MSP literature and practice is very limited. A key issue is that both narrowly conceive of 'conflict' in spatial terms. As Arbo and Thuy (2016) have argued, this is seldom an of contending parties being in direct conflict with one another, and more an issue of competing spatial claims being submitted to governance agencies. Furthermore, focusing on spatial competition avoids acknowledging more challenging forms of conflicts such as those concerned with the distribution of costs, benefits, rights, and obligations. Limiting MSP to spatial conflict management limits what it could achieve and prevents important discussions about other issues that should feature in plans (e.g. poverty alleviation, equity, justice, climate change adaptation, etc.). This may mean that MSP initiatives perpetuate more insidious conflicts that have shaped marine governance and created specific winners and losers in terms of the benefits and costs of management decisions. By failing to engage with conflict beyond spatial competition, MSP narrowly focuses on the final stages of policy implementation (e.g. allocating space to specific activities) and silences or excludes broader debates about how the benefits on the marine environment should be realised and by whom.

5.4 Is This Desirable, and What Can Be Done to Make MSP Matter?

The concept of MSP holds considerable transformative potential. This includes the possibility of addressing longstanding issues that have arisen from sectoral and fragmented approaches and the prospect of reducing the democratic deficit in marine governance. Academic evaluations indicate, however, that the translation of the MSP concept into practice fails to realise this potential. Failure to adopt more radical or progressive forms of planning means that MSP in practice leans towards preserving the *status quo* and, more than likely, producing the same winners and losers as the previous fragmented and sectoral regime (Bennett et al. 2019). This is not desirable and corrective actions should be developed and implemented by those interested in advancing progressive and radical forms of MSP. The key tensions outlined above provide opportunities to reclaim the potential of MSP. These tensions are interrelated, and productive action in one may have a positive impact on the others. Ideally, however, it would be preferable to develop actions that cut across all five tensions.

These tensions can be targeted through three interconnected actions: fostering greater stakeholder empowerment; politicizing MSP; and developing alternative and uncomfortable knowledge. To date, most MSP initiatives have tended to adopt tokenistic and power-blind forms of participation. Meaningful engagement cannot be achieved without acknowledging and addressing power asymmetries, especially those that prevent less powerful stakeholders from exercising an influence on decision-making (Greenwood and Van Buren III 2010). MSP initiatives need to be moved away from participation methods that ignore or reproduce these asymmetries and towards forms of engagement that recognise the uneven capacity across stakeholders to meaningfully engage with planning processes. To do this, MSP initiatives must start by recognising the different forms (Tafon et al. 2019) and mechanisms (Ramírez-Monsalve and van Tatenhove 2020) of power that can influence planning processes and outcomes, and by assessing stakeholder capacity to meaningfully engage with the planning initiative. Resources must then be provided to build stakeholder capacity before planning processes begin.

Capacity building will need to be targeted to the needs of specific stakeholders, but, given that MSP is here to stay, more general capacity-building initiatives should also be initiated. It may be useful to mirror initiatives from urban planning such as Planning Aid (RTPI 2020) and advocate planners (Flannery et al. 2016; Saunders et al. 2020; Tafon et al. 2018) that can provide support to stakeholders. Such intermediaries could focus on providing stakeholders with the necessary planning skills to make meaningful contributions to MSP processes. The capacity of planning teams to engage with stakeholders and to foster truly integrative planning processes should also be evaluated and addressed (Ansong et al. 2019; Vince and Day 2020).

There is a difference, however, between capacity building to engage with existing, skewed procesess and empowering stakeholders to change them. It is necessary, therefore, to develop mechanisms that facilitate stakeholder reflection about current processes and empower them to challenge existing discourses (van Tatenhove 2017). This can be done by politicizing MSP, which would entail debate about the very purpose of MSP and how it can be implemented in ways that serve a broader public good. Enabling deliberation within the limited remit of existing governance structures would probably fail to engender progressive changes. Mechanisms must be provided to enable stakeholders and governance institutions to engage in broader discussions about the structural and procedural changes needed to achieve more progressive MSP objectives. These discussions must include reflections on the purpose of MSP, how it can facilitate a break with past practices, and how to overcome structural barriers to transformative change. Reforming MSP is unlikely to feature highly on the political agenda and, therefore, different mechanisms of politicisation must be developed. This could be accomplished through, for example, long-term visioning exercises aimed at imagining radically different marine futures, supported by reflective processes for exploring and implementing the governance changes need to realise these visions. This could be facilitated by adopting a transition management approach to designing and implementing governance regime changes

(Kelly et al. 2018; Rudolph et al. 2020) and could incorporate more explicit processes for reflection and learning on an ongoing basis (Keijser et al. 2020). Any effort to change existing governance regimes must seek to deliberately include marginalised and excluded stakeholders (Tafon et al. 2021) and seek to empower them to engage meaningfully with these processes.

Empowering stakeholders to engage with and/or to politicise MSP regimes may mean that they will have to acquire the capacity to develop and mobilise alternative knowledge. By alternative knowledge, I am referring to knowledge that has not typically been captured by existing MSP processes and could include, for example, traditional and cultural knowledge, knowledge that illustrates the socio-ecological complexity of specific marine spaces, or uncomfortable knowledge (Rayner 2012) such as insights into corrupt planning practices, that have been excluded from planning processes. By producing and making use of alternative knowledge, stakeholders can begin to counter the prevailing discourses within marine governance. This may include, for example, countering how the marine problem is constructed (Ritchie and McElduff 2020), demonstrating to whose benefit and in whose interest existing problematisations serve (Ntona and Schröder 2020), or broadening the conceptualisation of social sustainability within MSP (Gilek et al. 2021).

The mechanisms of empowerment, politicisation, and knowledge production are clearly intertwined and can work together to rebalance the key tensions in MSP so that more progressive and novel forms are put into practice. Stakeholder empowerment will enable them to politicise MSP and counter processes that use participation to merely legitimise plans. Empowering them to develop and mobilise new or alternative knowledge will enable them to counter the assumed rationality of MSP and to better frame socio-political issues in ways that cannot be subsumed by the misapplication of geotechnologies. This new knowledge can also be developed in such ways that it can work with established geotechnologies to better illustrate the complexity of marine areas (St. Martin and Olson 2017). New knowledge about the 'marine issue' can be mobilised to develop progressive visions for the future of marine spaces and to foster real debate about how these might be realised in fair and just ways. However, none of these mechanisms will succeed if we fail to recognise that MSP is a concept whose time has come but that we need to develop alternative pathways to implementation for it to really matter.

5.5   Conclusion

The global uptake of MSP demands that attention is paid to understanding how it is being implemented and how it can be made better or to matter more. Evidence reported from recent evaluations indicates that MSP is not realising its transformative potential and that action needs to be taken to steer MSP towards something better. Focusing on key tensions may provide opportunities to insert different logic, knowledge, and power relations into ongoing and emerging MSP processes.

Action and research that focuses on empowering stakeholders, politicizing MSP processes, and developing alternative and uncomfortable knowledge, may provide opportunities to rebalance these tensions towards more novel and progressive forms of MSP.

References

Alexander K, Haward M (2019) The human side of marine ecosystem-based management (EBM): "sectoral interplay" as a challenge to implementing EBM. Mar Policy 101:33–38. https://doi.org/10.1016/j.marpol.2018.12.019

Ansong J, Calado H, Gilliland PM (2019) A multifaceted approach to building capacity for marine/maritime spatial planning based on European experience. Mar Policy 132:103422. https://doi.org/10.1016/j.marpol.2019.01.011

Arbo P, Thuy PTT (2016) Use conflicts in marine ecosystem-based management – the case of oil versus fisheries. Ocean Coast Manag 122:77–86. https://doi.org/10.1016/j.ocecoaman.2016.01.008

Aschenbrenner M, Winder GM (2019) Planning for a sustainable marine future? Marine spatial planning in the German exclusive economic zone of the North Sea. Applied Geogr 110:102050. https://doi.org/10.1016/j.apgeog.2019.102050

Bennett NJ, Cisneros-Montemayor AM, Blythe J, Silver JJ, Singh G, Andrews N, Calò A, Christie P, Di Franco A, Finkbeiner EM, Gelcich S (2019) Towards a sustainable and equitable blue economy. Nat Sustain 2(11):991–993. https://doi.org/10.1038/s41893-019-0404-1

Boucquey N, Martin KS, Fairbanks L, Campbell LM, Wise S (2019) Ocean data portals: performing a new infrastructure for ocean governance. Environ Plan D Soc Space 37(3):484–503. https://doi.org/10.1177/0263775818822829

Calado H, Pegorelli C, Frazão Santos C (2021) Maritime spatial planning and sustainable development. In: Leal Filho W et al (eds) Life below water. Springer, Cham, pp 1–11. https://doi.org/10.1007/978-3-319-71064-8_122-1

Campbell LM, St Martin K, Fairbanks L, Boucquey N, Wise S (2020) The portal is the plan: governing US oceans in regional assemblages. Maritime Studies 19(3):285–297

Carneiro G (2013) Evaluation of marine spatial planning. Mar Policy 37:214–229. https://doi.org/10.1016/j.marpol.2012.05.003

Clarke J, Flannery W (2020) The post-political nature of marine spatial planning and modalities for its re-politicisation. J Environ Policy Plan 22(2):170–183. https://doi.org/10.1080/1523908X.2019.1680276

Cohen PJ, Allison EH, Andrew NL, Cinner J, Evans LS, Fabinyi M, Garces LR, Hall SJ, Hicks CC, Hughes TP, Jentoft S (2019) Securing a just space for small-scale fisheries in the blue economy. Front Mar Sci 6. https://doi.org/10.3389/fmars.2019.00171

de Koning S, Steins N, van Hoof L (2021) Balancing sustainability transitions through state-led participatory processes: the case of the dutch north sea agreement. Sustainability 13(4):1–16. https://doi.org/10.3390/su13042297

Díaz S, Pascual U, Stenseke M, Martín-López B, Watson RT, Molnár Z, Hill R, Chan KM, Baste IA, Brauman KA, Polasky S (2018) Assessing nature's contributions to people. Science 359(6373):270–272

Douvere F (2008) The importance of marine spatial planning in advancing ecosystem-based sea use management. Mar Policy 32(5):762–771. https://doi.org/10.1016/j.marpol.2008.03.021

Douvere F, Ehler CN (2009) New perspectives on sea use management: initial findings from European experience with marine spatial planning. J Environ Manag 90(1):77–88. https://doi.org/10.1016/j.jenvman.2008.07.004

Dundas SJ, Levine AS, Lewison RL, Doerr AN, White C, Galloway AW, Garza C, Hazen EL, Padilla-Gamiño J, Samhouri JF, Spalding A (2020) Integrating oceans into climate policy: any green new deal needs a splash of blue. Conserv Lett 13:1–12. https://doi.org/10.1111/conl.12716

Eger SL, Stephenson RL, Armitage D, Flannery W, Courtenay SC (2021) Revisiting integrated coastal and marine management in Canada: opportunities in the Bay of Fundy. Front Mar Sci 8. https://doi.org/10.3389/fmars.2021.652778

Ehler CN (2018) Marine spatial planning: an idea whose time has come. In: Yates KL, Bradshaw CJA (eds) Offshore energy and marine spatial planning. Routledge, Oxon/New York, pp 6–17

Ehler CN (2020) Two decades of progress in marine spatial planning. Mar Policy 132:104134. https://doi.org/10.1016/j.marpol.2020.104134

Ehler C, Douvere F (2009) Marine spatial planning: a step-by-step approach. UNESCO, Paris

Fairbanks L, Boucquey N, Campbell LM, Wise S (2019) Remaking oceans governance: critical perspectives on marine spatial planning. Environ Soc Adv Res 10(1):122–140. https://doi.org/10.3167/ares.2019.100108

FAO (2020) The state of world fisheries and aquaculture 2020. Sustainability in action., FAO. https://doi.org/10.4060/ca9229en

Flannery W, McAteer B (2020) Assessing marine spatial planning governmentality. Marit Stud 19(3):269–284. https://doi.org/10.1007/s40152-020-00174-2

Flannery W, Ellis G, Ellis G, Flannery W, Nursey-Bray M, van Tatenhove JP, Kelly C, Coffen-Smout S, Fairgrieve R, Knol M, Jentoft S (2016) Exploring the winners and losers of marine environmental governance/marine spatial planning: cui bono?/?More than fishy business?: epistemology, integration and conflict in marine spatial planning/marine spatial planning: power and scaping/surely not all. Plan Theory Pract 17(1):121–151. https://doi.org/10.1080/14649357.2015.1131482

Flannery W, Healy N, Luna M (2018) Exclusion and non-participation in marine spatial planning. Mar Policy 88:32–40. https://doi.org/10.1016/j.marpol.2017.11.001

Flannery W, Clarke J, McAteer B (2019) Politics and power in marine spatial planning. In: Zaucha J, Gee K (eds) Maritime spatial planning. Palgrave Macmillan, London, pp 201–217

Flyvbjerg B (2001) Making social science matter. Cambridge University Press. https://doi.org/10.1017/CBO9780511810503

Flyvbjerg B, Landman T, Schram S (2012) Important next steps in phronetic social science. In: Flyvbjerg B, Landman T, Schram S (eds) Real social science: applied Phronesis. Cambridge University Press, Cambridge. https://doi.org/10.1017/CBO9780511719912

Flyvbjerg B, Landman T, Schram S (2016) Tension points: learning to make social science matter. Available at:. http://ssm.com/abstract=2721321. Accessed 8 May 2021

Foley MM, Halpern BS, Micheli F, Armsby MH, Caldwell MR, Crain CM, Prahler E, Rohr N, Sivas D, Beck MW, Carr MH (2010) Guiding ecological principles for marine spatial planning. Mar Policy 34(5):955–966. https://doi.org/10.1016/j.marpol.2010.02.001

Gerhardinger LC, de Andrade MM, Corrêa MR, Turra A (2020) Crafting a sustainability transition experiment for the Brazilian blue economy. Mar Policy 120:104157. https://doi.org/10.1016/j.marpol.2020.104157

Gilek M, Armoskaite A, Gee K, Saunders F, Tafon R, Zaucha J (2021) In search of social sustainability in marine spatial planning: a review of scientific literature published 2005–2020. Ocean Coastal Manage 208:105618. https://doi.org/10.1016/j.ocecoaman.2021.105618

Gimpel A, Stelzenmüller V, Töpsch S, Galparsoro I, Gubbins M, Miller D, Murillas A, Murray AG, Pınarbaşı K, Roca G, Watret R (2018) A GIS-based tool for an integrated assessment of spatial planning trade-offs with aquaculture. Sci Total Environ 627:1644–1655. https://doi.org/10.1016/j.scitotenv.2018.01.133

Gissi E, Fraschetti S, Micheli F (2019) Incorporating change in marine spatial planning: a review. Environ Sci Policy 92:191–200. https://doi.org/10.1016/j.envsci.2018.12.002

Greenwood M, Van Buren HJ III (2010) Trust and stakeholder theory: trustworthiness in the organisation–stakeholder relationship. J Bus Ethics 95(3):425–438

Guerreiro J, Carvalho A, Casimiro D, Bonnin M, Calado H, Toonen H, Fotso P, Ly I, Silva O, da Silva ST (2021) Governance prospects for maritime spatial planning in the tropical Atlantic compared to EU case studies. Mar Policy 123:104294. https://doi.org/10.1016/j.marpol.2020.104294

Gustavsson M, Morrissey K (2019) A typology of different perspectives on the spatial economic impacts of marine spatial planning. J Environ Policy Plan 21:1–13. https://doi.org/10.1080/1523908X.2019.1680274

Hassan D, Ashraf MAA, Alam MA (2019) Institutional arrangements for the blue economy: marine spatial planning a way forward. *journal of ocean and coastal economics* 6(2). https://doi.org/10.15351/2373-8456.1107

Hoegh-Guldberg O, Northrop E, Lubchenco J (2019) The ocean is key to achieving climate and societal goals. Science 365(6460):1372–1374. https://doi.org/10.1126/science.aaz4390

Janßen H, Bastardie F, Eero M, Hamon KG, Hinrichsen HH, Marchal P, Nielsen JR, Le Pape O, Schulze T, Simons S, Teal LR (2018) Integration of fisheries into marine spatial planning: quo vadis? Estuar Coast Shelf Sci 201:105–113. https://doi.org/10.1016/j.ecss.2017.01.003

Jentoft S (2017) Small-scale fisheries within maritime spatial planning: knowledge integration and power. J Environ Policy Plan 19(3):266–278. https://doi.org/10.1080/1523908X.2017.1304210

Jones PJ, Lieberknecht LM, Qiu W (2016) Marine spatial planning in reality: introduction to case studies and discussion of findings. Mar Policy 71:256–264. https://doi.org/10.1016/j.marpol.2016.04.026

Karnad D, St. Martin K (2020) Assembling marine spatial planning in the global south: international agencies and the fate of fishing communities in India. Marit Stud 19(3):375–387. https://doi.org/10.1007/s40152-020-00164-4

Katsanevakis S, Coll M, Fraschetti S, Giakoumi S, Goldsborough D, Mačić V, Mackelworth P, Rilov G, Stelzenmüller V, Albano PG, Bates AE (2020) Twelve recommendations for advancing marine conservation in European and contiguous seas. Front Mar Sci 7:1–18. https://doi.org/10.3389/fmars.2020.565968

Keijser X, Toonen H, van Tatenhove J (2020) A "learning paradox" in maritime spatial planning. Marit Stud 19(3):333–346. https://doi.org/10.1007/s40152-020-00169-z

Kelly C, Ellis G, Flannery W (2018) Conceptualising change in marine governance: learning from transition management. Mar Policy 95:24–35. https://doi.org/10.1016/j.marpol.2018.06.023

Kelly C, Ellis G, Flannery W (2019) Unravelling persistent problems to transformative marine governance. Front Mar Sci 6:1–15. https://doi.org/10.3389/fmars.2019.00213

Kirkfeldt TS, Andersen JH (2021) Assessment of collective pressure in marine spatial planning: the current approach of EU member states. Ocean Coast Manag 203:105448. https://doi.org/10.1016/j.ocecoaman.2020.105448

Lester SE, Stevens JM, Gentry RR, Kappel CV, Bell TW, Costello CJ, Gaines SD, Kiefer DA, Maue CC, Rensel JE, Simons RD (2018) Marine spatial planning makes room for offshore aquaculture in crowded coastal waters. Nat Commun 9(1):945. https://doi.org/10.1038/s41467-018-03249-1

Lombard AT, Dorrington RA, Reed JR, Ortega-Cisneros K, Penry GS, Pichegru L, Smit KP, Vermeulen EA, Witteveen M, Sink KJ, McInnes AM (2019) Key challenges in advancing an ecosystem-based approach to marine spatial planning under economic growth imperatives. Front Mar Sci 6. https://doi.org/10.3389/fmars.2019.00146

Luhtala H, Erkkilä-Välimäki A, Eliasen SQ, Tolvanen H (2021) Business sector involvement in maritime spatial planning – experiences from the Baltic Sea region. Mar Policy 123(June 2020):104301. https://doi.org/10.1016/j.marpol.2020.104301

Macpherson E, Urlich SC, Rennie HG, Paul A, Fisher K, Braid L, Banwell J, Ventura JT, Jorgensen E (2020) "Hooks" and "anchors" for relational ecosystem-based marine management. Mar Policy. https://doi.org/10.1016/j.marpol.2021.104561

McKinley E, Acott T, Stojanovic T (2019) Socio-cultural dimensions of marine spatial planning. In: Zaucha J, Gee K (eds) Maritime spatial planning. Palgrave Macmillan, London, pp 151–174

Merrie A, Keys P, Metian M, Österblom H (2018) Radical ocean futures-scenario development using science fiction prototyping. Futures 95:22–32. https://doi.org/10.1016/j.futures.2017.09.005

Morzaria-Luna H, Turk-Boyer P, Polanco-Mizquez EI, Downton-Hoffmann C, Cruz-Piñón G, Carrillo-Lammens T, Loaiza-Villanueva R, Valdivia-Jiménez P, Sánchez-Cruz A, Peña-Mendoza V, López-Ortiz AM (2020) Coastal and Marine Spatial Planning in the Northern Gulf of California, Mexico: consolidating stewardship, property rights, and enforcement for ecosystem-based fisheries management. Ocean Coast Manag 197(July):105316. https://doi.org/10.1016/j.ocecoaman.2020.105316

Noble MM, Harasti D, Pittock J, Doran B (2019) Linking the social to the ecological using GIS methods in marine spatial planning and management to support resilience: a review. Mar Policy 108:103657

Ntona M, Morgera E (2018) Connecting SDG 14 with the other sustainable development goals through marine spatial planning. Mar Policy 93:214–222. https://doi.org/10.1016/j.marpol.2017.06.020

Ntona M, Schröder M (2020) Regulating oceanic imaginaries: the legal construction of space, identities, relations and epistemological hierarchies within marine spatial planning. Maritime Studies 19(3):241–254

O'Hagan AM, Paterson S, Le Tissier M (2020) Addressing the tangled web of governance mechanisms for land-sea interactions: assessing implementation challenges across scales. Mar Policy 112:103715. https://doi.org/10.1016/j.marpol.2019.103715

O'Leary BC, Stewart BD, McKinley E, Addison PF, Williams C, Carpenter G, Righton D, Yates KL (2019) What is the nature and extent of evidence on methodologies for monitoring and evaluating marine spatial management measures in UK and similar coastal waters? A systematic map protocol. Environ Evid 8(1):1–9. https://doi.org/10.1186/s13750-019-0178-y

Páez DP, Bojórquez-Tapia LA, Ramos GCD, Chavero EL (2020) Understanding translation: co-production of knowledge in marine spatial planning. Ocean Coast Manag 190:105163. https://doi.org/10.1016/j.ocecoaman.2020.105163

Peters K (2020) The territories of governance: unpacking the ontologies and geophilosophies of fixed to flexible ocean management, and beyond: the territories of governance. Philos Trans R Soc B Biol Sci 375(1814):20190458. https://doi.org/10.1098/rstb.2019.0458rstb20190458

Pınarbaşı K, Galparsoro I, Borja Á (2019) End users' perspective on decision support tools in marine spatial planning. Mar Policy 108:103658. https://doi.org/10.1016/j.marpol.2019.103658

Piwowarczyk J, Gee K, Gilek M, Hassler B, Luttmann A, Maack L, Matczak M, Morff A, Saunders F, Stalmokaite I, Zaucha J (2019a) Insights into integration challenges in the Baltic Sea region marine spatial planning: implications for the HELCOM-VASAB principles. Ocean Coast Manag 175:98–109. https://doi.org/10.1016/j.ocecoaman.2019.03.023

Piwowarczyk J, Matczak M, Rakowski M, Zaucha J (2019b) Challenges for integration of the Polish fishing sector into marine spatial planning (MSP): do fishers and planners tell the same story? Ocean Coast Manag 181:104917. https://doi.org/10.1016/j.ocecoaman.2019.104917

Pomeroy R, Douvere F (2008) The engagement of stakeholders in the marine spatial planning process. Mar Policy 32(5):816–822. https://doi.org/10.1016/j.marpol.2008.03.017

Quesada-Silva M, Iglesias-Campos A, Turra A, Suárez-de Vivero JL (2019) Stakeholder participation assessment framework (SPAF): a theory-based strategy to plan and evaluate marine spatial planning participatory processes. Mar Policy 108. https://doi.org/10.1016/j.marpol.2019.103619

Ramírez-Monsalve P, van Tatenhove J (2020) Mechanisms of power in maritime spatial planning processes in Denmark. Ocean Coast Manag 198:105367. https://doi.org/10.1016/j.ocecoaman.2020.105367

Rayner S (2012) Uncomfortable knowledge: the social construction of ignorance in science and environmental policy discourses. Econ Soc 41(1):107–125. https://doi.org/10.1080/03085147.2011.637335

Reimer JM, Devillers R, Claudet J (2021) Benefits and gaps in area-based management tools for the ocean Sustainable Development Goal. Nat Sustain 4(4):349–357. https://doi.org/10.1038/s41893-020-00659-2

Rilov G, Fraschetti S, Gissi E, Pipitone C, Badalamenti F, Tamburello L, Menini E, Goriup P, Mazaris AD, Garrabou J, Benedetti-Cecchi L (2020) A fast-moving target: achieving marine conservation goals under shifting climate and policies. Ecol Appl 30(1):1–14. https://doi.org/10.1002/eap.2009

Ritchie H, Ellis G (2010) "A system that works for the sea"? Exploring stakeholder engagement in marine spatial planning. J Environ Plan Manag 53(6):701–723. https://doi.org/10.1080/09640568.2010.488100

Ritchie H, McElduff L (2020) The whence and whither of marine spatial planning: revisiting the social reconstruction of the marine environment in the UK. Maritime Studies 19(3):229–240

RTPI (2020) About Planning Aid England. Available at: https://www.rtpi.org.uk/planning-advice/about-planning-aid-england/. Accessed 17 May 2021

Rudolph TB, Ruckelshaus M, Swilling M, Allison EH, Österblom H, Gelcich S, Mbatha P (2020) A transition to sustainable ocean governance. Nat Commun 11(1):1–14

Said A, Trouillet B (2020) Bringing "deep knowledge" of fisheries into marine spatial planning. Marit Stud 19(3):347–357. https://doi.org/10.1007/s40152-020-00178-y

Sander G (2018) Ecosystem-based management in Canada and Norway: the importance of political leadership and effective decision-making for implementation. Ocean Coast Manag 163:485–497. https://doi.org/10.1016/j.ocecoaman.2018.08.005

Santos CF, Agardy T, Andrade F, Crowder LB, Ehler CN, Orbach MK (2018) Major challenges in developing marine spatial planning. 132:103248. https://doi.org/10.1016/j.marpol.2018.08.032

Santos CF, Agardy T, Andrade F, Calado H, Crowder LB, Ehler CN, García-Morales S, Gissi E, Halpern BS, Orbach MK, Pörtner HO (2020) Integrating climate change in ocean planning. Nat Sustain 3:505–516. https://doi.org/10.1038/s41893-020-0513-x

Santos CF, Agardy T, Allison EH, Bennett NJ, Blythe JL, Calado H, Crowder LB, Day JC, Flannery W, Gissi E, Gjerde KM (2022) A sustainable ocean for all. npj Ocean Sustainability 1(1):1–2

Saunders F, Gilek M, Ikauniece A, Tafon RV, Gee K, Zaucha J (2020) Theorizing social sustainability and justice in marine spatial planning: democracy, diversity, and equity. Sustainability 12(6):1–18. https://doi.org/10.3390/su12062560

Scarff G, Fitzsimmons C, Gray T (2015) The new mode of marine planning in the UK: aspirations and challenges. Mar Policy 51:96–102. https://doi.org/10.1016/j.marpol.2014.07.026

Schram SF (2004) Beyond paradigm: resisting the assimilation of phronetic social science. Polit Soc 32(3):417–433

Schutter MS, Hicks CC (2019) Networking the blue economy in Seychelles: pioneers, resistance, and the power of influence. J Polit Ecol 26(1):425–447. https://doi.org/10.2458/v26i1.23102

Schütz SE, Slater AM (2019) From strategic marine planning to project licences – striking a balance between predictability and adaptability in the management of aquaculture and offshore wind farms. Mar Policy 110:103556. https://doi.org/10.1016/j.marpol.2019.103556

Shucksmith RJ, Kelly C (2014) Data collection and mapping – principles, processes and application in marine spatial planning. Mar Policy 50:27–33. https://doi.org/10.1016/j.marpol.2014.05.006

Slater AM, Claydon J (2020) Marine spatial planning in the UK: a review of the progress and effectiveness of the plans and their policies. Environ Law Rev 22(2):85–107. https://doi.org/10.1177/1461452920927340

Smith G (2018) Good governance and the role of the public in Scotland's marine spatial planning system. Mar Policy 94:1–9. https://doi.org/10.1016/j.marpol.2018.04.017

Smith G, Brennan RE (2012) Losing our way with mapping: thinking critically about marine spatial planning in Scotland. Ocean Coastal Manage 69:210–216. https://doi.org/10.1016/j.ocecoaman.2012.08.016

Smith G, Jentoft S (2017) Marine spatial planning in Scotland. Levelling the playing field? Mar Policy 84:33–41. https://doi.org/10.1016/j.marpol.2017.06.024

Smythe TC, McCann J (2019) Achieving integration in marine governance through marine spatial planning: findings from practice in the United States. Ocean Coast Manag 167:197–207. https://doi.org/10.1016/j.ocecoaman.2018.10.006

Spijkers J, Merrie A, Wabnitz CC, Osborne M, Mobjörk M, Bodin Ö, Selig ER, Le Billon P, Hendrix CS, Singh GG, Keys PW (2021) Exploring the future of fishery conflict through narrative scenarios. One Earth 4(3):386–396. https://doi.org/10.1016/j.oneear.2021.02.004

St. Martin K, Olson J (2017) Creating space for community in marine conservation and management: mapping 'communities at sea'. In: Levin P, Poe M (eds) Conservation in the Anthropocene Ocean. Elsevier, pp 123–141

Stefansson G, Punt AE, Ruiz J, Putten IV, Agnarsson S, Daníelsdóttir AK (2019) Implementing the ecosystem approach to fisheries management. Fish Res 216:174–176. https://doi.org/10.1016/j.fishres.2019.04.014

Steinberg P, Peters K (2015) Wet ontologies, fluid spaces: giving depth to volume through oceanic thinking. Environ Plang D Soc Space 33(2):247–264. https://doi.org/10.1068/d14148p

Steins NA, Veraart JA, Klostermann JE, Poelman M (2021) Combining offshore wind farms, nature conservation and seafood: lessons from a Dutch community of practice. Mar Policy 126(July 2020):104371. https://doi.org/10.1016/j.marpol.2020.104371

Stelzenmüller V, Lee J, South A, Foden J, Rogers SI (2013) Practical tools to support marine spatial planning: a review and some prototype tools. Mar Policy 38:214–227. https://doi.org/10.1016/j.marpol.2012.05.038

Stelzenmüller V, Fernández TV, Cronin K, Röckmann C, Pantazi M, Vanaverbeke J, Stamford T, Hostens K, Pecceu E, Degraer S, Buhl-Mortensen L (2015) Assessing uncertainty associated with the monitoring and evaluation of spatially managed areas. Mar Policy:151–162. https://doi.org/10.1016/j.marpol.2014.08.001

Stelzenmüller V, Cormier R, Gee K, Shucksmith R, Gubbins M, Yates KL, Morf A, Aonghusa CN, Mikkelsen E, Tweddle JF, Peccu E (2021a) Evaluation of marine spatial planning requires fit for purpose monitoring strategies. J Environ Manage 278(P2):111545. https://doi.org/10.1016/j.jenvman.2020.111545

Stelzenmüller V, Gimpel A, Haslob H, Letschert J, Berkenhagen J, Brüning S (2021b) Sustainable co-location solutions for offshore wind farms and fisheries need to account for socio-ecological trade-offs. Sci Total Environ 776:145918. https://doi.org/10.1016/j.scitotenv.2021.145918

Stojanovic T, Gee K (2020) Governance as a framework to theorise and evaluate marine planning. Mar Policy 120:104115. https://doi.org/10.1016/j.marpol.2020.104115

Strickland-Munro J, Kobryn H, Brown G, Moore SA (2016) Marine spatial planning for the future: using public participation GIS (PPGIS) to inform the human dimension for large marine parks. Mar Policy 73:15–26

Surís-Regueiro JC, Santiago JL, González-Martínez XM, Garza-Gil MD (2021) An applied framework to estimate the direct economic impact of Marine Spatial Planning. Mar Policy:127. https://doi.org/10.1016/j.marpol.2021.104443

Tafon RV (2018) Taking power to sea: towards a post-structuralist discourse theoretical critique of marine spatial planning. Environ Plan C Polit Space 36(2):258–273. https://doi.org/10.1177/2399654417707527

Tafon RV (2019) Small-scale fishers as allies or opponents? Unlocking looming tensions and potential exclusions in Poland's marine spatial planning. J Environ Policy Plan, 1–12. https://doi.org/10.1080/1523908X.2019.1661235

Tafon R, Howarth D, Griggs S (2018) The politics of Estonia's offshore wind energy programme: discourse, power and marine spatial planning. Environ Plan C Polit Space:1–20. https://doi.org/10.1177/2399654418778037

Tafon R, Saunders F, Gilek M (2019) Re-reading marine spatial planning through Foucault, Haugaard and others: an analysis of domination, empowerment and freedom. J Environ Policy Plan 21(6):754–768. https://doi.org/10.1080/1523908X.2019.1673155

Tafon R, Glavovic B, Saunders F, Gilek M (2021) Oceans of conflict : pathways to an ocean sustainability PACT oceans of conflict : pathways to an ocean sustainability PACT. Plan Pract Res 37:1–18. https://doi.org/10.1080/02697459.2021.1918880

Toonen HM, van Tatenhove JPM (2013) Marine scaping: the structuring of marine practices. Ocean Coast Manag 75:43–52. https://doi.org/10.1016/j.ocecoaman.2013.01.001

Toonen HM, van Tatenhove JPM (2020) Uncharted territories in tropical seas? Marine scaping and the interplay of reflexivity and information. Marit Stud 19(3):359–374. https://doi.org/10.1007/s40152-020-00177-z

Trouillet B (2019) Aligning with dominant interests: the role played by geo-technologies in the place given to fisheries in marine spatial planning. Geoforum 107:54–65. https://doi.org/10.1016/j.geoforum.2019.10.012

Trouillet B (2020) Reinventing marine spatial planning: a critical review of initiatives worldwide. J Environ Policy Plan 22(4):441–459. https://doi.org/10.1080/1523908X.2020.1751605

van Tatenhove JPM (2017) Transboundary marine spatial planning: a reflexive marine governance experiment? J Environ Policy Plan 19(6):783–794. https://doi.org/10.1080/1523908X.2017.1292120

Vince J, Day JC (2020) Effective integration and integrative capacity in marine spatial planning. Marit Stud 19(3):317–332. https://doi.org/10.1007/s40152-020-00167-1

von Thenen M, Hansen HS, Schiele KS (2021) A generalised marine planning framework for site selection based on ecosystem services. Mar Policy 124:104326. https://doi.org/10.1016/j.marpol.2020.104326

Wright G (2015) Marine governance in an industrialised ocean: a case study of the emerging marine renewable energy industry. Mar Policy 52:77–84. https://doi.org/10.1016/j.marpol.2014.10.021

Wright G, Gjerde KM, Johnson DE, Finkelstein A, Ferreira MA, Dunn DC, Chaves MR, Grehan A (2019) Marine spatial planning in areas beyond national jurisdiction. Mar Policy 103384:103384. https://doi.org/10.1016/j.marpol.2018.12.003

Xu W, Liu Y, Wu W, Dong Y, Lu W, Liu Y, Zhao B, Li H, Yang R (2020) Proliferation of offshore wind farms in the North Sea and surrounding waters revealed by satellite image time series. Renew Sust Energ Rev 133:110167. https://doi.org/10.1016/j.rser.2020.110167

Yates KL, Schoeman DS, Klein CJ (2015) Ocean zoning for conservation, fisheries and marine renewable energy: assessing trade-offs and co-location opportunities. J Environ Manag 152:201–209. https://doi.org/10.1016/j.jenvman.2015.01.045

Young N, Brattland C, Digiovanni C, Hersoug B, Johnsen JP, Karlsen KM, Kvalvik I, Olofsson E, Simonsen K, Solås AM, Thorarensen H (2019) Limitations to growth: social-ecological challenges to aquaculture development in five wealthy nations. Mar Policy 104(April 2018):216–224. https://doi.org/10.1016/j.marpol.2019.02.022

Zuercher R, Motzer N, Magris RA, Flannery W (2022a) Narrowing the gap between marine spatial planning aspirations and realities. ICES J Mar Sci 79(3):600–608

Zuercher R, Ban NC, Flannery W, Guerry AD, Halpern BS, Magris RA, Mahajan SL, Motzer N, Spalding AK, Stelzenmüller V, Kramer JG (2022b) Enabling conditions for effective marine spatial planning. Mar Policy 143:105141

Chapter 6
The Past, Present and Future of Ocean Governance: Snapshots from Fisheries, Area-Based Management Tools and International Seabed Mineral Resources

Pradeep A. Singh and Fernanda C. B. Araujo

Abstract Ocean governance comprises the law of the sea as well as all related policy and normative dimensions that relate to the regulation of human activity at sea and increasingly places a strong focus on marine environmental protection and the conservation of marine resources, with the aim of ensuring a healthy and productive ocean while sustaining a resilient ocean-based economy. Premised on this observation, this chapter aims to reflect on the past, present and future of ocean governance using three case studies as snapshot examples, namely, fisheries at sea, marine area-based management tools and international seabed mineral resources. Put together, these three case studies will demonstrate how the law of the sea has evolved when considered from the dimension of ocean governance, particularly with respect to the challenge of protecting and preserving the marine environment through the sustainable use of marine resources.

6.1 Introduction

This chapter aims to provide some insights into the past, present and future of ocean governance using three case studies as snapshot examples, namely, fisheries at sea, marine area-based management tools and international seabed mineral resources. Put together, these three case studies will demonstrate how the law of the sea has

P. A. Singh (✉)
Institute for Advanced Sustainability Studies (IASS), Potsdam, Germany

Research Centre for European Environmental Law (FEU), University of Bremen, Bremen, Germany
e-mail: pradeep.singh@iass-potsdam.de

F. C. B. Araujo
Universidade de Brasilia, Brasilia, Brazil

© German Institute of Development and Sustainability (IDOS) 2023
S. Partelow et al. (eds.), *Ocean Governance*, MARE Publication Series 25,
https://doi.org/10.1007/978-3-031-20740-2_6

evolved when considered from the dimension of ocean governance, in particular with respect to the protection and preservation of the marine environment as well as the sustainable use and conservation of marine resources.

The United Nations Convention on the Law of the Sea (LOSC), adopted in 1982, is also known as the 'constitution for the oceans' due to its comprehensiveness in codifying the law of the sea into a multilateral treaty with legally binding effect (Koh 1982). The LOSC explicitly designated the various maritime zones (alongside with the associated legal rights and obligations that apply respectively) and established a dedicated part to the protection of the marine environment. Although the LOSC only took shape from the late twentieth century, the law of the sea is one of the oldest branches of international law, where States have often sought to exercise rights and exert their influence. The LOSC, consequently, has had the benefit of centuries of experience of human activity at sea and could be seen as an instrument that configures the main framework for global ocean governance. As a concept, ocean governance has not been precisely defined and its contour and relationship with the law of the sea remains unclear (Takei 2015). However, it is clear that ocean governance comprises the law of the sea as well as all related policy and normative dimensions that relate to the protection of the marine environment and the regulation of human activity at sea (Rothwell and Stephens 2016).

Accordingly, ocean governance appears to place a strong focus on marine environmental protection and the conservation of marine resources (Singh and Ort 2019), with the aim of ensuring a healthy and productive ocean while sustaining a resilient ocean-based economy. Premised on this observation, we begin with fisheries at sea as representative of a marine resource exploitation activity long before the conclusion of the LOSC and an important interest of State Parties that the LOSC sought to protect (though still barely effective for addressing overexploitation and conserving marine ecosystems). We then turn to area-based management tools as a marine conservation approach that has received increasing attention since the 1980s and in the current times. Finally, we consider the management of the international seabed mineral resources as example of an interest that sparked great debate during the negotiations of the LOSC and yet today still remains an activity for the future. Each case study will involve a brief historical analysis prior to 1982, as well as attempt to track developments since the LOSC was adopted and subsequently entered into force, and critically evaluate how things broadly stand today.

6.2   Fisheries at Sea: A Persistent Challenge

Fisheries lie among the very origins of the law of the sea. Since the early attempts of managing the oceans, fishing activities have been involved in the development of a series of instruments that try to harmonize the needs, interests and concerns at sea. Yet, fisheries regulations so far have been barely effective for the purposes of protecting fish stocks from overexploitation and the conservation of marine ecosystems, what makes it a persistently challenging activity for ocean governance.

The origins of international fisheries law are intertwined with the foundation of the law of the sea. The great conflict between the defenders of exclusive rights (*mare clausum*) and those who claim free exploitation (*mare liberum*) over marine resources and spaces dates back to colonial times of the sixteenth–seventeenth centuries, and agreements aimed at restricting access to certain maritime areas could be identified already in the Classical Age (Markus and Markus 2021). But even though there were already several conservation measures foreseen in fisheries legal norms by the mid-twentieth century, reversing fish stocks depletion only became the main concern of international fisheries regimes around the 1970s. Before that, the priorities of States were pretty much focused on the conquest of new fishing grounds or the development of means to guarantee production levels (Garcia et al. 2014; Markus 2018). This shift came after the serious environmental impacts caused by the significant increase on the size and capacity of fishing vessels, usually fostered by State subsidies (WTO 1999; Sakai et al. 2019), started to become evident, giving birth to a multitude of marine living resources protection-oriented regional and global instruments.

The adoption of LOSC was undoubtedly a cornerstone to international fisheries law. While maintaining the principle of "freedom of the seas" on the high seas, which concerns freedom of navigation, fishing and exploitation of resources, non-prejudicial passage in regions beyond the jurisdiction of States (Arts. 87 and 116), the LOSC assured to coastal States full sovereignty in Inland Waters and the Territorial Sea of up to 12 nautical miles.[1] Sovereign rights were accorded over the exploitation of natural resources in the Exclusive Economic Zone (EEZ) and the Continental Shelf that can extend up to 200 nautical miles (and in the case of the continental shelf, may extend even further pursuant to Article 76). In terms of the conservation and sustainable use of fishing resources, the LOSC detailed out rights and duties of coastal States in the EEZ. In this respect, States shall determine the total allowable catch of their living resources based on the best available scientific knowledge and in co-operation with the competent international organizations in order to achieve maximum sustainable yield (Articles 61-62). In addition, international cooperation is required, directly or through regional or subregional organizations, to manage shared, straddling, marine mammals, anadromous or catadromous stocks. In this process, economic and environmental factors must be considered, such as the economic needs of coastal fishing communities and developing States, as provided for in Article 61(3).

However, the main measures of the conservation strategy adopted for the EEZ (namely, "total allowable catches" and "maximum sustainable yield") are not only difficult to implement, as they are subject to the jurisdiction of coastal States and depend on high economic cost stocks assessments, but also tend to leave out relational analysis, such as bycatch and the impacts of marine pollution and other economic activities on biodiversity. Therefore, despite being known as the general legal

---

[1] Subject to the right of innocent passage of foreign vessels through these areas, as established from Articles 17 to 26 of the LOSC.

framework for international fisheries law, the LOSC lacks detailed and ambitious provisions applicable to all maritime spaces as well as a solution to the growing pressures on fish stocks, especially on the high seas (Birnie et al. 2009; Sands et al. 2018).

Since the LOSC was adopted, several norms and instruments to complement the regime applicable to marine fisheries have been developed. Of those pertaining to multilateral binding instruments, three agreements stand out. The first is the United Nations Agreement for the Implementation of the Provisions of the LOSC relating to the Conservation and Management of Straddling Fish Stocks and Highly Migratory Fish Stocks (UNFSA or simply the UN Fish Stocks Agreement), which was adopted in 1995 and came into force in 2001. The UNFSA aims to ensure long-term conservation and sustainable use of these fish stocks (Article 2). It further elaborates upon relevant provisions under the LOSC by setting out obligations both for areas beyond and under national jurisdiction, such as the need for applying a precautionary approach (Article 6) and by strengthening the role of regional and sub-regional fisheries organizations (RFMO) (see Articles 8-14 and 17(1)(2)). Specifically, it stresses on the need to consider the effects of other activities and environmental factors on target populations and associated ecosystems in fisheries assessments (Article 5(d)), as well as the relationships between biological characteristics and geographical particularisms and the impacts on living marine resources as a whole in determining conservation measures (Article 7.2), and to avoid adverse impacts on and ensure access to fisheries by small-scale and artisanal fish workers (Article 24(2)(b).

The other two global binding instruments were approved under the mandate of the UN Food and Agriculture Organization (FAO). On the one hand, the Agreement to Promote Compliance with International Conservation and Management Measures by Fishing Vessels on the High Seas (Compliance Agreement), adopted in 1993 and entered into force in 2003, aims to address the issue of compliance with international conservation measures in the high seas. In this sense, it requires flag States to take necessary measures to ensure that fishing vessels flying their flag do not engage activities that undermine international norms, such as the requirement of authorization to fish, the provision of sanctions and cooperation with other States to help identifying vessels engaged in such activities. On the other hand, the Agreement on Port State Measures to Prevent, Deter and Eliminate Illegal, Unreported and Unregulated (IUU) Fishing (PSMA), adopted in 2009 and in force since 2016, in turn, puts the spotlight on the point at which fish are landed, by providing, among others, that the local authorities can deny permission to entry into its port if they suspect that the vessel has engaged in IUU fishing (Article 9.1).

Apart from that, the contributions of the FAO to the development of international fisheries law through non-binding instruments also stand out.[2] In the last decades, the FAO has been striving to lead the settlement of the notion of sustainable

---

[2] The resolutions from the UN General Assembly, although less noticeable, have played an important role, too. On this subject, see the Chapter from Nakamura in this book.

fisheries. Notably, the most important means for that is the Code of Conduct, from 1995, which aims to promote responsible fisheries by providing principles, guidance and standards for its implementation (Article 2). Its Article 6 brings expressly the duty of States and users of bio-aquatic resources to conserve aquatic ecosystems as a result of their right to fish. The Code also gave birth to a series of plans of action, technical and international guidelines. One of them officially adopted what was called the "ecosystem approach to fisheries", which presupposes the need for fisheries management to associate fisheries concerns with conserving the structure, diversity and functioning of the biotic, abiotic and human components of ecosystems, aiming to promote convergence towards a more holistic and balanced approach through principles such as the precautionary approach, equity, stakeholder participation and ecosystem integrity (FAO 2003). Another important example comes from 2014, with the Voluntary Guidelines for Securing Sustainable Small-Scale Fisheries in the context of Food Security and Poverty Eradication (SSF Guidelines). The document was approved in order to guide public policies in the sector and ensure decent working conditions to this marginalized group through a human-rights approach. It recommends, among others, that States (especially developing countries) facilitate, train and support fishing communities to participate and assume responsibilities in the management of resources (FAO 2015).

Regional arrangements have also proven to be fruitful in the provision of norms concerning the international conservation and management of fish resources. Indeed, building on the political momentum for considerations on sustainability driven by multilateral summits such as the UN Conference on Environment and Development 1992 (the Earth Summit or Rio Conference), the UNFSA explicitly called for the establishment of subregional and regional management organizations or arrangements in order to improve fisheries governance (Harrison 2019, p. 80). Since then, many RFMOs (that were already long in existence) have revised their enabling conventions in order to adopt innovative approaches, such as the ecosystem approach to fisheries management and the requirement of undertaking periodic performance reviews (Harrison 2019). Alongside with the efforts of other related institutions, these reforms have brought about progressive legal frameworks capable of providing tools for the sustainable management of stocks, particularly in the case of tuna and tuna-like species RFMOs (Unterweger 2015).

The Common Fisheries Policy (CFP), under the scope of the European Union (EU), is also worth mentioning. The CFP defines various principles and management tools in the search for long-term sustainable fisheries, notably since its last extensive reform which entered into force in 2014.[3] The text provides for the adoption of a precautionary approach, as well as an ecosystem approach to fisheries management (see Art. 4(8 and 9)). In terms of conservation and sustainable exploitation measures, a wide range of options is listed, including input (e.g. multiannual plans and restrictions on the use of certain types of mesh or vessel sizes) and output (e.g. TAC and landing obligation) regulations, as well as market driven instruments,

---

[3] See Art. 2 of Regulation n. 1380/2013 of the European Parliament and of the Council.

such as economic incentives to fishing with low impact on the marine ecosystem and fishery resources (Art. 7), making fisheries one of the most regulated activities in the EU (Hadjimichael 2018).

Despite the developments in international fisheries law built upon the LOSC, international fisheries law is still deficient not only in substantive fisheries measures but also in terms of compliance. The LOSC, even after being complemented by the UNFSA, is essentially based on the enunciation of generic measures and objectives, relying on state practice to detailing and implementing them. However, there is low compliance by the States, either because they cannot afford the high costs of conservation measures, especially in developing countries, or because they give priority to other economic and political interests, in the case of industrialized countries (Molenaar 2019). On the other hand, although FAO has made much progress in regulatory terms, it is unable to overcome the reluctance of States to a large extent. As for regional fisheries bodies, the existing ones still leave some regions and species uncovered (e.g. the South-West Atlantic), as well as not all have the power of adopting legally binding conservation and management measures (Harrison 2019). Moreover, most of them still have not reached transparent, timely and effective decision-making mechanisms (Leroy and Morin 2018). Even the CFP has not been able to overcome the contradictions between ambitious declarations and state practice. The officially established EU target of achieving maximum sustainable yield exploitation rates for all fish stocks by 2020 (see Art. 2.2 from EU Regulation 1380/2013) was not achieved (European Commission 2020) while small scale fisheries fleet has been decreasing since the beginning of the new millennium (Lloret 2018), problems that experts link to the fact that the measures put into practice often fall considerably short of scientific recommendations and social concerns (Hadjimichael 2018; Lado 2016). Thus, there is an insistence on the application of traditional management techniques (e.g. gear and effort restrictions), with rare cases where measures that give due attention to the relationships among species are legally prescribed (Serdy 2018).

Therefore, fisheries at sea can be considered an example of how such a traditional activity can represent an ever-present challenge to ocean governance. If historically it was cause of conflicts primarily due to difficulties in regulating competing economic or geopolitical interests, the implementation of the increasingly important environmental protection measures and obligations suffers from the lack of political will or financial conditions by the States, as well as integration and coordination mechanisms for the many institutions and regulations that deal, direct or indirectly, with fisheries management.[4] As a result, international fisheries law has not been able to overcome the serious failures in addressing the negative impacts generated by fishing activities in the ocean (see FAO 2020 and WWF 2020). Area-based management tools, which are essentially multidimensional, have been increasingly prescribed by international norms to tackle such deficiencies. Nevertheless, they also

---

[4] For a profound incursion on fisheries governance norms and institutions and the practical interaction between regimes, see: Young 2011.

face problems to encompass all the complexity involved in achieving a good ocean governance.

6.3   Area-Based Management Tools: The Current Trend

Area-based management tools (ABMTs) gained momentum throughout the years as a useful tool not only in the broader global conservation agenda, but also for the protection of the marine environment. Their ability to mobilize a variety of legal regimes in specific areas to achieve a desired outcome turned ABMTs into an essential element in the ocean governance toolbox. Although ABMTs are an undeniable success in terms of adoption (particularly considering marine protected areas), they can be controversial. In fact, sensitive issues regarding biodiversity conservation, such as acknowledging all the complexity of ecosystems and properly taking into consideration social interests, are even more challenging in marine realities.

Conceptually speaking, ABMTs can cover a wide range of different legal measures. They operate by guiding determined spaces to pursue certain objectives, such as the protection and preservation of marine environment, the conservation of marine biodiversity, sustainable use of marine biodiversity components and revolving conflicts of use and interests in coastal and maritime zones. A study carried out by the World Conservation Monitoring Centre (WCMC), from the United Nations Environment Program (UNEP), mapped case studies related to seven ABMT, such as: integrated coastal zone management (ICZM), marine spatial planning (MSP), marine protected areas (MPAs), locally-managed marine areas (LMMA), MARPOL[5] special areas, particularly sensitive sea areas (PSSA) and fisheries closures (UNEP 2018). Nevertheless, since no global consensus on the definition of ABMT exists, we will focus here on the two examples which have been more significantly developed in international law: marine protected areas and marine spatial planning.[6]

The custom of protecting special places at sea by local communities exist for millennia (Laffoley et al. 2018). However, the creation of MPAs for environmental policy purposes is a recent development and mostly relies on the international regulatory framework for protected areas in general, since the LOSC does not mention them expressly. Protected areas were consecrated as an international commitment to spaces (terrestrial or marine) within the jurisdiction of the countries through the Convention on Biological Diversity (CBD) in 1992, which has become the main reference in the international arena for discussions and legal measures related to

---

[5] MARPOL is how the International Convention for the Prevention of Pollution from Ships, signed in 1973, together with its 1978 protocol, is better known.

[6] In this respect, it is important to note that MPAs have been much more widely integrated into international law and policies than MSP.

nature protection.[7] The CBD also provides a definition of a protected area, described as "a geographically defined area which is designated or regulated and managed to achieve specific conservation objectives". As highlighted by the *Ad Hoc* Technical Expert Group on Marine and Coastal Protected Areas, created by the Conference of the Parties of CBD at its fifth meeting, they can cover both coastal and offshore zones, with the effect of increasing the level of biodiversity protection within these areas set aside by law (Secretariat of the Convention on Biological Diversity 2004a, b, p. 7).

Over the past two decades, the spatial area covered by MPAs showed a ten-fold increase.[8] The LOSC appears to have contributed to this shift to some extent. Despite not explicitly mentioning MPAs, the LOSC strengthens coastal states capabilities to create such legal instrument, by granting them with sovereign rights in their territorial seas and EEZ for the purposes of managing and conserving natural resources and, at the same time, creating the duty to protect and preserve the marine environment. Most importantly, targets relating to the establishment of MPAs have been defined over the last decades. Back in 1992, Agenda 21 already dedicated Chap. 17 to push States to "undertake measures to maintain biological diversity and productivity of marine species and habitats under national jurisdiction", including the "establishment and management of protected areas" (see Article 17(7)). Goal 11 of the Aichi Biodiversity Targets and Goal 14 of the 2030 UN Agenda for Sustainable Development established quantitative and qualitative targets: they call on States to conserve, by 2020, at least 10% of coastal and marine areas through protected areas or other effective means consistent with national and international law.[9] This target could increase to 30% in the next global political commitment, i.e. the Post-2020 Global Biodiversity Framework, which is expected to be adopted in late 2022.[10] All these factors have helped push MPAs to become the core of ocean governance legal strategies today, essentially through domestic action.

---

[7] Before CBD, a few international conventions that mention species of marine protected areas can be listed: the 1971 Ramsar Convention established the list of Ramsar Sites (*Convention on Wetlands of International Importance Especially as Waterfowl Habitat* 1971, art. 2); in 1972, UNESCO introduced the concept of World Heritage Sites (UNESCO 1972, art. 4); in the 1990s, the Antarctic Specially Protected Areas and the Specially Managed Antarctic Areas were established by the 1991 Antarctic Protocol on Environmental Protection (*Protocol on Environmental Protection to the Antarctic Treaty* 1992, Annex V).

[8] According to the World Database on Protected Areas online platform. See: https://www.protectedplanet.net/marine

[9] The Zero Draft of the post-2020 Global Biodiversity Framework, which was released in July 2021, among the targets to be completed by 2030, calls for states to "ensure that at least 30 per cent globally of land areas and of sea areas, especially areas of particular importance for biodiversity and its contributions to people, are conserved through effectively and equitably managed, ecologically representative and well-connected systems of protected areas and other effective area-based conservation measures, and integrated into the wider landscapes and seascapes" (CBD/WG2020/3/3).

[10] For more information on the preparations for the Post-2020 Biodiversity Framework, visit: https://www.cbd.int/conferences/post2020

Indeed, in theory MPAs are very promising. MPAs work by establishing zones where different types and levels of human intervention are allowed or prohibited. Moreover, it is typically a multi-sector planning tool, instead of single-sector, enabling the application of rules to restrict different human activities at the same time. They have, then, the potential to encompass a comprehensive zoning approach (Singh & Ort 2019, pp. 48–49).

In addition, MPAs are intrinsically related to the ecosystem approach. The CBD bodies pioneered the development of ecosystem approach as a broad concept, encouraging its adoption as an approach that implies integrated and adaptive management techniques in order to adapt to the changing nature of a number of issues: the availability of scientific knowledge, the living systems themselves, the threats they suffer, as well as the multifaceted interests of those who use them (Secretariat of the CBD 2004a, b, pp. 1–4). The notion itself originated from practical experiences with the implementation of protected areas, which served, at the same time, to demonstrate that MPAs already provide many of the principles that make up the ecosystem approach and to call States to act upon protected areas failures and successes (CBD 1998).

The forthcoming binding international instrument under the LOSC on the conservation and sustainable use of marine biological diversity of areas beyond national jurisdiction (BBNJ Agreement) may improve MPAs regulation, which can be rather complicated when it comes to areas beyond the limits of national jurisdiction where no State may exercise sovereignty or sovereign rights. The pressing need for such an agreement was agreed in 2017 by the UN General Assembly (A/RES/72/249), after more than a decade of discussions within the *Ad Hoc* Open-ended Informal Working Group to Study Issues Relating to the Conservation and Sustainable Use of Marine Biological Diversity beyond Areas of National Jurisdiction. This legal instrument, whose draft text is under construction, was theme of three Intergovernmental Conferences and is expected to conclude its negotiations in 2023. ABMTs, in particular, MPAs, figure as one of the core components of the BBNJ Agreement, which may very well provide the necessary platform for the effective protection of the marine environment in areas beyond national jurisdiction.[11]

The BBNJ Agreement is also an opportunity to consolidate the role of ABMT in international law and to better delimitate the scope of MPAs. The draft of the treaty innovates when it provides for definitions for both ABMT and MPA, which can

---

[11] For more information on the negotiations towards the BBNJ Agreement, see: https://www.un.org/bbnj/

serve as a legal framework also for areas under national jurisdiction.[12] The definitions proposed elect the possibility of taking into consideration particular and cumulative impacts of different human activities in determined areas as the essential feature of ABMTs, as well as reaffirm MPAs as a species of ABMT that is oriented for long-term marine biodiversity conservation and sustainable use objectives.[13] Nonetheless, although there seems to be consensus on the desire to include in the Agreement a list of outcome-oriented objectives and to strengthen ecosystem approach, the use of best available science and of the traditional knowledge of local communities and indigenous peoples as basic requirements to the designation of any ABMT, the negotiations so far have not achieved significant outcomes on defining the creation, implementation, monitoring and reviewing processes and the bodies in charge of analyzing MPAs and other ABMT proposals (IISD 2019).

Aside from MPAs, marine (or maritime) spatial planning also stands out as a form of ABMT.[14] This tool is already institutionalized in more than 20 countries and is expected to cover at least one third of the surface area of world's EEZ in 2030 (Ehler et al. 2019, p. 1) as well as to be implemented in areas beyond national jurisdiction (Becker-Weinberg 2017). MSP is essentially a public planning process that brings together and maps out different impacts from human uses occurring in the same area, thereby permitting decision-making to restrict or foster ocean-based activities based on this geographic mapping (Ehler and Douvere 2009, p. 18; Zacharias 2014). Originating due to the exceeding demand for marine uses against space availability, it had its first legal foundation indirectly formed by the notion of integrated coastal zone management or ICZM (e.g. item 17.5 from Agenda 21) and by the LOSC provisions on the need for promoting peaceful uses of the sea (see the Preamble) and the regulatory competence of coastal States on supra-sectorial planning.[15] MSP aims to achieve social, economic and environmental results and has long been ascribed as a tool to implement an ecosystem-based approach *par*

---

[12] The revised text of November 27, in its Article 1 affirms that "ABMT means a tool, including a marine protected area, for a geographically defined area through which one or several sectors or activities are managed with the aim of achieving particular conservation and sustainable use objectives [and affording higher protection than that provided in the surrounding areas]" and that ""Marine protected area" means a geographically defined marine area that is designated and managed to achieve specific [long-term biodiversity] conservation and sustainable use objectives [and that affords higher protection than the surrounding areas]" (Intergovernmental conference on marine biodiversity of areas beyond national jurisdiction 2019).

[13] In the same sense of IUCN's guidelines (see Day 2012). Scovazzi (2011, p. 14) proposes a different definition when he considers MPA "an area of marine waters or seabed that is delimited within precise boundaries (including, if appropriate, buffer zones) and that is granted a special protection regime because of its significance for a number of reasons (ecological, biological, scientific, cultural, educational, recreational, etc.)", recalling note 11 of Decision VII/5 on marine and coastal biological diversity of the CBD's COP.

[14] Another ABMT that may gain value as a legal instrument in oceans governance for its integrative feature is the Ecologically or Biologically Significant Areas (EBSA), which prepares areas for the adoption of other management measures by describing spaces of ecological importance. To learn more about it, see: (Diz 2018).

[15] See Articles 56-58 of the LOSC.

excellence (Douvere 2008). The manner in which it has been concretized in current legal systems, however, is not so coherent in practice.

In the implementation of MSP under national and regional legal settings, economic considerations seem to have prevailed over environmental concerns. MSP is usually institutionalized under the context of promoting the development of a "blue growth".[16] This can be illustrated by the case of the Directive 2014/89/EU establishing a framework for MSP. Although it can be considered a milestone for an integrated long-term planning of the EU maritime space (Schubert 2018, p. 1021) – i.e., by aiming to promote the sustainable growth of maritime economies, the sustainable development of marine areas and the sustainable use of marine resources (art. 1(1)) – coastal zones have been left out of it (Cudennec 2015), while studies show that it has been implemented mainly to further economic purposes (Frazão Santos et al. 2014).

Accordingly, experiences in other countries demonstrate the need for better assessing of MSP social implications (Flannery et al. 2016; Flannery et al. 2018; Queffelec et al. 2021). The tool is an answer to deal with the growing interest in the exploitation of marine resources and space. However, MSP also attracts new users to a territory that was historically used essentially for fishing purposes. Therefore, the allocation of new activities at sea, even if formally stated to seek integration and adaptability, may end up legitimizing, just like some MPAs do, expropriations of vulnerable coastal communities whose livelihoods depend on artisanal fisheries, a phenomenon increasingly described in literature as ocean grabbing (Bennett and Govan 2015).

In fact, in terms of effectiveness, even the apparent success of MPAs remain highly controversial. Shortcomings have been pointed out by scientists regarding both the lack of reliable information and ecological and socioeconomic MPAs' potentialities. As for the latter, while there are studies showing that many public procedures behind the establishment of MPAs either do not take into consideration the rights, needs and interests of traditional coastal communities that are affected by the restrictive regimes they create or exclude them from the resources' management, being source of various conflicts and injustices (Araujo and Moita 2018; Barros et al. 2021; Sharma and Rajagopalan 2017), others reveal that when MPAs receive local support, these have the tendency to be more effective and successful (Bennett and Dearden 2014; Andrade and Rhodes 2012; Alder et al. 2002). With respect to the effectiveness of results from an environmental conservation perspective, many scientific studies endorse that closing off areas of the ocean to fishing and other extractive activities through MPAs do help species recover, especially those habitually under threat. Nevertheless, partially protected areas and the surroundings are overlooked by scientists, which makes it hard to conclude that fully protected areas are the best for marine biodiversity conservation (Dasgupta and Fensome 2018). Moreover, there is still the proliferation of the so-called "paper MPAs", i.e. those established in places that, instead of representing great biological importance,

---

[16] As recognized by UNESCO. See: http://msp.ioc-unesco.org/world-applications/overview/

are chosen simply because they have no economic importance and/or will unlikely implement any restrictions on exploitation or access (Rife et al. 2013).

The Chagos Marine Protected Area Arbitration helped in defining the role of MPAs in ocean governance, highlighting the need for balancing the competing rights at stake. In the dispute, Mauritius claimed that the creation of a MPA in the Chagos Archipelago by the United Kingdom violated Mauritian fishing rights, protected under the LOSC, among other agreements. In the decision rendered in 2015, the tribunal acknowledged that Part XII of the LOSC does not only apply to the prevention, reduction or control of marine pollution, but may also involve the creation of MPAs. In order not to violate the provisions of the LOSC, however, the coastal State must respect the rights and obligations of other States, which includes the duty to present a meaningful commitment to justify such a measure and after having explored other less restrictive alternatives. The tribunal declared that in establishing the MPA surrounding the Chagos Archipelago, the United Kingdom breached its obligations under Articles 2(3), 56(2), and 194(4) of the LOSC (PCA 2015, paras. 320; 538–541).

The ecosystem approach, which could be a guiding principle for the necessary adaptations of MPAs and the elaboration and implementation of new ABMTs, by its turn, does not have clearly delimited contours in international law. Notwithstanding the fact that it was the bodies of the CBD that most joined efforts to develop the ecosystem approach as a legal concept, there is no international consensus on its content and objectives yet (Engler 2015). This vagueness has been opening space for it to be appropriated by the discourse of ecosystems services economic valuation, which can reinforce chronic problems of ABMTs, instead of helping to overcome them (De Lucia 2018).

Summing up, the adoption of LOSC and the shift towards marine environmental protection has strongly stimulated the adoption of ABMT, especially with respect to MPAs and MSP. This trend would seemingly continue in the near future, given the increase in global political commitments (e.g. Goal 14 of the SDGs) and should also make some important strides in areas beyond national jurisdiction through the forthcoming BBNJ Agreement. That said, it is apparent that the mere existence of such political commitments is still far from guaranteeing the harmonic consideration of all rights and concerns involved in the establishment and implementation of ABMTs and to arrest the increase in the level of marine biodiversity loss.[17] It is expected that the forthcoming BBNJ Agreement would not seek to undermine any existing arrangements in areas beyond national jurisdiction, which most notably would include the dedicated regime established to administer the mineral resources of the international seabed.

---

[17] According to the Global Assessment Report on Biodiversity and Ecosystem Services (IPBES 2019), over one-third of marine mammals and nearly one-third of sharks, shark relatives, and reef-forming corals are threatened with extinction.

6.4   International Seabed Mineral Resources: Back to the Future

The deep seabed (of depths of 200 meters and beyond) is home to abundant mineral deposits with rich content of metals such as nickel, copper, cobalt, and manganese, amongst other critical metals. These deposits include polymetallic nodules, polymetallic sulphides and cobalt-rich ferromanganese crusts, which are known to exist in areas within the limits of national jurisdiction, as well as in areas beyond the limits of national jurisdiction (i.e. the international seabed). In the case of the latter, the framing of regulations to govern access to these resources as well as the sharing of financial and other economic benefits that are derived from their exploitation have been the subject of intense debates and ongoing negotiations for over half a century. In this respect, commercial mining activities are still yet to take place. With growing environmental concerns surrounding the harmful effects of seabed mining to the marine environment, and at the same time being one of the rare examples where a human activity is being thoughtfully regulated before it even commences, it remains to be seen how the regime and the legitimacy of its activities will shape up in the future.

The LOSC in Part XI classifies the seabed areas beyond the limits of national jurisdiction as the 'Area' and declares the mineral resources therein as the 'common heritage of mankind' (Articles 1(1)(1) and 136 of the LOSC). This declaration of the Area and its mineral resources as the 'common heritage of mankind', wherein the exploration for and exploitation of the mineral resources of the Area is to be carried out 'for the benefit of mankind as a whole' (Article 140(1) of the LOSC) through a single global regime, is consistently hailed as one of the greatest accomplishments of the LOSC (Lodge 2013). Essentially, the common heritage of mankind, now widely referred to as an established principle under international law, is considered as one of the foundational structures of the LOSC (Wolfrum 1983). Two salient provisions in the LOSC, both to be found outside of Part XI of the LOSC, confers strong support for this notion. First, the Preamble of the LOSC, which sets the tone for the entire instrument, gives stark effect to this declaration by affirming that "the area of the seabed and ocean floor and the subsoil thereof, beyond the limits of national jurisdiction, as well as its resources, are the common heritage of mankind, the exploration and exploitation of which shall be carried out for the benefit of mankind as a whole". Second, Article 311(6) of the LOSC unequivocally prescribes that there shall be no derogation from the "basic principle relating to the common heritage of mankind". Numerous provisions in Part XI, as will be explored in the coming paragraphs, also lend effect to the primacy of the 'common heritage of mankind' in the context of seabed mining activities in the Area.

However, in order to better comprehend the deep seabed mining regime for the Area that Part XI of the LOSC established, it is necessary to look beyond the LOSC and appreciate the historical developments that took place decades before the LOSC was adopted (White 1982). One particular fact to take cognizance of from the outset is that the LOSC, while concluded in 1982 after nearly a decade of multilateral negotiations, only came into force in 1994. This is due to the dissatisfaction of numerous developed countries specifically with respect to Part XI (Tanaka 2011). It is important to stress here that the LOSC was negotiated with a view of adoption as a 'package deal' (Treves 2008), and additionally, any State wishing to be a signatory to the instrument must accept it as a whole without exceptions or exemptions, which are otherwise known as reservations (UN DOALOS 1998). Since there were disagreements in relation to the deep seabed mining regime in Part XI, a significant number of States (mostly industrialized) were not inclined to ratify the LOSC. This deadlock was only resolved with the adoption of the 1994 Agreement relating to the implementation of Part XI of the LOSC (UN DOALOS 2016).

Although the existence of ocean minerals was already known since the 1860s when the HMS Challenger successfully collected polymetallic nodules from the seabed, the defining moment that gave rise to the strong political will to initiate the process to establish a mining regime for the international seabed only came a century later (Morgan 2011). This impetus was largely driven by John L. Mero's publication entitled 'Mineral Resources of the Sea' in 1965, which speculated the availability of abundant mineral resources on the seafloor that could be easily procured with assured profits (The Geological Society 2013). However, other contemporaneous events may have also played a role in propelling the creation of the international seabed mining regime. Most notably, the traditional practice of the freedom of the high seas was already under challenge since the 1940s (UN DOALOS 1998). Through a 1945 Proclamation by President Truman, the US unilaterally declared jurisdiction over non-living seabed resources up to the extent of the continental shelf. In contrast, newly independent and developing countries, in particular in South America, were more concerned with living resources (i.e. fisheries) and sought to extent their jurisdiction over fish stocks up to 200 nautical miles (as compared to the existing practice of coastal State jurisdiction of between 3 to 12 nautical miles). As state practice proliferated in this regard, the areas that were left as areas beyond national jurisdiction were substantially reduced. Thus, questions arose about how to regulate access to resources in areas that were beyond national jurisdiction. These questions mainly centred on the mineral resources in those areas, given that most coastal States (in particular newly independent and developing States) were content if their claims of 200 nautical miles of exclusive rights over fisheries were acceded to (thereby leaving them little cause for concern overfishing activities taking place outside their jurisdiction). Developed States – mainly concerned with offshore resources at this point in time – were equally content if their rights over the non-living seabed resources on their continental shelf were acknowledged in return.

In 1967, Ambassador Arvid Pardo (Malta) delivered a speech to the First Committee of the United Nations, expressing the urgent need to designate the Area and its mineral resources as the 'common heritage of mankind' in order to ensure that it is not exploited by rich and developed countries on a 'first come, first serve' basis (United Nations General Assembly 1967). This passionate plea gained widespread acceptance and formed the basis of two important UN General Assembly resolutions in 1970, which designated the international seabed and its mineral resources as the common heritage of mankind, that it should be developed 'for the benefit of mankind as a whole' and administered through an agreed international machinery (United Nations General Assembly, Resolutions 2749 and 2759 (XXV), 1970a, b). Another unrelated event to deep seabed mining that might have also propelled the demand for 'enclosing' the then open access feature of the international seabed is the publication of Garrett Hardin's 'The Tragedy of the Commons' in 1968, which resulted in increased attention towards the problems of open and unregulated access to a shared common resource (Hardin 1968). Shortly thereafter, in 1973, multilateral negotiations via the Third UN Conference on the Law of the Sea (UNCLOS III) commenced. The UNCLOS III culminated in 1982 with the conclusion of the LOSC (UN DOALOS 1998).

On the one hand, the conclusion of the LOSC brought an end to differing state practices in relation to the rights (and obligations) of coastal States over the maritime space in areas within national jurisdiction as well as provided legal clarity with respect to the rights (and obligations) of all States in areas beyond national jurisdiction. On the other hand, the LOSC has also received some criticism for affirming the claims of States that effectively 'territorialized' the seas and allowed States to disproportionately appropriate its commonly-owned resources through the exercise of sovereignty or sovereign rights (Constantinou and Hadjimichael 2020). Indeed, it has been observed that the speech delivered by Ambassador Pardo and the genesis of the 'common heritage of mankind' principle, as applicable to the Area and its mineral resources through the LOSC, specifically embodied a highly anthropocentric view and sense of entitlement over those resources with the primary intention of securing monetary gains (Constantinou and Hadjimichael 2020).

Part XI of the LOSC is dedicated to the Area and its mineral resources. It establishes the International Seabed Authority (ISA), headquartered in Kingston, Jamaica, to organize, manage and control the conduct of activities in the Area (defined as the exploration and exploitation of mineral resources in the international seabed area) (Articles 1(1)(1), 153(1), 156(1) and (4) of the LOSC). In particular, Part XI of the LOSC entrusts the ISA to establish a regulatory framework to administer the mineral resources of the Area (Article 157(1) of the LOSC) while simultaneously ensuring the effective protection of the marine environment from the harmful effects of mining activities (Article 145 of the LOSC). To this end, the LOSC authorizes the ISA to issue out contracts for mineral exploration (and in future, exploitation) activities, to supervise the conduct of such activities and ensure compliance, and to distribute the proceeds therefrom in an equitable manner through an appropriate mechanism (Articles 140 and 153 of the LOSC). The ISA comprises of three main organs: the Assembly, the Council and the Secretariat. The Assembly is the supreme organ of the ISA; all member States to the LOSC are *ipso facto* members of the Assembly (Article 156(2) of the LOSC). The Council is the executive organ of the ISA; the Assembly elects 36 member States to sit in the Council, which is entrusted with critical decision-making functions (Articles 161 and 162 of the LOSC). The Council is assisted by the Legal and Technical Commission, an advisory subsidiary body that provides recommendations to the Council on matters under its purview (Articles 163 and 165 of the LOSC). The Secretariat is the administrative organ of the ISA; it is led by the Secretary-General, which administers the day-to-day functions of the ISA pursuant to the instructions from the Council or Assembly, as the case may be (Article 166 of the LOSC).

Pursuant to its mandate, the ISA has developed three sets of regulations to govern the exploration of the three mineral resources of interest: polymetallic nodules (exploration regulations adopted in 1999, amended in 2013), polymetallic sulphides (exploration regulations adopted in 2010), and cobalt-rich ferromanganese crusts (exploration regulations adopted in 2012). As of June 2021, the ISA has issued 31 exploration contracts covering all three types of resources in various parts of the Area. It is to be noted that a majority of the existing contractors are either private actors or state agencies. These actors or entities operate under the sponsorship of a member State of the ISA. This concept of a sponsoring State is particularly pertinent, given that only States and international organizations are recognized as subjects of international law (and therefore, liable to responsibility for internationally wrongful acts). While contractors remain contractually liable for the conduct of their activities, which is enforceable under the domestic laws of the sponsoring State, the sponsoring State is exposed to responsibility under international law (Seabed Disputes Chamber, Advisory Opinion 2011).

Exploration contracts permit contractors to survey their contract areas, with a view of determining specific areas of interest to exploit, but do not permit the commercial harvesting of the resources. The commercial exploitation of the resources is to be conducted at a later stage, which entails a separate round of application, approval and award of a contract. Given that some exploration contracts have been in existence for approximately two decades, the present focus of the ISA is now shifted towards developing regulations to facilitate exploitation activities. Contrary to the earlier approach with exploration, the ISA is proceeding to develop one set of exploitation regulations that will govern the exploitation of all three types of mineral resources. The current draft exploitation regulations is at an advanced stage and is being considered by the Council (ISA 2019). Simultaneously, the ISA is also taking steps to develop the financial terms for exploitation, to design an appropriate mechanism to distribute the proceeds from activities in the Area in a fair and equitable manner, to study how activities in the Area could affect the economies of developing countries that depend on land-based mining sources, and to develop necessary standards and guidelines that would accompany the final regulations.

One crucial important area within the scope of responsibilities of the ISA is the adoption of necessary measures to ensure the effective protection of the marine environment from the harmful effects of mining activities (Singh and Hunter 2019). In this respect, the ISA is currently in the process of developing regional environmental management plans (or REMPs) to ensure that region-specific considerations are given effect to, and in particular to ensure that spatial and temporal measures are adopted, in order to ensure the effective protection of the marine environment. One particular feature of REMPs is the designation of "areas of particular environmental interest" (or APEIs) within the region. In designated APEIs, no mining activities are expected to take place, at least in the short term (i.e. 5 years), and these areas will be used for monitoring purposes as controlled areas. In the sole existing REMP at the moment, i.e. for the Clarion-Clipperton Zone (CCZ) of the Pacific Ocean, thirteen APEIs have been designated so far with the initial nine in 2021 and an additional four in 2021 (ISA 2021). However, it is pertinent to note that the REMP for the CCZ was only established after a significant amount of exploration contracts had already been awarded in the region, whereby APEIs had to be designated outside those contract areas and predominantly covered areas that were of lesser commercial interests, as opposed to truly representing areas in need of environmental protection (Wedding et al. 2013; Wedding et al. 2015; Dunn et al. 2018; Washburn et al. 2021).

Consequently, while APEIs are rightly accepted as a form of ABMT (Rayfuse 2020) and particularly as an exercise of MSP (McQuaid et al. 2020), their parity with MPAs (at least, when considered in a strict sense) may be open to debate. While its creation may have been guided by science, it is apparent that commercial mining interests would likely prevail over environmental considerations when it comes to the designation of APEIs. If this is the case, REMPs may be better termed as regional mining management plans, as opposed to environmental ones. That said, APEIs could play an important role in relation to short-term conservation efforts as well as for impact monitoring purposes in the CCZ region. Concurrently, efforts are ongoing to develop REMPs for other regions that are subject to increasing mining interests, namely, the Mid-Atlantic Ridge, Northwest Pacific Ocean and Indian Ocean (ISA 2021), and it remains to be seen how effective ABMT measures will be under these instruments. Finally, it is important to note that the Council of the ISA also has the powers to disapprove mining areas where substantial evidence indicates the risk of mining activities in those areas to cause serious harm to the marine environment (Article 162(2)(x) of the LOSC), which would then operate as a partial MPA (i.e. only closed to deep seabed mining activities, since the ISA has a narrow, sectoral mandate). To date, however, the ISA has not designated any of such "no mining" areas or formally considered any proposals to this effect, not least because the ISA is yet to define – in operational terms – what would amount to "serious harm" or the risk thereof. In this respect, it would be interesting to see how measures undertaken through the ISA, especially via REMPs or "no mining" areas, could be harmonized with efforts that could potentially be pursued with respect to MPAs and MSP in areas beyond national jurisdiction under the forthcoming BBNJ Agreement (Christiansen et al. 2022).

The international seabed mining regime represents a unique case study, in which a specific activity has been the subject of intense regulatory focus for decades, especially where no real activity has taken place to date. On the one hand, this may be seen as an application of the precautionary approach, whereby the conduct of an activity is postponed until its environmental implications are properly understood and can be effectively managed. On the other hand, it is apparent that economic and technological realities appear to have had a more controlling effect in hampering the conduct of seabed mining activities as opposed to the hitherto absence of detailed regulations. In any event, it is interesting to note that as the LOSC nears its 40th anniversary in 2022, deep seabed mining still remains an activity that is slated for the future, when and if at all the international community decides to permit such activities to take place, under what conditions, and at what price.

6.5 Conclusion

The LOSC is the bedrock of the law of the sea and ocean governance, providing the legal framework for jurisdiction, rights and responsibilities that binds its State parties and the conduct of their ocean affairs. Indeed, from the perspective of ocean governance, the LOSC functions to promote collective action, and cooperation among states as well as international and regional organizations. As states design their own policies for a national sustainable ocean economy, it would be wise to remember that a true definition of an ocean economy should not only consider the economic activities of ocean-based industries, but also the assets, goods and services of marine ecosystems as natural capital (OECD 2016). In other words, there are limits to growth, and overconsumption, pollution, as well as irresponsible or unsustainable practices need to be urgently arrested. A healthy, resilient and productive ocean is necessary to sustain human well-being, and consequently, all states should be held accountable for pollution and degradation of the marine environment that occurs within their jurisdiction or under their control in areas beyond national jurisdiction.

The three snapshot case studies covered in this chapter have shown that the LOSC plays a central role in shaping how the ocean is governed with a focus on marine environmental protection and resource management in terms of fisheries of sea, marine area-based management tools and the international seabed mineral resources. At the same time, certain limitations became clear when we turned our attention to specific developments in the law of the sea in our case studies, such as that the trajectories have not always converge into an integrated management (which requires more coherent consideration on all the relevant social, economic and environmental interests as well as the spatial and natural interactions at stake). Being a "living instrument" that is capable of being extended to address new uses, interests and concerns (Barrett 2016; Barnes 2016), the LOSC will continue to play an important role in overcoming the environmental threats and problems of the ocean. Indeed, apart from the fields of interest from the past, present and future that are already anticipated by the LOSC (three of which have been considered in this chapter), emerging themes such as marine genetic resources, offshore renewable energy, marine geoengineering, and ocean-climate nexus, among others, will also turn to the LOSC for solutions and innovation.

References

Alder J, Zeller D, Pitcher T, Sumaila R (2002) A Method for evaluating marine protected area management. Coast Manag 30(2):121–131. Viewed 25 July 2021. http://www.tandfonline.com/doi/abs/10.1080/089207502753504661

Andrade GSM, Rhodes JR (2012) Protected areas and local communities: an inevitable partnership toward successful conservation strategies? Ecol Soc 17(4):art14. Viewed 25 July 2021 http://www.ecologyandsociety.org/vol17/iss4/art14/

Araujo FCB, Moita E de AP (2018) The problems of under-inclusion in marine biodiversity conservation: the case of brazilian traditional fishing communities. Asian Bioethics Rev 10:261–278. Viewed 18 June 2019, https://rdcu.be/bchGp

Barnes R (2016) The continuing vitality of UNCLOS. In: Barrett J, Barnes R (eds) Law of the sea: UNCLOS as a living treaty. BIICL, London, pp 459–489

Barrett J (2016) The UN convention on the law of the sea: a "living" treaty? In: Barrett J, Barnes R (eds) Law of the sea: UNCLOS as a living treaty. BIICL, London, pp 3–37

Barros S, Medeiros A, Gomes EB (eds) (2021) Conflitos socioambientais e violações de direitos humanos em comunidades tradicionais pesqueiras no Brasil: relatório 2021, 2nd edn. Conselho Pastoral de Pescadores, Olinda. http://www.cppnacional.org.br/sites/default/files/publicacoes/Relatório%20de%20Conflitos%20Socioambientais%20em%20Comunidades%20Pesqueiras%20-%202021.pdf

Becker-Weinberg V (2017) Preliminary thoughts on marine spatial planning in areas beyond national jurisdiction. Int J Marine Coastal Law 32:570–588

Bennett N, Govan H (2015) Ocean grabbing. Mar Policy 57:61–68

Bennett NJ, Dearden P (2014) From measuring outcomes to providing inputs: governance, management, and local development for more effective marine protected areas. Mar Policy 50:96–110. Viewed 26 July 2021, https://linkinghub.elsevier.com/retrieve/pii/S0308597X14001353

Birnie P, Boyle A, Redgwell C (2009) International law and the environment, 3rd edn. Oxford University Press, New York

CBD (1998) Report of the workshop on the ecosystem approach. (UNEP/CBD/COP/4/Inf.9). CBD. Accessed 13 Feb 2021. https://www.cbd.int/doc/meetings/cop/cop-04/information/cop-04-inf-09-en.pdf

Christiansen S, Durussel C, Guilhon M, Singh P, Unger S (2022) Towards an ecosystem approach to management in areas beyond national jurisdiction: REMPs for deep seabed mining and the proposed BBNJ instrument. Front Mar Sci 9:720146. https://doi.org/10.3389/fmars.2022.720146

Constantinou C, Hadjimichael M (2020) Liquid entitlement: sea, terra, law, commons. Glob Soc. https://doi.org/10.1080/13600826.2020.1810642

Cudennec A (2015) Le cadre européen de la planification de l'espace maritime: illustration des limites de la méthode de l'intégration fonctionnelle. In: Nicolas BOILLET (ed) L'aménagement Du Territoire Maritime Dans Le Contexte de La Politique Maritime Intégrée. A. Pedone, Paris, pp 89–104

Dasgupta S, Fensome A (2018) The ups and downs of marine protected areas: examining the evidence. Mongabay series: conservation effectiveness. 25 January, viewed 19 April 2021. https://news.mongabay.com/2018/01/the-ups-and-downs-of-marine-protected-areas-examining-the-evidence/

Day JDN (2012) Guidelines for applying the IUCN protected area management categories to marine protected areas. IUCN, Gland

De Lucia V (2018) A critical interrogation of the relation between the ecosystem approach and ecosystem services. Review of European, Comparative & International Environmental Law 27(2):104–114. Viewed 1 May 2020, http://doi.wiley.com/10.1111/reel.12227

Diz D (2018) Marine biodiversity: opportunities for global governance and management coherence. In: Salomon M, Markus T (eds) Handbook on marine environment protection: science, impact and sustainable management. Springer, Cham, pp 855–870

Douvere F (2008) The importance of marine spatial planning in advancing ecosystem-based sea use management. Mar Policy 32:762–771

Dunn D, Van Dover C, Etter R, Smith C, Levin L, Morato T, Colaco A, Dale A, Gebruk A, Gjerde K, Halpin P, Howell K, Johnson D, Perez J, Ribeiro M, Stuckas H, Weaver P, Participants S (2018) A strategy for the conservation of biodiversity on mid-ocean ridges from deep-sea mining. Sci Adv 4(7):1–15

Ehler C, Douvere F (2009) *Maritime spatial planning: a step-by-step approach. Toward ecosystem-based management.* ICAM dossier no. 6, manual and guides no 153. Intergovernmental Oceanographic Commission UNESCO IOC. https://unesdoc.unesco.org/ark:/48223/pf0000186559

Ehler C, Zaucha J, Gee K (2019) Maritime/Marine Spatial Planning at the interface of research and practice. In: Maritime spatial planning: past present, future. Springer, Cham, pp 1–21

Engler C (2015) Beyond rhetoric: navigating the conceptual tangle towards effective implementation of the ecosystem approach to oceans management. Environ Rev 23(3):288–320. Viewed 14 May 2020, http://www.nrcresearchpress.com/doi/10.1139/er-2014-0049

European Commission (2020) Towards more sustainable fishing in the EU: state of play and orientations for 2021. In: Communication from the Comission to the European Parliament and the council COM/2020/248. European Commission, Brussels. https://eur-lex.europa.eu/legal-content/EN/TXT/?uri=CELEX%3A52020DC0248

FAO (2003) Fisheries management – 2: the ecosystem approach to fisheries. FAO Technical guidelines for responsible fisheries, FAO. Accessed 5 Jan 2020. http://www.fao.org/3/Y4470E/y4470e00.htm#Contents

FAO (2015) Voluntary Guidelines for Securing Sustainable Small-Scale Fisheries in the context of Food Security and Poverty Eradication, Rome. http://www.fao.org/documents/card/en/c/I4356EN

FAO (2020) The state of the world fisheries and aquaculture 2020. Sustainability in Action, Rome. https://doi.org/10.4060/ca9229en

Flannery W, Ellis G, Ellis G, Flannery W, Nursey-Bray M, van Tatenhove JPM, Kelly C, Coffen-Smout S, Fairgrieve R, Knol M, Jentoft S, Bacon D, O'Hagan AM (2016) Exploring the winners and losers of marine environmental governance/marine spatial planning: cui bono?/"more than fishy business": epistemology, integration and conflict in marine spatial planning/marine spatial planning: power and scaping/surely not all planning is evil?/marine spatial planning: a Canadian perspective/maritime spatial planning – "ad utilitatem omnium"/marine spatial planning: "it is better to be on the train than being hit by it"/reflections from the perspective of recreational anglers and boats for hire/maritime spatial planning and marine renewable energy. Plan Theory Pract 17(1):121–151. Viewed 22 March 2021, http://www.tandfonline.com/doi/full/10.1080/14649357.2015.1131482

Flannery W, Healy N, Luna M (2018) Exclusion and non-participation in marine spatial planning. Mar Policy 88:32–40. Viewed 22 March 2021, https://linkinghub.elsevier.com/retrieve/pii/S0308597X1730324X

Frazão Santos C, Domingos T, Ferreira MA, Orbach M, Andrade F (2014) How sustainable is sustainable marine spatial planning? Part I-linking the concepts. Mar Policy 49:59–65. https://doi.org/10.1016/j.marpol.2014.04.004

Garcia SM, Rice J, Charles A (2014) Governance of marine fisheries and biodiversity conservation: a history. In: Governance of marine fisheries and biodiversity conservation: interaction and coevolution. Willey, Oxford, pp 3–17

Hadjimichael M (2018) A call for a blue degrowth: unravelling the European Union's fisheries and maritime policies. Mar Policy 94(August):158–164. https://doi.org/10.1016/j.marpol.2018.05.007

Harrison J (2019) Key challenges relating to the governance of regional fisheries. In: Caddell R, Molenaar EJ (eds) Strengthening international fisheries law in an era of changing oceans. Hart Publishing, Oxford

IISD (2019) Summary of the third session of the intergovernmental conference (IGC) on the conservation and sustainable use of marine biodiversity of areas beyond national jurisdiction: 19–30 August 2019, IISD reporting services, 25, n. 218. Earth Negotiations Bulletin, IISD. Accessed 20 Dec 2019. http://enb.iisd.org/oceans/bbnj/igc3/

Intergovernmental conference on marine biodiversity of areas beyond national jurisdiction (2019) Advance, unedited version. UN. Viewed 12 April 2019, https://www.un.org/bbnj/sites/www. un.org.bbnj/files/revised_draft_text_a.conf_.232.2020.11_advance_unedited_version.pdf

IPBES (2019) Global assessment report on biodiversity and ecosystem services of the Intergovernmental Science-Policy Platform on Biodiversity and Ecosystem Services. Zenodo. Accessed 28 July 2021. https://zenodo.org/record/3831673

ISA (2019) Draft regulations on exploitation of mineral resources in the Area. ISBA/25/C/WP.1. https://isa.org.jm/files/files/documents/isba_25_c_wp1-e_0.pdf

ISA (2021) Environmental management plans. https://isa.org.jm/minerals/environmental-management-plan-clarion-clipperton-zone

Koh T (1982) A Constitution for the Oceans, Remarks by Ambassador Koh of Singapore, President of the Third United Nations Conference on the Law of the Sea. https://www.un.org/Depts/los/convention_agreements/texts/koh_english.pdf

Lado EP (2016) The common fisheries policy: the quest for sustainability. Wiley Blackwell, Oxford

Laffoley D, Baxter JM, Day JC, Wenzel L, Bueno P, Zischka K, Sheppard C (2018) Marine protected areas. In: World seas: an environmental evaluation. Academic, pp 549–569

Leroy A, Morin M (2018) Innovation in the decision-making process of the RFMOs. Mar Policy 97:156–162. Viewed 19 April 2021. https://linkinghub.elsevier.com/retrieve/pii/S0308597X17308369

Lloret J et al (2018) Small-scale coastal fisheries in European seas are not what they were: ecological, social and economic changes. Mar Policy 98:176–186. https://doi.org/10.1016/j.marpol.2016.11.007

Lodge M (2013) Common heritage of mankind. In: Freestone D (ed) The 1982 law of the sea convention at 30: successes, challenges and new agendas. Brill, Leiden/Boston, pp 59–68

Markus T (2018) Challenges and foundations of sustainable ocean governance. In: Salomon M, Markus T (eds) Handbook on marine environment protection: science, impact and sustainable management. Springer, Cham, pp 545–562

Markus T, Markus G (2021) The economics of the law of the sea. In: Markus T, Markus G (eds) Oxford research encyclopedia of environmental science. Oxford University Press. https://doi.org/10.1093/acrefore/9780199389414.013.423

McQuaid KA, Attrill MJ, Clark MR, Cobley A, Glover AG, Smith CR, Howell KL (2020) Using habitat classification to assess representativity of a protected area network in a large, data-poor area targeted for deep-sea mining. Front Mar Sci 7:1–21

Molenaar EJ, Caddell R (2019) International fisheries law: achievements, limitations and challenges. In: Caddell R, Molenaar EJ (eds) Strengthening international fisheries law in an era of changing oceans. Hart Publishing, Oxford

Morgan C (2011) Manganese nodules, again? Oceans'11: Mts/Ieee Kona, pp 1–6. https://doi.org/10.23919/OCEANS.2011.6106912

OECD (2016) Sustainable ocean for all: harnessing the benefits of sustainable ocean economies for developing countries. https://www.oecd-ilibrary.org/sites/bede6513-en/index.html?itemId=/content/publication/bede6513-en

PCA (2015) Chagos Marine Protected Area Arbitration. Viewed 18 June 2019. https://pca-cpa.org/en/cases/11/

Queffelec B, Bonnin M, Ferreira B, Bertrand S, Teles Da Silva S, Diouf F, Trouillet B, Cudennec A, Brunel A, Billant O, Toonen H (2021) Marine spatial planning and the risk of ocean grabbing in the tropical Atlantic. ICES J Mar Sci:1–13. Viewed 22 March 2021. https://academic.oup.com/icesjms/advance-article/doi/10.1093/icesjms/fsab006/6154827

Rayfuse R (2020) Crossing the sectoral divide: modern environmental law tools for addressing conflicting uses on the seabed. In: Banet C (ed) Law of the seabed. Brill, Leiden/Boston, pp 527–552

Rife AN, Erisman B, Sanchez A, Aburto-Oropeza O (2013) When good intentions are not enough... insights on networks of "paper park" marine protected areas: concerns regarding marine "paper parks". Conserv Lett 6(3):200–212. Viewed 25 July 2021. https://onlinelibrary.wiley.com/doi/10.1111/j.1755-263X.2012.00303.x

Rothwell D, Stephens T (2016) The international law of the sea, 2nd edn. Hart Publishing, Oxford

Sakai Y, Nobuyuki Y, Sumaila UR (2019) Fishery subsidies: the interaction between science and policy. Fish Sci 85(3):439–447. https://doi.org/10.1007/s12562-019-01306-2

Sands P, Peel J, Fabra A, Mackenzie R (2018) Principles of international environmental law, 4th edn. Cambridge University Press, Cambridge

Schubert M (2018) Marine spatial planning. In: Salomon M, Markus T (eds) Handbook on marine environment protection: science, impact and sustainable management. Springer, Cham, pp 1013–1024

Scovazzi T (2011) The conservation and sustainable use of marine biodiversity, including genetic resources, in areas beyond national jurisdiction: a legal perspective. Panel Discussion, 12th meeting of the UN open-ended informal consultative process on Oceans and the Law of the Sea, New York. Viewed 20 December 2019, http://www.un.org/depts/los/index.htm

Seabed Disputes Chamber Advisory Opinion (2011) Responsibilities and obligations of states sponsoring persons and entities with respect to activities in the area. Case No. 17 of ITLOS. https://www.itlos.org/fileadmin/itlos/documents/cases/case_no_17/17_adv_op_010211_en.pdf

Secretariat of the Convention on Biological Diversity (2004a) CBD guidelines: the ecosystem approach. Secretariat of the Convention on Biological Diversity, Montreal. https://www.cbd.int/doc/publications/ea-text-en.pdf

Secretariat of the Convention on Biological Diversity (2004b) Technical advice on the establishment and management of a national system of marine and coastal protected area. Technical series no.13. CBD, Montreal

Serdy A (2018) The international legal framework for conservation and management of fisheries and marine mammals. In: Salomon M, Markus T (eds) Handbook on marine environment protection: science, impact and sustainable management. Springer, Cham, pp 637–657

Sharma C, Rajagopalan R (2017) Aires marines protégées et droits fonciers des communautés de pêcheurs. Entre terre et mer : quel avenir pour la pêche ? 24(1):199–218

Singh P, Ort M (2019) Law and policy dimensions of ocean governance. In: Jungblut S, Liebich V, Bode-Dalby M (eds) YOUMARES 9 – the oceans: our research, our future. Proceedings of the 2018 conference for young marine researcher in Oldenburg, Germany. Springer, Cham, pp 45–56. https://doi.org/10.1007/978-3-030-20389-4

Singh P, Hunter J (2019) Protection of the marine environment: the international and National Regulation of deep seabed mining activities. In: Sharma R (ed) Environmental issues of deep-sea mining: impacts, consequences and policy perspectives. Springer, Cham, pp 471–503

Takei Y (2015) A sketch of the concept of ocean governance and its relationship with the law of the sea. In: Ryngaert C, Molenaar E, Nouwen S (eds) What's wrong with international law. Brill, Leiden/Boston, pp 48–62

Tanaka Y (2011) Protection of community interests in international law: the case of the law of the sea. In: Bogdandy, Wolfrum (eds) Max Planck yearbook of United Nations law, vol 15, pp 329–375

The Geological Society (2013) Treasures of the abyss. https://www.geolsoc.org.uk/Geoscientist/Archive/May-2013/Treasures-from-the-abyss

Treves T (2008) United Nations Convention on the Law of the Sea. https://legal.un.org/avl/pdf/ha/uncls/uncls_e.pdf

UN DOALOS (1998) The United Nations Convention on the Law of the Sea. https://www.un.org/depts/los/convention_agreements/convention_historical_perspective.htm

UN DOALOS (2016) Agreement relating to the implementation of Part XI of the United Nations Convention on the Law of the Sea of 10 December 1982: Overview. https://www.un.org/depts/los/convention_agreements/convention_overview_part_xi.htm

UNEP (2018) Applying marine and coastal area-based management approaches to achieve multiple sustainable development goal targets: summary for policy makers. UN Environmental regional seas reports and studies, 206, UNEP. Accessed 22 Nov 2019. https://www.unep-wcmc.org/resources-and-data/ocean-sdgs

United Nations General Assembly (1967) Twenty-second session: official records. First Committee, 1515th Meeting. https://www.un.org/depts/los/convention_agreements/texts/pardo_ga1967.pdf

United Nations General Assembly, Resolution 2749 (XXV) (1970a) https://undocs.org/en/A/RES/2749(XXV)

United Nations General Assembly, Resolution 2750 (XXV) (1970b) https://undocs.org/en/A/RES/2750(XXV)

Unterweger I (2015) International law on tuna fisheries management: is the Western and Central Pacific Fisheries Commision ready for the challenge? Nomos. https://doi.org/10.5771/9783845263915

Washburn T, Jones D, Wei C-L, Smith C (2021) Environmental heterogeneity throughout the clarion-Clipperton zone and the potential Representativity of the APEI network. Front Mar Sci 8

Wedding L, Friedlander A, Kittinger J, Watling L, Gaines S, Bennett M, Smith C (2013) From principles to practice: a spatial approach to systematic conservation planning in the deep sea. Proc R Soc B Biol Sci 280

Wedding L, Reiter S, Smith C, Gjerde Ķ, Kittinger J, Friedlander A, Gaines S, Clark M, Thurnherr A, Hardy S, Crowder L (2015) Managing mining of the deep seabed: contracts are being granted, but protections are lagging. Science 349(6244):144–145

White M (1982) The common heritage of mankind: an assessment. Case Western Reserve Journal of International Law 14(3):509–542

Wolfrum R (1983) The principle of the common heritage of mankind. https://www.zaoerv.de/43_1983/43_1983_2_a_312_337.pdf

WWF (2020) Living planet report 2020 – bending the curve of biodiversity loss. Almond REA, Grooten M, Petersen T (eds). Gland. https://f.hubspotusercontent20.net/hubfs/4783129/LPR/PDFs/ENGLISH-FULL.pdf

WTO (1999) On the environmental impact of fisheries subsidies: a short report by the Icelandic Ministry of Fisheries. WTO doc WT/CTE/W/111. World Trade Organization, Geneva

Young MA (2011) Trading fish, saving fish: the interaction between regimes in international law. Cambridge University Press, Cambridge. https://doi.org/10.1017/CBO9780511974526

Zacharias M (2014) Marine policy: an introduction to governance and international law of the oceans. Routledge, London. https://doi.org/10.4324/9780203095256

Chapter 7
The Diverse Legal and Regulatory Framework for Marine Sustainability Policy in the North Atlantic – Horrendograms as Tools to Assist Circumnavigating Through a Sea of Different Maritime Policies

Helena Calado, Marta Vergílio, Fabiana Moniz, Henriette Grimmel, Md. Mostafa Monwar, and Eva A. Papaioannou

Abstract Although considerable progress has been made in the management and planning of the marine environment, important gaps still exist in streamlining policies across governance levels, maritime sectors, and between different countries. This can hinder effective Maritime Spatial Planning (MSP) and prevent harmonious cross-sectoral cooperation, and importantly, cross-border or trans-boundary collaboration. These may in turn have serious implications for overall ocean governance and ultimately, marine sustainability. The North Atlantic presents an ideal

H. Calado (✉)
UAc/FCT/MARE - University of the Azores/Faculty of Science and Technology and Marine Environmental Science, Ponta Delgada, Portugal
e-mail: helena.mg.calado@uac.pt

M. Vergílio
Trisolaris Advanced Technologies, Lda., Ponta Delgada, Portugal

F. Moniz
FGF/UAc/FCT, Fundação Gaspar Frutuoso, University of the Azores/Faculty of Sciences and Technology, Ponta Delgada, Portugal

H. Grimmel
Independent Researcher, Zurich, Switzerland

Md. M. Monwar
Institute of Marine Sciences, University of Chittagong, Chittagong, Bangladesh

Australian National Centre for Ocean Resources and Security (ANCORS), University of Wollongong, Wollongong, Australia

E. A. Papaioannou
Independent Researcher, Athens, Greece

Present Affiliation: GEOMAR – Helmholtz Centre for Ocean Research Kiel, Kiel, Germany

© German Institute of Development and Sustainability (IDOS) 2023
S. Partelow et al. (eds.), *Ocean Governance*, MARE Publication Series 25,
https://doi.org/10.1007/978-3-031-20740-2_7

137

case-study region for reviewing these issues: North Atlantic countries have different governance structures, and as such, different approaches to marine policy. Therefore, for an effective marine management, cross-sectoral and cross-border MSP in the region, there is a need to review marine and maritime policies in order to identify differences and commonalities among countries. This chapter reviews major policies for the marine environment in the North Atlantic and assesses where differences between countries exist and at which governance level they are being created. Key research questions include: (i) Are there significant differences in marine policy between North Atlantic countries? Moreover, are there any substantial geographical/political differences? (ii) Are there differences in implementation of key policies? Such an analysis requires a sound framework for comparison among countries. To that end, the use of "horrendograms", a tool increasingly being used by the marine research and planning community to assess such issues, is adopted. Results indicate that key differences between countries are created primarily at a national level of marine governance. Although differences between countries exist, overall strategic targets are similar. For instance, whilst the political systems of certain North Atlantic countries may differ substantially, key objectives for major sectors, such as fisheries and conservation, are similar – even when such objectives are implemented at different levels. Findings from the study can enable targeted policy intervention and, as such, assist the development of future outlooks of ocean governance in the region. Results can also aid the development of future visions and scenarios for MSP in the Atlantic region.

Keywords Environmental legislation · Horrendogram · Maritime spatial planning (MSP) · North Atlantic · Ocean/marine governance · Ocean/marine policy

7.1   Introduction

7.1.1   The Need for Effective Marine Management and Governance

Maritime users and activities have pronounced impacts in the marine environment and their control is a fundamental aspect of maritime policy (Boyes and Elliott 2016). It is progressively being recognised that major global challenges such as overfishing, pollution, biodiversity and habitat degradation and loss, and the adverse impacts of climate change on the world's oceans, are frequently the result of ineffective marine and ocean governance (Crowder et al. 2006). Although considerable progress is lately taking place in novel, integrated approaches to marine and ocean management, obstacles still remain: marine and ocean management have historically focused on single-sector approaches resulting in numerous agencies having competencies for different issues. As such, institutions and organisations frequently have varied and non-comprehensive or limited mandates (Crowder et al. 2006; Durussel et al. 2019). Moreover, in the marine environment, political and jurisdictional borders and delineations seldom correspond to the limits of maritime activities and ecosystems. The previous may result in turn in considerable differences in the national environmental governance systems of countries bordering the same marine region (Kern and Gilek 2015; Carval and Jarno 2019). Different policy timescales between authorities, countries, institutions, and organisations give rise in turn to temporal mismatches between environmental problems and human institutions (Crowder et al. 2006). Most importantly, marine governance systems are largely shaped by environmental problems and institutions, and this situation may frequently result in different outcomes, despite common objectives (Kern and Gilek 2015).

There exists rising consensus that major challenges facing the marine environment are complex and multifaceted, beyond the capacity of a single sector or country to resolve (UNDP 2015; Zaucha 2014). To that end, cross-sectoral and cross-border cooperation, namely the communication, coordination or planning across spatial jurisdictions (regional, national, sub-national), encompassing both *vertical* (collaboration among different levels of government) and *horizontal* (i.e. nation to nation) dimensions of governance (Carneiro et al. 2017), is progressively being recognised as fundamental for the sound governance of the marine environment (Boyes and Elliott 2016; Van Tatenhove 2017; Morf et al. 2019).

However, marine governance systems' architectures remain largely fragmented across different sectors and governance levels combining national, regional and international governance (Gold et al. 2011; Kern and Gilek 2015). As a result, key policies relating to the marine environment are still lacking cross-sectoral and cross-border integration and coordination in many regions. Also, although international legal frameworks for dealing with some of the most pressing threats to the marine environment have emerged, additional effort of capacity-building is still required to implement these frameworks for many countries (UN 2017), including EU countries. Harmonising maritime policy across countries and ensuring maritime plans are coherent and coordinated constitute key objectives of major policy frameworks [e.g. Article 11 of the EU Maritime Spatial Planning (MSP) Directive text; European Parliament and Council 2014]. There thus exists a vital need for a detailed assessment of the different policy frameworks for the marine environment, to ensure a sound understanding of such frameworks, which in turn is crucial for the effective coordination and ultimately cooperation across different sectors, governance levels and countries (Carneiro et al. 2017; Rudd et al. 2018).

7.1.2   The North Atlantic Marine Region: Key Challenges and Opportunities

The need of a thorough assessment and comparison of marine policy frameworks is especially evident in the North Atlantic. Such an assessment and comparison would enable realising transboundary planning objectives, as also dictated by major policy provisions in the area. In the North Atlantic region, key policies specify the need for: (i) cooperation on transboundary issues; (ii) mechanisms for transnational consultations on marine spatial plans and issues arising from them; (iii) region-specific, tailor-made approaches to MSP for supporting the Ecosystem Based Approach (EBA); (iv) exchange of best practices and experiences with regard to MSP. For instance, the North-East Atlantic Environment Strategy of the OSPAR Commission for the Protection of the Marine Environment of the North-East Atlantic states that delivering these objectives requires consistency in assessment and monitoring methodologies and mutual compatibility of environmental targets (OSPAR 2010). To that end, policy harmonisation is set forward by key policy: OSPAR (2019a) for instance, stresses the need for contracting parties to harmonise policies and strategies relating with the prevention of maritime pollution.

There lately has been considerable progress in the review and assessment of various aspects relating to the policy and governance framework of the North Atlantic marine environment. Past studies include detailed reviews of the marine policy framework of individual countries (Boyes and Elliott 2014, 2016). Studies conducted within the framework of the Atlantic Ocean Research Alliance (AORA) (under the auspices of the Galway Statement on Atlantic Ocean cooperation), reviewed the role mandates play with respect to the implementation of Ecosystem-based Management (EBM), within and across jurisdictions in Canada, the US and the EU (Rudd et al. 2018). The CALAMAR project [Cooperation across the Atlantic for Marine Governance Integration, 2010–2011], developed a series of policy recommendations for improving integration of maritime policies and promoting transatlantic cooperation (Gold et al. 2011; Speer et al. 2011). Past projects (e.g. SIMNORAT – Supporting in the Northern European Atlantic) assessed a plethora of planning documents and concluded that heterogeneous spatial planning organisations are present in the region (Carval and Jarno 2019). Other projects in the wider region (Strong High Seas) also stressed the varying and non-comprehensive or limited mandates of authorities with reference to key maritime issues, notably Biodiversity Beyond National Jurisdiction (BBNJ) (Durussel et al. 2019).

The previous have generated considerable knowledge and a wealth of relevant information. However, a detailed assessment of the diverse and disparate marine policy and governance frameworks, encompassing multiple marine activities and maritime sectors, and the subsequent comparison between countries are largely missing. For the North Atlantic, such an assessment can help overcoming the following inherent difficulties: (i) different systems of marine policy (Gold et al. 2011; Rudd et al. 2018; Durussel et al. 2019), including a heterogeneous spatial planning organisation (Carval and Jarno 2019), which present challenges for a comparison between countries (especially with US and Canada); and (ii) varying degrees of maturity and progress with respect to implementation, even in the case of EU Member States (Marques et al. 2019). This requires a review of the different marine policy systems and a systematic assessment of commonalities and differences, especially at a national level.

7.1.3   Aims and Objectives: A Framework for Policy Comparison in the North Atlantic

The present study reviews the marine policy framework in the North Atlantic, while examining the compatibility of marine policies across different sectors and governance levels (international, regional, national). A key objective is to determine how national circumstances influence ocean governance, linking the implementation of regional initiatives and agreements of ocean management (Calado et al. 2018). Research builds on the expertise generated by past studies conducted in the wider North Atlantic region. As such, results from the analysis should be seen as complementary, in an "open dialogue" with respective findings from past (Boyes and Elliott 2014, 2016; Rudd et al. 2018) and ongoing studies.

In the present context, governance is understood as the sum of those policies, politics, administration and legislation pertaining to the marine environment, spanning from the global down to the local level of governance (Boyes and Elliott 2014, 2016). Regarding the mandate of competent institutions, this typically involves: "*an authorization to act in a particular way on a public issue*" which may include "*legally binding obligations as well as so-called soft law agreements, principles and declarations that are not necessarily legally binding*" (Rudd et al. 2018).

Key research questions include:

(i)   How are marine- and maritime- related topics treated within the policy frameworks in the North Atlantic? Are there important differences between/within countries in the North Atlantic? Are there any substantial geographical differences (e.g., EU vs non-EU)?

(ii)   Are there significant gaps in the implementation of key marine policies?

Such an analysis requires a methodical and systematic approach with attention to detail. For that matter, the use of "horrendograms", a tool increasingly being used by the marine research and planning community (Boyes and Elliott 2014, 2016) is adopted. Horrendograms constitute in essence comparisons between the organograms of the policy frameworks of countries under comparison. Main advantages include a methodical way of depicting relevant information, streamlining across different legislations, and importantly, allow for establishing where differences across policy frameworks are being created, and the essence of these differences. Meanwhile, such an approach enables a multi-sectoral assessment, not focusing on single sectors and themes, while enabling comparison between multiple countries. Such a framework can in turn disclose important information on the governance level where differences and commonalities exist, which in turn can enable targeted intervention, streamlining of relevant policies and ultimately promoting transboundary coordination of relevant activities.

7.2 Materials and Methods

7.2.1 The North Atlantic Marine Region

The North Atlantic marine region includes major administrative and jurisdictional units, including FAO Major Fishing Areas 21 (NW Atlantic) and 27 (NE Atlantic) (FAO 2015); OSPAR Regions V, III, IV (i.e. Wider Atlantic; Celtic Seas; Bay of Biscay and Iberian Coast respectively, OSPAR 2019b) and ICES Statistical Areas Xa, Xb; and XII (ICES 2019) (Fig. 7.1). The area borders some of the world's most indus- trialised nations and is home to a multitude of maritime uses and activities (Speer et al. 2011). Meanwhile, the region contains a wealth of natural resources and areas of high ecological diversity. Vital actions are required in dealing with the pronounced impacts of climate change in the region and their implications (Gold et al. 2011).

7.2.2 Comparing Marine Policy Across North Atlantic Countries

The present analysis is structured in three main phases (Fig. 7.2):

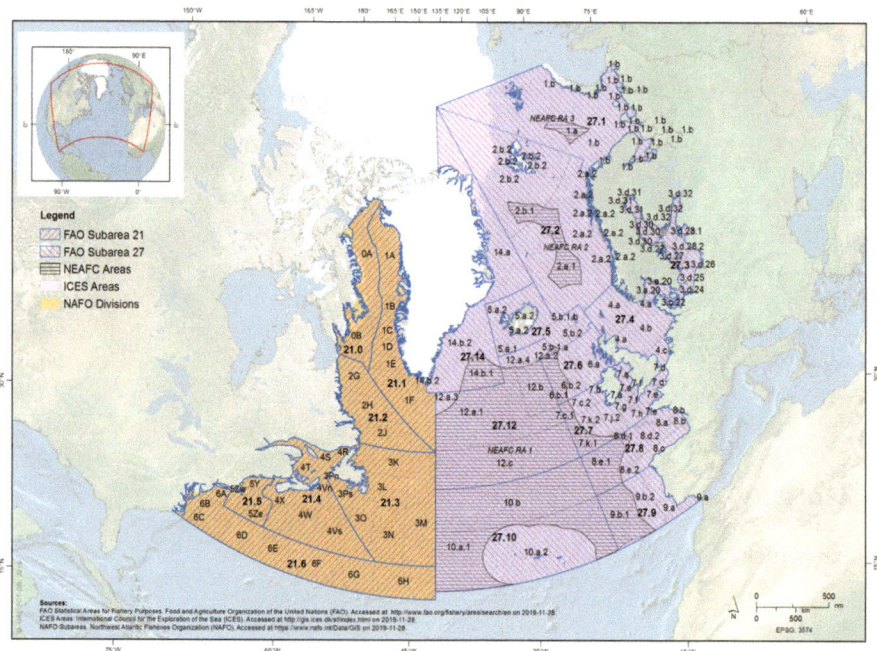

Fig. 7.1 Study area. (Source: Authors)

Fig. 7.2 Flowchart of methodological approach. (Source: Authors)

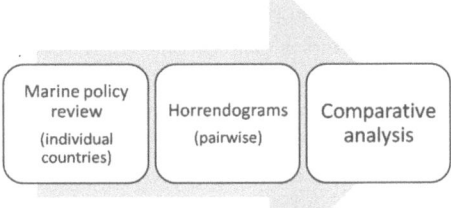

1. Review of international, regional and national legislation pertaining and influencing either directly or indirectly the marine environment in North Atlantic countries.
2. Development of an analytical framework that enables consistency in the comparison across countries: construction of horrendograms (after Boyes and Elliott 2014, 2016).
3. Comparison of countries' policies using horrendograms: assessment of the complexity of marine policies across different sectors and governance levels.

7.2.2.1 Marine Policy Review

An assessment of major national, regional and international legislation relating to the marine environment of the North Atlantic, management and governance of maritime activities and sectors is performed. Governance data are systematically gathered, collated and reviewed. Relevant data is obtained from major international (UN), regional/trans-national (OSPAR, EU) and national institutions and organisations (e.g. US National Oceanic and Atmospheric Administration, NOAA).

Key criteria for the selection of data include direct reference to the management, planning and governance of the marine environment, marine and maritime activities, users and sectors. Results from past and ongoing projects in the study area and scientific literature pertaining to the scope of the study are also addressed. Information is categorised to correspond to the respective marine governance levels, enabling the subsequent integration of information within the horrendograms framework. Collected data is subsequently validated by experts/officials at each country (e.g., practitioners at governmental agencies of environment and sea affairs).

7.2.2.2 Horrendograms

The horrendogram framework developed by Boyes and Elliott (2014, 2016) provides a suitable framework for analysis. Horrendograms summarize the marine policy framework for individual countries, streamlining and mainstreaming relevant information to enable the comparative analysis of regulatory frameworks and

ultimately establishing the major differences that exist between compared countries. Similar approaches have also been used in the framework of past projects in
the wider region (e.g. Strong High Seas project, Durussel et al. 2019).

The horrendogram for the UK developed by Boyes and Elliott (2014) is the
frame of reference for comparison between countries. The UK, has a robust tradition in MSP, while it being a unitary and island state, it has also made considerable
efforts to address the complex issue of streamlining legislation across different sub-
national levels (known as "devolved administrations" in an UK context). Importantly,
the UK has been instrumental in the development of EU environmental policy
(Boyes and Elliott 2016), and as such, enables comparison with non-EU Member
States. A pairwise horrendogram is also developed for the comparison of the US to
the Canadian marine policy framework.

For the horrendograms development, policies pertaining to the marine environment are placed in co-centric circles, following a clockwise pattern, and structured
along a vertical governance level. The centremost circle corresponds to international policy objectives and targets (e.g. UN conventions, laws and/or commitments) (Fig. 7.3). The following circle, i.e. the second circle from the centre,
represents the directives, policies or strategies of a regional (North Atlantic, such as
OSPAR) or trans-national level (e.g. EU) (Calado et al. 2018). As regulations usually have a stronger influence on policy than guidelines or recommendations

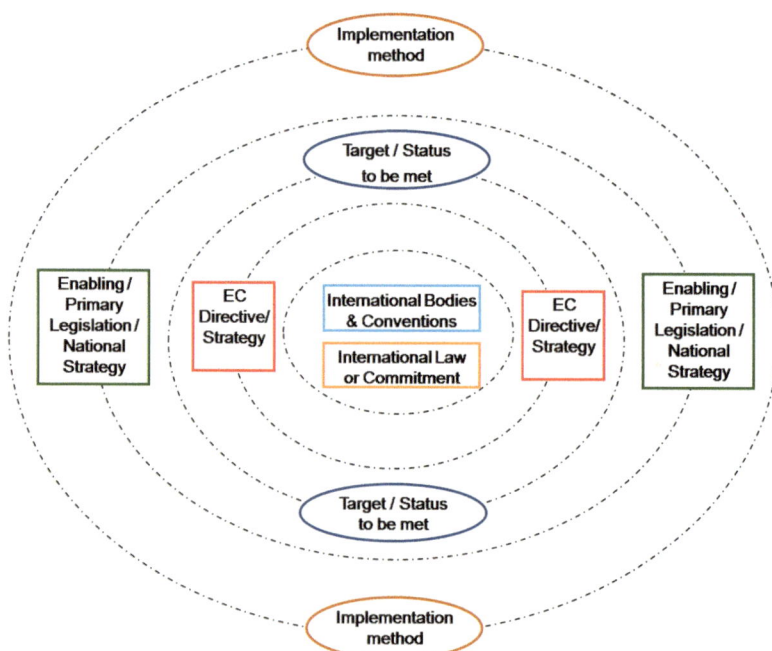

Fig. 7.3 Conceptual diagram of a horrendogram, describing the different categories across the
circles. (Diagram adapted after: Boyes and Elliott 2014)

(frequently termed "soft laws", Rudd et al. 2018; Durussel et al. 2019), they are foregrounded in the horrendogram framework. Different colours are used to represent differences in the approaches between compared countries and enable comparison: highlighted boxes in green denote a given country's unique legislation or policy and highlighted boxes in yellow an approach different to the one followed by the UK.

Policies are grouped in the following key categories, to correspond to major maritime users, activities, and sectors requiring particular attention in the context of cross-border and/or transboundary cooperation:

- Fisheries and aquaculture
- Food security
- Flood and risk assessment
- MSP
- Nature conservation
- Maritime cultural heritage
- Strategic Environmental Assessment (SEA)
- Environmental Impact Assessment (EIA)
- Shipping
- Ocean management
- Water quality environmental standards

This grouping and comparison enable the review and assessment of key policies that affect, either directly or indirectly, the management and governance of the marine environment.

7.2.2.3   Limitations of the Analysis

A comparative assessment of maritime and marine policies across different countries has inherent limitations. The present study seeks to identify major differences in the marine policy frameworks of North Atlantic countries, and it was accepted from the beginning that it would not entail an exhaustive comparison of all regulations, laws, directives, recommendations and other policies. Greenland has been excluded from the scope of the present study as governance data required for the analysis are scarce to locate and assess. Major political developments in the region are currently ongoing (Table 7.1) and their implications for key marine activities and maritime sectors are still unclear, and have not been incorporated in the framework of the present analysis. These include most notably BREXIT; after BREXIT and the end of the transition period, no major changes are expected to occur in the short- or medium- term in the UK's legal framework for the marine environment: Fundamental EU Directives are integrated as UK domestic law, while close cooperation in key sectors (e.g. fisheries) with the EU will continue. Moreover, consolidate impacts of change are time-consuming and anticipated to result in an enlarged time-elapse.

Table 7.1 Adaptations from the initial UK horrendogram of Boyes and Elliott (2014) – Additional international policies assessed for the purpose of the analysis

Policy	Canada	France	Ireland	Iceland	Portugal	Spain	UK	US	Notes
CBD Cartagena protocol	–	X	x	–	x	x	X	–	Included
CBD Nagoya protocol	–	X	–	–	x	x	X	–	Assessed, excluded[f]
HELCOM convention for the protection of the Baltic[a]	–	–	–	–	–	–	X	–	Assessed, excluded
UNEP and NOAA Honolulu strategy[b]	–	X	x	x	x	x	X	x	Included
UNESCO[c]	x	X	x	x	x	x	X	?	Included
UN FCCC – Paris agreement	x	X	x	x	x	x	X	x	Included
World network of biosphere reserves (WNBR); UN man and the biosphere (MAB) Programme[d]	x	X	x	–	x	x	X	x	Included
UN regional seas Programme (RSP) – Protection of the Arctic marine environment[e]	x	–	–	x	–	–	–	x	Included

x: Country member of respective legislation/policy; –: Country not member;?: Unclear status
[a]Other than the UK, no other N. Atlantic countries are members of HELCOM, thus the Convention was not included in the horrendogram comparison
[b]No evidence of the Honolulu Strategy influencing relevant Canadian national policy
[c]Following the US recently rejoining the UN Framework Convention on Climate Change (UNFCCC) Paris Agreement (2/2021), it has been speculated that it could also pledge to rejoin UNESCO, after leaving the Organization in 2017
[d]There are currently no designated Biosphere Reserves contained within the global WNBR network for Iceland (UNESCO 2018a)
[e]Although the Arctic Seas Regional Programme is not in the N. Atlantic it was included as it affects the MSP policy of three major countries in the area; it was hypothesized that the comparison within the horrendogram would disclose important information on the differences in the MSP process for those countries and the rest
[f]Assessed in the case of the UK-Portuguese pairwise comparison (c/f section 3.2.4)

7.3 Results and Discussion

7.3.1 Marine Policy Review

The UK horrendogram developed by Boyes and Elliott (2014) constitutes the basis for the analysis, with the present study building and extending on this seminal work. Adaptations to the original UK horrendogram result from the inclusion of recent (i.e. 2014–2019) policy developments in the field, the assessment of the legislative frameworks of other North Atlantic countries, and the subsequent addition and streamlining of relevant regulations within the horrendogram framework. Table 7.1 summarises key international policies that were assessed for the purpose of the present analysis and resulting adaptations to the original UK horrendogram.

The assessment of the policy frameworks of the North Atlantic countries enabled establishing (i) core regulations pertaining to the governance and management of the marine environment and key maritime sectors; and (ii) various instruments for the implementation of relevant policy. Amongst North Atlantic countries, Canada, Iceland, Ireland, the UK, and the US use Acts and Plans for regulating their marine environment and maritime sectors; Portugal and Spain use binding tools such as Law Decrees for governing marine resources and activities, while France utilises a set of different instruments for its MSP approach.

Canada has adopted an Ocean Act and individual Action Plans, but has no dedicated marine planning legislation. Iceland possesses an Ocean Policy and has not developed a dedicated integrated marine management framework. The marine policy framework in the US is established through numerous Acts, spanning the entire breadth of the country's Federal (>3 nm) and State (<3 nm) waters.

In France, marine policy is primarily comprised of Strategic Frameworks and Action Plans relevant to the marine environment, and the transposition of the EU MSP Directive to national law is ongoing (DIR 2017). In Portugal, the main policy framework for planning and management of the marine environment is established by the national law of Planning and Management of Maritime Space, adopted in 2014 prior to the EU MSP Directive, and subsequently entered into force with Law Decree 39/2015. In Spain, Royal Decree 363/2017 (Ministerio de Agricultura y Pesca, Alimentación y Medio Ambiente 2017) constitutes the national legislative framework for MSP, transposing the EU MSP Directive (European Parliament and Council 2014) into national law. In the UK, marine policy comprises three main themes: MSP, Marine Strategy and the Marine and Coastal Access Act (MCAA) (2009). The latter comprises the fundamental Act for marine policy, specifying regulations pertaining to fisheries, marine conservation, and setting the licensing and governance framework, further organising the administrative processes and competent authorities.

7.3.2 Horrendograms for the North Atlantic Countries

This section presents results from the comparison of the horrendograms for selected North Atlantic countries, summarising the major differences in their legislative frameworks for the marine environment. Horrendograms depicting the comparisons between UK and EU countries have been excluded on the grounds that key differences are mostly created at a national level. Figures 7.4, 7.5, 7.6, 7.7, 7.8 and 7.9 present the pairwise comparison between the frameworks of the UK and Iceland; the US and Canada; and the US and the UK. The respective horrendograms for

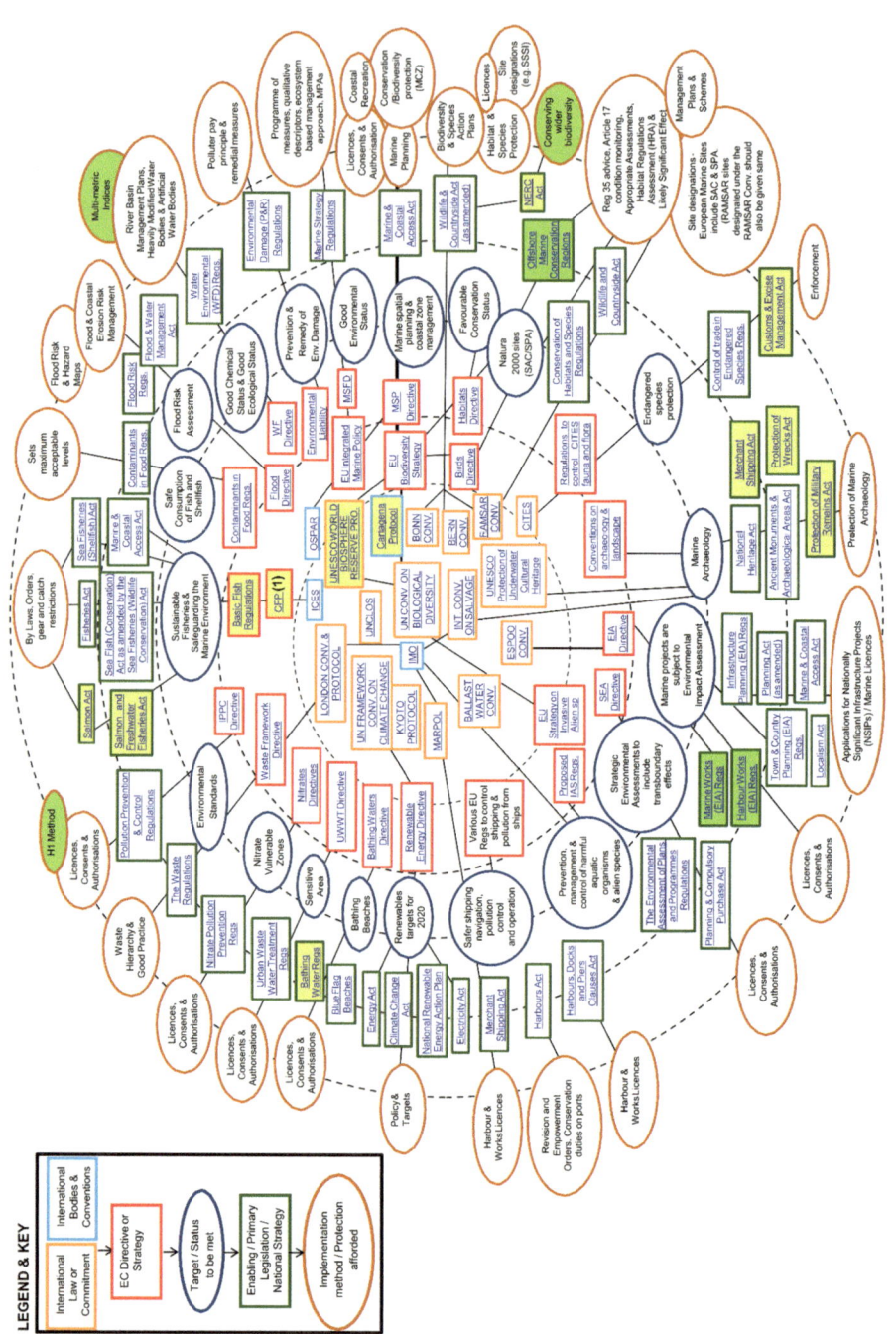

Fig. 7.4 **Horrendogram of the UK, when compared to the horrendogram of Iceland**. Showing international, European and UK legislation for the marine environment (after Boyes and Elliott 2014). Highlighted boxes in yellow show common approaches between the UK and Iceland and highlighted ones in green show legislative policies unique to the UK within the comparison at hand

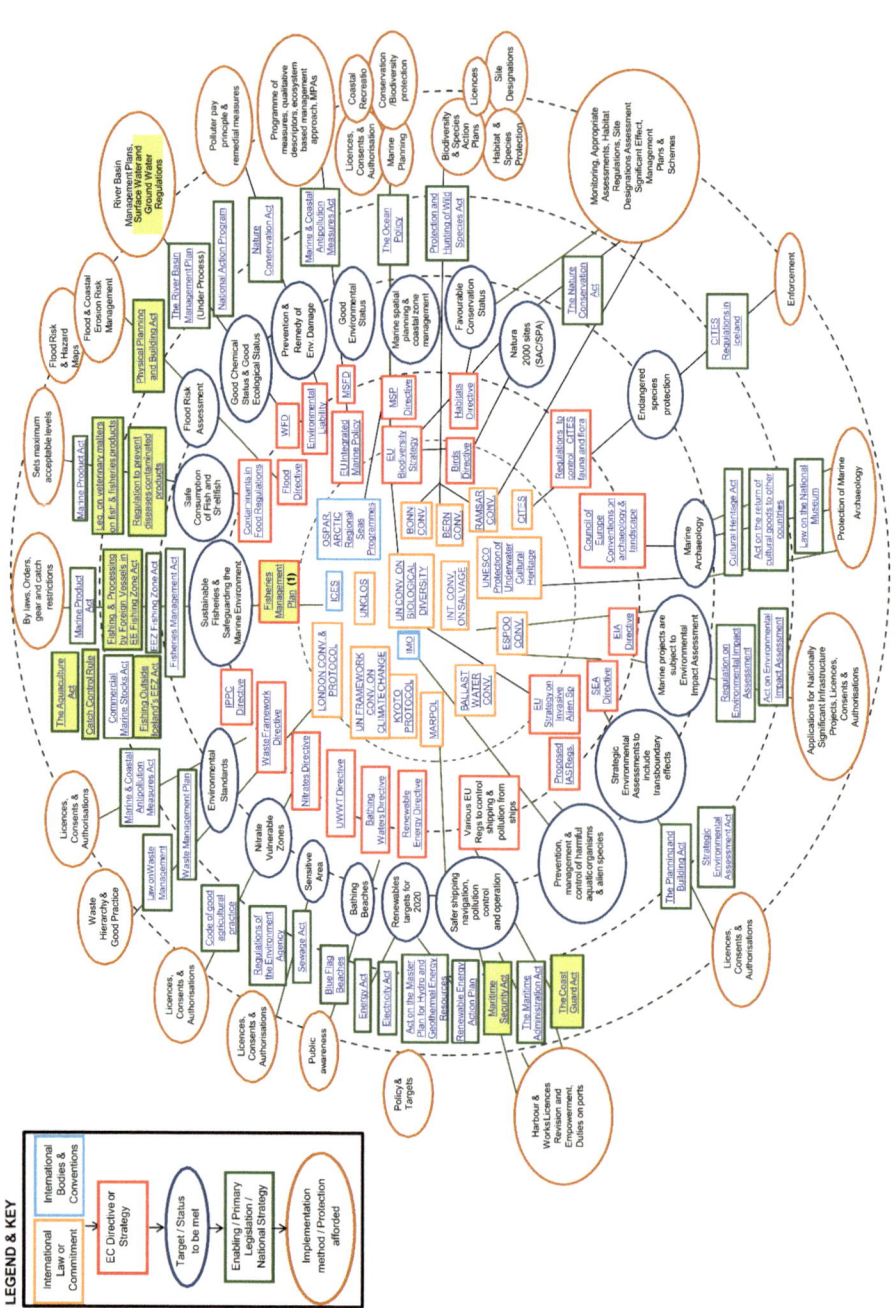

Fig. 7.5 Horrendogram of Iceland, when compared to the horrendogram of the UK. Showing international, European and Icelandic legislation for the marine environment (after Boyes and Elliott 2014). Highlighted boxes in yellow show common approaches between Iceland and the UK and highlighted ones in green show legislative policies unique to Iceland within the comparison at hand

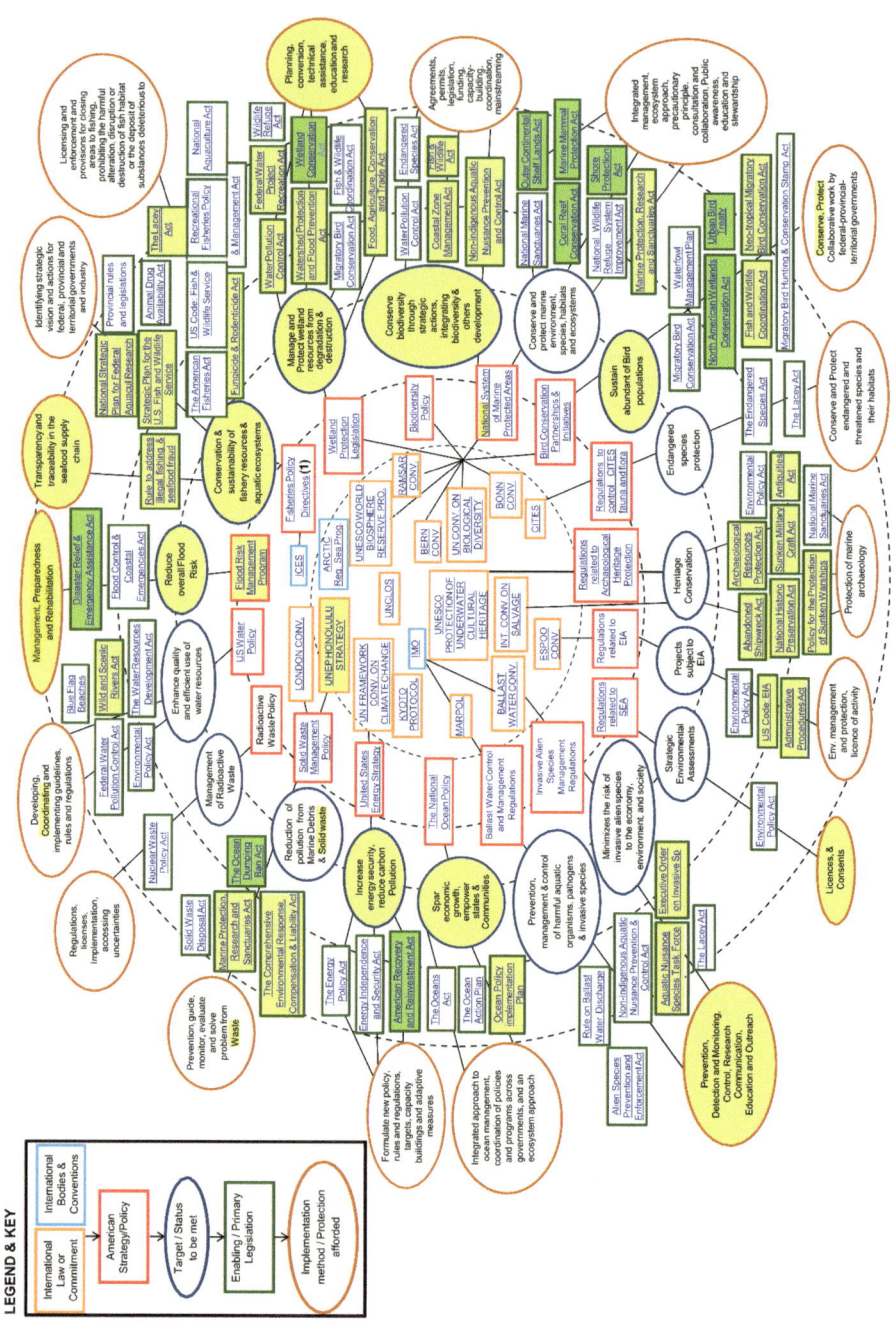

Fig. 7.6 **Horrendogram of the US, when compared to the horrendogram of Canada.** Showing international and US legislation for the marine environment (after Boyes and Elliott 2014). Highlighted boxes in yellow show common approaches between the US and Canada and highlighted ones in green show legislative policies unique to the US within the comparison at hand

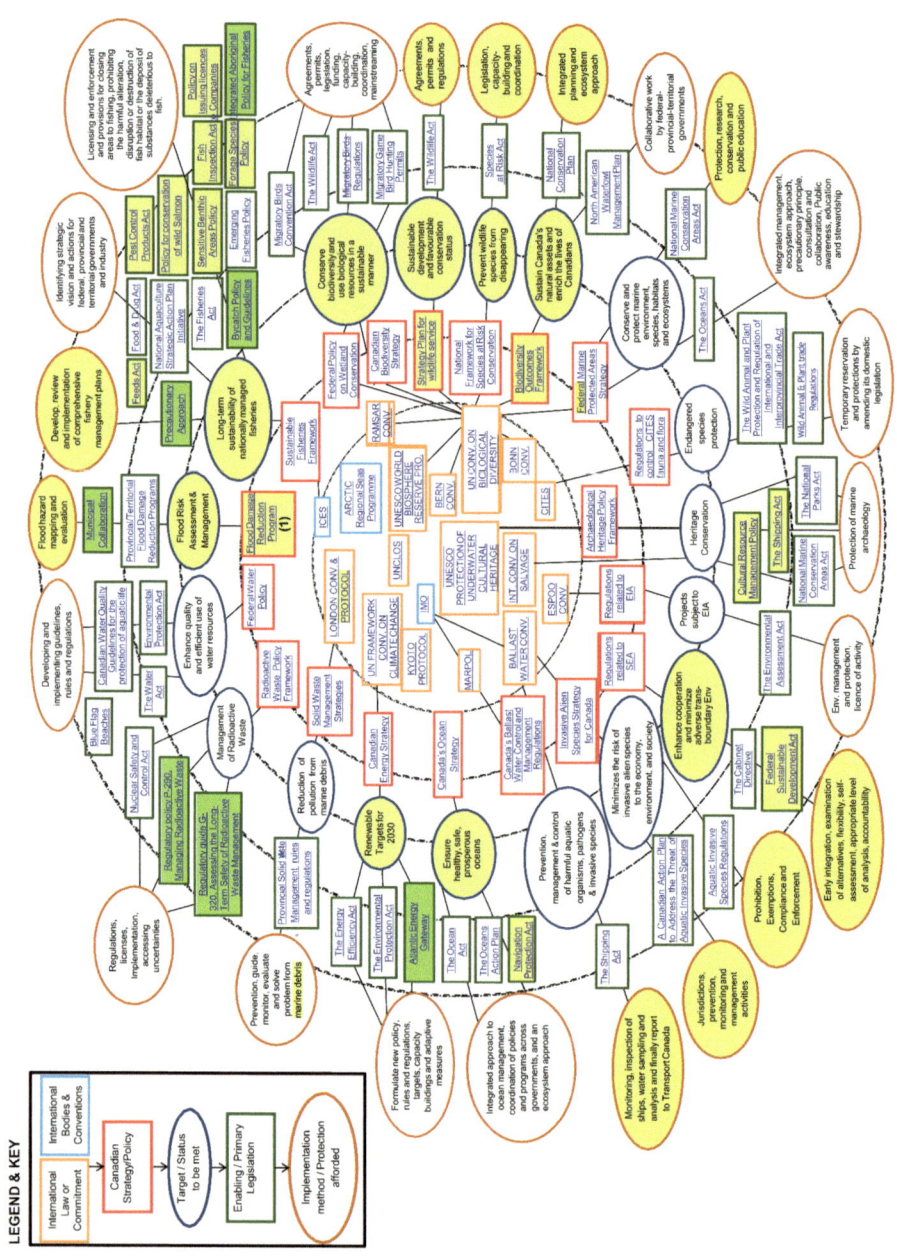

LEGEND & KEY

International Law or Commitment

International Bodies & Conventions

Canadian Strategy/Policy

Target / Status to be met

Enabling / Primary Legislation

Implementation method / Protection afforded

Fig. 7.7 **Horrendogram of Canada, when compared to the horrendogram of the US.** Showing international and Canadian legislation for the marine environment (after Boyes and Elliott 2014). Highlighted boxes in yellow show common approaches between Canada and the US highlighted ones in green show legislative policies unique to Canada within the comparison at hand

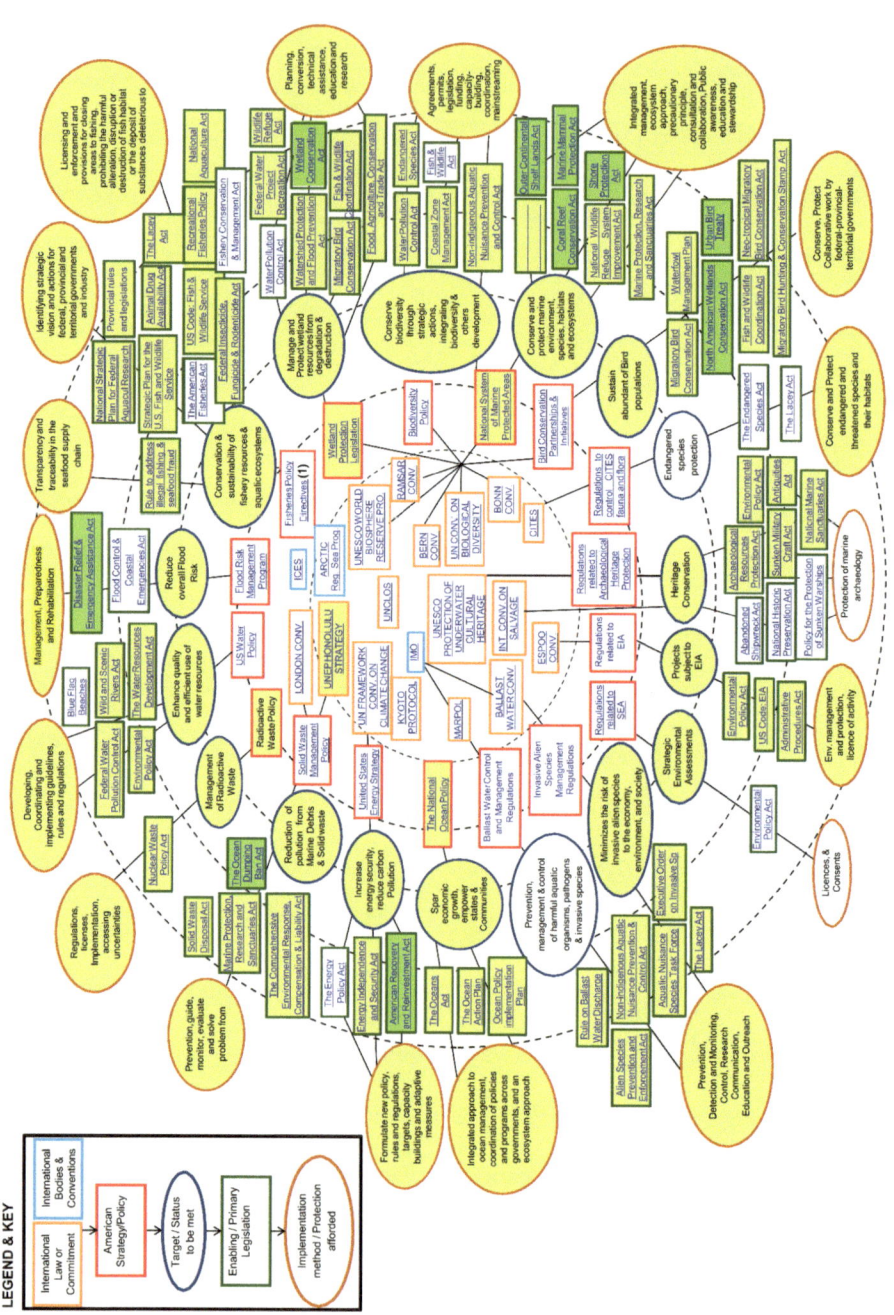

Fig. 7.8 Horrendogram of the US, when compared to the horrendogram of the UK. Showing international and US legislation for the marine environment (after Boyes and Elliott 2014). Highlighted boxes in yellow show common approaches between the US and the UK and highlighted ones in green show legislative policies unique to the US within the comparison at hand

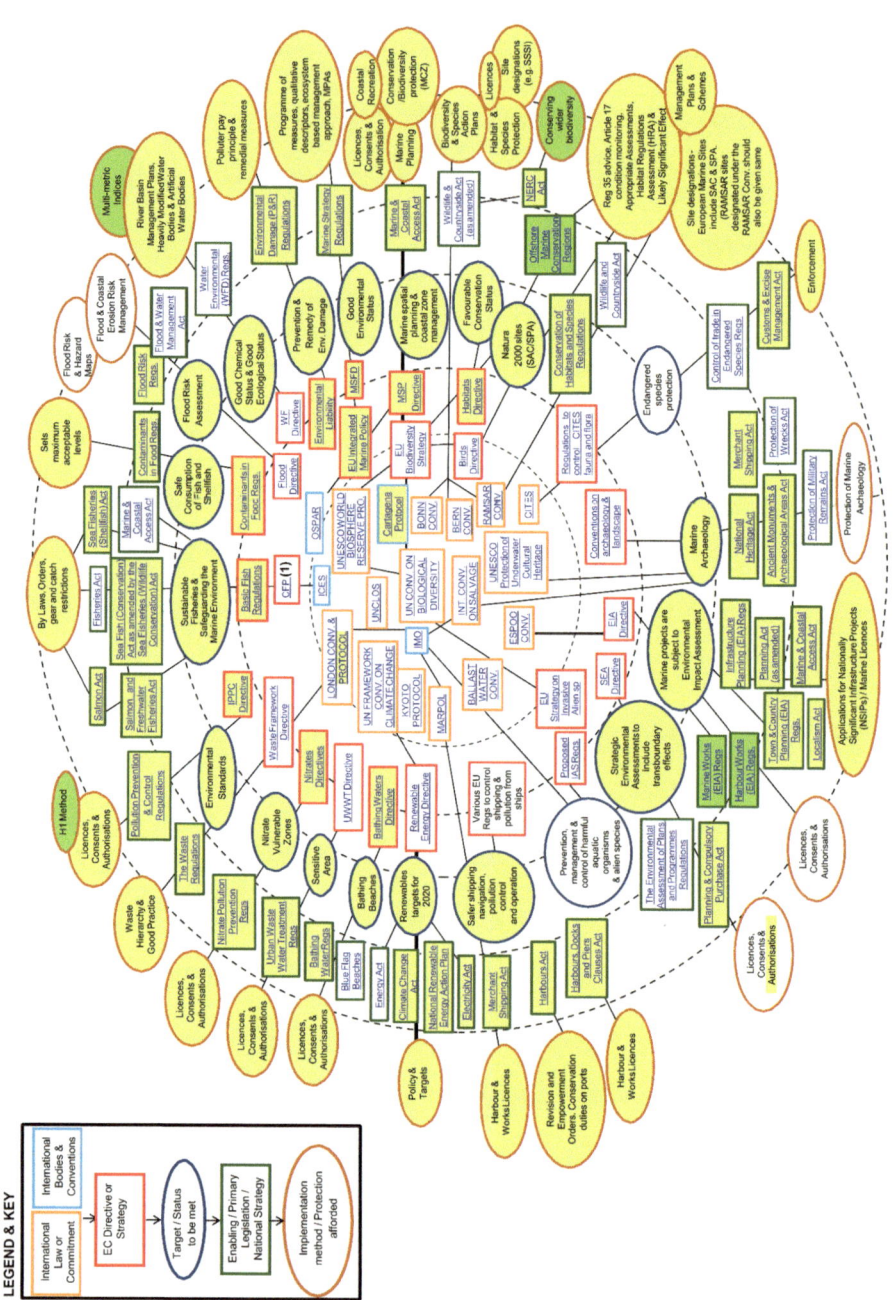

Fig. 7.9 **Horrendogram of the UK, when compared to the horrendogram of the US.** Showing international, European, and UK legislation for the marine environment (after Boyes and Elliot 2014). Highlighted boxes in yellow show common approaches between the UK and the US highlighted ones in green show legislative policies unique to the UK within the comparison at hand

Ireland, France, Portugal, and Spain are available at the website of the Geographical and Political Scenarios in Maritime Spatial Planning for the Azores and North Atlantic (GPS Azores) project.[1] Table 7.2 summarises the main national policies that were assessed in the scope of the present analysis and found to differ during the comparison between individual countries.

7.3.2.1 UK – Ireland

Ireland and the UK have similar marine policy frameworks, as shown by their respective horrendograms.[2] The content and color-coding of the boxes in the inner circles signify that international and regional/trans-national (North Atlantic/EU) marine policies are similar in scope and level of government implementation. Differences arise at a national and sub-national level (green boxes) and relate not as much to the scope of relevant policies, but mostly, the government implementation level (brown boxes). Notable differences are evident for the fisheries and aquaculture sectors. The UK has stringent and elaborate regulations for salmon fisheries [Salmon and Freshwater Fisheries Act 1975 (amended), UK Parliament, 1975] and specific policies and monitoring programmes regarding animal welfare, to ensure the safe consumption of fish and shellfish and the premium quality of final product. Ireland has specific regulations in place for aquaculture [Aquaculture (License Application) Regulations 1998, Statutory Instrument (S.I.) No. 236/1998], and has elaborated an environmental code of practice for aquaculture operators. These differences reflect the specificities of the two countries with respect to targeted and cultured species, the scale and size of fisheries' and aquaculture operators and the trade dimensions of final products. Other differences between the two countries involve the policy frameworks for marine conservation, marine heritage and EIA for key sectors and activities. The UK has dedicated regulations on offshore marine habitats and species [Conservation of Offshore Marine Habitats and Species Regulations 2017, S.I. 1013 of 2017)]; and elaborate regulations concerning marine works and harbours. Ireland has a dedicated Act on Planning and Development [Planning and Development Act 2000; 2018 (and amendments)] including several objectives relating to heritage. These differences also reflect the specificities of the two countries with respect to key maritime activities/uses. In the UK for instance, ports and harbours comprise vital assets for the local and national economy, with their ownership and governance framework being unique and showing distinct differences from port to port – with ownership and governance structure including private; municipal; or trust ports – and from the respective framework of Ireland.

With reference to the government implementation level, differences relate primarily to the use of specific objectives and implementation tools by relevant

---

[1] Analysis and Comparison of the Legal Frameworks of the N. Atlantic Countries Report, 55 pp. Available at: https://www.gpsazores.com/media/GPSAzores_Report_WP1-merged.pdf [Accessed: 2021/09/14].

[2] Ibid 1.

Table 7.2 Major national policies for the marine environment that were assessed for the purpose of the analysis

Country	Legislative and regulatory framework	Type of policy	Sector	Source
CN	Biodiversity Strategy 2006	National legislation (ratification of UN CBD)	Biodiversity	https://biodivcanada.chm-cbd.net/documents/canadian-biodiversity-strategy#TOClink
	Biodiversity Outcomes Framework 2006	Policy framework for Biodiversity Strategy (Council of ministers decision)	Biodiversity	https://biodivcanada.chm-cbd.net/documents/biodiversity-outcomes-framework
ES	State Maritime and Fisheries Law 2001 (3/2001); 2014 (amendment) (Provisions on IUU)	National legislation (Law)	Fisheries	
	Order ARM/2077/2010 of 27 July for the access control to port services of fishing vessels of third countries, transit operations, transhipment, import and export of fisheries products to prevent, deter and eliminate IUU (BOE num.185, of 31.07.2010)	Ministerial Order (Regulations)	Fisheries	https://www.boe.es/eli/es/o/2010/07/27/arm2077 (Available in Spanish)
FR	National Biodiversity Strategy (NBS) (2011)	National legislation (ratification of UN CBD)	Biodiversity	https://www.cbd.int/doc/world/fr/fr-nbsap-v2-en.pdf
	Public Health Code	National legislation (Law); and regulations	Bathing waters	Overview: http://baignades.sante.gouv.fr/baignades/editorial/en/controle/reglementation.html
IC	Fisheries Management Act 2006 (No. 38/1990)	National legislation (Act)	Fisheries	extwprlegs1.fao.org › docs › texts › ice3455
IE	Fisheries (amendment) Act 1995. No. 23. Regulations under the Act also set out the procedure for licensing for aquaculture	National legislation (Act)	Fisheries; Aquaculture	https://www.lawreform.ie/_fileupload/RevisedActs/WithAnnotations/HTML/en_act_19970023.htm
	Aquaculture (License application) Regulations 1998. S.I. No. 236 of 1998 (and amendments)	Statutory Instrument (Regulations)	Aquaculture	http://www.irishstatutebook.ie/eli/1998/si/236/made/en/print

(continued)

162

H. Calado et al.

Table 7.2 (continued)

Country	Legislative and regulatory framework	Type of policy	Sector	Source
	Planning and Development Act 2000 (and amendments); Planning and Development (Amendment) Act 2018	National legislation (Act)	Marine heritage; MSP	http://www.irishstatutebook.ie/eli/2000/act/30/enacted/en/html
	Harnessing our ocean wealth – An Integrated Marine Plan for Ireland Roadmap New Ways New Approaches New Thinking. 88 pp.	Strategy; Roadmap	MSP, Marine Planning	https://www.ouroceanwealth.ie/sites/default/files/sites/default/files/Publications/2012/HarnessingOurOceanWealthReport.pdf
PT	National Ocean Strategy 2013-2020 [Resolution of the Council of Ministers 12/2014, of 12 February]	National legislation (Resolution of the Council of Ministers)	Cross-sector	https://dre.pt/application/conteudo/572585 https://www.dgpm.mm.gov.pt/enm-11-13
	Law 17/2014, of 10 April (basis for the Policy of Planning and Management of the National Maritime Space)	National legislation (Law)	MSP	https://dre.pt/application/conteudo/25343987
	Resolution of the Council of Ministers55/2018, of 7 May (National Strategy for Nature Conservation and Biodiversity 2030)	National legislation (Resolution of the Council of Ministers)	Conservation and Biodiversity	https://dre.pt/application/conteudo/115226936
UK	Salmon and Freshwater Fisheries Act 1975 (amended)	National legislation (Act)	Fisheries (esp. inland, freshwater)	http://www.legislation.gov.uk/ukpga/1975/51
	Customs and Excise Management Act (1979)	National legislation (Act)	Incl. provisions on endangered species	http://www.legislation.gov.uk/ukpga/1979/2/contents
	Wildlife and Countryside Act (1981)	National legislation (AAct)	Incl. conservation of wild birds	http://www.legislation.gov.uk/ukpga/1981/69
	Natural Environment and Rural Communities ("NERC") Act 2006 c. 16	National legislation (AAct)	Incl. provisions on wildlife; sites of special scientific interest, etc.	http://www.legislation.gov.uk/ukpga/2006/16/contents

	National legislation (AAct)	Cross-sector	
Marine and Coastal Access Act (MCAA) 2009 c.23	National legislation (AAct)	Cross-sector	http://www.legislation.gov.uk/ukpga/2009/23/contents
Marine Policy Statement 2011 [HM Government, Northern Ireland Executive, Scottish Government & Welsh Assembly Government]	Policy framework for marine planning (for the purposes of section 44 of MCAA)	Marine Planning; MSP	https://www.gov.uk/government/publications/uk-marine-policy-statement
Conservation of Offshore Marine Habitats and Species Regulations 2017. S.I. 1013 of 2017	Statutory Instrument (RRegulations)	Marine conservation (habitats, species)	http://www.legislation.gov.uk/uksi/2017/1013/contents/made
US Magnuson–Stevens Fishery Conservation and Management Act (MSFCMA)	National Law	Fisheries, cross-sector	
American Fisheries Act	National Law	Fisheries	https://www.maritime.dot.gov/ports/american-fisheries-act/american-fisheries-act
Marine Protected Areas Executive Order 2000	Presidential Executive Order, 2000	MPAs	https://www.govinfo.gov/content/pkg/WCPD-2000-05-29/pdf/WCPD-2000-05-29-Pg1230.pdf

Policies include the ones found to differ in the comparisons between countries. Policies appear for each country in chronological order
Where: *CN* Canada, *ES* Spain, *FR* France, *ICE* Iceland, *IE* Republic of Ireland, *PT* Portugal, *UK* United Kingdom, *US* Unites States
IUU Illegal, Unreported and Unregulated fisheries. All sources last accessed: 2019/12/11

competent authorities. The UK has a substantial tradition in the development and implementation of Marine Plans (UK Marine and Coastal Access Act; HM Government, 2009), encompassing most maritime sectors, for all devolved administrations. In Ireland, the 2012 "Harnessing our Ocean Wealth" (HOOW) (MCG 2012) Strategic Vision for marine planning consisted a key development in the process of integrated, multi-sectoral maritime planning.

7.3.2.2 UK – Iceland

At the international level, a distinct difference in the marine policy framework between the UK and Iceland relates to the fact that Iceland has not designated Biosphere Reserves within the UN WNBR network (UNESCO 2018a) (although other similar concepts with a strong coastal dimension are present in the country, such as UNESCO Geoparks, e.g. the Reykjanes peninsula UNESCO Geopark, UNESCO 2018b). Again, major differences between the two primarily occur at a national government level (Figs. 7.4 and 7.5), especially regarding the management and governance of fish and fisheries. The UK, as a former EU Member State, has transposed many of the provisions of the Common Fisheries Policy (CFP) into domestic law – with a strong post-BREXIT co-operation stipulated in the EU-UK Trade and Cooperation Agreement- while Iceland has its own national Fisheries Management Act (1990) (Act No. 38/1980). Importantly, policies underline the different approaches to the management of fisheries followed by the two countries, with the UK showing particular attention to environmental protection while in Iceland, the main emphasis is on economic efficiency and resource sustainability (Paul et al. 2016) and management involves the use of economic, market-based incentives (i.e. Individual Transferable Quotas, ITQs) (Popescu and Poulsen 2012). Iceland has more thorough regulations concerning seafood product safety. There exists a bilateral agreement between the two countries concerning the management of fisheries, with Iceland conforming to several provisions of the CFP (European Economic Community and Republic of Iceland 1993). The UK policy framework is especially advanced with respect to flood risk assessment, with Iceland only recently developing a relevant flood directive. The two countries follow similar approaches as regards conservation measures, although the UK has a dedicated Customs and Excise Management Act (1979), with provisions on the protection of endangered species. Small differences also exist with respect to the government system of maritime heritage and shipping: In Iceland, fisheries play a centremost role in marine cultural heritage (Antonova and Rieser 2019) with museums and villages comprising key features, while maritime clusters are becoming progressively important structures for the promotion of blue bio-economy.[3] In the UK, maritime heritage encompasses a diversity of features, ranging from ports and harbours, sea-

---

[3] European Commission, 2019: Iceland and the blue bioeconomy: making the most from fish Available at: https://webgate.ec.europa.eu/maritimeforum/en/node/4449 [Accessed: 2021/09/07].

side resorts, and maritime archaeology,[4] reflecting the respective diversity of such maritime cultural heritage elements.

7.3.2.3   UK – France

The marine policy frameworks of France and the UK are similar at international and regional levels,[5] with differences primarily arising at a national level. These include provisions relating to biodiversity protection, with France having developed a dedicated National Biodiversity Strategy (NBS) (2011). France possesses numerous provisions and regulations for the fisheries and aquaculture sectors implemented through a series of laws, decrees, codes and catch restrictions. France has also developed a dedicated Public Health Code, with provisions pertaining among others, to fish catches. With reference to nature conservation, in France, strong emphasis is placed on the need for stakeholder's mobilisation and commitment for delivering the objectives of the National Biodiversity Strategy (2011). The UK MCAA requires a statement of public participation (SPP) where relevant stakeholders can be involved and influence the development of a particular marine plan.

7.3.2.4   UK – Portugal

The international dimensions of marine policy are similar in both countries,[6] except that Portugal unlike the UK, is not party to the London Protocol (LP 1996; entry into force: 2006). Instead, Portugal is a party of the London Convention (LC) reflecting the general case of the challenges in the presence of those two global treaties of similar scope (Hong and Lee 2015). Once again, main differences occur mainly on the national policy level. These involve the fisheries and aquaculture sectors, with the policy framework in Portugal having a focus on deep-sea fisheries and a system of regulatory concessions for aquaculture farms; while in the UK, as discussed earlier, there is particular attention given to salmon and freshwater fisheries. As in the case for other countries, these differences reflect once again the specificities of the fisheries sectors in the two countries, with reference to the targeted species and scale of fisheries operators.

An important difference between Portugal and other EU countries lies in the fact that Portugal pioneered the development of a National Ocean Strategy 2013–2020 [Directorate General for Maritime Policy DGPM)] that also integrates ecological status objectives (which is the reason why no specific ecological policies are shown in the horrendogram for Portugal under the respective category). Differences also

---

[4] Historic England, 2021. Available at: https://historicengland.org.uk/research/current/discover-and-understand/coastal-and-marine/ [Accessed:2021/09/07].

[5] Ibid. 1.

[6] Ibid. 1.

relate to nature conservation policies for rural communities, with the UK having a set of Acts, the Wildlife and Countryside Act 1981 and the Natural Environment and Rural Communities (NERC) 2006 Act, whereas in Portugal the specific topics are dealt within provisions of the Nature Conservation and Biodiversity Strategy. Different approaches are also followed between the two countries for coastal recreation, biodiversity and species protection and site designations, with the UK having specific measures and action plans for those matters, while in Portugal relevant provisions are within the framework of a sustainable use of natural resources, in the context of the National Strategy for Nature Conservation and Biodiversity. Portugal is currently developing a national Animal Protection Law where key aspects relating with nature and species protection will be dealt within. Differences also relate to key environmental policy, notably EIA and SEA: in Portugal the framework for EIA and SEA is established through a set of Decrees and Laws, while in the UK these are regulated through Acts. Differences also relate to the competent authorities for the implementation of relevant regulations. In the UK, relevant provisions are also framed within the Marine and Harbour work Regulations, Town and Country Planning Regulations, and the Localism Act. Shipping and Marine Renewable Energy (MRE) are other categories where differences in the two arise at a national government level. The two countries also have different implementation methods for key marine policies, most notably the Water Framework Directive and Urban Waste Water Treatment (UWWT) Directive.

The Nagoya protocol was not adopted with specificities for the marine environment. However, it is worth referring that in the Azores (Portugal), the Regional Legislative Decree 9/2012/A, of 20 March, was created inspired by the Nagoya Protocol, developing and regulating the legal regime for access and use of natural resources of the Azores for scientific purposes, including the access of marine resources (Calado et al. 2014).

7.3.2.5 UK – Spain

At an international level, the two countries adhere to the same marine policy provisions.[7] Differences occur at the national level for certain maritime sectors, as shown for major categories of the respective horrendograms. For the fisheries and aquaculture sector, a difference relates to the issue of Illegal, Unreported and Unregulated (IUU) fisheries, with specific provisions in Spanish Law [State Maritime and Fisheries Law 2014 amendment] and relevant regulations [Ministerial Order ARM/2077/2010] available (ClientEarth 2017). Spain also has elaborate regulations, in the form of Royal Decrees, regarding the safe consumption of fish and shellfish, whereas the UK appears to have the least amount of dedicated regulations specifically for that matter, with relevant provisions mostly integrated within the context of the Salmon and Freshwater Fisheries Act 1975. The two countries also

---

[7] Ibid. 1.

possess different legislative tools regarding the transposition into national law and implementation of the EU Water Framework Directive (WFD) (European Parliament and Council 2010), with Spain having a dedicated Water Act and Water Policy. Regarding nature conservation, the UK has pioneered the development and designation of offshore Marine Protected Areas (MPAs) (Joint Nature Conservation Committee, JNCC 2019). Spain has a national declaration for MPAs and one for the protection of animals. Spain has pioneered the issue of alien, invasive species, with a dedicated Law on the issue, absent from the UK and most other EU countries. A key difference amongst the two relates to the fact that in Spain the management and governance of coastal uses is mostly dealt with through Marine Laws.

7.3.2.6   US – Canada

The marine policy framework of the two countries might differ overall, but there are also distinct similarities (Figs. 7.6 and 7.7). At an international policy level, the main difference relates to the fact that the US has not ratified the UNCLOS and is not a party to the London Protocol. Differences between the two also emerge as a result of the Honolulu strategy (UNEP and NOAA 2016) that applies to the case of the US and not Canada. Similarities also occur at a national level and primarily stem from the fact that, in both countries, the legislative and regulatory framework is mostly framed by international commitments. Radioactive Waste and Energy Strategies are found in both horrendograms. Small differences occur with reference to the flood and risk assessment category, with the US having developed a risk management programme, while Canada has one for flood damage reduction. The most distinct difference between the two countries seems to be in the nature conservation sector. Overall, the US has a large number of complementary regulations, in the form of Acts, applying to nature protection and conservation while Canada has a more straight-forward and streamlined approach: For instance, different approaches apply with reference to the conservation and protection of bird fauna and marine species and habitats: in Canada, a Biodiversity Strategy and Biodiversity Outcomes Framework (2006), a Strategic Plan for Wildlife Service, and a Federal Marine Protection Areas Strategy (Fisheries and Oceans Canada 2005) frame the governance of the sector. In the US, Birds Conservation Partnerships and Initiatives aim to sustain abundance of bird populations specifically, while a National System of MPAs [Presidential Executive Order, 2000) also largely influences conservation objectives for marine habitats and species, notably marine mammals.

However, at the level of legislation implementation, Canada exhibits larger diversity in implementation methods and competent authorities engaged in the process. Differences also relate to fisheries and aquaculture sectors. Canada manages fisheries resources based on a precautionary approach; in the US, there is a strong focus on economic efficiency of the sector, with economic, market-based incentives existing for the management of certain stocks and species.

7.3.2.7   US – UK

At an international level, the two countries show similarities with respect to key marine and maritime policies (Figs. 7.8 and 7.9): both are parties to major international conventions (Ramsar; 2001 UCH Protection, Espoo, Kyoto,[8] MARPOL) with the main difference being that the US has not ratified the UNCLOS and London Protocol. Also, the US has not signed the CBD Cartagena Protocol, as described in the methods section. The most distinct differences arise at the national policy level, and involve different approaches followed by the two countries, most notably for fisheries, nature (marine) conservation, ocean management, water quality and environmental standards sectors. Fisheries regulations in the UK derive primarily from previously adhering to provisions of the EU CFP, while in the US the framework for the management and exploitation of fisheries is governed by a fisheries Policy Regulation primarily framed though the Magnuson-Stevens Fishery Conservation and Management Act (MSA) 1976 (and amendments) but also through provisions of the American Fisheries Act (1998). The two countries show similar approaches with respect to their general environmental protection legislation, as argued in past studies (Boyes and Elliott 2016). The US has specific regulations, such as the Wetlands Protection Legislation and an elaborate national system for MPAs. Relevant policy in the UK is shaped through the EU Integrated Maritime Policy (European Commission 2007), and the provisions of key legislation, such as the Habitats Directive (European Council 1992) and the Environmental Liability Directive (European Parliament and Council 2004), as have been transposed in national legislation. The two countries also reveal differences in legislation pertaining to SEA and shipping, with the US having a stronger focus on environmental protection, whilst UK policy foregrounding aspects of navigation safety and pollution prevention.

7.4   Conclusions

A comparative analysis of marine policy can take many shapes and forms, as no pre-defined methodological framework exists (Van Hoecke 2015; Calado et al. 2018). The present analysis did not seek to provide an exhaustive assessment of national policies and subsequent comparison between countries, but aimed at determining the most distinct differences in national approaches. Such an analysis comprises a snapshot of the most current policies, not integrating ongoing developments in the policy arena. However, horrendograms substantially aided the process of analysis across the different policy frameworks, readily highlighting where new efforts, in the form of future research, but also in assisting targeted policy intervention, are required. This in turn can aid cross-border coordination and

---

[8] As discussed, the US has initiated the process of withdrawal from the UNFCC Paris Protocol.

decision-making, having significant advantages for transboundary cooperation. The existing institutional platforms for cross border cooperation in Marine Governance, outside the EU, are still much dependent on the UNCLOS provisions and follow-up bodies such as the Regional Seas Conventions (RSC) or the Regional Fisheries Management Organizations (RFMO). de Grunt et al. (2018) highlight the role of the RSCs in the cross-border coordination of major maritime economic activities. Specific attention to the desirability and perceived challenges of such an increased role for the RSCs is also addressed by the authors, concluding that even these mechanisms are far from achieving high performances on their roles worldwide. Although the UN Ocean Decade may open new paths, the world's ocean coordination mechanisms are still far from those that exist on Climate Change or Biodiversity. New opportunities, as the hopes for closer marine research cooperation between Atlantic nations raised by the Belem and Galway Statements, need better linkages to these existing mechanisms in order to profit from already functioning channels.

The comparison of the different policy frameworks disclosed some crucial differences but also similarities in marine policy for North Atlantic countries. No major differences were highlighted by the horrendogram-based approach between countries at the level of international marine policy as suggested by the innermost circle of respective horredongrams, except for UNCLOS and CBD for the US. Major ongoing political developments, most notably the UK BREXIT are envisaged to result in marked differences in policy relating to the marine environment at the international government level in the future, with reference to resource management and access of fishing fleets within EEZs. This however will happen gradually and at a time-horizon greater than 5 years (e.g. the "adjustment period" stipulated in the EU-UK Trade and Cooperation Agreement for the fisheries sector). Differences in marine policy mostly arise at a national level, with EU Member States showing more similarities than their non-EU counterparts, due to the transposition of EU legislation into national laws. Distinct groups of countries, reflecting major approaches to marine legislation, appear to be present in the region: (i) Ireland, Iceland and the UK have a similar approach, with policy delivered mainly though Acts and Regulations; (ii) Portugal and Spain also show similarities, with marine policy delivered through the use of Law Decrees; (iii) The US and Canada, being federal states, also exhibit similarities, with both of them using Acts; and (iv) France shows a similar approach to Ireland, Iceland and the UK but also uses a set of binding tools for delivering relevant policy. Identifying these differences is the first step to overcome barriers in scaling up sustainability policies and goals.

Results suggested that the difference in implementation of relevant marine policy in the North Atlantic countries stem from the different national approaches to marine policy. For instance, France and UK have a more bottom-up approach while other countries, such as Portugal, exhibit a more top-down approach to marine policy and governance (Pinto et al. 2015; Calado et al. 2018). In certain countries, such as the US, marine affairs and maritime issues are dealt through a multitude of different laws and regulations, highly relevant in scope and complementary in nature (Crowder et al. 2006). Other North Atlantic countries (e.g. Canada) might have a more streamlined policy framework, but implementation might frequently involve several competent authorities and institutions, thus requiring attention in coordination.

There were many instances where major differences primarily resulted from the policy framework for specific maritime sectors: these included fisheries and aquaculture, marine conservation, and maritime cultural heritage. For fisheries and aquaculture, differences in the targeted and cultured species and the scale of operations (large vs small) often resulted in notable differences in national policy frameworks. For marine/coastal conservation, differences were also a result of different jurisdictions, i.e. transitional waters and the implications of different planning jurisdictions (terrestrial/marine). The previous denote that while harmonisation of policy between countries is essential, it is still crucial to consider local specificities especially for those sectors that exhibit the most pronounced differences. For ports, harbours and marine works, it is important to remember that the UK, which constituted our frame of reference for the present analysis, has a unique governance and ownership framework that may amplify differences, but also clearly echoes the need for taking into consideration such local specificities. With regard to MSP, the analysis also indicates that, even if not under the explicit designation of MSP, in all cases analysed, the spatial planning of marine spaces is supported by existing regulation or strategic tools, and not hindered by the existence of dissimilarities between States.

The study highlighted important tools and enablers in marine policy. Bilateral agreements between countries enable streamlining marine policy regulations and have a major role to play in transboundary cooperation. Good examples are the case of Iceland and the EU concerning major fisheries policies; or the Honolulu Strategy for the US and EU for marine litter. Results also highlighted the usefulness of dedicated Ocean Strategies (Portugal) and national Marine Plans (UK) in integrating different sectors and objectives. Such examples can constitute good examples and practices and guide the development of policies for other countries currently developing relevant legislation.

Future work will focus on reviewing and scrutinizing findings from the present study by members of the North Atlantic research and planning communities. Most importantly, future work aspires to integrate expert knowledge on the various issues raised by the present analysis. As such, the present study should be seen as a starting point for a constructive and open dialogue with members of the North Atlantic marine research and planning communities. This dialogue can be based on the already existing mechanisms as the RSC or RFMO thus taking advantage of existing dialogue channels. However, a more holistic and integrated approach is needed and that can be triggered under the opportunities opened by the UN Ocean Decade and subsequent actions.

Acknowledgements The study was conducted in the framework of the Geographical and Political Scenarios in Maritime Spatial Planning for the Azores and North Atlantic (GPS Azores) project (Ref: ACCORES-01-0145-FEDER-002 GPS Azores), financed by Azores 2020 Operational Programme (85% by FEDER and 15% by Regional Funds) and PADDLE project, which has received funding from the European Union Horizon 2020 Research and Innovation Programme 1.3.3. under grant agreement No. 734271. The authors would like to thank Fernando Lopes, Marie Bonnin, Ana Vitoria Magalhães, Evangelia Tzika and António Medeiros.

Conflict of Interest The authors have no conflict of interest to disclose. The views expressed are those of the authors and do not reflect official EU and national policy.

References

Antonova AS, Rieser A (2019) Curating collapse: performing maritime cultural heritage in Iceland's museums and tours. Marit Stud 18:103–114

Boyes SJ, Elliott M (2014) Marine legislation – the ultimate 'horrendogram': International law, European directives and national implementation. Mar Pollut Bull 86:39–47

Boyes SJ, Elliott M (2016) Brexit – the marine governance horredogram just got more horrendous! Mar Pollut Bull 111:41–44

Calado H, Pinto Lopes C, Fonseca C (2014) The Nagoya protocol and the regime on access to natural resources and the fair and equitable sharing of benefits in the Azores Autonomous Region. In: Chantal Ribeiro M (coord) 30 Anos da assinatura da Convenção das Nações Unidas Sobre o Direito do Mar: Proteção do Ambiente e o Futuro do Direito do Mar.Coimbra Editora, pp 485–516. ISBN 978-972-32-2203-6

Calado H, Monwar MMd, Tzika E, Magalhães AV, Grimmel H, Moniz F, Bonnin B (2018) Analysis and comparison of the legal frameworks of the North Atlanric countries. GPS Azores project deliverable, 40 pp

Carneiro C, Thomas H, Olsen S, Benzaken D, Fletcher S, Roldan SM, Stanwell-Smith D (2017) Cross-border cooperation in maritime spatial planning. In: Final report: study on international best practices for cross-border MSP. Publications of the European Union, Luxembourg, 109 pp. https://doi.org/10.2826/28939

Carval D, Jarno D (2019) Analysis of data needs and existing gaps –specifically relating to trans-boundary working. EU Project Grant No.: EASME/EMFF/2015/1.2.1.3/03/SI2.742089. Supporting implementation of Maritime Spatial Planning in the Northern European Atlantic (SIMNORAT). Shom. 112 pp

ClientEarth (2017) The control and enforcement of fisheries in Spain. 65 pp. Available at: https://www.documents.clientearth.org/wp-content/uploads/library/2017-09-29-the-control-and-enforcement-of-fisheries-in-spain-ce-en.pdf. Last accessed: 2019/12/11

Crowder LB, Osherenko G, Young OR, Airamı S, Norse EA, Baron N, Day JC, Douvere F, Ehler CN, Halpern BS, Langdon SJ, McLeod KL, Ogden JC, Peach RE, Rosenberg AA, Wilson JA (2006) Resolving mismatches in U.S. Ocean Governance. Science 313:617–618

de Grunt LS, Ng K, Calado H (2018) Towards sustainable implementation of maritime spatial planning in Europe: a peek into the potential of the Regional Sea Conventions playing a stronger role. Mar Pol 95:102–110

DIR (2017) Décret n° 2017-222 du 23 février 2017 Stratégie nationale pour la mer et le littoral

Durussel C, Wright G, Wienrich N, Boteler B, Unger S, Rochette J (2019) 'Summary for decision-makers – strengthening regional ocean governance for the high seas: opportunities and challenges to improve the legal and institutional framework of the Southeast Atlantic and Southeast Pacific', STRONG High Seas Project, 2019. Available at: www.prog-ocean.org/our-work/strong-high-seas/. Last accessed: 2019/11/20

European Commission (2007) Communication from the Commission to the European Parliament, the Council, the European Economic and Social Committee and the Committee of the Regions. An integrated maritime policy for the European Union. Brussels, 10.10.2007, COM (2007) 575 final. Available at: http://eur-lex.europa.eu/LexUriServ/LexUriServ.do?uri=COM:2007:0575:FIN:EN:PDF. Last accessed: 2019/12/11

European Council (1992) Council Directive 92/43/EEC of 21 May 1992 on the conservation of natural habitats and of wild fauna and flora. Available at: https://eur-lex.europa.eu/legal-content/EN/TXT/?uri=CELEX%3A31992L0043. Last accessed: 2019/11/03

European Economic Community and Republic of Iceland (1993) Agreement on fisheries and the marine environment between the European Economic Community and the Republic of Iceland. Available at: https://eur-lex.europa.eu/legal-content/EN/TXT/PDF/?uri=CELEX:21993A070 2(01)&from=EN. Last accessed: 2022/11/06

European Parliament and Council (2004) Directive 2004/35/CE of the European Parliament and of the Council of 21 April 2004 on environmental liability with regard to the prevention and remedying of environmental damage. Available at: https://eur-lex.europa.eu/legal-content/EN/TXT/?uri=celex%3A32004L0035. Last accessed: 2019/11/03

European Parliament and Council (2010) Directive 2000/60/EC of the European Parliament and of the Council establishing a framework for the Community action in the field of water policy ("EU Water Framework Directive). Available at: https://eur-lex.europa.eu/legal-content/EN/TXT/?uri=CELEX:32000L0060. Last accessed: 2019/11/03

European Parliament and Council (2014) Directive No 2014/89/EU of 23 July 2014 establishing a framework for maritime spatial planning. Available at: http://eur-lex.europa.eu/legal-content/EN/TXT/PDF/?uri=CELEX:32014L0089&from=PT. Last accessed: 2019/11/03

FAO (2015) Fisheries and Aquaculture Department. Major Fishing Areas. Available at: http://www.fao.org/fishery/area/search/en. Last accessed: 2019/10/31

Fisheries and Oceans Canada (2005) Canada's Federal Marine Protected Areas Strategy. 18 pp. Available at: https://waves-vagues.dfo-mpo.gc.ca/Library/315822e.pdf. Last accessed: 2019/12/11

Gold BD, Pastoors M, Babb-Brott D, Ehler C, King M, Maes F, Mengerink K, Müller M, Cunha TPE, Ruckelshaus M, Sandifer P, Veum K (2011) Integrated marine policies and tools working group CALAMAR project expert paper. 24 pp. Available at:. https://biblio.ugent.be/publication/2024604/file/2024611.pdf. Last accessed: 2019/11/20

Hong GH, Lee YJ (2015) Transitional measures to combine two global ocean dumping treaties into a single treaty. Mar Pol 45:47–56

International Council for the Exploration of the Sea (ICES) (2019). ICES Statistical Areas. Shapefile available through the ICES Spatial Facility: http://gis.ices.dk/sf/. Last accessed: 2019/12/12

Irish Government, Marine Coordination Group (MCG) (2012). Harnessing our ocean wealth – an integrated marine plan for Ireland roadmap new ways new approaches new thinking. 88 pp

Joint Nature Conservation Committee (JNCC) (2019) Offshore Marine Protected Areas (MPAs). Available at: http://archive.jncc.gov.uk/page-6895. Last accessed: 2019/12/10

Kern K, Gilek M (2015) Governing Europe's marine environment: key topics and challenges'. In: Gilek M, Kern K (eds) Governing Europe's marine environment: Europeanisation of regional seas or regionalisation of EU policies? Ashgate Publishing, pp 1–12

Marine and Coastal Access Act (MCAA) (2009). UK Public General Act. 2009 Chap. 23. Available at:. http://www.legislation.gov.uk/ukpga/2009/23. Last accessed: 2019/11/03

Marques M, Quintela A, Sousa LP, Silva A, Alves FL, Dilasser J, Ganne M, Cervera-Núñez C, Campillos-Llanos M, Gómez-Ballesteros M., Alloncle N, Giret O (2019) Coordination of sectorial policies. EU Project Grant No.: EASME/EMFF/2015/1.2.1.3/03/SI2.742089. Supporting Implementation of Maritime Spatial Planning in the European Northern Atlantic (SIMNORAT). Cerema – UAVR. 13 pp

Ministerio de Agricultura y Pesca, Alimentación y Medio Ambiente (2017) Real Decreto 363/2017, de 8 de abril, por el que se establece un marco para la ordenación del espacio marítimo. Available at: https://www.boe.es/eli/es/rd/2017/04/08/363. Last accessed: 2019/11/03

Morf A, Moodie J, Gee K, Giacometti A, Kull M, Piwowarczyk J, Schiele K, Zaucha J, Kellecioglu I, Luttmann A, Strand H (2019) Towards sustainability of marine governance: challenges and enablers for stakeholder integration in transboundary marine spatial planning in the Baltic Sea. Ocean Coast Manage 177:200–212

OSPAR (2019a) OSPAR guidelines for the preparation of draft OSPAR decisions, recommendations and other arrangements draft OSPAR background documents and other reports (OSPAR Agreement: 2019–01)

OSPAR (2019b) OSPAR regions. Shapefile available through the ICES spatial facility: http://gis. ices.dk/sf/. Last accessed: 2019/12/12

OSPAR Commission (2010) NE Atlantic environmental strategy. OSPAR Agreement 2010–03. 27 pp. Available at:. https://www.ospar.org/site/assets/files/1200/ospar_strategy.pdf. Last accessed: 2019/12/11

Paul M, Andersen JL, Aranda M, Fitzpatrick M, Goti L, Guyader O, Haraldsson G, Hatcher A, Hegland TJ, Le Floch P, Macher C, Malvarosa L, Maravelias CD, Mardle S, Murillas A, Nielsen RJ, Sabatella R, Smith ADM, Stokes K, Thoegersen T, Ulrich C (2016) A comparative review of fisheries management experiences in the European Union and in other countries worldwide: Iceland, Australia, and New Zealand. Fish Fish 17:803–824

Pinto H, Cruz AR, Combe C (2015) Cooperation and the emergence of maritime clusters in the Atlantic: analysis and implications of innovation and human capital for blue growth. Mar Policy 57:167–177

Popescu I, Poulsen K (2012) Icelandic fisheries – a review. Report requested by the European Parliament's Committee on Fisheries. Policy Department B: Structural and Cohesion Policies. 54 pp. Available at: https://www.europarl.europa.eu/RegData/etudes/note/join/2012/474540/ IPOL-PECH_NT(2012)474540_EN.pdf. Last accessed: 2019/12/10

Rudd MA, Dickey-Collas M, Ferretti J, Johannesen E, Macdonald NM, McLaughlin R, Rae M, Thiele T, Link JS (2018) Ocean ecosystem-based management mandates and implementation in the North Atlantic. Front Mar Sci 5:485. https://doi.org/10.3389/fmars.2018.00485

Speer L, Gonçalves E, Ardron J, Arico S, Auster P, Gianni M, Gjerde K, Laffoley D, Lodge M, Orbach M, Pomponi S, Rochette J, Unger S (2011) CALAMAR project expert paper – High Seas working group. 20 pp. Available at:. https://www.ecologic.eu/sites/files/project/2016/ documents/calamar_high_seas.pdf. Last accessed: 2019/12/05

UN (2017) The ocean and the sustainable development goals under the 2030 agenda for sustainable development – a technical abstract of the first global integrated marine assessment. Available at: https://www.un.org/depts/los/global_reporting/8th_adhoc_2017/Technical_Abstract_ on_the_Ocean_and_the_Sustainable_Development_Goals_under_the_2030_Agenda_for_ Susutainable_Development.pdf. Last accessed: 2019/10/30

UNDP (2015) Water and ocean governance programme contribution to realising the UNDP strategic plan 2014–2017, 44 pp

UNEP and NOAA (2016) The Honolulu Strategy: a global framework for prevention and management of marine debris. Available at: http://wedocs.unep.org/handle/20.500.11822/10670?show =full. Last accessed: 2019/10/30

UNESCO (2018a) Man and the Biosphere Programme biennial activity report 2016–2017

UNESCO (2018b) Reykjanes peninsula geopark. Available at: http://www.unesco.org/new/en/ natural-sciences/environment/earth-sciences/unesco-global-geoparks/list-of-unesco-global-geoparks/iceland/reykjanes/. Last accessed: 2019/12/12

Van Hoecke M (2015) Methodology of comparative legal research. Law and Method, 1–35.

Van Tatenhove JP (2017) Transboundary marine spatial planning: a reflexive marine governance experiment? J Environ Policy Plan 19:783–794

Zaucha J (2014) Sea basin maritime spatial planning: a case study of the Baltic Sea region and Poland. Mar Policy 50:34–45

Chapter 8
International Fisheries Law: Past to Future

Julia Nakamura

Abstract Ocean governance is a collective effort. It depends on the ability of all actors, from States to individuals, to work together upon common understandings, values and rules for use of the ocean. The contemporary Law of the Sea regime, as reflected in the 1982 United Nations Convention on the Law of the Sea (UNCLOS), provides a global legal order for the control of diverse activities carried out in the ocean, aiming to achieve balanced relationships among multiple users and scarce marine resources. International marine fisheries, in particular, are regulated by International Fisheries Law (IFL). More intensively in the last decades, IFL has contributed to ocean governance by harmonising social, economic and environmentally-sound standards for fisheries, setting out important parameters to support the potential of fisheries to sustainably operate in the ocean. This chapter draws on a historical narrative of IFL from 1994, when the UNCLOS entered into force, to mid-2022. It analyses selected legal developments at global and regional levels with a view to clarify how the contemporary IFL has developed and responded to the recurrent problems in fisheries at global and regional levels, addressing current and future needs.

8.1 Introduction

Ocean governance is a collective effort. It depends on the ability of all actors, from States to individuals, to work together to form common understandings, values and rules for use of the ocean. The contemporary Law of the Sea (LOS) regime, as reflected in the 1982 United Nations Convention on the Law of the Sea (UNCLOS 1982), provides a global legal order for the control of numerous and diverse activities carried out in the ocean, aiming to achieve balanced relationships among multiple users and scarce marine resources. International marine fisheries, in particular, are regulated by International Fisheries Law (IFL). This legal domain predates the

J. Nakamura (✉)
Law School, University of Strathclyde, Glasgow, UK
e-mail: julia.nakamura@strath.ac.uk

© German Institute of Development and Sustainability (IDOS) 2023
S. Partelow et al. (eds.), *Ocean Governance*, MARE Publication Series 25,
https://doi.org/10.1007/978-3-031-20740-2_8

UNCLOS and falls within the LOS regime (Molenaar and Caddell 2019). For most of IFL's history, that vision for balanced ocean governance has been hindered by industrialized activities, unsustainable fishing, and many other issues, whilst the regulation of fishing activities at sea fell short on marine environmental protection and conservation (Freestone and Makuch 1997; Barnes 2019), the protection of fishers at sea and their human rights (Van der Burgt 2012; Papanicolopulus 2018). Significant changes in international law broadly occurred from the 1970s onwards, especially on the need to protect and conserve the environment, ecosystems, habitats, and biodiversity (Birnie et al. 2009; Harrison 2015). The international community's concerns with social matters such as maritime safety and the human rights of fishers also brought about international legal developments relevant to IFL more recently (Politakis 2008; Morgera and Nakamura 2021; Nakamura 2022). Additionally, economic aspects in sustainable fisheries have been the subject of the World Trade Organization's (WTO) two-decades of ongoing negotiations (Chang 2003; WTO 2021), which finally resulted in the important and promising, yet unfinished, agreement on fisheries subsidies (Switzer and Lennan 2022). As such, the contemporary IFL of the last half-century has been progressively evolving and supporting social, economic and environmentally-sound standards for international marine fisheries.

The analysis of IFL's evolution over time helps one to clarify how this domain strengthens the global legal order for the oceans and fisheries sustainability (Garcia et al. 2014; Molenaar and Caddell 2019). The present chapter draws on a historical narrative of IFL's legal developments from 1994, when the UNCLOS entered into force, up until mid-2022. It analyses selected instruments, legal issues, and judicial cases, drawing on how the contemporary IFL has developed and responded to the recurrent problems in fisheries at global and regional levels, addressing current and future needs.

This chapter is structured in four parts. After this introduction, which includes a brief recapitulation of IFL history until contemporary times, the analysis outlines the legal developments in IFL, focusing on certain fisheries issues addressed in international legally binding and non-binding instruments adopted under the auspices of the United Nations (UN). This includes the Food and Agriculture Organization of the United Nations (FAO), as well as two international judicial cases. The third part of this chapter examines regional legal developments in IFL, particularly of regional fishery bodies (RFBs) whose constitutive instruments entered into force within the past twenty-five years, and highlighting issues addressed in certain RFBs' conservation and management measures (CMMs). Finally, this chapter's fourth part provides a brief conclusion.

8.1.1 From Past to Contemporary Times

The history of IFL dates back to the seventeenth century (Thornton 2004; Somos 2012), although the interests of coastal States were not as predominant on fisheries topics as they were for navigation and trade. In 1609, Hugo Grotius published *Mare*

Liberium, advancing the notions of inexhaustible resources of the sea, arguably considered property of no one *res nullius*, a common possession *res communis*, or public property *res publica*, subject to the free exploitation by all in an open sea – *mare liberum*. This doctrine was later challenged by scholars' claims, as in John Selden's publication of *Mare Clausum* in 1635, founded on the principle of sovereignty over the closed sea (Thornton 2004; Somos 2012). Since this early period, the central issue in IFL has been the intersection of, on one hand, the free uses by States of the *open* seas, and, on the other hand, the restricted uses by States of *closed* seas. In seeking balance between the two with respect to fishing activities in the ocean, IFL has played a key role in regulating inter-State cooperation in the sustainable use, management and conservation of fishery resources (Dagget 1934; Carroz and Savini 1979; Churchill and Lowe 1999, 279–289; Kaye 2001, 44–88; Garcia and Cochrane 2009, 447–453). However, throughout most of IFL's history, international marine fisheries lacked global regulation, or a treaty with universal participation establishing minimum obligations on sustainable use and management of marine fisheries across the ocean (Dagget 1934; Oda 1983). Bilateral and multilateral agreements predominantly regulated the exploitation by few States of fishery resources of their joint interest, and marine waters where States' fishing activities overlapped (Carroz and Savini 1979).

Significant changes in IFL marked the twentieth century as States progressively recognized international rules on the ocean use, initially as customary international law, and later by a global treaty (Jacobson 1985; Churchill and Lowe 1999, 279–289). At regional level, from the early 1900s and more intensively from the 1940s onwards, States began cooperating through RFBs to coordinate activities on scientific research, data collection and dissemination, and the management and conservation of fishery resources (Heck 1975; Sydnes 2001). A large portion of high sea areas were governed by RFBs up until the 1970s, whereas coastal States enjoyed narrow maritime zones (Churchill and Lowe 1999, 283). Between the 1930s and the late 1950s, important developments in IFL at the global level occurred through two attempts by States in seeking consensus on the adoption of a treaty that would broadly cover ocean governance, including the delimitation of maritime zones (Jacobson 1985; Churchill and Lowe 1999, 279–289; Boyle 2005). After these attempts, States extensively negotiated at the Third United Nations Conference on the Law of the Sea, from 1973 to 1982, and concluded the UNCLOS, which was adopted by consensus as an interlocking package deal (Boyle 2005). This treaty is the main foundation of the contemporary LOS regime (Koh 1982) and the contemporary IFL domain (Molenaar and Caddell 2019). The UNCLOS is a global treaty with 168 Parties (as of October 2022), which partially filled the missing elements of IFL by setting out a general international legal framework for marine fisheries and certain specific fisheries management requirements to be observed by States Parties (Barrie 1986; Molenaar and Caddell 2019). Notably, a larger extension of national fishery limits, from 12 to 200 nautical miles, was formalized by the UNCLOS, considerably reducing the RFBs' sphere of influence over the management of fishery resources (Churchill 1998).

8.1.2 From 1994 Onwards

The adoption of the UNCLOS was paramount to the international regulation of marine fisheries (Oda 1983; Barrie 1986; Miles and Burke 1989; Hey 1999), albeit still with many limitations as detailed later. This treaty contributed to clarifying the rights and duties of State Parties with respect to fishing, and the fisheries legal regime applicable to areas under national jurisdiction (AUNJ) (Attard 1987; Tsamenyi and Hanich 2012; Andreone 2015). These include the coastal States and archipelagic States' territorial sea of up to 12 nautical miles, and the exclusive economic zone (EEZ), which can be claimed by them and reach up to 200 nautical miles (UNCLOS 1982, Articles 3, 48, 57–58). The UNCLOS also clarifies, to some extent, the legal regime applicable to the high seas, whilst setting out the foundation for further international regulation of fish stocks straddling between maritime zones and/or migrating across long distances (Davies and Redgwell 1997; Hewison 1999; Nelson 1999).

A decade after the UNCLOS's adoption, the 1992 UN Conference on Environment and Development (UNCED) brought about significant influence to the evolution of IFL, fostering sustainability perspectives from other specialised fields of law (Proelss 2016). This resulted in a growing interaction of IFL with environmental law (Freestone and Makuch 1997; Juda 2002), trade law (Young 2011; Urrutia 2018; Churchill 2019) and human rights law (Van der Burgt 2012; Papanicolopulu 2018; Song and Soliman 2019). Examples important to the international fisheries context include the precautionary principle (Freestone 1999; Boyle 2005; Ebben 2011), the ecosystem approach (Molenaar 2002; Diz 2012; De Lucia 2015; Kenny et al. 2018), the principle of marine biodiversity conservation (Rengifo 1997; Diz 2012; Garcia et al. 2014), and more recently the human rights-based approach (Azmi et al. 2016; Jentoft and Bavinck 2019; Nakamura 2022). Fisheries-specific concepts such as total allowable catch (TAC), illegal, unreported and unregulated (IUU) fishing, fisheries co-management and small-scale fisheries (SSF) have also evolved through the adoption of improved international standards by influence of those principles and approaches. All these legal developments have contributed to a richer and more holistic legal landscape for international marine fisheries (Molenaar and Caddell 2019; Harrison 2017; Garcia et al. 2014; Palma et al. 2010, 54–92).

After the 1992 UNCED, other key legal developments occurred in IFL. The UNCLOS entered into force in 1994 and, as will be examined later in this chapter, subsequent international legally binding and non-binding fisheries instruments were adopted, complementing the treaty's provisions on fisheries. Due to this progressive evolution of IFL following 1994, this chapter examines selected IFL developments from that year until mid-2022. The next section dedicates the analysis to instruments adopted at global level, highlighting certain issues that are less explored in the IFL literature.

8.2 Global Legal Developments in International Fisheries Law

IFL is mainly concerned with marine capture fisheries (Molenaar and Caddell 2019), which can be distinguished by their main *manner* and *purpose* of conducting marine fishing (Thomson 1980). As such, it comprises coastal artisanal and/or subsistence small-scale fisheries (SSF), which utilise small boats and low-capital finance, make for the largest fisheries workforce, and around forty percent of global fisheries production (Purcell and Pomeroy 2015; Chuenpagdee and Jentoft 2018; Smith and Basurto 2019). Marine capture fisheries also encompass industrial commercial large-scale fisheries, which utilise big vessels and high-capital finance, serving the broader market, including regional and international trade (McCauley et al. 2018). Both categories are not defined by international law, and while they may generally fall under generic references to 'fishing' or 'fisheries' in IFL instruments, national fisheries legislation of certain countries specifically define or refer to SSF (Nakamura et al. 2021). As will be detailed later, SSF gained increasing attention by the international community in the last decade due to their pivotal role in the provision of nutritious food, jobs, culture and livelihoods in coastal communities of both developed and developing countries, and due to the need to tackle their vulnerabilities to social, environmental and economic stressors (Nakamura et al. 2021; Nakamura 2022). Associated to marine capture fisheries are also the activities concerning planning, development, management, conservation, monitoring, control, surveillance and enforcement (MCSE), fisheries trade, which are or can be regulated by IFL instruments (Kuemlangan 2009). The contemporary IFL framework is, therefore, very broad and constantly evolving (see Table 8.1).

The way IFL has unfolded at diverse governance levels follows how public international law has itself developed more generally. It is part of a State-centric horizontal system where law-making primarily derives from States' consent (Caddell 2019). Such a system represents the traditional way of international law-making, which depends on States' willingness to agree to negotiate, prepare and adopt international rules on a given fisheries issue. The implementation of international obligations and standards on fisheries management and conservation greatly relies on States' individual and collective efforts in both internalizing and operationalizing the relevant international instruments at national levels (Kuemlangan 2009). During the period from 1994 to mid-2022, innovation and technology have been advancing quickly and consequently affecting fisheries, in positive terms (with improved data monitoring tools and systems) and negative terms (by highly mechanised equipment harmful to the environment and destructive fishing gears). At the same time, as the global population continues to grow, the greater is the pressure on fishery resources, which, in turn, has become more vulnerable to and impacted by increasing climate-related environmental changes (Cisneros-Montemayor et al. 2019). The next subsections will therefore examine how IFL has evolved to meet these global concerns, analysing selected issues based on the UNCLOS's legal developments; the United

Table 8.1 Selected legally binding instruments and other guidance relevant to International Fisheries Law 1994 to mid-2022

Year	International instrument[a]	Historical event	Status[b]
1994	UN Law of the Sea Convention	Entry into force (EIF) (adopted in 1982)	168 Parties
	CCBST – Convention for the Conservation of Southern Bluefin Tuna	EIF (adopted in 1993)	6 Parties
	UNGA Resolutions 49/**28** Law of the Sea (LOS), 49/**116** Unauthorized Fishing, 49/**118** Fisheries By-catch and Discards, 49/**121** UN Conference on Straddling Fish Stocks and Highly Migratory Fish Stocks, and 39/**436** Large-scale Pelagic Drift-net Fishing	UNGA 49th Session	
1995	CCBSP – Convention on the Conservation and Management of the Pollock Resources in the Central Bering Sea	EIF (adopted in 1994)	6 Parties
	ATLAFCO – Convention on the Cooperation among African States bordering the Atlantic Ocean	EIF (adopted in 1991)	22 Parties
	FAO Code of Conduct for Responsible Fisheries (CCRF)	FAO Conf. 28th Session	
	FAO Rome Consensus on Sustainable Fisheries	Ministerial Conference	
	Kyoto Declaration on Sustainable Contribution of Fisheries to Food Security	International Conference	
	Cancun Declaration on Responsible Fishing	International Conference	
	CBD Jakarta Mandate on Marine and Coastal Biological Diversity	CBD CoP-2	
	UNGA Resolutions 50/**23** LOS, 50/**24** UN Fish Stocks Agreement (UNFSA), 50/**25** Large-scale Pelagic Drift-net Fishing; Unauthorized fishing; fisheries by-catch and Discards	UNGA 50th Session	
1996	IOTC – Agreement for Establishment of the Indian Ocean Tuna Commission	EIF (adopted in 1993	29 Parties
	UNGA Resolutions 51/**34** LOS, 51/**35** UNFSA, 51/**36** Large-scale Pelagic Drift-net Fishing; Unauthorized Fishing; fisheries By-catch and Discards	UNGA 51st Session	
1997	UNGA Resolutions 52/**26** Oceans and LOS, 52/**28** UNFSA, 52/**29** Large-scale Pelagic Drift-net Fishing; Unauthorized Fishing; and fisheries By-catch and Discards	UNGA 52nd Session	
1998	Protocol on Environmental Protection to the Antarctic Treaty	EIF (adopted in 1991)	37 Parties
	OSPAR – Convention for the Protection of Marine Environment of the North-East Atlantic	EIF (adopted in 1992)	16 Parties
	UNGA Resolutions 53/**32** Oceans and LOS, 53/**33** Large-scale Pelagic Drift-net Fishing; Unauthorized Fishing; fisheries By-catch and Discards; and other Developments	UNGA 53rd Session	

(continued)

Table 8.1 (continued)

Year	International instrument[a]	Historical event	Status[b]
1999	OSPESCA Agreement for the Cooperation between the Fisheries and Aquaculture National Authorities of the Inter-American Countries and the General Secretary of the American Integration System	EIF (adopted in 1999)	8 Parties
	Agreement on the International Dolphin Conservation Programme	EIF (adopted in 1998)	14 Parties
	Rome Declaration on the CCRF Implementation	Ministerial Meeting	
	UNGA Resolutions 54/**31** Oceans and the LOS, 54/**32** UNFSA, 54/**33** Results of the review by Commission on Sustainable Development of the Sectoral Theme of 'Oceans and Seas'	UNGA 54th Session	
2000	FAO IPOA-Seabirds, IPOA-Sharks and IPOA-Capacity	FAO Council	
	UNGA Resolution 55/**2** UN Millennium Declaration (Millennium Development Goals)	UNGA 55th Session	
	CBD Ecosystem Approach's Description and Operational Guidelines	CBD CoP-5	
	UNGA Resolutions 55/**7** Oceans and the LOS, 55/**8** Large-scale Pelagic Drift-net Fishing; Unauthorized Fishing; fisheries By-catch and Discards; and other Developments	UNGA 55th Session	
2001	UN Fish Stocks Agreement	EIF (adopted in 1995)	91 Parties
	RECOFI – Agreement for Establishment of Regional Commission for Fisheries	EIF (adopted in 1999)	8 Parties
	ACCOBAMS – Agreement on the Conservation of Cetaceans of the Black Sea, Mediterranean Sea and contiguous Atlantic Area	EIF (adopted in 1996)	24 Parties
	Inter-American Convention for Protection and Conservation of Sea Turtles	EIF (adopted in 1996)	16 Parties
	EUROFISH – Agreement for the Establishment of the International Organization for the Development of Fisheries in Easter and Central Europe	EIF (adopted in 2002)	13 Parties
	FAO IPOA-IUU	FAO Council	
	Reykjavik Declaration on Responsible Fisheries in the Marine Ecosystem	International Conference	
	UNGA Resolution 56/**12** Oceans and LOS, and 56/**13** UNFSA	UNGA 56th Session	

(continued)

Table 8.1 (continued)

Year	International instrument[a]	Historical event	Status[b]
2002	CRFM – Agreement Establishing Caribbean Regional Fisheries Mechanism	EIF (adopted in 2002)	17 Parties
	CITES Appendix II Inclusion of basking sharks, whale sharks, seahorses, giant date mussel, various species of dolphins	CITES CoP-12	In force for most Parties
	Johannesburg Declaration on Sustainable Development and Plan of Implementation	World Summit on Sustainable Developm.	
	UNGA Resolutions 57/**141** Oceans and LOS, 57/**142** Large-scale pelagic drift-net fishing; Unauthorized Fishing/IUU Fishing; and fisheries By-catch and Discards; and other Developments, and 57/**143** UNFSA	UNGA 57th Session	
2003	FAO Compliance Agreement	EIF (adopted in 1993)	45 Parties
	SEAFO – Convention on the Conservation and Management of Fishery Resources in the South East Atlantic Ocean	EIF (adopted in 2001)	7 Parties
	Agreement on the BOBP-IGO	EIF (adopted in 2003)	4 Parties
	Venice Declaration on the Sustainable Development of Fisheries in the Mediterranean	Ministerial Conference	
	UNGA Resolutions 58/**240** Oceans and LOS, 58/**14** Sustainable Fisheries, including through the UN Fish Stocks Agreement and related instruments	UNGA 58th Session	
2004	WCPFC – Convention for the Conservation and Management of Highly Migratory Fish Stocks in the Western and Central Pacific Ocean	EIF (adopted in 2000)	25 Parties
	CITES Appendix II Inclusion of great white sharks, humphead maori wrasse	CITES CoP-13	In force for most Parties
	ACAP – Agreement on the Conservation of Albatrosses and Petrels	EIF (adopted in 2001)	13 Parties
	UNGA Resolutions 59/**24** Oceans and LOS, 59/**25** Sustainable Fisheries, including through the UNFSA and related instruments	UNGA 59th Session	

(continued)

Table 8.1 (continued)

Year	International instrument[a]	Historical event	Status[b]
2005	FAO-SWIOFC – Resolution/Statutes of the South West Indian Ocean Fisheries Commission	SWIOFC Session	12 Members
	FAO/ILO/IMO Revised Code of Safety for Fishermen and Fishing Vessels FAO/ILO/IMO Revised Voluntary Guidelines for the Design, Construction and Equipment of Small Fishing Vessels	IMO-MSC 79th Session FAO-COFI 26th Session ILO 293rd Session	
	FAO Rome Declaration on Illegal, Unreported and Unregulated Fishing	Ministerial Meeting	
	Abuja Declaration on Sustainable Fisheries and Aquaculture in Africa	NEPAD Meeting	
	UNGA Resolutions 60/**30** Oceans and LOS, 60/**31** Sustainable Fisheries	UNGA 60th Session	
2006	FAO and CITES Secretariat Memorandum of Understanding (MoU)	Signed	Active
	UNGA Resolutions 61/**222** Oceans and LOS, 61/**105** Sustainable Fisheries	UNGA 61st Session	
2007	FCWC – Convention establishing the West Central Gulf of Guinea's Fishery Committee	EIF (adopted in 2007)	6 Parties
	CITES Appendix II Inclusion of European eels and CITES Appendix I Inclusion of sawfish	CITES CoP-14	In force
	ILO Recommendation Concerning the Work in the Fishing Sector (R199)	General Conference	
	UNGA Resolutions 62/**215** Oceans and LOS, 62/**177** Sustainable Fisheries	UNGA 62nd Session	
2008	Third Arrangement implementing the Nauru Agreement setting forth additional terms and conditions of access to the fisheries zones of the Parties	EIF (adopted in 2008)	8 Parties
	FAO International Guidelines for the Management of Deep Sea Fisheries in the High Seas	Technical Consultation	
	UNGA Resolutions 63/**111** Oceans and LOS, 63/**112** Sustainable Fisheries	UNGA 63nd Session	
2009	SADC – Southern African Development Community's Protocol on Fisheries	EIF (adopted in 2006)	12 Parties
	UNGA Resolution 64/**71** Oceans and LOS, 64/**72** Sustainable Fisheries	UNGA 64th Session	
2010	CACFish – Agreement on the Central Asian and Caucasus Regional Fisheries and Aquaculture Commission	EIF (adopted in 2009)	5 Parties
	CMS, States and Cooperating Partners MoU on the Conservation of Sharks	Signed	Active
	UNGA Resolution 65/**37** Oceans and LOS 'A', 65/**37** Oceans and LOS 'B', and 65/**38** Sustainable Fisheries	UNGA 65th Session	

(continued)

Table 8.1 (continued)

Year	International instrument[a]	Historical event	Status[b]
2011	FAO International Guidelines on Bycatch Management and Reduction of Discards	COFI 29th Session	
	UNGA Resolutions 66/**231** Oceans and LOS, 66/**68** Sustainable Fisheries	UNGA 66th Session	
2012	UNGA Resolution 66/**288** The Future We Want	UNGA 66th Session	
	SIOFA – Southern Indian Ocean Fisheries Agreement	EIF (adopted in 2006)	10 Parties
	SPRFMO – Convention on the Conservation and Management of High Seas Fishery Resources in the South Pacific Ocean	EIF (adopted in 2009)	15 Parties
	IMO International Convention on Standards of Training, Certification and Watchkeeping for Fishing Vessel Personnel (STCW-F)	EIF (adopted in 1995)	33 Parties
	IMO Cape Town Agreement of 2012 on the Implementation of the Provisions of the 1993 Protocol relating to the 1977 Torremolinos International Convention for the Safety of Fishing Vessels	Diplomatic Conference	Not in force 16 Parties
	SRFC - Convention on the Determination of Minimal Conditions for Access and Exploitation of Fishery Resources within the Maritime Areas under Jurisdiction of the Member States of the Sub-Regional Fisheries Commission	EIF (adopted in 2012)	7 Parties
	UNGA Resolution 67/**78** Oceans and LOS, 67/**69** Sustainable Fisheries	UNGA 67th Session	
2013	CITES Appendix II Inclusion of oceanic whitetip sharks, hammerhead sharks (scalloped-, great- and smooth-), porbeagle sharks, manta rays,	CITES CoP-16	In force for most Parties
	UNGA Resolutions 68/**70** Oceans and LOS, 68/**71** Sustainable Fisheries	UNGA 68th Session	
2014	FAO Voluntary Guidelines for Securing Sustainable Small-Scale Fisheries in the context of food security and poverty eradication	COFI 31st Session	
	FAO Voluntary Guidelines for Flag State Performance	COFI 31st Session	
	UNGA Resolutions 69/**245** Oceans and LOS, 69/**109** Sustainable Fisheries	UNGA 69th Session	

(continued)

Table 8.1 (continued)

Year	International instrument[a]	Historical event	Status[b]
2015	NPFC – Convention on the Conservation and Management of the High Seas Fisheries Resources in the North Pacific Ocean	EIF (adopted in 2012)	7 Parties
	UNGA Resolution 70/**1** 'Transforming Our World: the 2030 Agenda for Sustainable Development'	UN Sustainable Development Summit	
	UNGA Resolutions 70/**235** Oceans and LOS, 70/**226** UN Conference to Support Implementation of SDG 14, 70/**75** Sustainable Fisheries, and 69/**292** Development of an International Legally Binding Instrument under the UNCLOS on the Conservation and Sustainable Use of Marine Biological Diversity of ABNJ (BBNJ Agreement)	UNGA 70th Session	
2016	FAO Port States Measures Agreement	EIF (adopted in 2009)	74 Parties
	CITES Appendix II Inclusion of silky sharks, thresher sharks, devil rays, clarion angelfish, chambered nautilus.	CITES CoP-17	In force for most Parties
	UNGA Resolutions 71/**257** Oceans and LOS, 71/**123** Sustainable Fisheries, 71/**124** World Tuna Day, and 70/**303** Modalities for UN Conference to Support SDG14's Implementation	UNGA 71st Session	
2017	ILO Work in Fishing Convention No. 188	EIF (adopted in 2007)	20 Parties
	FAO Voluntary Guidelines for Catch Documentation Schemes	FAO Conf. 40th Session	
	UNGA Resolutions 71/**312** Our ocean, our future: call for Action, 72/**72** Sustainable Fisheries, 72/**73** Oceans and LOS, 72/**74** BBNJ Agreement	UNGA 72nd Session	
2018	FAO Voluntary Guidelines on the Marking of Fishing Gears	COFI 33rd Session	
	UN Declaration on the Rights of Peasants and Other People Working in Rural Areas	UNHRC 39th Session	
	UNGA Resolutions 73/**124** Oceans and LOS, 71/**123** Sustainable Fisheries, and 71/**124** World Tuna Day	UNGA 73rd Session	
2019	CITES Appendix II Inclusion of mako sharks (shortfin- and longfin-), giant guitarfish, wedgefish and teatfish species	CITES CoP-18	
	UNGA Resolutions 74/**19** Oceans and LOS, 74/**18** Sustainable Fisheries, and 73/**292** 2020 UN Conference to Support the Implementation of SDG 14	UNGA 74th Session	
2020	UNGA Resolutions 75/**239** Oceans and LOS, 75/**89** Sustainable Fisheries	UNGA 75th Session	

(continued)

Table 8.1 (continued)

Year	International instrument[a]	Historical event	Status[b]
2021	FAO COFI Declaration for Sustainable Fisheries and Aquaculture	COFI 34th Session	
	Agreement to Prevent Unregulated High Seas Fisheries in the Central Arctic Ocean	EIF (adopted in 2018)	10 Parties
2022	WTO Agreement on Fisheries Subsidies	Adopted (not yet in force)	
	Voluntary Guidelines for Transshipment	COFI 35th Session	

[a]titles of the instruments were shortened for better visualization
[b]As of consultations on official websites of the RFBs and on FAOLEX and ECOLEX, in October 2022

Nations General Assembly's (UNGA) fisheries-related non-binding resolutions; the instruments adopted under FAO's auspices; and two international judicial cases.

8.2.1 Law of the Sea Convention and the International Regulation of Fisheries

The UNCLOS has the ability to 'live' beyond its adoption, addressing persisting and emerging problems, as well as adapting to technological progress and social recognition of values (Barret and Barnes 2016; Molenaar and Caddell 2019, 3). This treaty's incorporation of generally accepted international rules and standards (GAIRS) make room for other IFL instruments to be interpreted and applied complementarily, arguably determining certain coastal States' obligations to manage and conserve their domestic fish stocks (Harrison 2017, 171–180). Notwithstanding, the UNCLOS has a limited approach to the regulation of fishing activities (Freestone and Makuch 1997). Fisheries management and conservation are more specifically addressed therein with respect to the EEZ, including the imposition of TAC, and use of best scientific evidence available for the conservation and management of fishery resources (UNCLOS 1982, Articles 61–62; Nakamura 2022). Beyond the EEZ's water column, in the high seas, States enjoy the freedom of fishing pursuant to their duty to cooperate with other States for the conservation and management of living resources (UNCLOS 1982, Articles 87(1)(e), 116–118). In all maritime zones, States have the general duty to protect and preserve the marine environment (UNCLOS 1982, Articles 192–237).

Despite the milestones achieved with the adoption of UNCLOS, many issues remained insufficiently addressed or imprecisely regulated, such as high seas fisheries and marine biodiversity conservation (Barrie 1986; Vicuña 1993). The Convention's provisions on straddling fish stocks and highly migratory fish stocks were further elaborated by the second implementing agreement relating to the UNCLOS, adopted in 1995, widely known as the UN Fish Stocks Agreement

(UNFSA 1995). This agreement is considered an adaptation or modification of the UNCLOS by subsequent practice (Buga 2015), building upon the 1992 Agenda 21 oriented vision of sustainable development and conservation (Agenda 21, Chapter 17). The UNFSA has gone beyond UNCLOS in various ways. It follows an ecosystem approach to fisheries (EAF) (Garcia et al. 2003) and expressly provides for the protection of 'biodiversity in the marine environment' (UNFSA 1995, Articles 3(1) and 5(b), (d), (f)). The UNFSA provides for the precautionary principle, outlining the measures to be taken in applying this principle (UNFSA 1995, Articles 5(c), (i) and 6(3), (4), (6)). It also addresses SSF by requiring the Parties to take into account the interests of 'artisanal and subsistence fishers' in their duty to cooperate under this agreement (UNFSA 1995, Article 6(5)). All these three elements are to be observed by the Parties in any maritime area, including in AUNJ (UNFSA 1995, Article 3(1)). Another important feature of the UNFSA is the principle of compatibility, according to which States are required to cooperate for ensuring coherent and non-conflictual conservation and management measures (CMMs) applicable in EEZ and adjacent high sea areas (UNFSA 1995, Article 7(2)).

In respect to high seas fisheries, the UNFSA has contributed to lift RFBs to their central role in the conservation and management of high seas stocks while enabling all States, including distant water fishing nations and RFBs' non-members, to enter into international fisheries, or at least challenge their potential exclusion from participating in fishing and fishing related activities in the areas governed by RFBs (Serdy 2016). The UNFSA contains far-reaching provisions on fisheries enforcement by States members of RFBs (Buga 2015), which hold the right to board and inspect any other State's vessels to ensure compliance with the applicable CMMs for stocks falling under the competent RFB area (UNFSA 1995, Article 21). With such provisions, the UNFSA was considered a pioneering legal instrument to move away from the primary control of flag State jurisdiction over fishing vessels on the high seas (Lodge and Nandan 2005). In turn, the control over fishing vessels by States other than the flag States, with respect to legal compliance with CMMs, rules on customs, immigration, sanitation and national security, is a matter that was (and continues to be) challenged by the increasing influence of port States' control and enforcement (Molenaar 2007).

Overall, the UNFSA has had a very constructive influence in IFL (Fresstone and Makuch 1997; Hayashi 1999; Bratspies 2001; Lodge and Nandan 2005) despite the lack of consideration for climate change (Pinsky et al. 2018). This is one of the gaps underpinning the debate around the UNCLOS's third implementing agreement i.e., the proposed international legally binding instrument on the conservation and management of marine biodiversity beyond national jurisdiction (BBNJ Agreement). Though most nations fish within their own EEZ, high seas fishing is a reality for some flag States and fishing entities, including China, Spain, Taiwan, Japan and South Korea (Kroodsma et al. 2018). The fact that high seas fisheries are regulated by existing international instruments, including the UNFSA and regional CMMs, has raised difficult questions vis-à-vis its inclusion in the proposed BBNJ Agreement (Barnes 2016). While such an agreement could conflict with the activities of RFBs already in place, it could nevertheless be an alternative or complementary tool to help them tackle poorly regulated, weakly enforced and unsustainable high seas fishing. At most, it could regulate discrete high seas stocks and other aquatic species and/or areas not regulated by RFBs (Barnes 2016) and which may be impacted by fisheries industry directly or indirectly through abandoned, lost or otherwise discarded fishing gears.

In relation to treaty-monitoring mechanism for UNCLOS, it is important to note that the UN Secretary-General performs functions through the Division for Oceans Affairs and Law of the Sea (DOALOS), which serves as the Secretariat of UNCLOS. One of the functions of the UN Secretary-General, pursuant to the UNCLOS, is to convene the meetings of State parties to the Convention (UNCLOS 1982, Article 319(2)(e)), but the matters dealt in such occasions have not focused on fisheries issues *per se* (Tarassenko and Tani 2012). Legal developments on fisheries have rather been showcased through several high-level UN conferences and meetings, as seen below.

8.2.2 Other Legal Developments Through High-Level UN Conferences and Meetings

The main legal sources produced, under the UN auspices and which integrate the IFL framework, are the LOS-related and fisheries-related resolutions adopted at the UNGA annual meetings (Harrison 2011; Caddell 2019). The UNGA resolutions are non-binding instruments, but they hold law-making importance by influencing activities of States, regional and international organizations in numerous issues, including fisheries management and conservation (Caddell 2019). Notably, specific concerns with large-scale driftnet fishing came to force in the early 1990s, and by the mid-2000s, UNGA began to address the negative impacts caused on deep-sea vulnerable marine ecosystems (VMEs) by unsustainable bottom fishing practices (Caddell 2019). From 1994 onwards, there have been at least two UNGA resolutions per year, one addressing the broad LOS theme, which resonates more closely with the UNCLOS, and others addressing certain fisheries topics, which concern those issues of the UNFSA and related fisheries instruments (Caddell 2019). In this context, important contributions of the UNGA resolutions to IFL developments have addressed three key fisheries issues, highlighted by UNGA in the past twenty-five years, relating to: (i) unauthorised fishing, (ii) fisheries by-catch and discards and (iii) artisanal and subsistence small-scale fisheries.

The matter of 'unauthorised fishing' was introduced in UNGA's discussions in 1994. The initial concern was with the detrimental impact caused by fishing in AUNJ, especially in developing countries, and the duty of flag States with respect to duly implementing, controlling and enforcing their fishing authorisation schemes (UNGA Resolution 49/1161994). Only in 1999, the UNGA Resolution 54/32 expressly mentioned concern with IUU fishing, as reflected previously at the regional level (Serdy 2016), and it provided for FAO's mandate to develop what came later to be the International Plan of Action (IPOA)-IUU. Since then, the general treatment of 'unauthorised fishing' was also associated with IUU fishing practices on the high seas as well as the numerous related activities concerning compliance with international CMMs. From 2003 onwards, an IUU-fishing dedicated section was included in the fisheries-specific resolution, deepening the discussions on this topic and referring not only to the related activities by States and FAO, but also including WTO's efforts and cooperation through the International Maritime Organization (IMO). The recurrent appearance of these issues have influenced the adhesion of States to the Agreement to Promote Compliance with International Conservation and Management Measures by Fishing Vessels on the High Seas (Compliance Agreement 1993) and the Agreement on Port State Measures to Prevent, Deter and Eliminate Illegal, Unreported and Unregulated Fishing (PSMA 2009), both adopted under FAO auspices, as well as have fostered inter-agency cooperation to implement them. The recently adopted UNGA Resolution 75/89 reflects the various additional matters which have been included in the topic of IUU fishing throughout the past years. These comprise concerns with effective flag States' jurisdiction, control and enforcement over the vessels flying their flag, port States' measures and control, maritime safety and decent labour conditions, landings and catch reporting and associated data-sharing, the importance of trade and market-related measures, public and private ecolabelling schemes, and the linkage between illegal fishing and transnational organised crime (UNGA Resolution 75/892021).

The 'fisheries by-catch and discards' issue was also introduced in UNGA's discussions in 1994 (UNGA Resolution 49/1181994). While by-catch concerns non-target species caught incidentally, the problem of discards applies to any species subject to 'oceans wasting' (Gillespie 2002). Discards occur for all sorts of reasons such as lack of space to keep the species on board the fishing vessel, non-profitability of the species, which lead one to discard the species overboard instead of landing it or bringing it to the shore. Certain species of marine mammals, sharks, sea turtles and seabirds have nevertheless acquired special protection in other international instruments, including multilateral environmental agreements through time. Associated debates were then generally improved in the international fora and, since 2003, a 'fisheries by-catch and discards' dedicated section has been fostering activities to reduce and combat these problems, including catch by lost or abandoned gear and post-harvest losses, with particular attention to juvenile fish (UNGA Resolution 58/142003). The UNGA Resolution 75/89 includes the concern with impacts by large-scale fish aggregating devices, the importance of electronic monitoring, standardised data collection and reporting protocols, conservation of non-target species incidentally harvested, minimizing sea turtles and seabirds by-catch and increasing post-release survival of these species (UNGA Resolution 75/892021).

Finally, SSF issues, which do not enjoy a specific section in the fisheries-related UNGA resolutions yet, are worth highlighting for their increasing importance and limited coverage by IFL literature. Particular attention to SSF by UNGA was made in 2003, highlighting the impacts of directed and non-directed shark catch fisheries on shark populations and related species, taking into account the nutritional and socio-economic considerations 'particularly as they relate to small-scale, subsistence and artisanal fisheries and communities' (UNGA Resolution 58/142003). In 2005, UNGA acknowledged the importance of the fisheries sector 'including small-scale and artisanal fisheries' to developing countries, in respect of the need to eliminate fisheries subsidies that contribute to IUU fishing and fishing overcapacity (UNGA Resolution 60/312005). In 2006, the 'participation of small-scale fishery stakeholders' in policy development and fisheries management strategies were emphasized, and FAO was mandated to develop guidance for enhancing the contribution of SSF to poverty alleviation and food security (UNGA Resolution 61/1052006), later resulting in the adoption of the 2014 Voluntary Guidelines for Securing Sustainable Small-Scale Fisheries in the Context of Food Security and Poverty Eradication (SSF Guidelines 2014). The SSF Guidelines are the first comprehensive international instrument dedicated to the full SSF value-chain, and have significantly contributed to strengthen the recognition, protection and empowerment of small-scale fishers, their human rights, and SSF sustainability (Morgera and Nakamura 2021; Nakamura 2022). The SSF Guidelines arguably hold normative significance, despite their non-binding nature, and can produce law-making effects at international, regional and national levels of governance (Nakamura 2022). While a specific section in the UNGA resolutions has not yet been fixed, the consideration of SSF needs and the mandate of FAO to develop guidelines for this fisheries subsector illustrate the growing importance given by UNGA to SSF. The UNGA Resolution 75/89 includes the concerns with SSF access to fishery resources and markets, capacity development and technical support to SSF, participation of SSF stakeholders in policy development and fisheries management, the recognition of SSF's important role and need for support to their long-term environmental, economic and social sustainability (UNGA Resolution 75/892021).

In addition to these specific issues of IUU fishing, by-catch and SSF, it is worth noting the UNGA Resolution that established the UN Open-ended Informal Consultative Process on Oceans and the Law of the Sea in 1999 (UNGA Resolution 54/331999), which has been meeting annually for the international review of ocean affairs and generating important instruments and discussions, on which DOALOS has been producing relevant reports as well (de La Fayette 2006). The UNGA Resolution 70/1, in turn, provides for the Sustainable Development Goals (SDGs), setting out SDG14, entirely dedicated to the conservation and sustainable use of the oceans and marine resources, contemplating ten targets, four of which directly related to fisheries, tackling overfishing and IUU fishing (SDG 14.4), harmful fisheries subsidies (SDG 14.6), sustainable fisheries in Small Island Developing States and least developed countries (SDG 14.7), and SSF access to marine resources and markets (SDG 14B) (UNGA Resolution 70/12015). Of particular relevance to IFL and ocean governance is SDG 14C, aimed at enhancing conservation ad sustainable use of oceans and their resources by implementing international law as reflected in the UNCLOS. While SDG14 is the evident SDG related to IFL, other SDGs are particularly important in addressing key social and environmental issues (e.g., hunger, gender, decent work, climate change) that affect fisheries, especially SSF (Said and Chuenpagdee 2019; Morgera and Nakamura 2021).

8.2.3 FAO Complementary Instruments

Since the establishment of FAO in 1945, the organization has been facilitating the cooperation among its members with respect to the appropriate use, management, development and conservation of world fisheries (Harrison 2011). It acts as the principal body for developing IFL and promoting the implementation of UNCLOS's provisions on fisheries (Harrison 2011; Boyle and Chinkin 2007, 126–128). FAO's initiatives involve providing technical support through the elaboration and improvement of the international standards for fishing, and do not depend on the provision of a clear mandate delegated by the UNCLOS (e.g., CCRF 1995; Edeson 1996). Nonetheless, the Convention has strengthened the ability of FAO to perform such activities. The aforementioned UNCLOS's rule of reference or GAIRS, addressing the conservation of living resources and protection of the marine environment in any maritime zone (UNCLOS 1982, Articles 61(3), 119(1)(a) and 197), arguably encourage coastal States to follow best practices when developing CMMs (Harrison 2017, 171), many of which are oriented by FAO's guidance. From the mid-1990s onwards, numerous FAO guidelines on fishing and fishing related activities as well as two legally binding instruments complemented the UNCLOS.

Two of such instruments were adopted under Article XIV of FAO's Constitution, namely the Compliance Agreement and the PSMA, whose provisions respectively bind 45 and 74 Parties, including the EU (as of October 2022). The advantage of being adopted at different times in IFL history is that each can resonate with the interests of governments, which neither the UNCLOS nor UNFSA may have captured before. The United States, for example, is not a Party to the UNCLOS, but has ratified the UNFSA, the PSMA and the Compliance Agreement. In turn, Libya and Turkey are non-Parties to the UNCLOS and the UNFSA, but both countries are Parties to the PSMA. As such, these States, while non-Parties to the UNCLOS, are bound by those other IFL instruments, which provide more detailed rules on, for instance, the conservation and management of straddling fish stocks and highly migratory fish stocks, port States measures and flag States jurisdiction on the high seas. Despite their limited participation, both the PSMA and the Compliance Agreement play significant roles in IFL and, as seen further below, many regional initiatives have to some extent addressed their requirements in relevant CMMs.

In the same year the UNFSA was adopted, 1995, FAO Members adopted the Code of Conduct for Responsible Fisheries (CCRF 1995). In spite of its voluntary nature, this code reflects rules already provided in legally binding instruments, including the UNCLOS and the Compliance Agreement (CCRF 1995, Article 1). The use of the CCRF in providing more precise or detailed meanings of the obligations contained in legally binding instruments arguably strengthens the CCRF's capacity of generating a normative effect or influence (Barnes 2006 at 253; Harrison 2017, 180). The objectives of this code include providing guidance for the implementation of other international legal instruments and standards of conduct for all persons in the fisheries sector (CCRF 1995, Article 2(a)(j)). In form, the CCRF resembles one of a general regulatory framework, providing an improved set of provisions that elaborates on those present in the UNCLOS's primary rules. In doing so, the CCRF clearly provides for an EAF (CCRF 1995, Articles 6.1–6.7, 6.9), the precautionary principle (CCRF 1995, Article 6.5), and takes into account the interests of the SSF (CCRF 1995, Article 6.18). It has a broader scope than the UNFSA, applying to any aquatic species subject to fishing and fishing related activities, and not solely to the straddling and highly migratory fish stocks. It also covers concerns falling under national sovereignty such as fisheries management and operations in AUNJ (CCRF 1995, Articles 7–8), aquaculture (CCRF 1995, Article 9) and the interaction between fisheries and coastal management (CCRF 1995, Article 10), setting important international standards or GAIRS in IFL.

With more than twenty-five years of implementation, the CCRF has been significantly influential in IFL developments. Similar to the UNCLOS, it may be hard to find an international fisheries instrument adopted in the course of the last two decades which does not mention the CCRF either explicitly or in replicating certain CCRF's provisions. The aforementioned two FAO agreements have in fact intrinsic relationships with the CCRF, sharing common provisions and concepts (Moore 1999 at 91–93). While the Compliance Agreement forms an integral part of the CCRF (CCRF 1995, Article 1.1), the PSMA provides the regulatory stream that expands on the Port States duties of Article 8.3 of the CCRF. The influence that the CCRF has on IFL developments is also perceived in the numerous declarations adopted at international conferences and ministerial meetings, which reinforce the importance of applying the CCRF (e.g., Kyoto Declaration 1995; Rome Declaration 1999 and Reykjavik Declaration 2001). The UNGA recognises the CCRF and other international related instruments, including FAO's IPOAs), as setting out the 'principles and global standards of behaviour for responsible practices for conservation of fisheries resources and the management and development of fisheries' (UNGA Resolution 75/892021, Preamble).

The FAO has also developed several technical guidelines for responsible fisheries to further clarify and guide the implementation of the CCRF provisions with respect to a specific matter (Kuemlangan 2009). Despite holding no formal legal status, these instruments have an important role in the development of customary law in IFL by reproducing the set of internationally elaborated principles based on which States are expected to follow in their domestic practices (Barnes 2006, 254; Kuemlangan 2009). Notably, FAO's IPOA aimed at preventing, deterring and eliminating IUU fishing has created the parameters necessary for what further came to be the PSMA. These voluntary instruments have also addressed certain matters that have been for long neglected or insufficiently covered by IFL. For instance, with the adoption of the SSF Guidelines in 2014, bringing international recognition and attention to a fisheries subsector that has been widely suffered from marginalisation and vulnerability (Béné et al. 2010; Chuenpagdee and Jentoft 2011; Purcell and Pomeroy 2015). These FAO instruments consolidate common understandings about a given subject in fisheries, filing gaps in the international legal regime of fisheries, to which States and judicial bodies may use for evidence of what IFL stands for (Edeson 1999). It is on the latter that the next subsection turns to, analysing two selected international cases to illustrate how IFL sources can be interpreted by an international court.

8.2.4 *Judicial Interpretation of IFL in Selected International Cases*

Although international judicial decisions, within contentious cases, apply strictly to the parties of the relevant dispute, these decisions set a precedence to guide international judges in deciding future cases, thereby being significant for any State (Harrison 2007). International judicial decisions may also be of an advisory nature, non-binding to the party requesting the opinion, but also serves States in interpreting and applying international legal instruments. While the merits of the cases are not put in scrutiny, this subsection examines how certain IFL instruments were interpreted or considered by the international judicial body constituted by the UNCLOS (UNCLOS 1982, Annex VI) – the International Tribunal for the Law of the Sea (ITLOS or Tribunal). Disputes concerning the interpretation or application of the UNCLOS provisions must be initially resolved consensually, and, if such consensus is not reached, may be referred to through the ITLOS by any Party, pursuant to all Parties having declared the Tribunal as their preferred means of settlement (Churchill 2007, 387). Parties may also declare their preference for other dispute settlement mechanisms, including the International Court of Justice and arbitral tribunals constituted in accordance with UNCLOS requirements (UNCLOS 1982, Article 287(1)). Due to the limited space left in this chapter, however, only two selected ITLOS cases will be examined: an advisory opinion and a contentious case.

The ITLOS 2015 *Sub-Regional Fisheries Commission (SRFC) Advisory Opinion* (ITLOS Case No. 21) was the first advisory opinion delivered by the full Tribunal (Freestone 2016). It clarified the SRFC's four questions related to the exercise of fishing in the EEZs of SRFC's member States by fishing vessels flying the flag of the EU member States, with which the SRFC have concluded fishing access agreements. In respect of IFL instruments, the Tribunal noted the importance of the definition of IUU fishing provided by the IPOA-IUU, highlighting that it 'draws up within the framework of the [CCRF]', was 'subsequently incorporated and reaffirmed in article 1(e) of the [PSMA]' and 'has also been included in decisions of some regional fisheries management organizations, (…) the national legislation of a number of States and the law of the [EU]' (ITLOS Case No. 21, Para 92). This reference indicated the Tribunal's view of the importance of the IUU fishing definition in the IPOA-IUU, which played 'an important role in the context of the consideration of the obligations borne' within the SRFC Convention's area of application (ITLOS Case No. 21, Para 95). The definition of 'unregulated fishing' in particular helped the Tribunal to clarify the duty of the coastal State to 'have in place national management and conservation measures and policies in relation to fishing resources' within its EEZ (ITLOS Case No. 21, Para 114).

The Tribunal answered the SRFC's four questions based on interpretation of the UNCLOS, especially the provisions on the EEZ, as well as on relevant international cases. Other specific IFL instruments were discussed by the Tribunal. For instance, the ITLOS noted that the bilateral fisheries access agreements concluded by the SRFC member States provided for the obligation of the flag State to ensure compliance with CMMs of the International Commission for the Conservation of the Atlantic Tunas (ICCAT) (ITLOS Case No. 21, Para 96), and it referred to the EU Common Fisheries Policy's definition of 'Union fishing vessel' for arguing on the liability aspects of the case (ITLOS Case No. 21, Paras 165–174). Such references were quite limited, but Judge Paik's separate opinion elaborated further on the relevance of IFL instruments, particularly those non-legally binding, noting that 'the post-UNCLOS normative developments as a whole (…) are relevant to the present case as to the state and direction of international fisheries law on this question'. Judge Paik emphasised the reason for the Tribunal to look carefully into such legal developments as a means to clarify what constitutes the generally accepted international regulations, procedures and practices or GAIRS, 'not because they are binding upon States as either treaty law or customary law, but rather because they are indicative of such regulations, procedures and practices' (Separate Opinion, Para 27).

In turn, the ITLOS 2014 *M/V Virginia G (Panama v Guinea Bissau)* case (ITLOS Case No. 19) generated important views by the Tribunal and Judges on certain IFL instruments. Notwithstanding other matters dealt by the Tribunal, the key point for the interpretation of IFL instruments was addressed in respect of the competence to exercise regulatory jurisdiction over the bunkering activities (i.e., provision of gas and oil) in support of foreign vessels fishing in Guinea Bissau's EEZ. The Tribunal clarified the need of such activities to have a 'direct connection to fishing' in order to fall under the list of matters on which the coastal State, in the exercise of its sovereign rights to explore, exploit, conserve and manage its EEZ living resources, is entitled to adopt laws and regulations (ITLOS Case No. 19, Paras 207–215). The Tribunal concluded that 'coastal States have jurisdiction to regulate the bunkering of foreign vessels fishing in their [EEZs] and to provide for the necessary enforcement measures', which include the boarding, inspection and arrest of vessels concerned (ITLOS Case No. 19, Paras 264–265).

Notably, ITLOS expressly affirmed that, in reaching such conclusion, it was 'also guided by the definitions of "fishing" and "fishing related" activities in several of the international agreements' (ITLOS Case No. 19, Para 216). The Tribunal cited various examples of IFL instruments, including the PSMA, the revised SRFC Convention, the North Pacific Anadromous Fish Commission (NPAFC)'s Convention, the South East Atlantic Fisheries Organisation (SEAFO)'s Convention, the Southern Indian Ocean Fisheries Agreement (SIOFA), the Western and Central Pacific Fisheries Commission (WCPFC)'s Convention, and the Commission for the Conservation of Southern Bluefin Tuna (CCSBT)'s Convention. Based on these IFL instruments, the Tribunal concluded that the bunkering of foreign fishing vessels in

Guinea Bissau's EEZ, including the supply of fuel to fishing vessels, is comprised by these instruments' definition of 'fishing related activities' (ITLOS Case No. 19, Paras 216–219). This part of the judgement is an important example of how IFL instruments, to which the Parties of the dispute are not necessarily bound by, may be used to guide the Tribunal's reasoning.

According to Judge Gao, the Tribunal's decision was a pioneering and progressive step which might be regarded as 'breaking new ground in international case law' by determining that such bunkering activities connected to fishing vessels do not fall under the category of freedom of navigation, allowing for coastal States to regulate on and take enforcement measures against them (Separate Opinion, Paras 11–12). Judge Ndiaye, in turn, recalled the role of the UN system's specialised agencies to 'concern themselves with the technical details under the chapter headings established by the Convention [UNCLOS]', referring to instruments drawn up under the auspices of the FAO, expressly mentioning the CCRF, the IPOA-IUU, the Compliance Agreement and the PSMA (Dissenting Opinion, Para 179). Such instruments were again referred by Judge Ndiaye as examples at the global level of the 'extensive regulation of fishing and related activities in the EEZ', as well as many other IFL instruments of regional scope (Dissenting Opinion, Paras 209–215).

These two ITLOS cases, particularly the Judges' separate opinions highlighted above, strengthen the legal force of the overall IFL framework, which can be used to guide the resolution of future cases or dispute resolutions in other international adjudicatory and arbitral forums. These cases demonstrate how international jurisprudence can also contribute to the development of IFL and the interpretation of relevant IFL instruments, including non-binding ones, which consist of a large part of the IFL domain.

8.3 Regional Regulation of Marine Fisheries

A substantive part of IFL is produced at regional or multilateral levels through RFBs, of which there are now about 50 (Løbach et al. 2020). Inter-State cooperation through RFBs, for the management of straddling and highly migratory fish stocks in AUNJ and beyond, existed years before the UNCLOS's entry into force (Heck 1975). Yet, when comparing the contexts before and after the EEZ concept was codified, the number of RFBs has doubled (Sydnes 2001, 355). Since 1994, over fifteen of the RFBs constitutive instruments have entered into force (see Table 8.1). Historical trends in the RFBs from the years before the negotiation of UNCLOS to the period following the 1990s, have been characterised as moving from 'loose, mainly advisory regional commissions which had multi-species responsibilities and relatively limited powers' into further being predominated by 'the establishment of several species-specific institutions' (Barston 1999, 341–342). Despite the multiple RFBs currently in place, there remain regions on the high seas and species, including high seas discrete species, which are not governed by an RFB, a regulatory gap that could be filled by the proposed BBNJ Agreement (Barnes 2016).

The functions of RFBs vary, but the main feature distinguishing those referred to as regional fisheries management organizations and/or arrangements (RFMO/As) is their competence to establish legally binding CMMs, as opposed to a mandate focused on scientific research, coordinative and/or developmental (Caddell 2019; Harrison 2019; Sydnes 2001). Most of RFBs have a purely advisory role (Løbach et al. 2020). In general, the constitutive instruments of RFMO/As provide for their competence to adopt CMMs that may be binding on their members pursuant to applicable procedures (Harrison 2019; Molenaar 2019). These CMMs contribute to the regional regulatory framework of IFL by, for instance, regulating issues not covered by the UNFSA. IFL instruments of RFMO/As therefore include their constitutive instrument, binding on the parties, and the CMMs, which may be binding on member States or not, depending on the State member's acceptance of the CMM (Harrison 2019). An interesting point of debate is the differentiated opt-out procedures adopted by RFMO/As, which often pose constraints on members objecting to a given CMM. The restrictions vary and may include an additional requirement for members to justify their objecting reasons and/or present alternative measures, or a detailed procedure by which members' objections, reasons and alternative measures are also subject to the judgement of a review panel (Harrison 2017, 183–184). Another important discussion concerns the legal personality and capacity of such organisations, which entitle them to exercise rights and powers on various fisheries issues in the international fora (Manoa 2016).

The next subsections examine selected RFBs created under the auspices of FAO and other selected RFMOs outside the UN system.

8.3.1   RFBs Created Under FAO's Auspices

A key contribution to IFL from FAO, in the exercise of the powers provided by FAO's Constitution (Articles VI(1)(2) and XIV), is the creation of RFBs, which have been supporting the preparation, adoption and implementation of CMMs for fisheries resources falling under their areas of competence (Barnes et al. 2006, 10). The RFBs established by a legally binding instrument originate from FAO's competence to approve conventions and agreements. These RFBs include the Asia-Pacific Fishery Commission (APFIC), the General Fisheries Commission for the Mediterranean (GFCM), the Indian Ocean Tuna Commission (IOTC) and the Regional Commission for Fisheries (RECOFI). Another set of RFBs are those created by non-binding instruments adopted by FAO's Conference and Council, both with competence to establish regional commissions for the purpose of advising on the formulation and coordinated implementation of policy, as determined by FAO's Constitution (Article VI(1)) or for the purpose of studying and reporting on matters pertaining to the purpose of the Organization (Article VI(2)). Those RFBs include the Western Central Atlantic Fishery Commission (WECAFC), the South West Indian Ocean Fisheries Commission (SWIOFC) and the Fishery Committee for the Eastern Central Atlantic (CECAF).

Most of these RFBs were established prior to 1994, but their main contributions to IFL stem from their practical operation through regular meetings to report, discuss, share data, best practices, concerns, activities, decide on institutional arrangements, programmes of works and to adopt recommendations towards bettering the sustainable utilization, management, development of living resources of the respective areas falling under their competence. As mentioned earlier, the main difference among these RFBs is the normative nature of their recommendations, which can be legally binding on the members that have accepted them i.e., not objected, pursuant to the decision-making procedures laid out in their constitutive instruments. In this respect, the latest compilation of CMMs issued by two of FAO's RFMOs provides useful insights into their alignment with global IFL developments. These RFMOs are the GFCM and the IOTC, whose respective recommendations and resolutions, if adopted by a qualified majority of two-thirds votes, become legally binding on members except for those who make a timely objection to the proposed measure (GFCM Agreement 1949, Article 13, IOTC Agreement 1993, Article IX(1)–(7)). These RFMOs have the membership of two countries in common, France and Japan, as well as the EU. Thus, if considering the number of members that each hold, a total of 54 members are legally bound by CMMs applicable in their areas of competence.

Such CMMs have significantly strengthened the IFL's framework in addressing a range of contemporary issues and even reinforcing States' obligations, which previously relied on non-binding instruments. For instance, the management and conservation of sharks and ray species, which were partially covered by the non-binding IPOA-Sharks, currently correspond to legally binding CMMs for the members of both the GFCM (Recommendations GFCM/42//2018/2, GFCM/36/2012/3) and the IOTC (Resolutions 19/03, 18/02, 17/05, 13/05, 13/06, 12/09). The former has generally addressed all sharks and rays through strict management measures (e.g., prohibitions on removal of shark fins on-board vessels, on retaining, transhipping or landing shark fins, on beheading and skinning of specimens on-board and before landing) (Recommendation GFCM/42//2018/2, Para 4) and specific conservation measures (e.g., obligations to ensure a high protection to certain species, which must be released unharmed and alive, to the extent possible) (Recommendation GFCM/42//2018/2, Para 6). The IOTC, in turn, has adopted general CMMs for all sharks (e.g., retention by the fishing vessel of all parts of sharks, except its head, guts and skins, to the point of landing) (Resolution 17/05, Para 2), and special conservation measures for certain shark species (e.g., blue sharks, whale sharks and thresher sharks) (Resolutions 18/02,13/05, 12/09 and 19/03). In a similar manner, both GFCM and IOTC have reflected the IPOA-Seabirds in their CMMs on reducing the incidental bycatch of these species in longline fisheries (Recommendation GFCM/35/2011/3, Resolution 12/06), and they have also each established a list of vessels presumed to have carried out IUU fishing respectively in their areas of competence (Resolution 18/03, Recommendation GFCM/33/2009/8), which supports both the implementation of the IPOA-IUU fishing and the PSMA.

Numerous other issues addressed by the CMMs of IOTC and GFCM indicate their evolution in respect to emerging concerns outlined in the analysed global IFL framework. Progressive examples from the IOTC were the measures on non-entangling and the use of biodegradable fishing aggregated devices within their detailed management plan procedures (Resolutions 19/02, 15/09), while from the GFCM an important recent measure included the establishment of a fisheries restricted area in the Jabuka/Pomo Pit are in the Adriatic Sea for the purpose of protecting VMEs and essential fish habitats for demersal stocks (Recommendation GFCM/41/2017/3). These two specific issues align with IFL's contemporary concerns and the overall contribution to respectively minimise the detrimental impacts caused by destructive fishing gears and to protect and conserve coastal and marine areas, including fragile ecosystems and habitats. Even though these measures suggest an important step forward, there seems not to be sufficient integration of certain matters such as those concerning sustainable SSF. In this respect, however, the GFCM has taken the initiative by adopting a non-binding resolution, which calls for the support to accelerate the implementation of the SSF Guidelines (Resolution GFCM/40/2016/3).

8.3.2   Other RFBs Outside the UN System

There is a range of other RFBs, including RFMOs (e.g. NAFO; CCAMLR; SEAFO; ICCAT) which have been created throughout the last decades and their works have generated what likely constitutes the largest part of IFL sources. As anticipated, the legal developments and contributions of RFMOs to IFL stem from their constitutive instruments as well as their evolution through time, by adoption of amendments to these constitutive instruments and/or of updated CMMs based on their most recent meetings and performance reviews. The present section sheds light on the South Pacific Regional Fisheries Management Organization (SPRFMO), whose constitutive instrument entered into force in 2012 (SPRFMO Convention 2009). The SPRFMO's Convention has gained deserved attention for providing an improved legal framework for international fisheries management, suggesting higher IFL standards for regional rules, with innovative decision-making procedures concerning their member's adoption of CMMs, as well as provision for compulsory dispute settlement mechanisms (Harrison 2019; Caddell 2019; Schiffman 2013).

In respect of the substantive issues dealt with by the SPRFMO, some key provisions of its Convention are worth noting. The SPRFMO Convention requires its Parties, Commission (SPRFMO Convention 2009, Article 6) and subsidiary bodies (Articles 6(2) and 9(1)) to apply the precautionary approach and the EAF (Article 3(1)(b) and (2)(a)). It also requires them to apply principles of transparency, accountability and inclusion in adopting CMMs (Article 3(1)(a)(i)), and the proportionality principle in the establishment of sanctions that are adequate in severity as to avoid illegal fishing (Article 3(1)(a)(ix)). The SPRFMO Commission's technical committee is not only required to monitor the implementation and compliance with CMMs, but also to review such implementation as well as review the implementation of cooperative measures for MCSE (Article 11(2)(a),(c)). Moreover, the

SPRFMO's Convention expressly refers to VMEs in both considerations which its Scientific Committee and the CMMs adopted by SPRFMO's members are required to observe (Articles 10(c) and 20(1)(d)). Additionally, it follows the UNFSA provision on the duty to cooperate for the establishment of CMMs, taking special account to the need to avoid adverse impacts on, and ensure access to fisheries by, 'subsistence, small-scale and artisanal fishers and women fish workers, as well as indigenous people' in developing States' SPRFMO members and their territories and possessions (Article 19(2)(b)).

8.4 Conclusion: From Past to Future in International Fisheries Law

As Ottenheimer noted, back in the early 1970s, '[l]egal policy in general and legal fisheries policy in particular must choose between giving priority to potentialities for change in an evolving future or to determinants for stability in an unchangeable past' (Ottenheimer 1973). These remain the underlying options in contemporary IFL, though the need for more improvements in this domain appears to reveal States' reliance on the second choice. As seen in this chapter, some progress has been made at both global and regional levels, but this analysis was limited in the face of the numerous RFBs, relevant international instruments, judicial and arbitral cases. Recent IFL literature highlighted several key issues that have to some extent been leading ongoing and future developments in IFL. Such matters include the consideration of fisheries and related issues in the proposed BBNJ Agreement, furthering the application of the precautionary principle to new and exploratory fisheries management, in light of increasing population and fish food demand, as well as climate change and climate variability threats (Molenaar and Caddell 2019). Notably, the latter issue has fallen short in the RFMO arena (Rayfuse 2019). The present chapter narrated some other important developments of the recent past in contemporary IFL, particularly with respect to SSF issues, which are not sufficiently explored by IFL scholars.

In following the trend of integration, enhanced cooperation and coherence in ocean governance, numerous institutions interested in bettering the uses of marine living resources more generally have also acquired an interest in fisheries issues, therefore, being important drivers of IFL development. They include regional development and/or economic bodies, which have had issues of weak coordination and overlap with fisheries management due to the political, cultural and economic diversity of the region (e.g., West Central Atlantic and the Gulf regions) (Barston 1999, 343). It is also worth mentioning the growing interaction between IFL and other specialized legal regimes. For instance, the inclusion of aquatic species commercially exploited by the fisheries sector in Appendix II of the Convention on International Trade in Endangered Species of Wild Fauna and Flora (CITES 1973), has required improved coordination between government authorities involved in

CITES implementation and the fisheries sector (Nakamura and Kuemlangan 2020). SSF, in turn, raises important linkages between the SSF Guidelines and international human rights standards, including the recently adopted 2018 UN Declaration on the Rights of Peasants and Other People Working in Rural Areas, which explicitly applies to small-scale fishers (Morgera and Nakamura 2021; Nakamura 2022).

Despite the efforts taken by the international community in bringing global marine fisheries to an improved state of healthy, resilient and restored fish stocks, the status of currently recognised fish stocks remains alarming (FAO 2022). In dealing with the fisheries crisis, it is fundamental that the IFL is interpreted and applied by taking due consideration of all the existing IFL instruments at global and regional levels. IFL sets out the minimum standards of permissible action in fisheries management and outlines the principles guiding such management (Kaye 2001, 1–2). However, the regulation of international fisheries needs to advance faster and more effectively to live up to a growing global population, increasing demand for seafood protein and often unpredictable environmental changes. At the regional level, adequate incentives for RFMOs to fully embrace their roles as 'custodians of regional fish stocks' as well as mechanisms to hold them accountable for their CMMs (Barnes et al. 2006) remain key suggestions for future developments in IFL. As Ottenheimer's put it, '[s]urely our hopes lie not with yesterday, but tomorrow' (Ottenheimer 1973), and States would need to take that first choice more incisively to allow the promising developments that have occurred in the recent past of contemporary IFL to resonate better in the future.

References

International Instruments

Agenda 21: Programme of Action for Sustainable Development, adopted at the United Nations Conference on Environment and Development, in Rio de Janeiro, 3 to 4 June 1992, UN Doc A/Conf.151/26 (Agenda 21)

Agreement for the Establishment of the General Fisheries Commission for the Mediterranean, adopted 24 September 1949, Rome, entered into force 20 February 1952 (GFCM Agreement)

Agreement for the Establishment of the Indian Ocean Tuna Commission, adopted 25 November 1993, Rome, entered into force 27 March 1996 (IOTC Convention)

Agreement for the Implementation of the Provisions of the United Nations Convention on the Law of the Sea relating to the Conservation and Management of Straddling Fish Stocks and Highly Migratory Fish Stocks, adopted 4 August 1995, in New York, entered into force 11 December 2001, 2167 UNTS 3 (UNFSA)

Agreement on Port State Measures to Prevent, Deter and Eliminate Illegal, Unreported and Unregulated Fishing, adopted 22 November 2009, in Rome, entered into force 5 June 2016 (PSMA)

Agreement to Promote Compliance with International Conservation and Management Measures by Fishing Vessels on the High Seas, adopted 29 November 1993, in Rome, entered into force 24 April 2003 (Compliance Agreement)

Code of Conduct for Responsible Fisheries, adopted at the 28th Session of FAO Conference, in Rome, 31 October 1995, Resolution 4/95 FAO Conference (CCRF)

Convention on the Conservation of Antarctic Marine Living Resources, adopted 20 May 1980, in Canberra, entered into force on 7 April 1982, 19 ILM 841 (CCAMLR Convention)

Convention on Future Multilateral Cooperation in the North-West Atlantic Fisheries, adopted 24 October 1978, in Ottawa, entered into force 1 January 1979, 1135 UNTS 369 (NAFO Convention)

Convention on International Trade in Endangered Species of Wild Fauna and Flora, adopted 3 March 1973, in Washington, entered into force 1 July 1975, 993 UNTS 243 (CITES)

Convention on the Conservation and Management of Fishery Resources in the South East Atlantic Ocean, adopted 20 April 2001, in Windhoek, entered into force 13 April 2003 (SEAFO Convention)

Convention on the Conservation and Management of High Seas Fishery Resources in the South Pacific Ocean, adopted 14 November 2009, in Auckland, entered into force 24 August 2012 (SPRFMO Convention)

International Convention for the Conservation of Atlantic Tunas, adopted 14 May 1966, in Rio de Janeiro, entered into force 21 March 1969, 673 UNTS 63 (ICCAT Convention)

Kyoto Declaration, adopted at the International Conference on the Sustainable Contribution of Fisheries to Food Security, in Kyoto, 4 to 9 December 1995 (Kyoto Declaration)

Reykjavik Declaration on Responsible Fisheries in the Marine Ecosystem, adopted at the Reykjavik Conference on Responsible Fisheries in the Marine Ecosystem, Reykjavik, 1 to 4 October 2001 (Reyjavik Declaration)

Rome Declaration on the implementation of the code of conduct for responsible fisheries, Adopted by the FAO ministerial meeting on fisheries, Rome, 10-11 March 1999 (Rome Declaration)

United Nations Convention on the Law of the Sea, adopted 10 December 1982, in Montego Bay, entered into foce 16 November 1994, 1883 UNTS 397 (UNCLOS)

United Nations General Assembly Resolution 49/116, Unauthorized Fishing in Zones of National Jurisdiction and its Impact on the Living Marine Resources of the World's Oceans and Seas, adopted at the 49th Session of the UNGA, 19 December 1994, A/RES/49/116 (UNGA Resolution 49/116)

United Nations General Assembly Resolution 49/118, Fisheries by-catch and discards and their impact on the sustainable use of the world's living marine resources, adopted at the 49th Session of the UNGA, 19 December 1994, A/RES/49/118, (UNGA Resolution 49/118)

United Nations General Assembly Resolution 54/33, Results of the review by Commission on Sustainable Development of the sectoral theme of 'Oceans and Seas': international coordination and cooperation, adopted at the 54th Session of the UNGA, 24 November 1999, A/RES/54/33 (UNGA Resolution 54/33)

United Nations General Assembly Resolution 58/14, Sustainable fisheries, including through the 1995 Agreement for the Implementation of the Provisions of the United Nations Convention on the Law of the Sea of 10 December 1982 relating to the Conservation and Management of Straddling Fish Stocks and Highly Migratory Fish Stocks, and related instruments, adopted at the 58th Session of the UNGA, 24 November 2003, A/RES/58/14 (UNGA Resolution 58/14)

United Nations General Assembly Resolution 60/31, Sustainable fisheries, including through the 1995 Agreement for the Implementation of the Provisions of the United Nations Convention on the Law of the Sea of 10 December 1982 relating to the Conservation and Management of Straddling Fish Stocks and Highly Migratory Fish Stocks, and related instruments, adopted at the 60th Session of the UNGA, 29 November 2005, A/RES/60/31 (UNGA Resolution 60/31)

United Nations General Assembly Resolution 61/105, Sustainable fisheries, including through the 1995 Agreement for the Implementation of the Provisions of the United Nations Convention on the Law of the Sea of 10 December 1982 relating to the Conservation and Management of Straddling Fish Stocks and Highly Migratory Fish Stocks, and related instruments, adopted at the 61st Session of the UNGA, 8 December 2006, A/RES/61/105 (UNGA Resolution 61/105)

United Nations General Assembly Resolution 70/1, Transforming our world: the 2030 Agenda for Sustainable Development, adopted at the 70th Session of the UNGA, 25 September 2015, A/RES/70/1 (UNGA Resolution 70/1)

United Nations General Assembly Resolution 75/89, Oceans and the law of the sea: sustainable fisheries, including through the 1995 Agreement for the Implementation of the Provisions of the United Nations Convention on the Law of the Sea of 10 December 1982 relating to the Conservation and Management of Straddling Fish Stocks and Highly Migratory Fish Stocks, and related instruments, adopted at the 75th Session of the UNGA, 18 December 2021, A/RES/75/89 (UNGA Resolution 75/89)

Voluntary Guidelines for Securing Sustainable Small-scale fisheries in the context of food security and poverty eradication, adopted at the 31st Session of the FAO Committee on Fisheries, in Rome, June 2014 (SSF Guidelines)

International Judicial Cases

Request for an Advisory Opinion Submitted by the Sub-Regional Fisheries Commission (SRFC) ITLOS Reports, Case 21, Advisory Opinion of 2 April 2015

The M/V 'Virginia G' Case (Panama v Guinea-Bissau) ITLOS Reports, Case 19, Judgment of 14 April 2014

Journal Articles, Books and Chapters

Andreone G (2015) The exclusive economic zone. In: Rothwell DR, Oude Elferink AG, Scott KN (eds) The Oxford handbook of the law of the sea. Oxford University Press, Oxford

Attard DJ (1987) The exclusive economic zone in the law of the sea convention. In: Attard DJ (ed) The exclusive economic zone in international law. Claredon Press, Oxford

Azmi K, Hanich Q, Vrahnos A (2016) Defining a disproportionate burden in transboundary fisheries: lessons from international law. Mar Policy 70:164–173

Barnes R (2006) The convention on the law of the sea: an effective framework for domestic fisheries conservation? In: Barnes R, Freestone D, Ong DM (eds) The law of the sea: progress and prospects. Oxford University Press, Oxford

Barnes R (2016) The proposed LOSC implementation agreement on areas beyond National Jurisdiction and its impact on international fisheries law. Int J Mar Coast Law 31:583–619

Barnes R (2019) The pursuit of good regulatory design principles in international fisheries law. In: Van Erp J, Faure M, Nollkaemper A, Philipsen N (eds) Smart mixes for transboundary environmental harm. Cambridge University Press, Cambridge

Barnes R, Freestone D, Ong DM (2006) The law of the sea: progress and prospects. In: Barnes R, Freestone D, Ong DM (eds) The law of the sea: Progress and prospects. Oxford University Press, Oxford

Barrett J, Barnes R (2016) The UN convention on the law of the sea: a "living" treaty? The British Institute of International and Comparative Law, London

Barrie GN (1986) Fisheries and the 1982 United Nations law of the sea convention. Acta Juridica 43–50

Barston R (1999) The law of the sea and regional fisheries Organisations. Int J Mar Coast Law 14:333–352

Béné C, Hersoug B, Allison EH (2010) Not by rent alone: analysing the pro-poor functions of small-scale fisheries in developing countries. Dev Policy Rev 28:325–358

Birnie P, Boyle A, Redgwell C (2009) International law and the environment. Oxford University Press, Oxford

Boyle A (2005) Further development of the law of the sea convention: mechanisms for change. Int Comp Law Q 54:563–584

Boyle A, Chinkin C (2007) The making of international law. Oxford University Press, Oxford

Bratspies R (2001) Finessing king Neptunes: fisheries management and the limits of international law. Harv Environ Law Rev 25:213–258

Buga I (2015) Between stability and change in the law of the sea convention: subsequent practice, treaty modification, and regime interaction. In: Rothwell DR, Oude Elferink AG, Scott KN (eds) The Oxford handbook of the law of the sea. Oxford University Press, Oxford

Caddell R (2019) International fisheries law and interactions with global regimes and processes. In: Caddell R, Molenaar EJ (eds) Strengthening international fisheries law in an era of changing oceans. Hart Publishing, Oxford

Carroz JE, Savini MJ (1979) The new international law of fisheries emerging from bilateral agreements. Mar Policy 3:79–98

Chang SW (2003) WTO disciplines on fisheries subsidies: a historic step towards sustainability? J Int Econ Law 6:879–921

Chuenpagdee R, Jentoft S (2011) Situating poverty: a chain analysis of small-scale fisheries. In: Jentoft S, Eide A (eds) Poverty mosaics: realities and prospects in small-scale fisheries. Springer, Dordrecht

Chuenpagdee R, Jentoft S (2018) Transforming the governance of small-scale fisheries. Mar Stud 17:101–115

Churchill R (1998) Legal uncertainties in international high seas fisheries management. Fisheries Research 37:225–237

Churchill R (2007) The jurisprudence of the International Tribunal for the law of the sea relating to fisheries: is there much in the net? Int J Mar Coast Law 22:383–424

Churchill R (2019) International trade law aspects of measures to combat IUU fishing and unsustainable fishing. In: Caddell R, Molenaar EJ (eds) Strengthening international fisheries law in an era of changing oceans. Hart Publishing, Oxford

Churchill R, Lowe V (1999) The law of the sea. Manchester University Press, Manchester

Cisneros-Montemayor A, Cheung WWL, Ota Y (2019) Predicting future oceans: sustainability of ocean and human systems amidst global environmental change. Elsevier, Amsterdam

Dagget AP (1934) The regulation of maritime fisheries by treaty. Am J Int Law 28:693–717

Davies PGG, Redgwell C (1997) The international legal regulation of straddling fish stocks. Br Year Book Int Law 67:199–274

De La Fayette L (2006) The role of the UN in international oceans governance. In: Barnes RA, Freestone D, Ong DM (eds) The law of the sea: progress and prospects. Oxford University Press, Oxford

De Lucia V (2015) Competing narratives and complex genealogies: the ecosystem approach in international environmental law. J Environ Law 27:91–117

Diz D (2012) Fisheries management in areas beyond national jurisdiction. Martinus Nijhoff, Leiden

Ebben T (2011) The implementation of the precautionary principle into international Fishery law: a move towards green fisheries. New Zealand J Environ Law 15:113–146

Edeson W (1996) Current legal developments: food and agriculture organization of the UN. The International Journal of Marine and Coastal Law 11:233–238

Edeson W (1999) Closing the gap: the role of soft international instruments to control fishing. Aust Year Book Int Law 20:83–104

Freestone D (1999) International fisheries law since Rio: the continued rise of the precautionary principle. In: Boyle A, Freestone D (eds) International law and sustainable development: past achievements and future challenges. Oxford University Press, Oxford

Freestone D (2016) International Tribunal for the law of the sea, case 21: request for an advisory opinion submitted by the sub-regional fisheries commission (SRFC). Asia-Pac J Ocean Law Policy 1:131–138

Freestone D, Makuch Z (1997) The new international environmental law of fisheries: the 1995 United Nations straddling stocks agreement. Yearb Int Environ Law 7:3–51

Garcia SM, Cochrane KL (2009) From past management to future governance: a perspective view. In: Cochrane KL, Garcia SM (eds) A Fishery Manager's guidebook. FAO and Wiley-Blackwell

Garcia SM, Rice J, Charles A (2014) Governance of marine fisheries and biodiversity conservation: a history. In: Garcia SM, Rice J, Charles A (eds) Governance of marine fisheries and biodiversity conservation: interaction and co-evolution. Wiley, Hoboken

Gillespie A (2002) Wasting the oceans – searching for principles to control bycatch in international law. Int J Mar Coast Law 17:161–193

Harrison J (2007) Judicial law-making and the developing order of the oceans. Int J Mar Coast Law 22:283–302

Harrison J (2011) The contribution of the food and agriculture organization to international fisheries law. In: Harrison J (ed) Making the law of the sea: a study in the development of international law. Cambridge University Press, Cambridge

Harrison J (2015) Actors and institutions for the protection of the marine environment. In: Rayfuse R (ed) Research handbook on international marine environmental law. Edward Elgar Publishing, Cheltenham

Harrison J (2017) Saving the oceans through law: the international legal framework for the protection of the marine environment. Oxford University Press, Oxford

Harrison J (2019) Key challenges relating to the governance of regional fisheries. In: Caddell R, Molenaar EJ (eds) Strengthening international fisheries law in an era of changing oceans. Hart Publishing, Oxford

Hayashi M (1999) The straddling and highly migratory fish stocks agreement. In: Hey E (ed) Developments in international fisheries law. E Kluwer Law International, The Hague

Heck CB (1975) Collective arrangements for managing ocean fisheries. Int Organ 29:711–743

Hewison G (1999) Balancing the freedom of fishing and coastal state jurisdiction. In: Hey E (ed) Developments in international fisheries law. E Kluwer Law International, The Hague

Hey E (1999) The fisheries provisions under the LOS convention. In: Hey E (ed) Developments in international fisheries law. E Kluwer Law International, The Hague

Jacobson JL (1985) International fisheries law in the year 2010. Louisiana Law Rev 45:1161–1200

Jentoft S, Bavinck M (2019) Reconciling human rights and customary law: legal pluralism in the governance of small-scale fisheries. J Leg Plur Unoff Law 51:271–291

Juda L (2002) Rio plus ten: the evolution of international marine fisheries governance. Ocean Dev Int Law 33:109–144

Kaye SM (2001) International fisheries management. Kluwer Law International, The Hague

Kenny AJ, Campbell N, Koen-Alonso M, Pepin P, Diz D (2018) Delivering sustainable fisheries through adoption of a risk-based framework as part of an ecosystem approach to fisheries management. Mar Policy 93:232–240

Kroodsma DA, Mayorga J, Hochberg T, Miller NA, Boerder K, Ferretti F, Wilson A, Bergman B, White TD, Block BA, Woods P, Sullivan B, Costello C, Worm B (2018) Tracking the global footprint of fisheries. Science 359:904–908

Kuemlangan B (2009) Legal aspects. In: Cochrane KL, Garcia SM (eds) A Fishery Manager's guidebook. FAO/Wiley-Blackwell, Rome/Hoboken

Lodge MW, Nandan SN (2005) Some suggestions towards better implementation of the United Nations agreement on straddling fish stocks and highly migratory fish stocks of 1995. Int J Mar Coast Law 20:345–479

Manoa PE (2016) The contribution of tuna regional fisheries management organisations to international law. DPhil thesis, Australian National Centre for oceans resources and security (ANCORS). University of Wollongong

McCauley DJ, Jablonicky C, Allison EH, Golden CD, Joyce FH, Mayorga J, Kroodsma D (2018) Wealthy countries dominate industrial fishing. Sci Adv 4:1–9

Miles EL, Burke WT (1989) Pressures on the United Nations convention on the law of the sea of 1982 arising from new fisheries conflicts: the problem of straddling stocks. Ocean Dev Int Law 20:343–358

Molenaar EJ (2002) Ecosystem-based fisheries management, commercial fisheries, marine mammals and the 2001 Reykjavik declaration in the context of international law. Int J Mar Coast Law 17:561–595

Molenaar EJ (2007) Port state jurisdiction: toward a comprehensive, mandatory and global coverage. Ocean Dev Int Law 38:225–257

Molenaar EJ (2019) Participation in regional fisheries management organizations. In: Caddell R, Molenaar EJ (eds) Strengthening international fisheries law in an era of changing oceans. Hart Publishing, Oxford

Molenaar EJ, Caddell R (2019) International fisheries law: achievements, limitations and challenges. In: Caddell R, Molenaar EJ (eds) Strengthening international fisheries law in an era of changing oceans. Hart Publishing, Oxford

Moore G (1999) The code of conduct for responsible fisheries. In: Hey E (ed) Developments in international fisheries law. E Kluwer Law International, The Hague

Morgera E, Nakamura J (2021) Shedding a light on the human rights of small-scale fishers: complementarities and contrasts between the UN Declaration on peasants' rights and the small-scale fisheries guidelines. In: Alabrese M, Bessa A, Brunori M, Giuggioli PF (eds) The UN Declaration on peasants' rights. Routledge, Abingdon. Available at SSRN: https://ssrn.com/abstract=3850133 (Forthcoming)

Nakamura J (2022) Legal reflections on the small-scale fisheries guidelines: building a global safety net for small-scale fisheries. Int J Mar Coast Law 37:31–72

Nakamura J, Chuenpagdee R, El Halimi M (2021) Unpacking legal and policy frameworks: a step ahead for implementing the small-scale fisheries guidelines. Mar Policy 129:1–9

Nelson D (1999) The development of the legal regime of high seas fisheries. In: Boyle A, Freestone D (eds) International law and sustainable development: past achievements and future challenges. Oxford University Press, Oxford

Oda S (1983) Fisheries under the United Nations convention on the law of the sea. Am J Int Law 77:739–755

Ottenheimer GR (1973) Patterns of development in international Fishery law. Can Yearb Int Law 11:37–47

Palma MA, Tsamenyi M, Edeson W (2010) Promoting sustainable fisheries: the international legal and policy framework to combat illegal, unreported and unregulated fishing. Martinus Nijhoff, Leiden

Papanicolopulus I (2018) International law and the protection of people at sea. Oxford University Press, Oxford

Pinsky ML, Reygondeau G, Caddell R, Palacios-Abrantes J, Spijkers J, Cheung WWL, Malin L (2018) Preparing Ocean governance for species on the move. Science 360:1189–1191

Politakis GP (2008) From tankers to trawlers: the International Labour Organization's new work in fishing convention. Ocean Dev Int Law 39:119–128

Proelss A (2016) Fisheries. In: Morgera E, Kulovesi K (eds) Research handbook on international law and natural resources. Edward Elgar Publishing, Cheltenham

Purcell SW, Pomeroy RS (2015) Driving small-scale fisheries in developing countries. Front Mar Sci 2:1–7

Rayfuse R (2019) Addressing climate change impacts in regional fisheries management organizations. In: Caddell R, Molenaar EJ (eds) Strengthening international fisheries law in an era of changing oceans. Hart Publishing, Oxford

Rengifo A (1997) Protection of marine biodiversity: a new generation of fisheries. Rev Eur Comp Int Law 6:313–321

Said A, Chuenpagdee R (2019) Aligning the sustainable development goals to the small-scale fisheries guidelines: a case for EU fisheries governance. Mar Policy 107:1–7

Schiffman HS (2013) The South Pacific regional fisheries management organization (SPRFMO): an improved model of decision-making for fisheries conservation? J Environ Stud Sci 3:209–216

Serdy A (2016) The new entrants problem in international fisheries law. Cambridge University Press, Cambridge

Smith H, Basurto X (2019) Defining small-scale fisheries and examining the role of science in shaping perceptions of who and what counts: a systematic review. Front Mar Sci 6:1–19

Somos M (2012) Selden's Mare Clausum: the secularisation of international law and the rise of soft imperialism. J Hist Int Law 14:287–330

Song AM, Soliman A (2019) Situating human rights in the context of fishing rights – contributions and contradictions. Mar Policy 103:19–26

Sydnes AK (2001) Regional Fishery organizations: how and why organizational diversity matters. Ocean Dev Int Law 32:349–372

Tarassenko S, Tani I (2012) The functions and role of the United Nations secretariat in ocean affairs and the law of the sea. Int J Mar Coast Law 27:683–699

Thomson D (1980) Conflict within the fishing industry. ICLARM 3:3–4

Thornton H (2004) Hugo Grotius and the freedom of the seas. Int J Marit Hist 16:17–38

Tsamenyi M, Hanich Q (2012) Fisheries jurisdiction under the law of the sea convention: rights and obligations in maritime zones under the sovereignty of coastal states. Int J Mar Coast Law 27:783–793

Urrutia O (2018) Combating unregulated fishing through unilateral trade measures: a time for change in international fisheries law? Vic Univ Wellingt Law Rev 49:671–695

Van der Burgt N (2012) The contributions of international fisheries law to human development: an analysis of multilateral and ACP-EU fisheries instruments. Brill, Leiden

Vicuña FO (1993) Toward an effective management of high seas fisheries and the settlement of the pending issues of the law of the sea. Ocean Dev Int Law 24:81–92

Young MA (2011) Trading fish, saving fish: the interaction between regimes in international law. Cambridge University Press, Cambridge

Technical Papers, Reports and Other Online Sources

FAO (2022) The State of World Fisheries and Aquaculture 2022. Towards Blue Transformation, Rome. https://doi.org/10.4060/cc0461en

Garcia SM, Zerbi A, Aliaume C, Do Chi T, Lassere G (2003) The ecosystem approach to fisheries. Issues, terminology, principles, institutional foundations, implementation and outlook. FAO fisheries technical paper. No. 443. Rome, FAO. 71p

Koh T (1982) A constitution for the oceans. UN archive. Available at https://www.un.org/Depts/los/convention_agreements/texts/koh_english.pdf. Accessed 24 Apr 2021

Løbach T, Petersson M, Haberkon E, Mannini P (2020) Regional fisheries management organizations and advisory bodies. FAO fisheries and aquaculture technical paper. No. 651. Rome, FAO. https://doi.org/10.4060/ca7843en

Nakamura J, Kuemlangan B (2020) Implementing the convention on international trade in endangered species of wild Fauna and Flora (CITES) through national fisheries legal frameworks: a study and a guide. FAO legal guide. No. 4. Rome, FAO. https://doi.org/10.4060/cb1906en

Switzer S, Lennan M (2022) The WTO's agreement on Fisheries Subsidies. 'It's good, but it's not quite right' (23 June 2022). Available at https://oneoceanhub.org/the-wtos-agreement-on-fisheries-subsidies-its-good-but-its-not-quite-right/. Accessed 29 Oct 2022

WTO (2021) Fisheries Subsidies: draft consolidated Chair text. Doc. TN/RL/W/276/Add.1. Available at https://docs.wto.org/dol2fe/Pages/SS/directdoc.aspx?filename=q:/TN/RL/W276A1.pdf&Open=True. Accessed 4 Aug 2021

Chapter 9
Managing Land Sea Interactions: Case Studies of Coastal Governance in Four EU Member States

Paul Lawlor and Daniel Depellegrin

Abstract Under the Marine Strategy Framework Directive, EU member states are committed to delivering Good Environmental Status in EU marine and coastal areas but the risk of damage from land based pollutants is rising, along with increased economic uses and activities in marine and coastal areas. While it is accepted that land sea interactions need to be managed, and uses and activities in our marine and coastal areas must be regulated, the complexity and dynamic nature of land sea connections create challenges for governance systems. This chapter reviews the marine and coastal management systems in operation in Ireland, Romania, Spain and France. Using relevant case studies at national, sub national and local level, we assess their capacity to manage complex and dynamic land sea interactions. We further examine their ability to achieve integrated, multiscalar and cross sectoral governance of their marine and coastal areas. Recommendations to assist EU member states who are developing marine and coastal governance systems are also provided.

Keywords Evolutionary Governance Theory (EGT) · Land Sea Interactions (LSI) · Marine Strategy Framework Directive (MSFD) · Good Environmental Status (GES) · Intergrated Coastal Zone Management (ICZM) · Features and Mechanisms of the Ocean and Coastal Governance

P. Lawlor (✉)
School of Architecture, Building & Environment, Technological University Dublin, Dublin, Ireland
e-mail: paul.lawlor@TUDublin.ie

D. Depellegrin
Landscape Analysis and Management Laboratory, Department of Geography, University of Girona, Girona, Spain

S. Partelow et al. (eds.), *Ocean Governance*, MARE Publication Series 25,
https://doi.org/10.1007/978-3-031-20740-2_9

9.1 Introduction

Occupying the interface between marine and terrestrial areas, coastal zones are highly diverse and truly unique multifunctional natural areas that are critical habitats for endangered species which accommodate more than 60% of the worlds population (O'Connor et al. 2009, p. 923) and provide significant ecosystem services (Ramesh et al. 2015, pp. 85–86). Despite widespread recognition of their environmental sensitivity and crucial ecological role, pressures on coastlines are increasing due to growing human populations and economic activities on the landward side in addition to climate induced changes such as sea level rise, higher sea temperatures and more frequent and intense weather events on the seaward side (ibid, 2015, pp. 85–86). These complex and interconnected land sea interactions (LSI hereafter) have the potential to undermine the ecological health of coastal areas and their ability to fulfil their many important roles. Yet managing LSI is a challenging task and there is concern that existing governance frameworks, instruments and mechanisms that are in place in coastal areas are insufficient to ensure the sustainable use of coastal and marine resources (Van Assche et al. 2020, p. 2). The intense pressures that coastal areas face and their ineffective management systems has led commentators to conclude that coastal zones are *'arguably the most transformed and imperilled social ecological system on earth (which) are characterised by pervasive unsustainable practices'* (Ramesh et al. 2015, p. 86). Thus, in order to ensure sustainable ocean governance, better management of the land sea interface is required.

The need to manage LSI and address the unsustainable use of our coastal and marine resources is recognised in the requirements of the 1982 United Nations Convention of the Law of the Sea (UNCLOS) and by the adoption of EU Member States of the Marine Strategy Framework Directive (MSFD hereafter), which commits them to achieving Good Environmental Status (GES hereafter) in marine and coastal environments. There is an appreciation that effective governance systems are needed to manage the complex interrelated factors that influence the environmental quality of marine and coastal areas (Schlüter et al. 2020, p. 1) but the historical regulation of land and sea as separate entities and the governance of coastal areas in accordance with terrestrial models pose challenges (Partelow et al. 2020, p. 2) to the delivery of the required systems. The need for 'fit for purpose' coastal and marine governance systems has led to much debate among scientists and environmental managers on *'effective policy mixes and regulatory instruments to facilitate integrated forms of multiscalar and cross sectoral governance across ecologically diverse marine spaces'* (Van Assche et al. 2020, p. 2). The continuing implementation of the MSFD has brought this issue into sharp focus and noting the diversity of terrestrial and marine planning systems throughout the EU, an examination of how LSI are handled in marine and coastal management regimes in European countries is both timely and necessary.

Using the perspective of Evolutionary Governance Theory (EGT), this research attempts to inform the previously mentioned debate among scientists and environmental managers on what are the most effective *'policy mixes and regulatory instruments'* for managing LSI in the EU and facilitating the *integrated forms of multiscalar and cross sectoral governance across ecologically diverse marine spaces* that are urgently required. EGT is considered to be a suitable lens for this approach as it is presents an understanding of governance as a radically evolutionary and constantly changing process that is influenced by the interplay of actors, institutions, knowledges and systems of sense-making (natural, technological, infrastructural), materialities and interest formations in any community, in any location and at any point in time (Van Assche et al. 2020, p. 3). The chapter begins with a brief review of how the issue of LSI has been dealt with at EU level and it continues with an examination of the institutional mechanisms and measures that are currently being used to manage LSI in the marine and coastal governance regimes in 4 EU member states (Ireland, Romania, Spain and France). The effectiveness of these institutional mechanisms and measures for delivering improved environmental outcomes is considered and the findings of the research are used to draw lessons for the future implementation of MSFD in achieving GES in the coastal and marine areas of the EU.

9.2 Background to EU Level Regulatory Frameworks for Managing Land Sea Interactions

Concerns arising from the pollution of coastal and marine waters from land based sources are well established. The 1982 United Nations Convention of the Law of the Sea (UNCLOS) includes a specific requirement for States (under Article 194) to put measures in place to deal with pollution of the marine environment including pollutants arising from land-based sources (Kidd et al. 2019, p. 247). It is likely that the inclusion of LSI in UNCLOS was influenced by the emergence of ICZM – (also known as ICM or Integrated Coastal Management) which focuses on the need for integrated planning and management of human relationships with the coastal and marine environment. The ICZM approach is considered to have been particularly influential in focussing attention on LSI in Europe and elsewhere in the mid 1990s where it was recognised as a *'mechanism to reduce the deterioration of coastal areas, and progress the sustainable use of coastal resources in Europe'* (Falaleeva et al. 2011, p. 787). A range of European countries participated in an ICZM Demonstration Programme in 1996 which examined the approach and its suitability for national level implementation in Member States. The findings from this Programme later informed the Communication to the Council and the European Parliament entitled "Integrated Coastal Zone Management: A Strategy for Europe" (COM (2000) 547 final) which identified the 8 principles of ICZM (Table 9.1). According to Kidd et al. 4 of these principles refer specifically to core areas of LSI consideration – Principles 1 & 5 (which focus on interactions within and between natural systems and human activities) and Principles 7 & 8 (which relate to governance arrangements) (Kidd et al. 2019, p. 249).

Table 9.1 The 8 principles of Integrated Coastal Zone Management (ICZM)

	ICZM principles
1	A broad overall perspective (thematic & geographic) to take into account the interdependence and disparity of natural systems and human activities with an impact on coastal areas
2	A long-term perspective which will take into account the precautionary principle and the needs of present and future generations
3	Adaptive management during a gradual process which will facilitate adjustment as problems and knowledge develop. This implies the need for a sound scientific basis concerning the evolution of the coastal zone
4	Local specificity and the great diversity of European coastal zones, which will make it possible to respond to their practical needs with specific solutions and flexible measures
5	Working with natural processes and respecting the carrying capacity of ecosystems, which will make human activities more environmentally friendly, socially responsible and economically sound in the long run
6	Involving all the parties concerned (economic and social partners, the organisations representing coastal zone residents, non-governmental organisations and the business sector) in the management process, for example by means of agreements and based on shared responsibility
7	Support and involvement of relevant administrative bodies at national, regional and local level between which appropriate links should be established or maintained with the aim of improved coordination of the various existing policies. Partnership with and between regional and local authorities should apply when appropriate
8	Use of a combination of instruments designed to facilitate coherence between sectoral policy objectives and coherence between planning and management

Source EC (2002a, b)

The ICZM Communication was influential as it led to a 2002 recommendation by the European Commission (EC hereafter) EC (2002a, b) which encouraged Member States to prepare ICZM strategies (Falaleeva et al. 2011, pp. 787–788). However, the recommendation was not binding and as a result, its impact on governance was limited as only a small number of larger EU Member States (France, Spain and Germany) adopted it (Shipman and Stojanovic 2007, p. 378).

The Marine Strategy Framework Directive in 2008 (MSFD) (Directive 2008/56/EC) also addresses LSI as it requires member states to maintain GES (Bellas 2014, p. 16) by protecting and preserving the marine environment, restoring altered ecosystems, and preventing and reducing inputs into the marine environment by phasing out pollution. A subsequent review of the first implementation phase of MSFD acknowledged the work of member states in completing initial assessments of the environmental status of their marine and coastal areas. However, it stated that greater co-ordination of monitoring programmes and measures was needed along with full implementation of the EU's legislative framework for dealing with land based sources of pollution. The review also called for more systemic efforts to achieve ICZM (EC 2014a, b). The adoption of the 2014 MSP Directive is seen as significant to LSI management as it not only requires LSI to be taken into account (under article 6) but it also provides member states with the choice of using the MSP process or the ICZM approach to manage LSI in their coastal areas (Kidd et al. 2019, p. 248). According to O'Hagan, the key issue for member states following their adoption of the MSP Directive became the management of LSI as they had to ensure that the implementation of the MSP Directive in their coastal and marine areas was coherent with other relevant processes related to LSI at member state level (such as spatial planning) (O'Hagan et al. 2020, p. 4).

Therefore, there is a clear understanding at EU and member state level that LSI must be effectively managed to achieve good marine and coastal environmental quality. It is also understood that the complexity of LSI and their dynamic nature is creating major problems for management approaches. In response to these concerns, the MSP Expert Group (who advise the European Commission) developed a framework that recognises LSI as the synergies created from land-sea natural processes (Fig. 9.1) and land sea economic activities (SUPREME 2015). The framework also includes guidance for the management of these synergies by recommending that MSP Authorities (as well as other stakeholders) should address LSI in a two

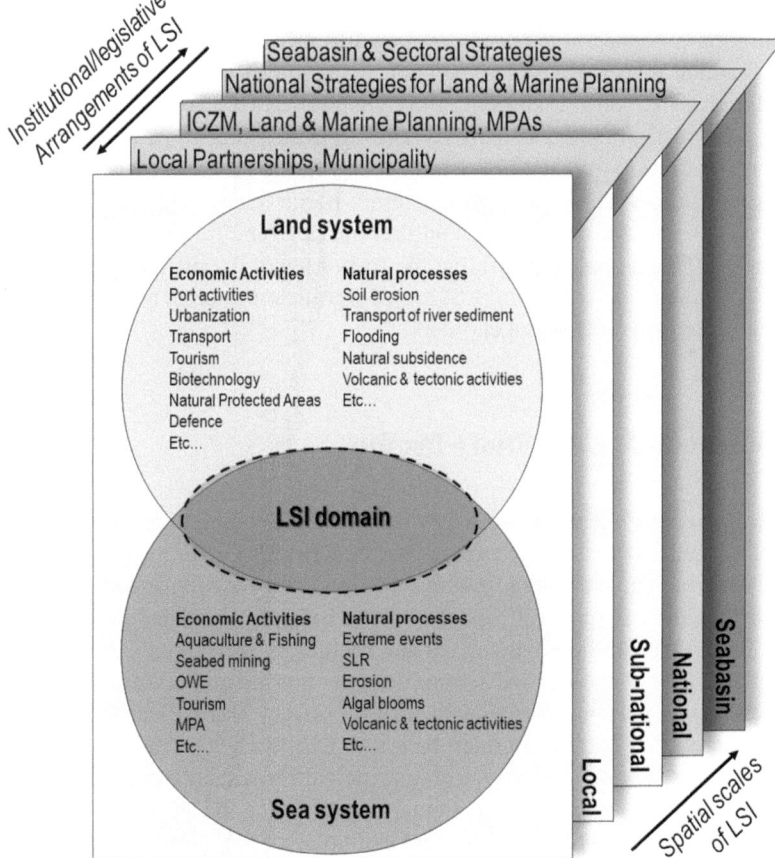

Fig. 9.1 LSI framework presenting land and sea systems and the relevant legislative/institutional arrangements relevant across spatial scales. (Adapted from EC 2017; SUPREME 2015)

phase process that involves understanding the dynamics involved and by identifying institutional arrangements/mechanisms that are most suited to managing them. While the framework acknowledges that different institutional mechanisms and measures are available for this purpose, no advice is offered on which of these mechanisms and measures should be used. Instead, it states that member states should choose institutional mechanisms and measures that are most suitable to the government context that they will be implemented in. ICZM is also included in the Framework as a management option (although it is referred to as ICM (Integrated Coastal Management)). In addition, it is made clear that LSI processes can be managed at various spatial scales such as local scale (e.g., local partnerships of municipalities and interest groups), sub-national scale (e.g., regional territorial planning), national scale (e.g., national and sectoral strategies) and seabasin scale (e.g., European seabasin strategies, cross-border cooperation protocols). Once again, no reference is made to the governance scales that are most appropriate for managing LSI as it is left to individual Member States to devise appropriate spatial scales for LSI planning and management.

Noting the guidance in the MSP Framework, this research seeks to evaluate the extent to which 4 member states (Ireland, Romania, Spain and France) have followed the guidance on investigating the dynamics of LSI in their jurisdictions. In addition, the institutional mechanisms and measures that each of these member states have chosen to manage LSI within their marine and coastal governance systems are considered along with their overall effectiveness and suitability to their respective government contexts. Given that the deadline for achieving GES under MSFD was 2020, it is anticipated that the responses of the different member states to the EU guidance on managing LSI are of significant interest to all MSP authorities, practitioners and other stakeholders.

9.3   Methods & Case Study Profiles

This research seeks to draw lessons from how LSI are being managed in a range of different marine and coastal governance systems from diverse European geographic areas, all of which are striving to achieve GES to comply with the MSFD. A total of 4 case studies were purposefully selected from Ireland, Romania, Spain and France in order to investigate the policy mixes and regulatory instruments that are in place for managing LSI in the EU and to explore how marine and coastal areas are governed at national, sub-national and local levels. Data was collected by reviewing earlier research that had been undertaken into LSI in each of the case study areas and by carrying out one interview with a principal researcher from each of the four selected case study areas between January and May 2020. A total of 7 questions were put to each principal researcher and examples of the questions are as follows:

- What are the features of coastal governance in the case study area
- Describe the barriers to coastal governance in the case study area

- What are the enablers for coastal governance in the case study area
- Describe the mechanism (or mechanisms) that are used to manage land sea interactions in the case study area

The responses given to the interviews were transcribed manually by the researcher during and immediately after the interviews and a manual qualitative assessment of the information given by each respondent was carried out. A thematic analysis of the data was then undertaken to see if common themes could be identified in each of the case studies based on the interview responses. The approach enabled a comparative analysis to be completed of the experiences of Member States in managing LSI and marine resources at all governance levels. The results of the comparative analysis were subsequently used to examine the link between governance and environment quality and to draw lessons for future marine and coastal governance. The selected case studies (Fig. 9.2) are as follows.

Case Study 1: Ireland (Atlantic Ocean and Irish Sea). The first case study considers the coastal and marine governance system for the extensive maritime area and 5800 km coastline in the Republic of Ireland (O'Hagan and Cooper 2002, p. 547). The governance system which is concentrated at national level is described as highly centralised and sectoral in its approach with at least 34 different government departments, agencies, and bodies with responsibilities for estuarine, coastal, and marine management across different territorial scales. Regional and Local Authorities tend to have a limited role in coastal and marine governance due to doubts about their own legal jurisdiction (O'Hagan et al. 2020, p. 10). However, changes have taken place since 2016 with the launch of the national marine planning framework (in July 2021) and the establishment of a national coastal change management strategy group to consider the development of an integrated coastal change strategy. Nonetheless, a strong land-sea divide remains in the Irish marine and coastal governance structure with very little integrative national legislation (O'Hagan et al. 2020, p. 10). In addition, there is no formal role for coastal communities and other non statutory stakeholder groups.

Case Study 2: Romania (Black Sea). In the second case study, the Romanian approach to coastal and marine governance on the semi-enclosed Black Sea is examined. Like Ireland, coastal and marine governance in Romania is centralised at the national level in the Ministry of the Environment. No regional or local authorities in Romania have marine or coastal management responsibilities and coastal communities are not involved in marine and coastal governance. The Black sea is classified as a vulnerable marine ecosystem and its governance is complicated as it is bordered by two EU Member States (Romania and Bulgaria) and four non EU Countries (Russia, Ukraine, Georgia and Turkey) – two of whom (Russia and Ukraine) are engaged in an interstate conflict (Vaidanu et al. 2020, p. 1). Despite these challenges, there have been Black Sea cooperation initiatives between bordering countries to improve its management and they include the preparation of a Strategic Action Plan in 2009 (Vaidanu et al. 2020, p. 3).

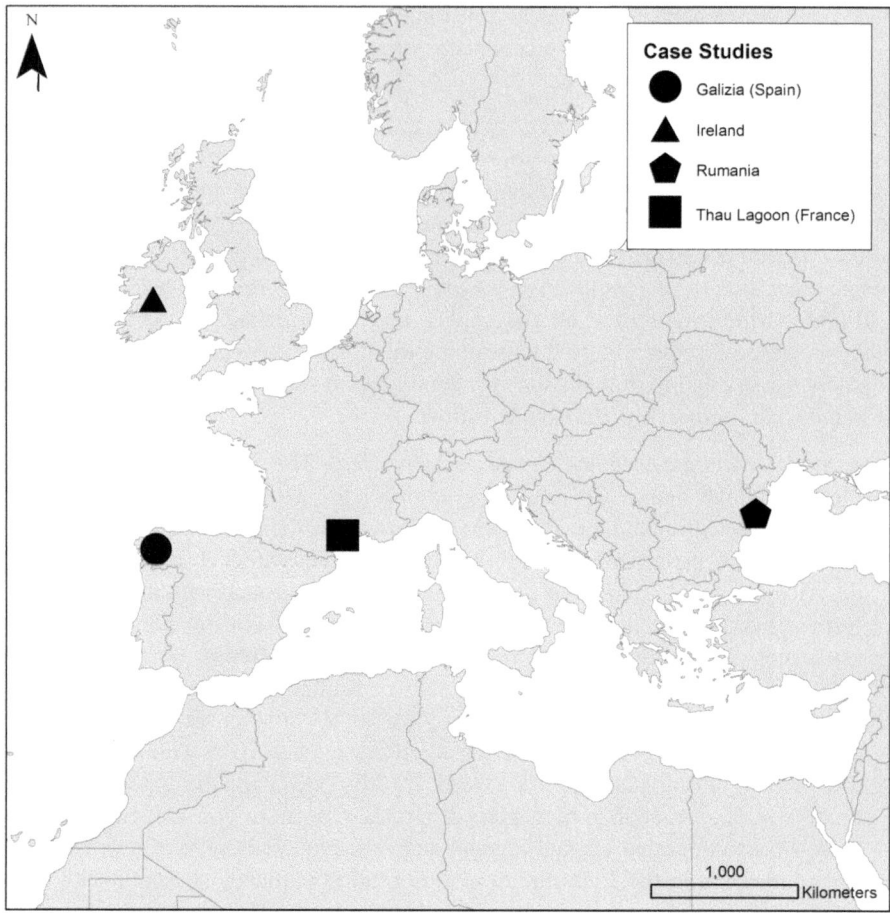

Fig. 9.2 Map of the case study areas. (Source: Authors)

Case Study 3: Galizia (Spain, Atlantic Sea). The third case study is focused on the regional (sub-national) governance of coastal and marine areas in Galicia. The area is comprised of 10 municipalities and 10% of its 136,000 population rely on coastal/marine activities such as fishing, aquaculture and seafood processing for their livelihoods. With respect to governance, central government has responsibility for marine and coastal areas at the national level while resource management (fisheries/aquaculture) and land and coastal planning are handled at the regional level by the autonomous Galician government (Pineiro-Antelo et al. 2020, p. 2) through a Coastal Management Plan (POLGA). All muncipal level plans must adhere to the provisions of the POLGA. A notable feature of the region is that coastal and marine management is traditionally carried out in collaboration with Galician fishermen's guilds which are associations comprising fishermen and shellfish gatherers.

Case Study 4: Thau Lagoon (France, Mediterranean). The fourth case study considers local (sub-national) level coastal and marine governance in the Thau Lagoon, which is a stream-fed semi-enclosed lagoon connected to the Mediterranean Sea in the Languedoc-Roussillon region of France. Economic activities such as oyster farming and fishing take place in the lagoon while the surrounding area accommodates viticulture, horticulture and livestock farming. Tourism is also significant and urbanisation is creating further environmental pressures on the lagoon. The comprehensive governance structure in the Thau Lagoon involves the participation of stakeholders at all levels (community organisations, local municipalities, regional and state/national bodies) but these arrangements led to responsibilities for key issues (such as water quality) being spread across many organisations and stakeholders. To improve coordination and decision making between the different levels of governance, a brokering organisation (with multi disciplinary staff) called Syndicat Mixte du Bassin de Thau (SMBT) was created at the regional level (Daniell et al. 2020, p. 7).

9.4 Presentation, Analysis and Discussion of Results

A total of seven themes were identified from the interview responses; the influence of the EU, features of marine and coastal governance in Member States, opportunities for and barriers to effective governance, mechanisms of governance, the relationship between governance and environmental outcomes and the application of evolutionary governance theory. Insights across the four case studies are presented in aggregate below, with specific examples given from each case study.

9.4.1 The Influence of the EU on Evolving Coastal Governance Structures

The research findings reveal that overall, the EU has had a positive impact on coastal and marine governance as each of the four member states that were the subject of investigation have either devised or are in the process of developing mechanisms to deliver coastal and marine governance in response to their obligations as member states under the Marine Strategy Framework Directive. However, the research also revealed that prior to the adoption by member states of the MSFD in 2016, the level of engagement between the EU and member states in the area of marine and coastal governance has been somewhat variable as some (such as Spain and France) adopted the (non binding) EC Recommendation on ICZM in 2002 and others (Ireland and Romania) did not (Shipman and Stojanovic 2007, p. 378). This variable level of engagement has had clear implications on how the coastal governance systems of the member states have evolved – as the countries who engaged with marine and coastal management in 2002 (Spain and France) are now much further advanced than those who did not (Ireland and Romania).

The research has illustrated that the approaches to marine and coastal governance structures and systems in the Spanish and French case studies have evolved over an extended period of time thereby enabling them to be adapted and more focussed on achieving better environmental outcomes for their marine and coastal areas. In the case of Spain, the researchers stated that the path towards integrated coastal management began in the 1990s with land use and planning laws relating to coastal areas being adopted at regional and national level in 1995, 2002 and 2007 and a coastal management plan being approved for Galicia in 2011. Despite the progress made, the researchers for the Spanish case study noted that the integration of ICZM policies on a vertical scale (between national, regional and local level) had yet to take place. With respect to France, the evolution of the governance system for coastal and marine areas (as shown in the Thau Lagoon) is demonstrated by the constant adaptation of administrative boundaries and governance arrangements that have taken place to take account of multiple changes within the lagoon and deliver specific environmental outcomes such as improved water quality.

In contrast to the Spanish and French case studies, there were no integrated marine and coastal governance structures in place in Ireland or Romania prior to their adoption of the Marine Strategy Framework in 2016. In the case of Ireland, the researchers expressed concern (at the time of the research in 2020) that the legislation, the policies and mechanisms being devised to give effect to MSP seemed to have been rushed and did not appear to have been 'road tested' or assessed for their suitability to the governance structure in which responsibilities for coastal and marine areas were fragmented (by a range of different government departments/ministries and supporting agencies). The researchers from Ireland used the example of the linear approach that has been applied in the UK to test policies to demonstrate this point. The UK linear approach involves the development of a green paper on a particular issue, which (after due consideration) progresses to a white paper and finally to leglisation. This linear approach provides for a logical evolution in the development of policy which enhances understanding and promotes confidence among stakeholders. However, this logical evolution (or road testing) of policy was not evident in Ireland with respect to marine and coastal governance. Similar concerns were expressed by the researchers who undertook the case study of Romania. As a result of this lack of 'road testing' of policies and mechanisms, the researchers in Ireland and Romania were less confident that the legislation, mechanisms and policies to support marine and coastal governance would have the capacity to manage LSI and deliver the required improvements to the marine and coastal environment.

9.4.2   The Features of Marine and Coastal Governance in the Case Study Areas

Two distinct types of marine and coastal governance systems were observed. In both France and Spain, the marine and coastal governance systems provide for comprehensive devolution with active participation by authorities and agencies at the national, regional and local levels. This presents a strong contrast with the centralised Irish and the Romanian systems that are confined to national level only and have no meaningful roles afforded to authorities and agencies at regional, local or community levels. The results also show that the French and the Spanish systems have been evolving since their adoption of the (non binding) EU recommendation to prepare ICZM strategies in 2002 by incorporating additional governance 'layers'. The more recent modifications to the French and Spanish systems have included the development of partnerships with coastal communities and oyster farmers (in France) and the reorganisation of coastal governance (in Spain) to integrate Fishing Guilds and other local actors as a means of achieving community level involvement in marine and coastal governance. In contrast, there is no evidence of marine and coastal governance layers being developed below the national level in the Irish or Romanian systems.

The influence of the ICZM approach in the evolution of the coastal and marine governance systems in the case study areas were also considered. This was measured by assessing the extent to which the governance systems of each case study area adhered to the 8 principles of ICZM. It was significant to note that the governance system in the French case study seemed to adhere to all 8 principles of ICZM. In the Spanish case study, 7 out of the 8 ICZM principles were reflected in their approach to marine and coastal management. The one lacking principle was using a combination of instruments to facilitate coherence between sectoral objectives. The findings indicated that the Irish and Romanian approaches adhered to the least number of ICZM principles – with just 3 principles reflected in their marine and coastal governance systems.

The high level of adherence to the ICZM principles in both Spain and France reflects the fact that both of these countries actively engaged with the ICZM approach since the EC recommended its adoption in 2002. Similarly, the low level of adherence to ICZM principles by Ireland and Romania is also understandable as neither of these countries (like many other EU member states at that time) are considered to have engaged in ICZM in a meaningful way (Shipman and Stojanovic 2007, p. 378). The research also demonstrated that there is a positive relationship between the rate of adherence to the ICZM principles and the environmental outcomes for marine and coastal areas. In both the French and Spanish case studies, the researchers were confident that the marine and coastal governance systems had either achieved (or were achieving) improvements in marine and coastal environments. In contrast, the Irish and Romanian researchers were not confident their respective marine and coastal governance systems had the capacity to deliver an improvement in environmental outcomes.

9.4.3 Opportunities for Effective Marine and Coastal Governance

The development of coastal and marine governance systems in each of the case study areas has had a number of positive impacts that have been beneficial to managing LSI and achieving GES. All researchers reported that there are higher levels of awareness of their marine and coastal environments. Heightened awareness is also leading to positive changes. In Romania, demands for participatory management (from sectoral partnerships and NGO's) are emerging, and there has been a move away from hard engineering solutions to coastal protection. In Ireland, the adoption of a National Marine Planning Framework and the opportunities to participate in its preparation were both seen as positive developments and it was acknowledged that there has been a significant increase in the number of new data sets for the marine and coastal environment. However, the above positive impacts did not (at the time of the research in 2020) have any discernible influence on the development of the Irish and Romanian coastal and marine governance systems.

There were also higher levels of awareness in France and Spain of the need to achieve good marine and coastal environment status and this change is believed to have influenced the provision of an extra 'layer' in their governance systems for non statutory stakeholders which has led to community and non statutory stakeholder groups being assigned decision making roles in marine and coastal management. As a result of this change, actions are being undertaken by community and non statutory stakeholder groups in both countries that enable the conservation and improvement of the marine and coastal environments in their respective areas. Examples of the actions undertaken in Galicia (Spain) include the provision of better signposting, engaging in the cleaning and maintenance of coastal amenities and changing access arrangements to preserve and improve the environment. There is also evidence from the French case study that allocating tasks to the community and non statutory stakeholders in the management of the lagoon has led to innovations in comanagement that included the development of a pollution tracking project which provided citizens with a digital means to indicate geolocalised pollution points.

9.4.4 Barriers to Effective Marine and Coastal Governance

The research revealed that despite their varied backgrounds and differing legislative contexts, there are strong similarities in the barriers faced by member states when attempting to manage LSI and govern their marine and coastal areas. In all cases, there is a fragmentation of responsibilities for coastal and marine areas among a range of different government departments/ministries and supporting agencies. A recurring theme of the research is the significant number of diverse government departments (or ministries) and agencies in all member states that either had (or still have) sectoral functions and responsibilities for marine and coastal areas. The research reveals that poor co-ordination of decision making by the government departments (or ministries) and agencies with marine and coastal responsibilities has led to fragmented approaches to governance as many pursue their own sectoral objectives (often using a range of governance mechanisms to do so) with little or no regard for holistic objectives like GES. In addition, all member states have struggled to achieve the integration of the policies that are designed to improve coastal and marine areas across all levels of governance (national, regional and local).

Given that all member states have experience of fragmented marine and coastal governance, the results of the research provide an insight into how each member state has responded to this issue. It was noted that fragmented responsibilities does not appear to have led to wholesale reform of existing governance structures for marine and coastal areas in any of the case study areas. In Ireland, Romania and France, the focus was very much on improving communication and engagement between the key authorities with marine and coastal responsibilities in order to coordinate their management efforts. However, there are notable differences in the mechanisms used to improve co-ordination. In France, a very effective brokerage organisation (Syndicat Mixte du Bassin de Thau (SMBT hereafter)) was established at the regional level to improve co-ordination and decision making of authorities with coastal and marine functions at different levels of government. With respect to Spain, it was acknowledged in the research that action is needed to address the fragmentation of responsibilities in marine and coastal functions. However, like France, there were examples of effective joint coastal and marine governance arrangements (such as the Atlantic Islands Natural Park in Galicia (Ons, Cíes, Sálvora and Cortegada)) that could provide guidance on managing LSI. In Ireland, the marine coordination group was established. This group was comprised of an interdepartmental committee in which high level departmental officials engage in matters of mutual interest in marine and coastal areas as a means of improving communications between government departments with coastal and maritime functions. However, the lack of oversight of the marine coordination group (who don't produce reports) means that its effectiveness is difficult to gauge. Romania adopted a similar approach to Ireland with the establishment of an inter ministry committee but its impact on improving co-ordination between stakeholders is unclear.

A lack of integrated data sets has also been identified as a barrier to marine and coastal governance in both Ireland and Romania, despite the acknowledgement that effective governance relies on good quality data. According to the researchers in both cases, the governance arrangements impose two strong influences on the type of data sets that are collected. Firstly, data sets are normally aggregated at national level only, as there are no regional or local authorities in either case who engage in data collection. Secondly, centralised governance systems generally lead to the collection of fragmented data sets as individual government departments/ministries focus on their own sector specific objectives, and gather sector specific data sets, that tend to be more limited in their application and use. An example (from Ireland) of a sector specific data set would include information on fisheries being collected by the Department of Agriculture, Food and the Marine. This contrasts with Spain and France where devolved governance systems have enabled the collection of more integrated data sets with local 'specifity' and which are also used to devise ecologically-based performance criteria for local marine and coastal areas. The high number of administrative staff and low number of technical staff (with scientific backgrounds) in Irish government departments and Romanian ministeries with coastal and marine responsibilities is also believed to amplify the difficulties with integrating data sets. This offers a sharp contrast to France where the SMBT brokering organisation has a multi disciplinary staff complement.

The Romanian researchers also drew attention to the issues arising from data being collected to different data baselines and standards by EU member states and non EU member states with borders on the Black Sea. This has created significant problems for governance as the data cannot be reliably used for comparative purposes or for devising (or for monitoring) performance standards for key criteria such as water quality. The recent departure of Britain from the EU also has the potential to create similar divisions between Ireland (a Member State) and the UK (a non EU country from January 2021). Despite the issues with respect to data collection standards, there appears to be potential to address these matters through existing transboundary bodies such as the Black Sea Commission and the British Irish Council, both of whom can be used to deliver common data collection standards and more effective transboundary governance of coastal and marine areas. The Romanian researchers also identified a lack of continuity at government level and insufficient political will to take action and address shortcomings as barriers to progress in marine and coastal governance.

9.4.5   Governance Mechanisms

Notable differences could be seen in the mechanisms used in the devolved marine and coastal governance systems of France and Spain and the more centralised systems of Ireland and Romania. The regional, local and community level authorities in the case study areas in France and Spain were using area based plans in order to manage LSI and achieve improved outcomes for their marine and coastal environments. The area based plan for the Thau Lagoon (in France) were also based on holistic objectives which are comprised of prescriptive theme based performance criteria for constituent elements of the marine and coastal environment. The theme based performance criteria (which were devised by using data sets collected at local level) are also used to overcome the difficulties created by administrative boundaries, unify the management approaches of the different authorities and create partnerships among statutory and non-statutory stakeholders (such as coastal communities and other interests such as oyster farmers). Similar partnership arrangements were in place in Galicia in Spain where local development strategies are focused on the preservation and improvement of the environment. The Irish and Romanian case studies provide a sharp contrast to the French and Spanish area based plan approach. In both Ireland and Romania, national level strategies focussed on non prescriptive high level objectives were under development (in 2020). As there are no regional or local authorities with coastal and marine responsibilities in

Ireland or Romania, it is not possible for either of these countries to engage in data collection or prepare and implement area plans (with theme based performance criteria) for marine and coastal areas below national level.

9.4.6 The Relationship Between Governance and Environmental Outcomes

There was a consensus among the researchers that comprehensive and effective marine and coastal governance systems can achieve the goal of GES. However, striking differences could be seen in the perceptions of researchers on the effectiveness of the current governance arrangements in each of the case study areas. With respect to the French and Spanish case studies, the researchers appeared convinced that the governance structures have either led to (or are leading to) an improvement in the quality of the coastal and marine environment in their subject areas and that the interactions between land and sea were being managed more effectively. As a result, the researchers in the French and Spanish case study areas had a high level of confidence that the overall objective of GES could be achieved.

In contrast, the Irish and Romanian researchers were not convinced that the governance arrangements for their countries would lead to improved environmental outcomes for their marine and coastal areas. While it was acknowledged that the Irish and Romanian systems were a work in progress and that it was too early to comment on whether they had achieved an improvement in marine and coastal environmental quality or not, both sets of researchers were of the view that the governance pathways for delivering effective marine and coastal governance were not clear. This view arose from the fact that in both cases, no obvious attempts seemed to have been made in either Ireland or Romania to assess the suitability of the MSP policy mixes and mechanisms to the existing governance structures that they were being introduced into. There was also a concern among Irish and Romanian researchers that both of these countries were persisting with centralised approaches to marine and coastal governance (confined to national level only) that had been abandoned by France and Spain in favour of more devolved governance systems.

9.4.7 The Application of Evolutionary Governance Theory (EGT)

Noting the complexity of LSI and the difficulties that arise in attempting to manage them, the capacity of EGT as an approach to analyse marine and coastal governance approaches in the four case study areas was considered. The results of the research confirm the consensus view among interview respondents that the EGT perspective provided a useful lens to explore and understand *"governance and governance transformation against the background of co-evolutions of all constituent parts of governance"* (Van Assche et al. 2020, p. 1). All respondents also agreed that it led to an enhanced understanding of coastal and marine governance pathways in each case study area. In addition, there was an appreciation that EGT was an effective conceptual framework of analysis for exploring the management of LSI in different EU member states.

In the Irish case study, EGT was considered to be an informative approach *"which allowed the researchers to review past ocean and coastal governance in Ireland and apply this experience when looking forward" (Researcher in the Irish case study, January, 2020)*. The Romanian researchers found that the EGT approach was useful *"for looking at the journey that Romania has been on – from its transition from a country heavily influenced by the USSR to an EU member state and for reviewing what has happened in the country in recent years and understanding the stage that the country is currently at" (Researcher in the Romanian case study, January, 2020)*. In the Spanish case study, the EGT perspective was considered to be an effective means *"of exploring the interactions between the different levels of government and their position in the new system of actors created in the coastal zone" (Researcher in the Spanish case study, February, 2020)*. The researchers involved in the French case study described EGT as a constructive approach for analysing marine and coastal management as it helped to reveal the failures of previous governance systems (many of which relied on physical water boundaries) in the Thau Lagoon.

9.4.8 Commonalities Between Approaches to Governing Marine and Coastal Areas in the EU

While MSFD has been adopted by all EU member states (since at least 2016) and all member states are commited to delivering the common desired goal of GES in marine and coastal environments, the research demonstrates that a degree of harmonisation of governance approaches to managing LSI and governing marine and coastal areas can be discerned in the four member states under study. This harmonisation is occurring despite the fact that the land use (and marine) planning systems differ significantly between the four case study areas. According to the research findings, two different types of marine and coastal governance systems can be identified. The first of these systems (found in both Ireland and Romania) has strongly centralised governance arrangements that are concentrated at the national level with fragmented responsibilities for government departments/ministries/agencies and no responsibilities for managing marine and coastal resources afforded to non statutory stakeholders. The strongly centralised systems also appear to rely on national level strategies and data sets as well as non prescriptive high level objectives to deliver GES in marine and coastal areas. The second type of system (that can be found in France (and to a lesser extent Spain)) has devolved marine and coastal governance arrangements with good coordination among stakeholders at all levels (national, regional, local and community). The devolved systems tended to use area based plans with theme based performance criteria (devised from local data sets) to realise GES. Co-management of marine and coastal resources between statutory and non-statutory stakeholders at community level is also a feature of the devolved systems of France and Spain.

9.5 Conclusions and Recommendations

Under MSFD, EU member states are committed to delivering GES in marine and coastal areas by managing LSI and regulating all uses and activities in their marine and coastal areas. While it is understood that comprehensive marine and coastal governance systems are needed to govern LSI and manage marine and coastal areas, the physical diversity of maritime areas and coastlines combined with the complex and dynamic relationship between the land and the sea present major challenges to achieving this. Recognising these difficulties, the MSP expert group in 2017 proposed a framework for addressing LSI that called for MSP Authorities (and other stakeholders) to engage in a two phase process that reflects the complexity of the task. The first phase of the process involves the development of an understanding of the dynamics involved in LSI in their jurisdiction and the second phase requires member states to identify institutional mechanisms to manage LSI that are most suited to their individual marine and coastal governance frameworks. This section of the research reviews the investigation of LSI in each case study area as well as the mechanisms and measures that were used to manage them. Conclusions are drawn on the effectiveness of the mechanisms and measures introduced to deal with LSI and marine and coastal management, while recommendations for future governance are provided.

(i) *The extent to which member states have investigated the dynamics of LSI in their jurisdictions*

The research reveals that the French case study (from the Thau Lagoon) has undertaken the most in depth investigation into LSI. This has been achieved by developing a devolved marine and coastal governance system comprising of sub national authorities (such as the SMBT) with multidisciplinary (i.e., technical and administrative) staff who engaged in the collection of local level 'holistic' data sets that are focussed on ecological themes. The holistic data sets were then analysed to ascertain the 'impact chain' of land based activities on marine and coastal areas by identifying the most ecologically harmful activities and devising measures to either mitigate or avoid them altogether. The specific local data sets are also used to devise performance criteria for key environmental indicators in the marine and coastal environment (such as water quality). A similar approach was followed in Spain where local level theme based data sets were gathered by authorities who devised local development strategies designed to preserve and improve the marine and coastal environment. In Ireland and Romania, the centralised governance systems were dominated by national level stakeholders with sectoral interests. This was also reflected in the data sets collected which were aggregated at national level and often had a sectoral focus. The absence of data sets with local specificity then made it more difficult to determine the 'impact chain' of land based activities on marine and coastal areas or to identify and take action on the most harmful terrestrial activities. Matters are further complicated in Romania as the national level data sets that exist on the marine and coastal environment (of the Black Sea) are not directly compa-rable with the data sets collected by the non EU member states that border the Black Sea.

Recommendation 1: Best Practice Guidance on Data Collection

It is strongly recommended that best practice guidance is produced at the EU level on collecting and recording holistic theme based data sets (at national, regional and local level) in order to underpin integrated approaches to managing LSI and marine and coastal resources. It is also recommended that common standards for data col-lection and recording are agreed between EU and non member states (who share borders with the EU) in order to ensure effective monitoring of shared marine and coastal resources.

(ii) *The institutional mechanisms and measures that each of these member states have chosen to manage LSI within their marine and coastal governance systems*

The four case study areas revealed that two distinct types of marine and coastal governance systems can be discerned from the research – devolved systems and centralised systems. Both France and Spain provide examples of devolved marine and coastal governance systems which afford decision making roles to stakeholders at national, sub national/regional, local and community level. There was also evi-dence (from France) to demonstrate that these devolved systems had higher levels of co-ordination between stakeholders and more integrated governance approaches to managing marine and coastal areas. This was achieved by creating a regional brokering organisation with multi disciplinary staff to coordinate land, water, sea and biodiversity planning and to facilitate interactions between statutory stakehold-ers and community level groups. Centralised marine and coastal governance sys-tems can be found in Ireland and Romania. These systems are confined to national level only as there are no competent authorities and agencies involved at regional, local or community levels. The research results have shown that a prominent feature of centralised systems is weak coordination of sectoral interests (many of whom have fragmented responsibilities) and an absence of devolved governance layers which enable sub national, local and community level stakeholders to participate in management, decision making and data collection.

The type of marine and coastal governance system also exerts a strong influence on the governance mechanisms that are used to deliver improved environmental outcomes for marine and coastal environments. The devolved French and Spanish systems are focussed on area based plans as a means of managing marine and coastal resources more effectively. This was demonstrated in the Thau Lagoon in France and in the local development strategies in Galicia, Spain where the area plans and the local development strategies at local levels have a strong environmental emphasis. This is particularly the case in the Thau Lagoon in France where prescriptive theme based performance criteria for constituent elements of the marine and coastal environment are included as targets of the area based plan. These performance criteria are compiled using the local area specific data sets and they are used to integrate the management approaches of all Authorities (statutory and non statutory) and overcome the difficulties created by administrative boundaries. The research also revealed that the devolved marine and coastal governance systems provided for greater participation at all levels of governance (from national to community level) and a higher level of coordination and engagement among statutory and non statutory stakeholders. The development of a community level of governance has also led to the formation of effective partnerships and co-management innovations between statutory authorities and community based stakeholders. Centralised marine and coastal governance systems (such as those found in Ireland and Romania) rely on national level strategies with high level aims and objectives. National level stategies (and objectives) afford little or no participation to statutory and non statutory stakeholders at regional, local and community levels in managing LSI and marine and coastal resources.

Recommendation 2: Prepare best practice guidance on coordinating the management of LSI

The research has demonstrated that best practice examples are available on coordination mechanisms that can be used to ensure integrated approaches to managing LSI and marine and coastal governance (such as the brokering organisation with multi disciplinary staff in the Thau Lagoon case study in France). It is recommended that best practice guidance should be prepared at EU level to illustrate how integrated marine and coastal governance can be achieved.

Recommendation 3: Engaging in participative management with coastal communities and non statutory stakeholders

It has been shown that significant benefits can be derived from involving coastal communities and / or non statutory stakeholders in the management of marine and coastal areas. These benefits include stakeholder groups (such as Fishermans Guilds (Spain) and oyster farmers (France)) undertaking stewardship roles by monitoring environmental quality and enabling the development of innovative co-management techniques between statutory authorities and non statutory stakeholders. It is recommended that EU member states should undertake proactive measures to involve coastal communities and non statutory stakeholders in their coastal and marine governance systems in order to realise these valuable benefits.

(iii) *The overall effectiveness of these mechanisms and measures*

The effectiveness of the different governance mechanisms and measures for managing LSI and maritime activities and for delivering GES for marine and coastal areas was considered. The results revealed that the researchers who worked on the

Thau Lagoon (France) and Galicia (Spain) case studies were confident that the devolved governance arrangements that were in place in these areas were proving effective and that they were leading to improvements in marine and coastal environments. In contrast, the researchers who carried out the case studies in Ireland and Romania stated there was no evidence that the coastal governance systems in these countries were leading to marine and coastal environmental improvements. The Irish and Romanian researchers also shared a lack of confidence in the capacity of their marine and coastal governance systems to deliver GES as the pathways for doing so were unclear.

Recommendation 4: Revise the current methodology for assessing the effectiveness of marine and coastal governance

There is evidence in the research which appears to show that some member states have introduced mechanisms and measures to comply with EU requirements on MSP and MSFD without carrying out the necessary due diligence to ascertain whether the adopted mechanisms and measures are suitable to existing governance systems. To address this issue, the methodology by which marine and coastal governance approaches are being assessed at EU level (i.e., the assessment procedure of measures adopted by member states) should be reviewed to ensure that the effectiveness of the approaches being followed by member states and their suitability to their different governance contexts is fully assessed.

Recommendation 5: Introduce tiered deadlines for compliance with GES

It is clear from the research that the marine and coastal governance systems of some member states are more advanced than others with respect to managing LSI and delivering GES for marine and coastal areas. As member states should be encouraged to road test the suitability of different measures to their differing governance contexts, staggered deadlines for compliance with GES should be considered at the EU level. This would enable member states to find the most effective measures that would suit their governance systems rather than rushing in changes to their systems that are unlikely to realise their desired environmental outcomes.

References

Bellas J (2014) The implementation of the marine strategy framework Directive: shortcomings and limitations from the Spanish point of view. Mar Policy 50:10–17

Daniell KA, Plant R, Pilbeam V, Sabinot C, Paget N, Astles K, Steffens R, Barreteau O, Bouard S, Coad P, Gordon A, Ferrand N, Le Meur P-Y, Lejars C, Maurel P, Rubio A, Rougier J-E, White I (2020) Evolutions in estuary governance? Reflections and lessons from Australia, France and New Caledonia. Mar Policy 112:103704. https://doi.org/10.1016/j.marpol.2019.103704

EC (2002a) Communication from the Commission to the Council and the European Parliament on integrated coastal zone management: a strategy for Europe COM/2000/0547 final. https://eur-lex.europa.eu/legal-content/EN/TXT/?uri=CELEX:52000DC0547

EC (2002b) Recommendation of the European Parliament and of the Council concerning the implementation of Integrated Coastal Zone Management in Europe. Off J Eur Communities, L 148/24, 6.6.2002. https://eur-lex.europa.eu/legal-content/EN/TXT/PDF/?uri=CELEX:3200 2H0413&from=EN

EC (2008) Directive 2008/56/EC of the European Parliament and of the Council of 17 June 2008 establishing a framework for community action in the field of marine environmental policy (Marine Strategy Framework Directive). https://eur-lex.europa.eu/legal-content/EN/TXT/?uri =CELEX:52000DC0547

EC (2014a) Report from the Commission to the Council of the European Parliament; The first phase of implementation of the Marine Strategy Framework Directive (2008/56/EC) The European Commission's assessment and guidance/* COM/2014/097 final

EC (2014b) DIRECTIVE 2014/89/EU establishing a framework for maritime spatial planning. European Commission, Brussels

EC (2017) Maritime spatial planning: addressing land-sea interaction a briefing paper. Web: https://www.msp-platform.eu/sites/default/files/20170515_lsibriefingpaper_1.pdf. Accessed 20 Oct 2020

Falaleeva M, O'Mahony C, Gray S, Desmond M, Gault J, Cummins V (2011) Towards climate adaptation and coastal governance in Ireland: integrated architecture for effective management? Mar Policy 35:784–793

Kidd S, Jones H, Jay S (2019) Taking account of Land-Sea interactions in marine spatial planning. In: Zaucha J, Gee K (eds) Maritime spatial planning – past, present & future. Palgrave Macmillan

O'Connor MC, Lymbery G, Cooper JAG, Gault J, McKenna J (2009) Practice versus policy-led coastal defence management. Mar Policy 33:923–929. https://doi.org/10.1016/j. marpol.2009.03.007

O'Hagan AM, Cooper JAG (2002) Spatial variability in approaches to coastal protection in Ireland. J Coast Res SI 36:544–551

O'Hagan AM, Paterson S, Tissier ML (2020) Addressing the tangled web of governance mechanisms for land-sea interactions: assessing implementation challenges across scales. Mar Policy 112:103715. https://doi.org/10.1016/j.marpol.2019.103715

Partelow S, Schlüter A, Armitage D, Bavinck M, Carlisle K, Gruby RL, Hornidge A-K, Tissier L, Martin P, Jeremy B, Song AM, Sousa LP, Văidianu N, Assche V, Kristof. (2020) Environmental governance theories: a review and application to coastal systems. Ecol Soc 25(4):19. https:// doi.org/10.5751/ES-12067-250419

Pineiro-Antelo M d 1 A, Felicidades-Garcia J, O'Keeffe B (2020) The FLAG scheme in the governance of EU coastal areas. The cases of Ireland and Galicia (Spain). Mar Policy 112:103424

Ramesh R, Chen Z, Cummins V, Day J, D'Elia C, Dennison B, Forbes DL, Glaeser B, Glaser M, Glavovic B, Kremer H, Lange M, Larsen JN, Le Tissier M, Newton A, Pelling R, Purvaja E, Wolanskin. (2015) Land-Ocean interactions in the coastal zone: past. Present Futur Anthropocene 12(2015):85–98

Schlüter A, Assche V, Kristof H, Anna-Katharina V, Nataşa. (2020) Land Sea interactions and coastal development; an evolutionary governance theory perspective. Mar Policy 112:103812

Shipman B, Stojanovic T (2007) Facts, fictions, and failures of integrated coastal zone Management in Europe. Coast Manag 35:375–398

Strategic Action Plan for the Environmental Protection and Rehabilitation of the Black Sea (2009) Available from: http://www.blacksea-commission.org/_bssap2009.asp. Accessed 26 Sept 2017 Sofia, Bulgaria

SUPREME (2015) Recommendation on how to perform analysis of land-sea interactions, combining MSP and ICZM in the considered project area. Initial definitions. Deliverable No 1.3.7. Web: https://www.pap-thecoastcentre.org/razno/C_137_LSI_initial%20description.pdf. Accessed 20 Nov 2020

Văidianu N, Tătui F, Ristea M, Stănică A (2020) Managing coastal protection through multi-scale governance structures in Romania. Mar Policy 112:103567. https://doi.org/10.1016/j.marpol.2019.103567
Van Assche K, Hornidge A-K, Schlüter A, Vaidianu N (2020) Governance and the coastal condition: towards new modes of observation, adaptation and integration. Mar Policy 112:103413

Part III
Thematic Analyses

Chapter 10
Sustainable Seafood Consumption: A Matter of Individual Choice or Global Market? A Window into Dublin's Seafood Scene

Cordula Scherer and Agnese Cretella

Abstract Seafood consumption is considered a key element for food security and for nutrition related policies. However, seafood is often not easily accessible or perceived as a popular option even by those living in close proximity to the sea, especially in the western world. Common culprits are usually identified as a lack of specialized shops, culinary knowledge or as the disconnection with local coastal cultural heritage. This is, for instance, the case in Ireland: Irish waters provide a great diversity of seafood and yet, its domestic consumption remains unusually low for an island nation. Most of Ireland's seafood is exported to other countries, whilst the Irish stick to the popular salmon, cod and tuna; a consumption habit that has obvious sustainability externalities. This contribution aims to unpack the issues connected to seafood consumption in Ireland's coastal capital Dublin and offers a window into the city's seafood scene. Data presented were gained within Food Smart Dublin, a multidisciplinary research project designed to encourage a behavioural shift of consumption towards more sustainable local seafood. The project's purpose was to reconnect Dublin's society with their tangible and intangible coastal cultural heritage by rediscovering and adapting historical recipes. The paper thus connects past, present, and future perspectives on the topic. First, the past is explored by delineating the potential of marine historical heritage in stimulating sustainable seafood consumption with the reintroduction of traditional Irish recipes. The present offers a data snapshot on consumption patterns towards seafood gathered from

C. Scherer (✉)
Trinity Centre for Environmental Humanities, Trinity College Dublin, Dublin, Ireland
e-mail: cscherer@tcd.ie

A. Cretella
Trinity Centre for Environmental Humanities, Trinity College Dublin, Dublin, Ireland

Department of Philosophy and Communication, University of Bologna, Bologna, Italy
e-mail: agnese.cretella@unibo.it

© German Institute of Development and Sustainability (IDOS) 2023
S. Partelow et al. (eds.), *Ocean Governance*, MARE Publication Series 25,
https://doi.org/10.1007/978-3-031-20740-2_10

233

structured online questionnaires results from the Food Smart Dublin project. Respondents offered insights into their relationship with the sea, on the frequency with which they consume seafood and the obstacles they see in consuming more of it. Finally, these perspectives delineate possible future scenarios and recommended governance actions to support policymakers in designing a better and more sustainable seafood system.

10.1 Introduction

10.1.1 The Irish Context

Ireland is an island nation with an extensive, indented coastline of over 7000 km, and 10 times more territory under the sea than on land. This provides ideal habitats for great coastal biodiversity and creates a vast range of seafood. Ireland's fishing grounds are among the richest in Europe and yet seafood is often overlooked in shaping the country's modern culinary identity. In the past, seafood played a pivotal role for the inhabitants of the island. There is evidence that shellfish such as oysters, scallops and cockles, fish like cod, whiting, wrasse and ling and all kinds of seaweed were consumed by the hunter-gatherers that first arrived and settled at Irish shores over 10,000 years ago (O'Sullivan and Breen 2007). These are species that can still be found in the Irish waters today and that are commercially exploited. With the advent of farming in the Neolithic period, Irish ancestors turned away from the sea and seafood became less essential food for survival. With different invasions and trades came different food cultures and seafood saw a rise and fall through the centuries with the arrival of the Beaker people, the Celts, the Vikings, the Normans and the English (O'Sullivan and Breen 2007; McMahon 2020).

In present day Ireland, people consume seafood just below the average European amount which is surprising given the richness of seafood at the doorstep. Some call this phenomenon the "sea blindness" of the Irish as the diversity of marine food does not seem much appreciated. Instead seafood like salmon, cod and tuna, top predators that could be regarded as the tigers and lions of the sea, are the regular items of the Irish seafood diet. These predators occupy the top trophic level of the marine food web and are heavily overfished while most of Ireland's treasures, such as lobster, herring and mussels are exported to other European and Asian countries who seem to have more appreciation of these local products.

Before March 2020, when the COVID-19 pandemic forced the gastronomy sector in Ireland to its knees and when restaurants were operating normally, over half of the seafood consumption took place outside the domestic setting. Reasons for not cooking seafood at home were often the lack of recipes and restricted availability (Scherer and Holm 2020). But alternative seafood to the traditional fish'n'chips such as mussels, seaweed or less-known fish were also considered unpalatable and a certain lack of knowledge led to insecurity around cooking a delicious seafood meal at home.

Indeed, knowledge amongst the Irish public on local and seasonal fish and seafoods from lower trophic levels is limited and incomplete. The sustainability of the fisheries is confusing and highly complex and without unifying certificates many consumers as well as hospitality professionals feel unsupported and discouraged from buying sustainable seafood. This seafood illiteracy was not always so severe in Ireland. The island nation has hundreds of years of experience in sourcing and cultivating food from the sea. Only in the nineteenth century local knowledge seemed to slip away when conflict, political indifference and economic abandonment led to a decline and neglect in coastal activity (O'Sullivan and Breen 2007). The Great Famine marked a key event of change in Irish food. Due to the massive reduction in population, the workforce and the knowledge was not available to produce food locally. Consequently, less food was grown on Irish fields and more was imported from abroad. This led to great changes in produce and therefore consumption with a strong influence of the world market and increased commercialisation (Clarkson and Crawford 2001).

In the last couple of decades or so, an appreciation of diverse seafood is gently resurging due to celebrity chefs introducing novel, healthy trends. These celebrity chefs promote the preparation of seafood on TV cooking shows and give workshops on sourcing and purchasing fresh local and sustainable seafood across the country. This strengthens the confidence of Ireland's citizens in past seafood knowledge and spurs curiosity. The recent COVID-19 pandemic also seems to have contributed to the incentive to appreciate local products, cook at home and reconnect with a more territorial, local cuisine, grounded in coastal habitats.

This contribution explores and presents findings of the multidisciplinary research project Food Smart Dublin. The project was designed to revive Ireland's sustainable seafood practices in an innovative dialogue between past knowledge, present palates and future interaction with Irish waters focusing on Ireland's coastal capital Dublin. This article's main objective is to investigate whether the rediscovery of cultural/culinary heritage could incentivise sustainable seafood consumption by Dublin's society. At the same time, it explores if there are obstacles that prevent citizens who live in such close proximity to the sea to eat more locally-sourced, sustainable seafood.

To this backdrop data are presented from structured online questionnaires on consumption patterns of seafood among Dublin's society and the participants' relationship with their surrounding sea. The results are discussed from the perspective if and how the rediscovery of historical seafood recipes can help with Dublin's image as a sustainable seafood city. The chapter concludes with recommendations on governance actions to support policymakers in designing a better and more sustainable seafood system on Ireland's East coast.

10.1.2   Seafood and Its Environmental Agency

Fishing arguably remains the oldest means of food gathering humans still practice on a global scale today. For centuries, the ocean was a distant place for many and the human-ocean relationship was not thought about in great detail (Brennan et al. 2019). Over the last one and a half centuries, anthropogenic use of the oceans increased dramatically with the exploitation for its oil and gas, wind and wave power, increased transport, recreation and of course intensified fisheries. Given the preference for certain seafood species and the industrialisation of fishing, stocks of the most commercially valuable species have become seriously depleted in the early twenty-first century (Pauly et al. 1998, 2002, 2003).

This is not without reason. Seafood provides important sources of employment and nutrition, especially in low-income countries, and is highly traded, both globally (Gephart and Pace 2015) and regionally (Belton et al. 2018). Hundreds of millions of people rely on seafood for their livelihood, culture, and food and nutrition security (FAO 2018). And yet, the real value of seafood is not well understood, protected or integrated into global food security and nutrition policy considerations (e.g. Béné et al. 2015). Moreover, food sourcing from the ocean in the last decades has mostly focused on exploiting top predators such as salmon, tuna, cod and haddock. The vast amounts of potential food at lower trophic levels such as filter feeders and algae are not as popular, despite being already harvested as economically viable and nutritious products.

The Food from the Oceans report (EU 2017), which was subsequently endorsed by the EU Group of Chief Scientific Advisors as the foundation for a range of recommendations posed a central question: *'How can more food and biomass be obtained from the oceans in a way that does not deprive future generations of their benefits?'* The scientific evidence in answering this question clearly points to act sustainably by increasing seafood production and consumption at lower trophic levels as a way to bring about such an increase in biomass. Moreover, the greatest and most feasible potential for expansion globally identified in The Food from the Oceans report lies in mariculture of herbivore filter feeders such as mussels and oysters and cultivated algae/seaweed for direct human consumption – or for a more ecologically-efficient source of feed for farmed marine carnivores (such as salmon). Another point addressed in the same report is that ocean-derived protein should play an increasingly important role globally to fulfil the UN Framework Convention on Climate Change. The challenge we are facing is a shift in consumption habits.

10.1.3   Food Systems and Consumption Behaviour

Food is a highly complex system, with social, economic and ecological components. It contributes significantly to greenhouse gas emissions and plays a key role in driving climate change. Our behaviour towards food, what we eat, how we eat it, and how we dispose of it too influences our health, food security, soil degradation and water quality. Around one third of global greenhouse emissions comes from the food system. The UN's Food and Agriculture Organisation estimates the annual financial cost of wasted food to be €900 billion in economic costs and an additional €800 billion in social costs (FAO 2018).

Food insecurity and sustainability are among the most significant global challenges faced by humanity in the twenty-first century. Ensuring safe, nutritious and sufficient food for a growing global population of close to ten billion people is a challenge exacerbated by increasing urbanisation and political instability that requires an interdisciplinary approach locally, nationally and regionally. The future of planet Earth is determined by our actions, our behaviour as consumers and as citizens (Holm 2014). The lasting COVID-19 pandemic has highlighted the necessity to study impacts and identify vulnerabilities within the food system and has provided opportunities for governments, international bodies, industries, small-scale actors, and civil society to respond, adapt, and build resilience to future shocks to the food system. Investing in food-based solutions while interlinked with agriculture, specifically targets the food supply chain that is highly dependent on individual behaviour change (IPCC 2014). To change how our society consumes food, we must first change people's routines, habits and norms.

Many people from countries with a developed economy in the global north have changed their attitude towards food during pandemic related lockdown and recent peer-reviewed publications show shifts in consumer behaviour (Kaiser et al. 2021; Lam 2021; Love et al. 2021). Some of these results are positive indeed – Love and co-authors (2021) reported more home cooking and from scratch while food waste decreased and grow-your-own food increased substantially. This shows scope for a changing attitude towards food and consumption behaviour.

While the literature on theoretical models of consumer behaviour is large and complex (Jackson 2005), environmental education emerged as one of the primary strategies to effect behaviour change (Williamson et al. 2018) although the authors point out that evidence suggests it is less effective alone than paired with other techniques. Within educational approaches, it is important to distinguish between different types of knowledge that may be useful in an intervention, such as the what, why, and how related to a behaviour (Kaiser and Fuhrer 2003). Against this backdrop the seminal work of Shove (2010) has vocally criticised the over-attention and public investment on individual consumer behaviour rather than on the economic structures and policies that would allow sustainable living.

In principle, humans find it extremely difficult to change established behaviour, even though we know the negative consequences that await us if this change is not taking place. One point that can help with keeping these good habits is to re-introduce the totemic value that food had before modern mass-production reduced it to its economic value. The ecosystem's agency needs to take centre stage when dealing with the planet's resources as human preferences. This means that practices and actions are the main drivers of global environmental change in the twenty-first century. But this cannot come solely from the bottom up, i.e. society, it must also be implemented in government policies. It is crucial, therefore, to promote pro-environmental behaviour throughout. Holm and co-authors (2015) argue that in order to accomplish this, we need to move beyond rational choice and behavioural decision theories, which do not capture the full range of commitments, assumptions, imaginaries, and belief systems that drive those preferences and actions. Disciplines of the humanities such as history, anthropology, psychology, and philosophy can provide deep insights into human motivations, values, and choices.

To this end Holm et al. (2013) developed the Global Change Research (GCR), a framework aimed at an integrated conception of human agency and the planetary environment combining different knowledges for a "radical interdisciplinarity". Within this framework, the humanities are seen as an ally of the natural sciences meaning that greater attention is paid to the bio-geophysical dimensions of the social sciences and to ecological approaches in the humanities, while developing concepts, theories and research that aim to form fields enabling transnational studies.

Food studies lend themselves perfectly to such approaches. Within those, seafood can play an important role in building sustainable lifestyles and circular, fair food systems, creating a more resilient global system against climate change, helping to improve biodiversity and reduce pollution (Olson et al. 2014). This can be achieved not only by providing important sources of employment and nutrition across the globe, but also through increased ocean literacy (Tran et al. 2010) which is defined as '...*an understanding of the ocean's influence on you – and your influence on the ocean*' according to the most popular definition by the National Oceanic and Atmospheric Administration (2013). An ocean literate person has knowledge on how the oceans work, the anthropogenic impact on them, is able to develop critical attitudes towards topics such as unsustainable and sustainable fisheries and the generally human-ocean relationship (Brennan et al. 2019). At the same time, it is important to remember that beyond individual attitudes the current food system is fully embedded in the global economy, in what has been defined as a Corporate Food Regime (McMichael 2005). As any other kinds of commodities food prices and markets are now established internationally, whilst food often travels around the globe following capitalistic dynamics. Per contra, the concept of Food Sovereignty "is at once a slogan, a paradigm, a mix of practical policies, a movement and a utopian aspiration" (Edelman 2014, p. 960), which aims to contrast such corporate system by fighting for equal redistribution of food, land and water.

10.2   Methods

This section presents data from the Food Smart Dublin research project, including historical information from archival data, as well as seafood consumption data gathered from a structured online questionnaire. It details the methodologies applied to the archival research, the basis on how historical recipes were selected and how the online questionnaire was constructed. Participants taking the online questionnaire offered their perspectives on their relationship with the sea, on the frequency with which they consume seafood and the obstacles they see in consuming more of it. On the basis of these perspectives possible future scenarios are discussed and recommended governance actions to support policymakers in designing a better and more sustainable seafood consumption system are explored.

10.2.1 Food Smart Dublin

The Food Smart Dublin project[1] was based on a multidisciplinary and trans-sectoral approach that applied methodologies in the humanities and natural sciences. This was to integrate and intertwine insights from history, social sciences, food policy and marine ecology and to apply a trans-sectoral concept of knowledge exchange involving academia, businesses, NGOs and the general public. In a wider context, the project implements ideas of the 'Humanities for the Environment' approach (Holm et al. 2015) in a transactional effort to increase sustainable seafood consumption of locally sourced food from lower trophic levels. Specifically, the framework builds on archival and folkloristic research of historical, local seafood recipes of the Dublin coastal communities to document the city's forgotten knowledge of local seafood. A selection of ten historical recipes, following the seasons through the year, were cooked in an appetising, innovative way by professional chefs. The old and new recipes were published on the project's website and promoted on social media with a link to a structured online questionnaire to respond to. An effort was also made to ensure the selected recipes were from a time prior to the Great Famine for reasons given in 1.1.

10.2.2 Data Collection

Historical, local seafood recipes were searched for engaging general search engines such as Google and Wikipedia. More specific software and internal search engines were used to search the archives of national institutions like the National Library Ireland Archives, the National Folklore Collection, The School Collection at Dúchas.ie. Several specialised websites dealing with local maritime and food history of Dublin and Dublin Bay Biosphere were also utilized. Keywords included 'fish', 'fishing', 'seafood', 'Irish boats', 'coastal living', 'Irish diets' 'Dublin Bay', 'catch', 'dinner', 'coastal activity', 'shellfish'. More specific words around seafood included 'lobster', 'salmon', 'cod', 'limpets'. A total of just over 190 seafood recipes were found from seven main sources providing suitable material. These consisted of actual printed cookbooks, observations of the natural history of Dublin,

---

[1] Food Smart Dublin was funded by the Irish Research Council and carried out at the Trinity Centre for Environmental Humanities in Trinity College, Dublin between 2019 and 2021.

handwritten manuscripts and letters from family and estate papers. All sources were written in English. The majority of the recipes were on salmon, cod, oysters and lobster. Lamprey, turbot and eel were also prominent. Less common were recipes on ray, weaver and limpets. All seafood recipes originated from a time period between the early 1690s and mid-1840s (Box 10.1).

Box 10.1: List of Main Sources of Historical, Local Seafood Recipes and References

Source name	Document type	Author	Publisher	Year	Reference
Mary Cannon's Commonplace Book – an Irish kitchen in the 1700s	Printed book	Marjorie Quarton	Lilliput Press	1700–1707	https://www.lilliputpress.ie/product/mary-cannon-commonplace-book-an-irish-kitchen-in-the-1700s
The Townley Hall papers	Handwritten manuscripts	Ce Bradell; Janc Bury	National Library Ireland	c.1840, 1702	Ms 16,844 – 16,846; Ms 9563
The Art of Cookery made plain and Easy	Scanned e-book	Hannah Glasse	Internet archive – public domain	1777 2nd edition	https://archive.org/details/TheArtOfCookery
The Lady's companion: or, Accomplish'd Director in the whole art of Cookery	Scanned e-book	A Lady 'Ceres'	National Library of Australia	1767	https://catalogue.nla.gov.au/Record/3197172
Smythe Family of Barbavilla, Collingstown, Co. Westmeath XXVI Recipes and Miscellaneous	Handwritten manuscripts	Several authors	National Library of Ireland	c. 1690	MS 41,603/2/1-2
A new system of domestic cookery	Scanned e-book	Maria E. Rundell	Internet archive – public domain	1807	https://archive.org/details/newsystemofdomes01rund/page/n4
An essay towards a natural history of the county of Dublin	Scanned e-book	John Rutty	Google books – public domain	1772	https://play.google.com/books/reader?id=u3FbAAAAQAAJ&pg=GBS.PP1&hl=en

Once organised, transcribed and logged, the recipes were selected in a collaborative manner between the researchers and chefs in constant dialogue around the concepts of *suitability, seasonality* and *sustainability*, explained in detail in the next paragraphs.

These following concepts are merely a methodology we adopted to select the appropriate seafood and recipes. Respondents of the structured online questionnaire were presented with the recipes and seafood chosen on the basis of these indicators and were thus not asked to provide feedback on them.

Suitability

The suitability of the dish/seafood was concerned primarily with the history of the marine creature in Irish tradition (e.g. lobster, oyster or hake), and also with the level of difficulty in making the dish, i.e. it required no special skills to be cooked in the domestic setting with ordinary kitchen tools and average cooking skills. Affordability was an additional element considered under suitability, i.e. aimed for participants to be able to comfortably incorporate the dish into their weekly diet based on an average Irish income and time availability.

Seasonality

Recipes were selected utilizing different perspectives and viewpoints of seasonality. For instance, avoiding certain seafood during spawning season was not the only aspect considered, also because it is not always a straightforward choice. Many chefs and fishmongers would agree that some seafood is only available during spawning season as that is when they become more active, and are accessible for the boats that catch them, or when they taste better. Moreover, when the seafood is landed in higher numbers, it usually also goes down in price and is, therefore, more affordable. Some seafood is seen as a delicacy when in roe and preferred by some such as the opaque scallop with its orange 'coral'. Sometimes the ethical imperative to avoid seafood during their spawning season is in contrast with their availability, affordability and taste. With the recipes selected for the Food Smart Dublin project, the seafood was generally not considered for a certain month when they were known to be spawning or when they were known to be 'spent'. This is a term used for seafood that just spent all their fat and protein content into egg production during spawning season which makes their flesh watery and soft.

Sustainability

Sustainability is not a concept, but rather an on-going process with three core elements that are intrinsically linked: economic growth, social inclusion and environmental protection (Purvis et al. 2019). We argue that the dimension of ethics is the fourth element crucial to harmonise the other three (in agreement with Suhonen and Sutinen 2014). This implies that a commitment to sustainability in and of itself does not prescribe a unique fixed future state of the world but leaves a dynamic leeway of options. The path to sustainable development is value-based and multi-dimensional (Kaiser 1997), often accompanied by trade-offs between basic values, affecting all other dimensions. Ultimately, there are many different definitions of sustainability, depending on what perspective one looks at it.

Likewise, the sustainability of seafood can vary significantly depending on how and where the seafood has been caught or farmed. Many single species are caught or farmed in different ways and by different methods. The recipes for the Food Smart Dublin project were chosen based on current data at the time of the project, knowledge and sustainability advice (2019–2021). The choice may have been different if the project would have been in the UK or another country or carried out at a different time. Key elements that were considered under the sustainability aspect were the harvesting methods of Irish fishing boats and the overall health of the respective stock and their resilience to other factors such as climate change.

From May 2020, when the archival research was complete and appropriate recipes were identified, the Food Smart Dublin team published a total of ten recipes on their website and one per calendar month until May 2021 (except from July 2020 and January 2021). The recipes were advertised through a variety of channels, including the project's Twitter, Facebook and Instagram accounts and a Mailchimp newsletter advertising every new recipe. The University's communications office helped to produce a short video clip for each recipe which was also published on the project's website and advertised on YouTube.[2] When the first recipe was published to kick off this part of the research, a press release was given and a piece about the collaborative work between the chefs and researchers was published in both, a national and a local newspaper. The initial plan also incorporated feedback on the historical seafood recipes via in-person tastings events in the seafood restaurant owned by the lead-chef involved. The surveys were planned to be conducted face to face during these events in order to engage with as many people as possible. However, with the COVID-19 pandemic and lockdown restrictions, the project had to be completely reconfigured to be carried out remotely with the transfer of all outreach and recipe promotion to an interactive online format. The recipes were advertised up to three times a week via the project's social media channels and participants were incentivised to engage with the online questionnaire by offering the chance to win a seafood voucher from the project's sustainable fishmonger partner each month. The structured online questionnaire was aimed at social media users from all backgrounds and ages. To try and engage people without an interest in seafood consumption fun facts and nutritional benefits gained by consuming food from the sea were also posted.

10.2.3   Structured Online Questionnaires

The data for the present research were collected as part of the structured online questionnaires conducted in connection with the historical recipes between May 2020 and May 2021. The structured online questionnaires focused on the respondents' experience with the specific recipes, as well as offering insights into

[2] https://bit.ly/3tS7XLN

consumption patterns regarding seafood, each participant's relationship with the sea and on suggestions to further improve sustainable seafood availability. The structured online questionnaire was built around three main blocks of information. In the first section respondents were presented with seven questions about their age, profession, income, gender, nationality, residency in Ireland and education. In the second section three questions investigated the respondents' relationship with seafood by asking about the frequency in consumption, the main obstacles for consuming more, and how they perceived their relationship with the sea. In the last sections, respondents could leave detailed feedback on the monthly recipe (therefore this last section was readapted each month) by answering seven questions regarding the sourcing of the seafood, rating the difficulty and taste of the recipe, for whom they had cooked it, the likeness of cooking it again, two final open questions on the pros and cons of the recipe and a last open comment box.

The pandemic and consequential lockdown affected the number and quality of responses, due to survey and screen fatigue. Nevertheless, structured online questionnaires were designed and delivered online using Google Form with the option for respondents to contribute even if they did not cook the recipes. In this case, the questionnaire consisted of only one multiple choice question aimed at understanding the reasons for why participants did not cook the recipe. The given choices for this were the following: "I could not find [seafood of the month] in shops"; "the recipe looked too complicated to cook"; "I did not want to kill animals"; "I am allergic to some of the ingredients used in the recipe"; "[seafood of the month] is too expensive"; "cooking the recipe would have taken too much of my time"; "other" (open answer).

10.3 Results

In total, 33 online questionnaires were completed. Eighteen of these were completed by participants who had cooked the dish and responded to the whole block of questions outlined in Sect. 2.3, while 15 responded to the question related to not having cooked the recipes. Among the reasons indicated by the latter group not having cooked the historical recipes the open responses included: "It is a bit oil rich" (with two entries), or "I did not know how it would taste". These open responses indicated a variety of reasons behind the choice of not cooking seafood. The structured online questionnaires from respondents who cooked the recipes constitute a more in-depth observation of participants' behaviour and attitude towards seafood. First of all, an equal representation of females (50%) and males (50%) were given with 72% residing in Dublin, 21% in Antrim and 7% in Cork. Respondents also indicated working in a wide variety of different professions, including radiochemist, software engineer, IT manager, company secretary, homemaker etc. The most represented category of respondents were chefs with 16% of the total responses.

Respondents were generally very positive with their feedback in rating the recipes (Fig. 10.1). The majority responded with "very good" and "excellent" when

Overall, how would you rate the recipe?

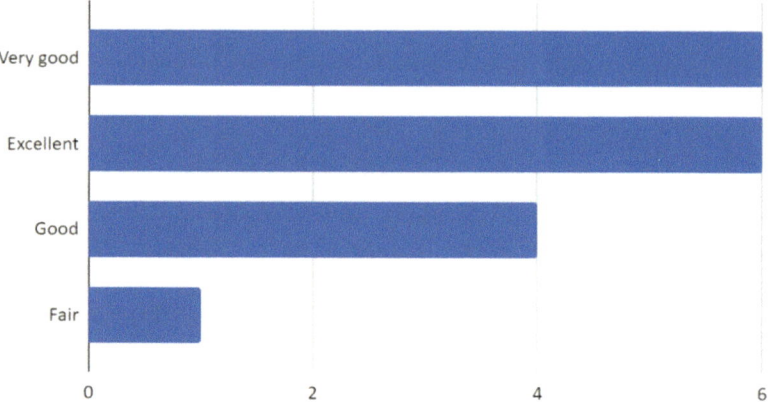

Fig. 10.1 Overall recipe ratings

How likely are you to prepare this recipe again?

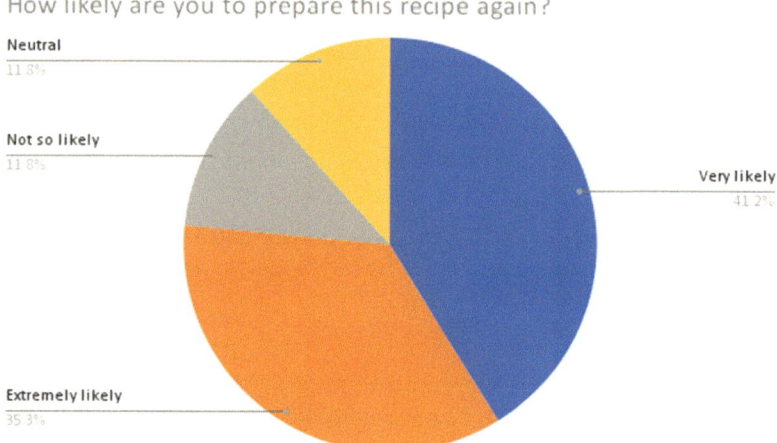

Fig. 10.2 Likelihood of repeated purchase

asked about their cooking experience. There were no negative responses. Likewise, in Fig. 10.2, the question "How likely are you to cook the recipe again" was responded to in a positive manner. More than 75% of respondents indicated that it would be "extremely" or "very likely" that they would cook the recipe again. This suggests a general enthusiasm for their cooking experience. Interestingly, almost 90% of participants sourced their seafood from fishmongers: in Fig. 10.3. "SSI" stands for "sustainable seafood Ireland" a local fishmonger owned by Food Smart Dublin's official partner chef who gave a 20% discount on the prevailing seafood. Only 6% of respondents purchased their seafood in supermarkets. This low number could potentially be explained by the different seafood types advertised in

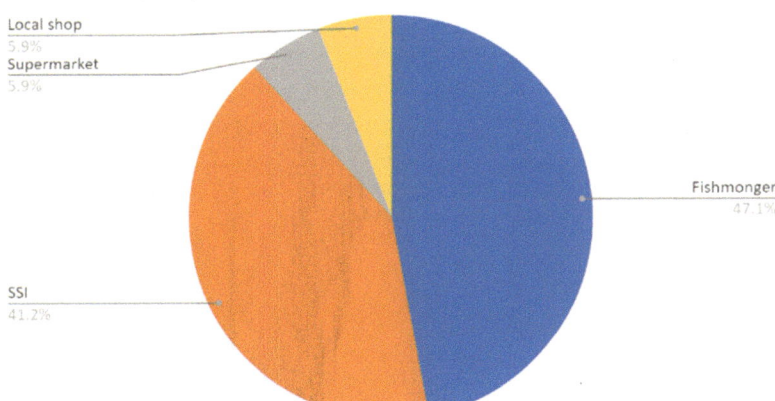

Fig. 10.3 Seafood sources by purchase location

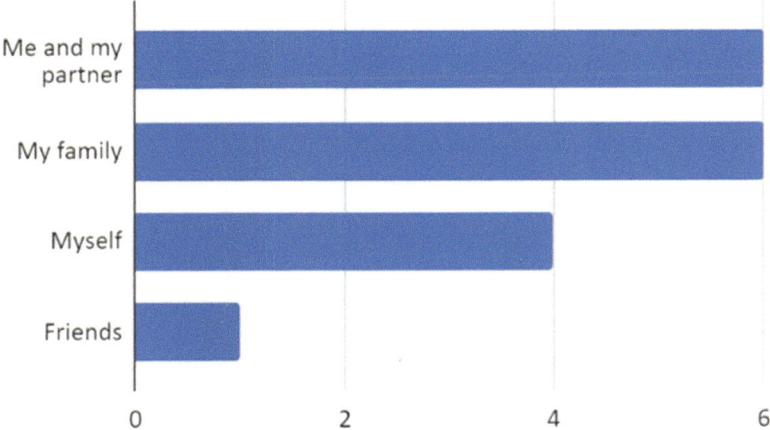

Fig. 10.4 Response regarding cooking partner

the Food Smart Dublin recipes that were not so common, and thus not necessarily accessible in supermarkets where seafood is limited to a few standard options.

When looking at whom the participants cooked the dish for (Fig. 10.4), only one respondent had cooked the recipe for friends, whilst others cooked it for themselves, family or partners. This is likely a consequence of the COVID-19 pandemic and associated lockdown in 2020 but also shows how home cooking encompasses a wide variety of family status. The questionnaire also investigated how strongly respondents felt connected to the sea. Figure 10.5 shows that most participants seem to have a strong connection with their marine surroundings, although 14.2% felt neutral about it or not very much connected. Respondents also proved to be

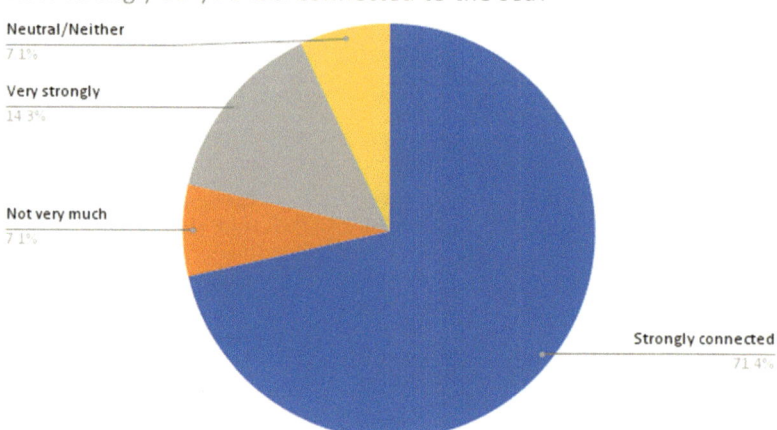

Fig. 10.5 Responses on connection to the sea

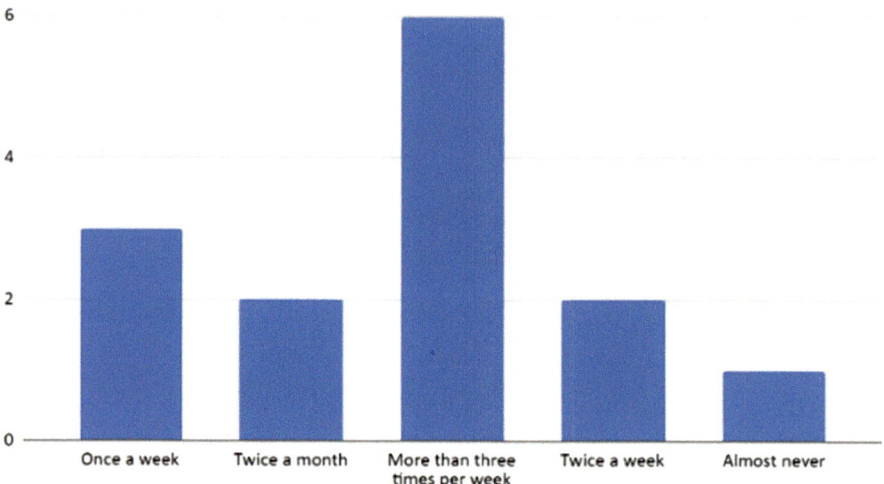

Fig. 10.6 Responses on the frequency of seafood consumption

enthusiastic about seafood: most of them consume seafood more than three times a week, or at least two or one time per week as shown in Fig. 10.6. Finally, although most respondents did not see obstacles in consuming more seafood (Fig. 10.7), some indicated a lack of knowledge and a lack of specialised shops as main bottlenecks in allowing them to consume seafood more regularly.

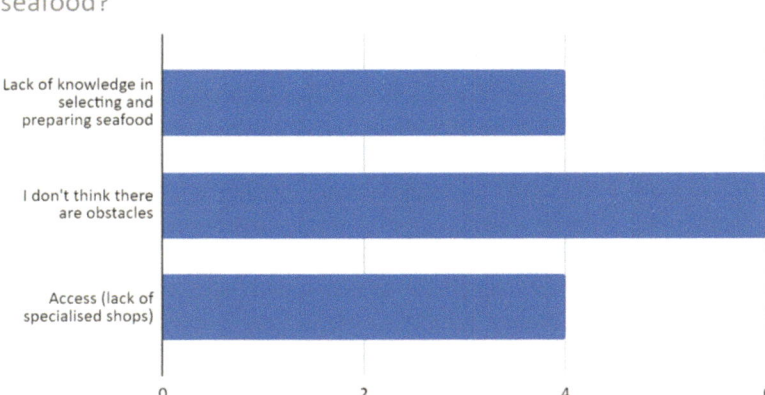

What are the main obstacles for you in consuming more seafood?

Fig. 10.7 Main obstacles for seafood consumption

10.4 Discussion

Although the number of responses is small, the results of the structured online questionnaires can still highlight important and novel insight into the participants' consumption patterns of seafood and their relationship with their surrounding sea. For instance, the outlined responses highlight two relevant points to encourage a more widespread consumption of seafood in Dublin. Firstly, people from Dublin often do not have the knowledge and skills when it comes to cooking seafood but the results of the project's structured online questionnaires demonstrate that they are eager to learn and that there is an interest. At the same time, this interest encompasses a wide variety of social, family, and gender status.

Instead, rather than a lack of skills, the main obstacle to a more sustainable seafood consumption is the actual inaccessibility to the varied seafood Irish waters have to offer. The questionnaire on the recipe with megrim, a deep-water flatfish, common in Irish waters, but not a traditional Irish fish *per se* demonstrates that only a few respondents were able to find the fish in their local fishmongers, not to mention the complete absence of it in supermarket chains. This is in contrast to Irish fishers who are well familiar with the fish as it is one of the most valuable fish exports.

Megrim is not an exception in this disconnection between Irish fisheries and local seafood consumers. Herring once brought wealth and fortune to Ireland, yet this traditional fish deeply rooted in Ireland's history, is rarely found at the fishmongers counter nowadays and completely absent from the fish counter at supermarkets. It is still fished, but since there is no local demand most catches are exported most profitable to the French market.

Interestingly, almost 90% of the participants sourced their seafood from local fishmongers or specialized shops, partly, because the local seafood in the recipes is simply not available in supermarkets. However, given that supermarkets constitute the main source of food for Irish consumers, we argue that actions to increase access to local seafood should be taken in this direction without creating obstacles for local fishmongers. One alternative route here could be to establish a supply link between the local fishmonger and the fresh counter in supermarkets so that access to locally sourced, sustainable seafood, is more readily available. Another point worth contemplating on is whether a push in the direction of high quality local, sustainable seafood is needed rather than just higher fish consumption overall. This however, would make seafood an elite food and would defeat the purpose of affordable locally sourced alternative protein available for people across demographic levels. Arguably, this indicates that the focus on individual consumption behaviour may be a misplaced effort if the overall structure obstructs the possibility for consumers to purchase sustainable seafood options. This echoes Shove's argument (2010) discussed in the introduction, describing how, the research into changing individual behaviour has become the main driver in public policy design around environmental sustainability. Shifting the focus to the market and the policies which sustain it could be a more effective strategy to achieve sustainable food systems.

10.5   Conclusions and Policy Recommendations

New insights across economics, anthropology, political science, humanities and natural sciences, have transformed global understandings of human behaviour and decision-making. Public, private, and non-governmental sectors are changing as a result. Policy is improved by commissioned 'behavioural insights teams' or 'nudge units' to apply novel insight into human decision making (OECD 2017). Product marketers are upgrading their approaches to pipeline development, advertising, and sales. All of these efforts are clearly not leading to real change if the market is still structured to consider food as a mere commodity. We believe, that in Ireland and particularly in its capital Dublin, the lack of access to specialized shops contributes to the "sea blindness" its society suffers.

Food political economists have long criticised the corporate-driven system which has transformed food into a simple commodity (McMichael 2005) where prices are established and imposed internationally. Our analysis shows how seafood is also increasingly exposed to the same dynamics on the island of Ireland, to the point that entire species, once popular seafood choices, disappear from local markets.

The concept of Food Sovereignty aims to contrast such mechanism with more culturally adapted control of food production and distribution (Rosset 2008). The re-introduction of the totemic value that food had before modern mass-production reduced it to its economic value could potentially help with this aim. However, such a vision has to be coordinated at multiple levels of governance and involve a range of actors from the bottom-up and on the individual level, but also from the top-down with governmental coordination. Consumer behaviour change is best effected with joined-up actions, addressing society as a whole and its structures rather than individuals. Taxation and legislation are key ways to drive change, while European policies in agriculture and fisheries can offer great opportunities for developing robustness and sustainability in food production.

Public policies can play a determinant role in shaping the future of seafood in Ireland: they can support small-scale fisheries, improve access to seafood in public spaces (such as markets and supermarkets), promote healthy and sustainable seafood options via public procurement in places like hospitals or governmental offices; and even increase seafood literacy by supporting appropriate actions in schools and other educational institutions. All of these potential actions however have to deal with the profit over people attitude, which often diverts policies towards highly competitive and overexploited forms of fishing and farming, leaving no space for a more humanistic, people centred, approach to local resources.

The Food Smart Dublin project used the radical transdisciplinarity approach for environmental education in which humanities and natural sciences are intertwined to pay greater attention to the history and coastal cultural heritage of seafood and its value in Ireland while not losing sight of the ecological and bio-geophysical dimensions. The identification of historical seafood recipes to encourage the (re)-discovery of the rich and diverse seafood from local Irish waters, accompanied by the ecological, historical and sustainability information seems to have appealed to people in supporting to change their consumption patterns towards more sustainable seafood ways. However, this humanistic, people centred approach needs to be spread wider and more holistically included in policy design.

The COVID-19 pandemic has changed society, including how we view, value and utilise the natural world around us. Examining the effects of the global crisis and the subsequent 'lockdown' on the natural world gives us the chance to examine the role of food resources in (re-)building healthy and sustainable communities. A post COVID-19 world cannot remain "business as usual". Facing the emergency status that we put our planet in, it is now more important than ever to co-create policies that promote healthy, culturally appropriate sustainable food systems.

References

Belton B, Bush SR, Little DC (2018) Not just for the wealthy: rethinking farmed fish consumption in the Global South. Glob Food Sec 16:85–92

Béné C, Barange M, Subasinghe R et al (2015) Feeding 9 billion by 2050 – putting fish back on the menu. Food Sec 7:261–274. https://doi.org/10.1007/s12571-015-0427-z

Brennan C, Ashley M, Molloy O (2019) A system dynamics approach to increasing ocean literacy. Front Mar Sci 6:360. https://doi.org/10.3389/fmars.2019.00360

Clarkson LA, Crawford EM (2001) Feast and famine: food and nutrition in Ireland 1500-1920. Oxford University Press, Oxford. 336pp

Edelman M (2014) Food sovereignty: forgotten genealogies and future regulatory challenges. J Peasant Stud 41(6):959–978

European Commission (2017) Directorate-General for Research and Innovation, Group of Chief Scientific Advisors. Food from the oceans: how can more food and biomass be obtained from the oceans in a way that does not deprive future generations of their benefits? Publications Office. https://data.europa.eu/doi/10.2777/66235

FAO (2018) The State of World Fisheries and Aquaculture 2018 - Meeting the sustainable development goals. Rome, Italy. Published by: Food and Agriculture Organization of the United Nations. Licence: CC BY-NC-SA 3.0 IGO.

Gephart JA, Pace ML (2015) Structure and evolution of the global seafood trade network. Environ Res Lett 10:125014

Holm P (2014) Anthropocene humanities. In: Conroy J, Kelleher M (eds) Restating the value of the humanities. Published Under the Auspices of the Humanities Serving Irish Society Consortium, Dublin

Holm P, Goodsite ME, Cloetingh S, Agnoletti M, Moldan B, Lang DJ, Leemans R et al (2013) Collaboration between the natural, social and human sciences in global change research. Environ Sci Policy 28:25–35. https://doi.org/10.1016/j.envsci.2012.11.010

Holm P, Adamson J, Huang H, Kidran L, Kitch S, McCalman I et al (2015) Humanities for the environment – a Manifesto for research and action. Humanities 4(4):977–992

IPCC (2014) Summary for policymakers. In: Edenhofer O, Pichs-Madruga R, Sokona Y, Farahani E, Kadner S, Seyboth K, Adler A, Baum I, Brunner S, Eickemeier P, Kriemann B, Savolainen J, Schlömer S, von Stechow C, Zwickel T, Minx JC (eds) Climate change 2014: mitigation of climate change. Contribution of working group III to the fifth assessment report of the Intergovernmental Panel on Climate Change. Cambridge University Press, Cambridge/New York

Jackson T (2005) Motivating sustainable consumption: a review of evidence on consumer behavior and behavioral change. Sustainable Development Research Network, Guildford

Kaiser M (1997) Fish-farming and the precautionary principle: context and values in environmental science for policy. Found Sci 2:307–341

Kaiser FG, Fuhrer U (2003) Ecological behaviour's dependency on different forms of knowledge. Appl Psychol 52(4):598–613

Kaiser M, Goldson S, Buklijas T, Glickman P, Allen K, Bardsley A, Lam ME (2021) Towards post-pandemic sustainable and ethical food systems. Food Ethics 6:4. https://doi.org/10.1007/s41055-020-00084-3

Lam ME (2021) United by the global COVID-19 pandemic: divided by our values and viral identities. Humanit Soc Sci Commun 8:31. https://doi.org/10.1057/s41599-020-00679-5

Love DC, Allison EH, Asche F, Belton B, Cottrell RS, Froehlich HE et al (2021) Emerging COVID-19 impacts, responses, and lessons for building resilience in the seafood system. Glob Food Secur 28. https://doi.org/10.1016/j.gfs.2021.100494

McMahon JP (2020) The Irish cook book. Phaidon Press Limited, London

McMichael P (2005) Global development and the corporate food regime. Res Rural Sociol Dev 11:265–299

National Oceanic and Atmospheric Administration (2013) Ocean literacy: the essential principles and fundamental concepts of ocean sciences for learners of all ages version 2, a Brochure resulting from the 2-week on-line workshop on ocean literacy through science standards. National Oceanic and Atmospheric Administration, Silver Spring

O'Sullivan A, Breen C (2007) Maritime Ireland: an archaeology of coastal communities. The History Press Ltd., Stroud, p 256

OECD (2017) Behavioural insights and public policy: lessons from around the world. OECD Publishing, Paris. https://doi.org/10.1787/9789264270480-en

Olson J, Clay PM, Pinto da Silva P (2014) Putting the seafood in sustainable food systems. Mar Policy 43:104–111

Pauly DV, Christensen V, Dalsgaard J, Froese RM, Torres FC Jr (1998) Fishing down marine food webs. Science 279(5352):860–863. https://doi.org/10.1126/science.279.5352.860

Pauly D, Christensen V, Guénette S, Pitcher TJ, Sumaila UR, Walters CJ et al (2002) Towards sustainability in world fisheries. Nature 418:689–695

Pauly D, Alder J, Bennett E, Christensen V, Tyedmers P, Watson R (2003) The future for fisheries. Science 302:1359–1361

Purvis B, Mao Y, Robinson D (2019) Three pillars of sustainability: in search of conceptual origins. Sustain Sci 14:681–695. https://doi.org/10.1007/s11625-018-0627-5

Rosset P (2008) Food sovereignty and the contemporary food crisis. Development 51(4):460–463.

Scherer C, Holm P (2020) FoodSmart City Dublin: a framework for sustainable seafood. Food Ethics 5:7. https://doi.org/10.1007/s41055-019-00061-5

Shove E (2010) Beyond the ABC: climate change policy and theories of social change. Environ Plan A 42(6):1273–1285

Suhonen J, Sutinen E (2014) The four pillar model-Analysing the sustainability of online doctoral pro-grammes. TechTrends 58(4):81–88.

Tran LU, Payne DL, Whitley L (2010) Research on learning and teaching ocean and aquatic sciences. NMEA Spec Rep 3:22–26

Williamson K, Satre-Meloy A, Velasco K, Green K (2018) Climate change needs behavior change: making the case for behavioral solutions to reduce global warming. Rare, Arlington

Chapter 11
Marine Governance as a Process of Reflexive Institutionalization? Illustrated by Arctic Shipping

Jan P. M. van Tatenhove

Abstract The objective of this chapter is to give insight in marine governance challenges, illustrated by Arctic shipping. To do this, this chapter presents a theory of marine governance as reflexive institutionalization, in which the structural properties of marine governance arrangements are (re)produced in interactions between governmental actors, maritime sectors and civil society actors within the structural conditions of the networked polity at sea. Based on an analysis of the institutionalization of shipping governance arrangements of three (possible) Arctic shipping routes; The Northwest Passage (NWP), the Northeast Passage and Northern Sea Route (NEP/NSR), and the Transpolar Sea Route (TSR) the following question will be answered, "What are the enabling and constraining conditions of marine governance as reflexive institutionalization?" In other words, what are the possibilities for public and private actors to challenge discursive spaces and to change the rules of the game, in order to find solutions for environmental, spatial, economic, and social problems at the Arctic Ocean? The analysis shows forms of institutionalization as structural reflectiveness in which the dominant discourse 'shipping is allowed in the Arctic' is not challenged. However, this form of reflectiveness showed how actors, such as China and Russia, are able the use rules from different institutional settings to strengthen their position.

11.1 Introduction

The increase of maritime activities in oceans and seas results in environmental pollution and increasing conflicts between these activities (Halpern et al. 2008; Van Tatenhove 2013). Marine ecosystems are under pressure not only from maritime activities, but also from land-based activities (Schlüter et al. 2019). Examples include eutrophication caused by (coastal) agriculture, wastewater treatment

J. P. M. van Tatenhove (✉)
Centre for Blue Governance, Aalborg University, Aalborg, Denmark

Van Hall Larenstein University of Applied Sciences, Leeuwarden, The Netherlands
e-mail: jan.vantatenhove@hvhl.nl

© German Institute of Development and Sustainability (IDOS) 2023
S. Partelow et al. (eds.), *Ocean Governance*, MARE Publication Series 25,
https://doi.org/10.1007/978-3-031-20740-2_11

facilities, industries and discharges from ports, plastic pollution, toxic and chemical pollution and atmospheric deposition from several sources, threatening of biodiversity by pollution of industries, tourism and maritime traffic. Besides spatial conflicts and environmental pollution, oceans, seas and coastal areas are impacted by the consequences of climate change, ranging from threatening land-based activities and coastal communities by sea-level rise to the possibilities of new shipping routes, due to the melting of sea ice in the Arctic Ocean (Eguíluz et al. 2016).

This chapter presents a social scientific analysis to understand processes of institutionalization and governance illustrated by the case of Arctic shipping. With the opening of shipping routes, due to a diminishing of the extent and volume of Arctic sea ice (Keil 2018) the governing of Arctic shipping has become a timely and relevant political, social and scientific topic. From a governance perspective, Arctic shipping is interesting, because it is regulated by not only a patchwork of governance structures and regulations, but navigation takes place both in the territorial waters of Arctic states (Russia, Canada, USA (Alaska), Norway and Denmark (Greenland)) and on the high seas. Within their Exclusive Economic Zones and territorial waters, coastal states may take the necessary steps to prevent passage which is not innocent[1] (UNCLOS, art. 25 (1), and suspend temporarily (…) the innocent passage of foreign ships (UNCLOS, art. 25 (3)), or levy charges for specific services (UNCLOS art 26 (2)). Beyond the territorial waters, on the high seas, national states do not have the ability to control, to monitor, and to govern environmental, spatial, social and economic processes at sea has diminished.[2]

To understand and explain the governance challenges of shipping in the Arctic this chapter will discuss and analyse the process of institutionalization of Arctic shipping governance arrangements. The focus will be on the governing capacity of these governance arrangements by looking at the dynamics of the shipping industry, the different forms of authority and the possibility of actors to change the rules of the game, to question discourses, and to mobilize resources. To understand the institutionalization of Arctic shipping governance arrangements in Sect. 11.2, a conceptual framework is developed in which marine governance is understood as a process of reflexive institutionalization. Chhotray and Stoker (2009) define governance as the rules of collective decision making in settings where there are a plurality of actors or organisations, and where no formal control system can dictate the terms of the relationship between these actors and organisations. Marine governance "involves a process of negotiation between, on the one hand, nested general

---

[1] Passage is innocent so long as it is not prejudicial to the peace, good order or security of the coastal State. Such passage shall take place in conformity with this Convention and with other rules of international law. (UNCLOS, art. 19 (1)).

[2] Areas Beyond National Jurisdiction (ABNJ) are open to all states. States have the freedom of navigation, to lay submarine cables and pipelines, to construct artificial islands and other installations permitted under international law, freedom of fishing and freedom of scientific research (art 87 UNCLOS). Every state, whether coastal or lan-locked, has the right to sail ships flying its flag on the high seas (at. 90 UNCLOS). At the high seas, states shall cooperate with each other in the conservation and management of living resources (art 118, UNCLOS).

institutions operating at several levels, and on the other hand, state actors, market parties and civil society organizations. This process leads to a sharing of competences for policymaking to govern activities at sea and control their consequences" (van Leeuwen and van Tatenhove 2010; Van Tatenhove 2013: 289). The conceptual framework consists of the concepts of governance arrangements, institutionalization and reflexivity. Core to reflexive institutionalization is that actors are capable of challenging discursive spaces and have the capacity to change the rules of the game in the processes of structuration (morphogenesis) and stabilization (morphostasis). In Sect. 11.3, the case of Arctic shipping is described and analysed. The main question for the case is: what are the enabling and constraining conditions for reflexive institutionalization? In other words, what are the possibilities for public and private actors organized in Arctic shipping governance arrangements to challenge discursive spaces and to change the rules of the game, in order to find solutions for environmental, spatial, economic, and social problems at the level of a regional sea (the Arctic Ocean) and the high seas? In Sect. 11.4, conclusions will be drawn.

11.2 Marine Governance as Reflexive Institutionalization

11.2.1 Marine Governance

In general, marine governance is the capacity of state actors, representatives of maritime sectors (market actors) and civil society actors (NGOs, coastal communities) in marine governance arrangements to govern maritime activities and their consequences (Van Tatenhove 2013). Marine governance encompasses the interplay of policy-making processes (in governance arrangements), politics (the power relations and dynamics between the public and private actors involved) and polity (the institutional setting in which policies and politics take place). This interplay of policy, politics and polity results in specific processes of institutionalization.

A *marine governance arrangement* refers to the way a policy domain, in this case Arctic shipping is temporarily shaped in terms of substance and organization (Liefferink 2006b; Van Tatenhove 2013; van Tatenhove et al. 2020). Substance refers to discourses, resulting in distinct policy and regulatory goals, whereas organization refers to the types of actors involved, the rules of the game (instruments, procedures, division of tasks), and the available resources. The structure of an Arctic shipping governance arrangement can be analysed along four dimensions; **actors and coalitions**; the unequal division of **resources**, formal and informal **rules** of the game and **discourses**[3] (Van Tatenhove et al. 2000).

---

[3] A discourse is the specific ensemble of ideas, concepts and categorizations through which meaning is given to physical and social realities (Hajer 1995).In this chapter, discourses refer to the ideas and concepts related to the development of Arctic shipping (now and in the future).

Marine governance arrangements do not develop in a vacuum. Their specific design and way of institutionalization is the result of the interplay of interactions between interdependent actors in policy practices and processes of political modernisation (van Tatenhove and Leroy 2000; Van Tatenhove et al. 2000; Arts et al. 2006; van Tatenhove 2019). Political modernization refers "to the shifting relationships between the state, market and civil society in political domains of society – within countries and beyond – as a manifestation of the 'second stage of modernity', implying new conceptions and structures of governance" (Arts and Van Tatenhove 2006: 29).

This raises several questions, such as who are the actors at sea? What is the institutional setting of marine governance arrangements? How can we understand the interactions between different governmental actors and the maritime industry? To understand the role of public and private actors at sea, the specific dynamics of maritime sectors, and the way coalitions of maritime actors and governmental actors are nested and embedded in a multilevel and multiple actor institutional setting, I introduce the following concepts: 'maritime regime complex' (Raustiala and Victor 2004; Keohane and Victor 2011; Colgan et al. 2012), 'network state' (Castells 2009), and 'networked polity' (Ansell 2000).

Raustiala and Victor (2004: 279) define a regime complex as "an array of partially overlapping and non-hierarchical institutions governing a particular issue area (Raustiala and Victor 2004). Inspired by this definition, a *maritime regime complex* is an array of organizations, institutions and coalitions of actors which govern a maritime sector and its (sectoral) activities. Maritime sectors, such as fishing, aquaculture, shipping/navigation, deep-sea mining, oil and gas, tourism, etc., are characterised by specific institutional dynamics, reflecting the different levels at which sectoral activities are organized and regulated. The relations and interactions between public (governmental) and private (non-governmental) actors in marine regime complexes are shaped by prevailing discourses, the expectations of actors involved and the institutional rules of that specific regime complex. Maritime regime complexes can be placed on a continuum running from fully integrated institutional arrangements at one extreme to highly fragmented collection of arrangements at the other (Keohane and Victor 2011).

Due to the fragmentation and dispersal of authorities at sea, the role of (nation) states and state authority should be reconceptualised. According to Jessop (2004) political authorities are becoming involved in all aspects of meta-governance in which (…) the role of the state has shifted from the direct governance of society to the 'meta-governance' of the several modes of intervention and from command and control through bureaucracy to the indirect steering of relatively autonomous stakeholders (Bevir and Rhodes 2011) (204). Additionally, states share sovereignty with other actors, such as the European Union (EU) and the United Nations (UN). According to Beck and Grande (2007) (32) the state is in a process of transformation in which it "is not replaced or suppressed entirely, but it is integrated in a variety of ways into new international regimes and organizations, new supranational institutions, new forms of regionalism, and the like". This is what they call reflexive modernization of statehood which leads to "the emergence of a plurality of diverse new forms of transnational governance beyond the nation-state" and "the increasing role of private actors in solving collective problems and producing public goods" (2007: 32–33). This new form of statehood is what (Castells 2009, 2010) calls the (emerging) *network state* which is 'characterised by shared sovereignty and responsibility between different states and levels of government; flexibility of governance procedures; and greater diversity of times and spaces in the relationship between governments and citizens compared to the preceding nation-state' (Castells 2009).

Ansell (2000) defines the *networked polity* (or institutional setting) as a governance structure in which both state and societal organization is vertically and horizontally disaggregated (as in pluralism), but linked together by cooperative exchange (as in corporatism). To understand the institutional setting of marine governance I define the *maritime networked polity* as the institutional setting of governance in which (emerging) network states, regime complexes and societal actors (NGOs, communities) are positioned vis-a-vis each other in a multi-level governance setting, while horizontally linked to each other in interactions of conflictual and/or cooperative exchange. The nature of these interactions is guided by institutional rules and discourse. Characteristic for the maritime networked polity is its embeddedness. Rules systems and regulations of different governmental levels come together at the level of regional seas, in what DiMento and Hickman (DiMento and Hickman 2012) (8 and 115) call clusters (the collection of international environmental institutions, regimes and complexes) (van Tatenhove 2016) (166).

With the introduction of the concepts of maritime regime complex, network state and maritime networked polity we can now define marine governance more specifically. *Marine governance* refers to the ability and capacity of network states, maritime regime complexes (the institutional order and dynamics of maritime sectors), NGOs and (coastal and marine) communities – organised in (marine) governance arrangements – to change the rules of the game, to mobilize resources and discourses, in order to govern maritime activities and their consequences in a specific maritime networked polity.

11.2.2 Reflexive Institutionalization

In general, institutionalization refers to the phenomenon whereby patterns arise in people's actions, fluid behaviour gradually solidifies into structures, and those structures in turn structure behaviour (Arts et al. 2006). Institutionalization is the ongoing process of patterning, preservation, construction, organisation and deconstruction of day-to-day activities and interactions in institutions (van Tatenhove and Leroy 2000). The concept incorporates the development of structures, stabilisation and change: institutions, no matter how stable they appear at first sight, are subject to continual change and adjustment, deconstruction and reconstruction.

More specifically, *institutionalization* is the process of production and reproduction of governance arrangements, in which the rules of the games are (re)produced in interaction within the context of long-term processes of societal and political transformation (political modernization) (Van Tatenhove et al. 2000; Arts and Van Tatenhove 2006; Liefferink 2006a). In other words, change induced by political modernization provides a structural focus on change because of the changing relations between state, civil society and market. Change stimulated by day-to-day interactions is strategic, focusing on the arguments (discourses), rules and resources actors use in interactions to define problems and to find solutions.

Analytically, two sub-processes of institutionalization can be distinguished: structuration and stabilization, in which the content and the organization of governance arrangements are (re)produced in interaction within the context of long-term processes of societal and political change. "Structuration refers to the (re)production of content and organisation of a policy domain in interaction, whereas stabilisation refers to the 'preservation of contents and organisation in specific policy concepts and arrangements" ((van Tatenhove and Leroy 2000): 19–20). The interplay of stabilization and structuration resemble the distinction made by Archer between morphogenesis and morphostasis (Archer 2010a, b, 2014). In her morphogenetic approach Archer refers to those processes which tend to elaborate or change a system's given form, structure or state (Buckley in Archer 2010a, b: 274). More specific morphogenesis is a process of structuration, which is the gradual formation and production of structural properties of a governance arrangement in interaction. Specific forms of interaction within relations of interdependency result in accepted rules of the game, discourses and the availability and division of resources. Morphostasis refers to processes in a complex system that tend to preserve these unchanged (Archer 2010a, b: 274). In this process of stabilization, institutionalized governance arrangements constrain agency (the involved actors) into adopting certain discourses, rules and resources. The institutionalization of marine governance arrangements (as the ordering of a specific maritime policy field in terms of actors/ coalitions, resources, rules and discourses) is the result of the interplay of contextual processes of structural political and social change (political modernisation), and problem-oriented renewal of policy making and decision-making by agents in day-to-day practices (policy innovation).

The institutionalization of maritime policies, politics and governance arrangements can be understood from the perspective of "reflexive modernization of statehood, which leads to the emergence of a plurality of diverse forms of transnational 'governance beyond the nation-state'" (Beck and Grande 2007) (6). This also contains what Saskia Sassen calls bordering capabilities by which actors shape (bordered) spaces transversal to traditional state borders (Sassen 2006, 2009, 2013, 2015). Her central thesis is "that opening of traditional national borders may, in fact, strengthen a range of transversal bordering capabilities—transversal in the sense that these capabilities cut across traditional borders and enter and exist deep inside national institutional spaces" (Sassen 2009): 596). These bordering capabilities can be mobilized for a broad range of dynamics, including some with scale-up potentials that can unsettle the territorial authority of the state. Sassen states that territory, as an analytic category, cannot be confined to its national instantiation, even if this is the dominant one. Whilst Sassen (2013: 31) in her work on cities argues that transversally bordered spaces entail the making of distinct, albeit elementary territories and jurisdictions inside nation-states, moving to the marine realm the focus also goes beyond or outside a single nation state. In the marine realm (at the level of regional seas and the high seas), all governmental and non-governmental actors, have transversal bordering capabilities, which are related to the ability to steer and control cross-border flows of resources (money, goods and information).

Reflexive institutionalization is a process of structuration (morphogenesis) and stabilization (morphostasis) in which the structural properties of marine governance arrangements are (re)produced in interactions between governmental actors, maritime sectors and civil society actors within the structural conditions of the networked polity at sea. Reflexivity refers to the capacity of actors to govern and to induce change (i.e., to change the processes of structuration and stabilization) by challenging the existing discursive spaces of marine governance arrangements (performative mobilization), and to activate and to use rules and resources from different rule systems and layers of government. In this sense, the dynamic process of institutionalization is driven by agency (reflexivity) and structural conditioning (networked polity and related power structure). Reflexive institutionalization at sea is not planned and designed, but is what Beck and Grande (2007: 6) call "institutionalized improvisation". Van Tatenhove (2017) distinguished three modes of reflexivity, representing different extents: structural and performative reflectiveness and reflexivity. *Structural reflectiveness* refers to the ability of actors to use rules and resources from different institutional settings within a given discursive space of a policy domain, but actors are not able to change the rules of the game. The dominant form of mobilization of actors is action-oriented within an existing governance setting. The conditions remain relatively unchanged (morphostasis). *Performative reflectiveness* refers to the ability of actors to challenge the discursive space of a governance arrangement (performative mobilization, (Pestman 2001)). This could result in for example alternative discourses, and related new coalitions, rules and resources existing side by side with the existing governance arrangement, but existing institutional rules and power relations (polity) are not challenged. *Reflexivity* refers to the situation when actors both challenge the existing discursive space of a policy domain, and are able to change the institutional rules (structural congruence, (Boonstra 2004)), which thus refers to a process of morphogenesis (structural and cultural elaboration).

The case study of Arctic shipping will analyse how to increase the institutional capacity and ability of governmental and non-governmental actors within different institutional settings at sea (networked polity) to act, to govern and to get involved in processes of governance, in order to find solutions for environmental, spatial, economic, and social problems. In the process of institutionalization, new governance arrangements are (re-)produced. Forms of reflexivity are the motor of change in this process of institutionalization.

11.3   The Institutionalization of Arctic Shipping

In this section, I describe and analyse the case of Arctic shipping. Section 11.3.1, presents some general characteristics of Arctic shipping, such as the accessibility of navigation in the Arctic region, and the different shipping routes, which are possible with diminishing sea ice covering, followed in Sect. 11.3.2 by the institutional governance setting of the Arctic. Aim of Sect. 11.3.3 is to reconstruct the institutionalization of different shipping governance arrangements of the three main Arctic shipping routes. The analysis focuses on the specific interplay of and interactions between actors within the shipping regime complexes related to forms of the network state, the guiding discourses and specific rules and resources within the networked polity related to each of the shipping routes. The analysis will give insight into different types of Arctic shipping governance arrangements, the processes of institutionalization of Arctic shipping related to the different Arctic routes and the enabling and constraining conditions for reflexive institutionalization of shipping, e.g., the possibilities of different actors to change the rules of the game and to challenge the dominant discursive spaces.

11.3.1   Arctic Shipping

The extent and volume of Arctic sea ice is diminishing (Keil 2018) (see Fig. 11.1). This opens up possibilities for navigation in the Arctic region. There are different forms of navigation in the Arctic,[4] such as liner shipping,[5] bulk shipping (liquid and dry), specialised shipping (LNG and reefer[6]) and cruise shipping. Three Arctic shipping routes are emerging (see Fig. 11.2). The Northwest Passage (NWP) connects the Atlantic Ocean with the Pacific Ocean via Canada's Arctic Archipelago. The Northeast Passage (NEP) also connects the Atlantic Ocean with the Pacific Ocean, but from Northwest Europe around the North Cape and along the coasts of Eurasia and Siberia through the Bering Street. Part of the NEP is the Northern Sea Route (NSR), which runs from Kara Strait (a small passage between Russia and Novaya Zemlya) to the Bering Strait.[7] In contrast to the NWP and the NEP, the Transpolar Sea Route (TSR) or Trans-Arctic route is high seas and does not run in the territorial waters of Arctic states. The TSR also connects Europe and Asia but is much shorter than the coastal NWP, NEP and NSR routes.

---

[4] www.worldshipping.org; visited 15/01/2020.

[5] Liner shipping is the service of transporting goods by means of high capacity via transit regular routes on fixed schedules (for example containerships and roll-on/roll-off ships).

[6] Reefer is a specialized ship to carry frozen products (fish and meet) (https://pame.is/index.php/projects/arctic-marine-shipping/).

[7] The main difference between the NSR and the NEP is that the latter comprises the Barents Sea and provides access to the port of Murmansk (Buixadé Farré et al. 2014).

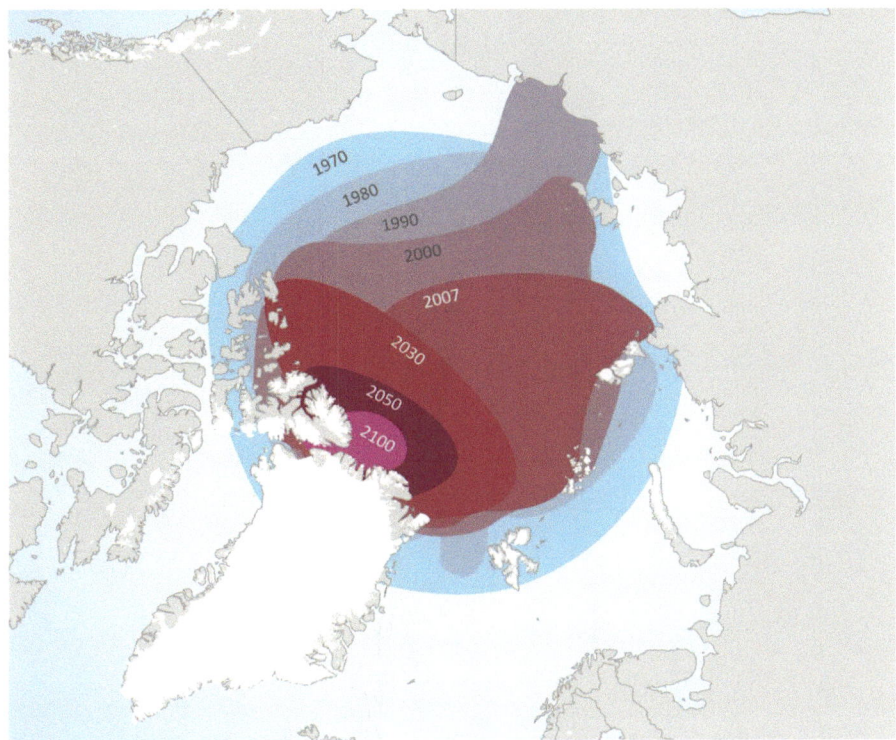

Fig. 11.1 Diminishing Arctic sea ice (1970–2100). (Source: Humpert and Raspotnik 2012)

Because the TSR passes outside territorial waters, it is of special geopolitical importance. A fourth possible route is the Arctic Bridge Route (linking the port of Murmansk in Russia with the port of Churchill in Canada via Iceland) (Humpert and Raspotnik 2012). This route will not be discussed in this chapter, because it does not connect the Pacific with the Atlantic Ocean.

Despite discussions about opening up the Arctic for navigation and other maritime activities, it will be very challenging in the near future, due to harsh weather conditions, free floating sea ice, remoteness, lack of communication and SAR (Search and Rescue) capabilities (Humpert and Raspotnik 2012; Buixadé Farré et al. 2014; Dyrcz 2017). Humpert and Raspotnik estimated that during summer (July – September) the maritime accessibility of the Arctic will increase (see Table 11.1). The Ice-free period along the Arctic's main shipping routes is expected to increase from 30 days (2010) to more than 120 days (2050). "However, free-floating ice in summer will remain a serious threat to navigation, and widespread ice in winter will continue to obstruct passage by most ships" (Buixadé Farré et al. 2014) (p. 321).

The harsh condition in the Arctic ocean requires technical innovation, what Buixadé Farré et al. (2014: 313) refer to as 'winterization': addressing the challenges unique to sub-zero environments (e.g., icing, snow, rain and fog).

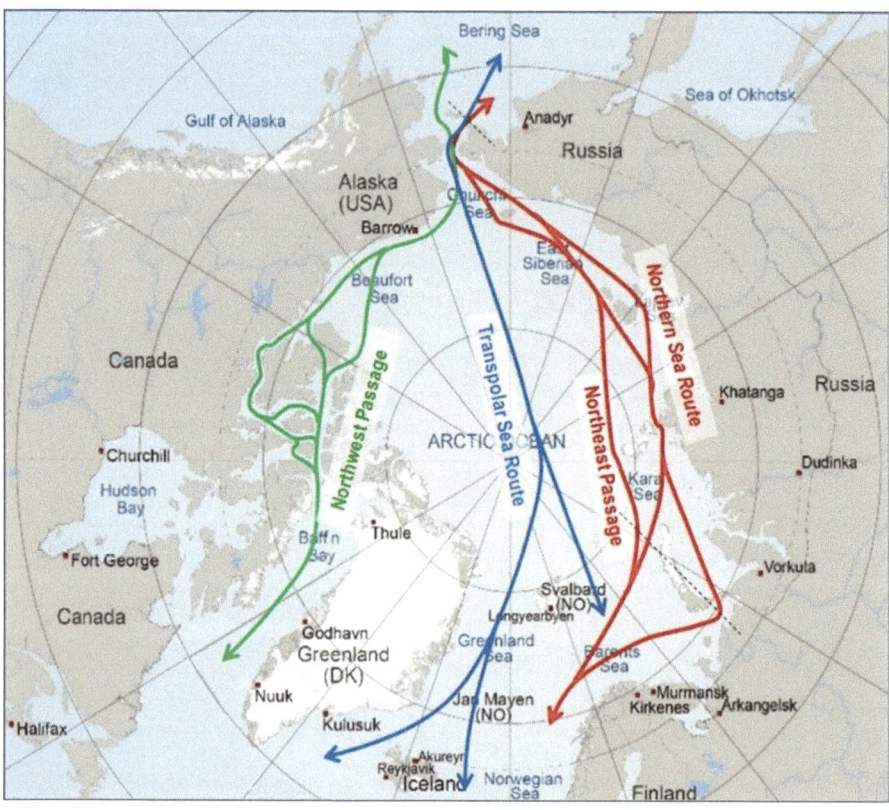

Fig. 11.2 Arctic shipping routes. (Czeslaw Dyrcz 2017)

Table 11.1 Maritime accessibility in 2000–2014 and 2045–2059 for Type A vessels (light icebreaker) in the period July–September

Route	Length (km)	% accessible, 2000–2014	% accessible, 2045–2059	Accessibility change (%) relative baseline
Northwest Passage	9324	63%	82%	30%
Northern Sea Route	5169	86%	100%	16%
Transpolar Sea Route	6960	64%	100%	56%
Arctic Bridge	7135	100%	100%	0%

Source: Humpert and Raspotnik (2012: 288)

Winterization solutions are building structures resistant to low temperatures, anti-freezing measures, the procurement of freezing-resistant supplies, etc.

In general, Artic shipping routes are much shorter than the Suez Canal/Malacca and Panama Canal routes (Østreng et al. 2013) (p. 50). However they will not substitute existing shipping routes, but will be supplementary and providing additional capacity.

11.3.2   The Networked Polity of Arctic Shipping

The networked polity of Arctic shipping is the institutional setting in which the emerging Arctic network state, consisting of formal and informal institutions, such as the Arctic Five,[8] the Arctic Council, the Northern Dimension,[9] the Nordic Council,[10] the UN (UNCLOS and IMO, shipping regime complexes and other nongovernmental actors (e-NGO's, indigenous communities) are positioned vis-à-vis each other. Key actors in the emerging network state of shipping are the Arctic Council and IMO.

The **Arctic Council** (established in 1996 with the Ottawa Declaration) is a high-level intergovernmental forum to provide a means for promoting cooperation, coordination, and interaction among the Arctic States, with the involvement of the Arctic indigenous communities and other Arctic inhabitants on common Arctic issues, in particular issues of sustainable development and environmental protection in the Arctic. The Arctic Council has decision-making power in which also non-Arctic States want to participate (Koivurova 2013; Smits et al. 2014, 2017). The Council consists of the eight Arctic States[11] and six organisations representing Arctic indigenous peoples.[12] Observer status in the AC is open to non-Arctic states, intergovernmental and inter-parliamentary organisations with a global and/or regional constituency, and NGOs that the Council determines as potential contributors to its work.[13] The primary role of observers is to observe the work of the Arctic Council,

---

[8] The Arctic 5 are the five Arctic littoral states, namely Russia, Norway, Denmark (Greenland), Canada, and the USA (Alaska).

[9] The Northern Dimension is an intergovernmental platform of cooperation between the EU, Russia, Norway, and Iceland. The Northern Dimension was launched in 1997 by Finland to emphasise the interdependence between the EU and Russia, Norway, Iceland, and the Baltic States (non-EU Member States at that time).

[10] The Nordic Council is an inter-parliamentary coalition between the Nordic countries, which include Denmark, Finland, Iceland, Norway, Sweden, the Faroe Islands, Greenland, and Åland. Parliamentarians of all Nordic countries are taking place in the Council and decide upon issues after which they call on the governments of the Nordic countries to implement these.

[11] Canada, Denmark (including Greenland and the Faroer Islands), Finland, Iceland, Norway, the Russian Federation, Sweden, and the USA (Alaska).

[12] the Arctic Athabaskan Council, Aleut International Association, Gwich'in Council International, Inuit Circumpolar Council, Russian Association of Indigenous Peoples of the North, and Saami Council   (https://arctic-council.org/index.php/en/about-us/permanent-participants,   visited 28/01/2020).

[13] See   (https://arctic-council.org/index.php/en/about-us/arctic-council/observers   visited 28/01/2020) for the list of observers.

and to contribute to the work of one of the six Working Groups[14] of the Arctic Council. The AC is increasingly an "active regional organization" (Buixadé Farré et al. 2014). An example is the Arctic Marine Shipping Assessment (AMSA) of 2009 (Arctic Council 2009), which recommendations resulted in the first two binding circumpolar treaties: *Cooperation on Aeronautical and Maritime Search and Rescue in the Arctic* (2011) and the *Agreement on Cooperation on Marine Oil Pollution Preparedness and Response* (2013).

The UN International Maritime Organization (**IMO**) regulates shipping by setting standards and regulations about safety, security, efficiency and environmental responsibility. Examples of IMO regulations are *International Convention on the Prevention of Pollution from Vessels* (MARPOL), the *International Convention on the Safety of Life at Sea* (SOLAS), *International Convention on the Control of Harmful Anti-Fouling Systems on Ships* (Anti-fouling Convention), *International Convention for the Control and Management of Ships' Ballast Water and Sediments* (Ballast Water Management), *International Convention on Oil Pollution Preparedness, Response and Co-operation* (OPRC). To improve the safety of shipping in the Arctic and to reduce the impact of shipping on the environment IMO's International Maritime Safety Committee established in July 2014 the *International Code for Ships Operating in Polar Waters* (The Polar Code).[15] The Polar Code covers all shipping related matters in Arctic and Antarctic waters, ranging from ship design, construction and equipment, operational and training concerns, search and rescue to the protection of the environment and eco-systems of the Polar Regions.[16]

The rationale of the Polar Code is that sustainable Arctic shipping is based on two pillars; human safety and environmental protection (Keil 2018). The environmental pillar of the Code consists of binding requirements and regulations relating to oil, invasive species, sewage, garbage and chemicals and defines three categories of ships.[17] The ship safety pillar formulates binding requirements and regulations concerning equipment, design & construction, operations & manning with the aim "to provide for safe ship operation and the protection of the Polar environment by addressing risks present in Polar waters and not adequately mitigated by other instruments".[18] The implementation and the enforcement of the Polar Code will

---

[14]There are six Working Groups of the Arctic Council: Arctic Contaminants Action Program (ACAP); Arctic Monitoring and Assessment Programme (AMAP); Conservation of Arctic Flora and Fauna (CAFF); Emergency Prevention, Preparedness and Response (EPPR); Protection of the Arctic Marine Environment (PAME); Sustainable Development Working Group (SDWG) (https://arctic-council.org/index.php/en/about-us/working-groups visited 28/01/2020).

[15]The Polar Code is developed as a complement to existing documents, as the new 14th Chapter of the SOLAS Convention and entered into force 01/01/2017.

[16]http://www.imo.org/en/MediaCentre/HotTopics/polar/Pages/default.aspx, visited 28/01/2020.

[17]http://www.imo.org/en/MediaCentre/HotTopics/polar/Documents/How%20the%20Polar%20Code%20protects%20the%20environment%20%28English%20infographic%29.pdf

[18]http://www.imo.org/en/MediaCentre/HotTopics/polar/Documents/Polar%20Code%20Ship%20Safety%20-%20Infographic_smaller_.pdf

have implications for a diversity of actors, such as ship-owners, assurance companies, trainers, operators, surveillance and controlling agencies, etc. The Polar Code defines a new stage of Arctic shipping, because it will both constrain navigational operations in the Arctic through binding requirements, while at the same time it is an expression of the dominant discourse of sustained Arctic shipping, by stimulating and enabling shipping activities, as it contributes to shape the necessary information, communication and material infrastructures that support shipping activities. According to Keil (2018: 46) does the Polar Code not conclude, that "Arctic shipping is too dangerous or risky (…) and should therefore not take place", but it an expression of a dominant discourse that "Arctic shipping is seen universally as an activity that can be conducted sustainably (…)". "Arctic shipping is considered to be capable of interacting with the natural environment, Arctic communities, and business interests in a way that enables these assets to co-exist over time without threatening the existence of nature, societies or businesses; thus, their relationship is regarded as fundamentally sustainable".

The Arctic networked polity does not replace or suppress nation states, but states are positioned besides the shipping regime complexes and the emerging Arctic network state, consisting of actors and institutions with conflicting interests and jurisdictions, such as UNCLOS (binding international law regulating shipping, and rules related to territorial claims, etc.), The Arctic Council (to promote environmental protection and sustainable development in the Arctic, IMO (regulating environmental and safety issues related to shipping), and the Arctic Five. This fragmented networked shipping polity sets the scene in which Arctic shipping governance arrangements institutionalize.

11.3.3   The Institutionalization of Arctic Shipping in the Three Shipping Routes

This sub-section gives an analysis of the institutionalization of shipping governance arrangements of the Northeast Passage NEP (including the Northern Sea Route, NSR), the Northwest Passage (NWP) and the Transpolar Sea Route (TSR). For each Arctic route, a shipping governance arrangement is constructed, consisting of shipping regime complexes, network states and NGOs, discourses, resources and rules. The dominant discourse in all three governance arrangements is that shipping is allowed in Arctic waters and can be sustainable under certain circumstances and conditions (see Sect. 11.3.2). The main rule supporting this discourse is the Polar Code, which is a crucial condition for sustainable Arctic shipping, because it addresses "(…) present in polar waters and not adequately mitigated by other instruments of the Organization" (Polar Code 2017: 5).[19]

---

[19] https://edocs.imo.org/FinalDocuments/English/MEPC68-21-ADD.1(E).doc.

11.3.3.1 The NEP/NSR Shipping Governance Arrangement

According to Buixadé Farré et al. (2014), the NEP is the most practicable route in the Arctic both as a corridor for the transport of natural resources and as a shorter route for transit shipping. Although it has the highest potential for transit shipping and transporting resources, there will be serious challenges for container shipping, because they operate under a just-in-time regime, which relies on predictability and precise schedules. Bulk cargo ships do not require such a regime; therefore, it is more likely that bulk-cargo ships can deal with the variability of the NEP. However despite potential for bulk resource transport there remain significant physical and logistic limitations (shallow bathymetry, see also Arctic Council 2009). For example, the shallow depths of the NEP/NSR make it impossible for the new generation ultra large container ships (ULCS) to transit. These ships will prefer the Suez Canal Route.

The dominant shipping regime complex of the NEP/NSR consists of the coalitions and infrastructures related to tankers (oil and LNG), general cargo shipping, and icebreakers. Of the 207 transits (between 2011 and 2015), 45% were tankers and 17% general cargo.[20] During the winters (January–April 2017–2019), shipping activities take place mainly to the west of the Kara Sea.[21] Actors involved are shipping companies (Sovcomflot (Russian), transportation of crude oil and LNG; Murmansk Shipping Company (partly Russian), oil transportation, transhipment and exploration; Nordic Bulk Carriers (Danish), dry bulk shipping), insurance companies, shipbuilders, icebreaker assistants, port authorities, flag, port and coastal states, and interested states like Russia and China, but also EU member states, such as Norway, Denmark, the Netherlands, Germany, etc. Examples of the needed infrastructure are harbour facilities (repair, maintenance, storage, processing industries, refineries, etc.), Search and Rescue facilities, and hinterland infrastructures (rail and roads).

Specific for the networked polity of the NSR shipping governance arrangement is the special position of Russia. Although most Russian regulations are consistent with international law and requirements (UNCLOS, IMO, and AC (SAR)), the country has adopted rules "pertaining to vessels operating in the NSR that contain certain provisions that go beyond international rules and standards (for example, inspections, requirements for ice pilots and transit fees)" (AC 2009: 119). This is reflected in the Russian Arctic Strategy in which Russia sees the utilization of the NSR as a national integrated transport and communication system to safeguard Russian interest in the Arctic (Buixadé Farré et al. 2014: 308). For Russia, the NSR is a national integrated transport and communication system to safeguard Russian interest in the Arctic, and has developed a framework that obligates all ships to

---

[20] https://pame.is/index.php/projects/arctic-marine-shipping

[21] In 2018: 278 transits (124 tankers; 34 LNG tankers; 59 icebreakers; 34 containerships; 26 general cargo and 1 SAR). In 2019: 426 transits (144 tankers; 118 LNG tankers; 86 icebreakers; 65 containerships; 1 bulk and 1 SAR)(Centre for High North Logistics Nord University, 2019).

request permission to access the NSR, and to deny passage for political reasons (for example, in 2013, Russia denied three times the requests made by Greenpeace's icebreaker Arctic Sunrise to enter the NSR).

An important resource for the future viability of the NEP/NSR is the availability and accessibility of ports. The current availability of Russian ports for repairs and maintenance is scare. Of the 18 marine ports in the Russian Arctic, 11 are in poor condition and located in regions with sparse land transportation infrastructure, only 4 ports (Murmansk, Arkhangelsk, Vitino and Kandalaksha) are in fairly good condition (Buixadé Farré et al. 2014) (313). (Liu et al. 2021) showed that the implementation of the Arctic strategy has not promoted the cargo throughput of ports along the NSR during 2003–2012. According to the authors the development of ports along the NSR route is restricted by the low level of economic development, foreign trade and a lag of Russian transportation infrastructure.

Russia is responsible for the coordination of SAR activities along the NSR. Although Russia has invested in the creation of 10 SAR centres along the NSR, substantial parts of the NSR lie outside the coverage of these centres, making Russian icebreakers the only potential respondents to a SAR request (Buixadé Farré et al. 2014) (314–315).

11.3.3.2 The NWP Shipping Governance Arrangement

The NWP shipping governance arrangement consists of different shipping regime complexes in the Canadian Arctic, consisting of actors and infrastructures related to "community re-supply; bulk shipments of raw materials, supplies and exploration activity for resource development operations; and tourism" (AC 2009: 113).

The Canadian St Roch realized the first complete transit from west to east in 1942, followed by the oil-tanker Manhattan in 1969. During the period 1969–1990, there were only 30 complete transits. In 2012, 30 vessels transited through the NWP, while in 2014 only 17 vessels managed the full transit. In 2013 for the first time, a large bulk carrier transited the NWP.[22] These figures point out that there will be no commercial shipping on a regular basis to transit the NWP from west to east, aside from a few small specialty cruise operators (AC 2009: 114). Except from cruise ship tourism it is not expected that the NWP will be a viable trans-Arctic route in the nearby future, "due to seasonality, ice conditions, a complex archipelago, draft restrictions, chokepoints, lack of adequate charts, insurance limitations and other costs" (AC 2009: 114). According to the Arctic Council there will be an increase in destinational shipping in the Canadian Arctic driven by increasing demand for seasonal re-supply activity, expanding resource development and tourism (AC 2009: 114).

---

[22] https://www.enr.gov.nt.ca/en/state-environment/73-trends-shipping-northwest-passage-and-beaufort-sea (visited 31/01/2020).

The dominant shipping regime complex of the NWP is related to cruise and expedition shipping, and consists of tour operators, cruise-ship owners, expedition leaders, AECO (Association of Arctic Expedition Cruise Operators), SAR facilities, shipbuilders, tourists, scientists and states (Arctic and non-Arctic). Inaccessible destinations such as the North Pole, Northwest Passage and the Northern Sea Route are increasingly open for the public. Between 1984 and 2004, 23 commercial cruise ships accomplished transits of the Northwest Passage; seven commercial tours were planned for 2008 alone. The Arctic tourism industry ranges from relatively small expedition style vessels that hold less than 200 people, to large luxury cruise liners that can hold 1000 or more. According to Cajaiba-Santana et al. (2020) cruise ship tourism in the Arctic is based on the "expedition" model of Arctic cruising (Cajaiba-Santana et al. 2020), involving small vessels (between 20 and 500 passengers). Expedition cruise tourism is about "shore landings and exploration using rubber boats, quality environmental and historical interpretation of biodiversity, landscapes, historical remains and current use, remote and exclusive wilderness experience, minimal environmental and social impact, human safety and flexibility depending on dynamic weather and sea-ice conditions" (Van Bets et al. 2017) (p. 1585).

Most of the passenger vessel traffic takes place along the Norwegian coast, the coasts of Greenland,[23] Iceland and Svalbard. Though there was some passenger vessel traffic in the Canadian Arctic and Alaska, those numbers were small in comparison to the higher traffic areas. Important destinations in the NWP organized by Polar Cruises[24] are Spitsbergen (Svalbard), from Kangerlussuaq (Canada) to Nome (Alaska) and from Greenland to the Bering Sea, the west and east coasts of Greenland (west and east), and Baffin Island.[25]

Arctic cruise shipping is facing ambiguity of rules and institutional voids (Cajaiba-Santana et al. 2020), such as a lack of central authority governing the sector, a lack of regulatory power by AECO, inconsistencies related to the multi-jurisdictional and transnational operating context, and gaps related to for example licencing and Polar Code training requirements, the lack of models for insurance and assessment, and the chartering of uncharted waters.

---

[23] Cruise ship traffic off the coast of Greenland is increasing rapidly. Between 2006 and 2007, port calls into Greenland increased from 157 to 222 cruise ships. The number of port calls in 2006 combined for a total of 22,051 passengers, this increased to a total of 110,567 passengers for all Greenland's harbours in 2018 (http://bank.stat.gl) almost doubling Greenland's total 2018 population of 56,171.

[24] https://www.polarcruises.com/arctic (visited 14/02/2020).

[25] Polar cruises has also cruises in the NEP: from Norway to Alaska, from Nome (Alaska to Murmansk), and from Tromsø to Nome (Alaska); Iceland; Newfoundland and Labrador; Russian Far East and Scotland/Ireland.

11.3.3.3   The TSR Shipping Governance Arrangement

The Transpolar Sea Route is a mid-ocean route and is shorter than the NWP and NEP. Because the TSR has a multitude of possible navigational routes, it is more interesting for bulk shipping (which follows less predictable schedules) then for liner shipping (which are dependent on regular routes and fixed schedules). According to (Humpert and Raspotnik 2012) (294) the challenge for Arctic shipping is not primarily technological, but economic. The lack of schedule reliability and variable transit time along the Arctic shipping routes is a major obstacle for the development of the TSR. Also, navigation at the TSR remains an unviable option in the near future due to climate conditions and economic uncertainties. To become economically profitable a different kind of economic optimization needs to be developed, taking into account "the lack of economic hubs, the cost associated with different types of Arctic shipping and uncertainties with regard to investments for special equipment and insurance" (Humpert and Raspotnik 2012: 301).

Compared to the other Arctic shipping routes, the TSR involves only limited legal uncertainties and controversies, because it lies outside the EEZs of Arctic states and is therefore subject to UNCLOS and to High Seas regulations. The TSR is mainly a potential route, now only navigated by icebreakers, but it is expected that the TSR could become the dominant Arctic route for bulk shipping in the second half of the twenty-first century. China is anticipating this development by investing in Iceland and by establishing free trade negotiations between China and Iceland in 2009 (Stanley 2012 in Humpert and Raspotnik 2012: 289). China prefers the TSR to avoid Russian territorial waters. By establishing a strategic partnership with Iceland (strategically located in the Northern Atlantic), Iceland may become an important trans-Arctic shipment hub. This would strengthen the geopolitical role of China as a "near-Arctic state" and as "a stakeholder".

11.3.3.4   Similarities and Differences in the Development of Artic Shipping Routes

The three Artic shipping governance arrangements shows similarities and differences. An important similarity is the dominant discourse, which frames shipping as a legitimate activity in the Arctic, with the related assumption that navigation can be sustainable under the condition of an effective implementation and enforcement of the rules of the Polar Code. However, the way sustainability is defined and implemented is dependent on the specific characteristics of each of the shipping governance arrangements. Table 11.2 summarizes the differences and similarities between the three Arctic Shipping governance arrangements.

Table 11.2 Differences and similarities in the development Arctic shipping governance arrangements

	Regime complexes	Rules	Resources	Networked polity
NEP/ NSR	Coalitions and infrastructures related to Tanker and Cargo shipping Embedded in global trade networks	UNCLOS; Polar Code; SAR (AC); Russian law	Accessibility of ports and hinterland infrastructure; control over SAR and icebreakers	Russia; China, AC; EU
NWP	Coalitions and infrastructures related to Cruise and expedition shipping, mainly regional	UNCLOS; Polar Code; SAR (AC); self-regulation cruise sector	Control over SAR activities and facilities; Accessibility of communities, destinations.	Canada, Svalbard (Norway), Greenland (Denmark) cruise and expedition sector
TSR	No regime complexes, future possibilities related to bulk shipping	UNCLOS; AC	Investments (in hubs/ ports); China's Polar Silk Route[a]	AC, China, Island

[a]In its Artic Policy China states that the Polar Silk Route "facilitates connectivity and sustainable economic and social development of the Arctic", by opening up an economic passage between China and Europe through the seas northern of Russia (Tianming et al. 2021)

11.4 Conclusions

This chapter presented a social scientific analysis to understand the process of reflexive institutionalization of Arctic shipping, by analysing three different governance arrangements related to three Arctic shipping routes (the NEP/NSR, NWP and the TSR), in terms of regime complexes, networked polity, resources, institutional rules and discourses.

The main questions of this article were: 'What are the enabling and constraining conditions for a reflexive institutionalization of Arctic shipping?', 'How do Arctic shipping governance arrangements in the three shipping routes institutionalize?', and 'Can we speak of reflexive institutionalization? In other words, are the governmental and non-governmental actors involved able to challenge and change the discursive space of Arctic shipping (performative reflectiveness), to use rules from different institutional settings, without changing the rules of the game (structural reflectiveness) or to change both the rules of the game and the discursive space of Arctic shipping (reflexivity)?

An important motor of the institutionalization of Arctic shipping is the framing of shipping as a legitimate activity in the Arctic under the conditions of sustainable shipping. Although this discursive space is challenged by some NGOs (Extinction Rebellion and Ecohustler),[26] the actors within the shipping regime complexes and

---

[26] Marianne Brooker in the Ecologist. The journal for the post-industrial Age, 21st February 2020.

governmental actors embrace this dominant discourse. The differences in processes of institutionalization are related to the multi-level characteristics of the networked polity and the role of states in each of the three governance arrangements. While the NWP is regional oriented, the NEP/NSR and TSR are embedded in global navigation and trade discourses and networks. Despite the fact that Russia tries to define the NSR shipping governance arrangement as a regional arrangement governed by specific Russian national rules, China's Arctic Policy will make this governance arrangement global. The future global economic role of China and the preferred routes by the Chinese government, ship-owners and investors will affect the institutionalization of the NSR and the TSR governance arrangements. Both cases are examples of institutionalization as structural reflectiveness; the discursive space of Arctic shipping is not challenged, but core actors, such as China and Russia, are able the use existing rules from different institutional settings, not only to strengthen their position in these governance arrangements, but also to influence their specific institutionalization.

Arctic shipping is at the beginning of its development. Depending on ice and weather conditions, some shipping routes will be more realistic in the future then others. This makes the future institutionalization and the type of reflexivity of Artic shipping governance arrangements difficult to predict. Theoretically, one can state that, the future institutionalization of Arctic shipping governance arrangements is affected by the Arctic governance setting. In this regionalized networked polity (van Tatenhove 2016), states are part of the Arctic network state (consisting of UNCLOS, the Arctic Council, the Commission on the Limits of the Continental Shelf (CLCS), the Arctic Five, Permanent Participants and Permanent Observers), which is in continuous interaction with the actors within the shipping regime complexes, NGOs and Arctic (indigenous) communities. Both the Artic network state and the shipping regime complexes are characterised by institutional ambiguity (van Leeuwen et al. 2012), which gives actors the possibility to negotiate and apply the rules and resources from different institutional settings. Whether this will increase the governance capacity of actors to develop sustainable solutions for Arctic shipping will be an important question for the future.

References

Ansell C (2000) The networked polity: regional development in Western Europe. Governance 13(2):279–291. https://doi.org/10.1111/0952-1895.00136

Archer MS (2010a) Morphogenesis versus structuration: on combining structure and action. Br J Sociol 61(SUPPL. 1):225–252. https://doi.org/10.1111/j.1468-4446.2009.01245.x

Archer MS (2010b) Routine, reflexivity, and realism. Sociol Theory 28(3):272–303. https://doi.org/10.1111/j.1467-9558.2010.01375.x

Archer MS (ed) (2014) Late modernity. Trajectories towards Morphogenic Society. Springer Berlin Heidelberg, New York/Dordrecht/London. https://doi.org/10.1007/978-3-319-03266-5

Arctic Council (2009) Arctic marine shipping assessment 2009 report. Arctic Council

Arts B, Van Tatenhove J (2006) Political modernisation. In: *Institutional dynamics in environmental governance*. Springer Netherlands, Dordrecht, pp 21–43. https://doi.org/ 10.1007/1-4020-5079-8_2

Arts B, Leroy P, van Tatenhove J (2006) Political modernisation and policy arrangements: a framework for understanding environmental policy change. Pub Org Rev 6(2):93–106. https://doi. org/10.1007/s11115-006-0001-4

Beck U, Grande E (2007) Cosmopolitan Europe. Polity Press, Cambridge/Malden

Bevir M, Rhodes RAW (2011) The stateless state. In: Bevir M (ed) The sage handbook of governance. SAGE Publications Ltd, London, pp 203–217

Boonstra FG (2004) Laveren tussen regio's en regels. Verankering van beleidsarrangementen rond plattelands ontwikkelingen in Noordwest-Friesland, de Graafschap en Zuidwest Salland. Available at: http://hdl.handle.net/2066/67439

Buixadé Farré A et al (2014) Commercial Arctic shipping through the northeast passage: routes, resources, governance, technology, and infrastructure. Polar Geogr 37(4):298–324. https://doi. org/10.1080/1088937X.2014.965769

Cajaiba-Santana G, Faury O, Ramadan M (2020) The emerging cruise shipping industry in the arctic: institutional pressures and institutional voids. Ann Tour Res 80:102796. https://doi. org/10.1016/j.annals.2019.102796

Castells M (2009) Communication power. Oxford University Press, Oxford

Castells M (2010) *End of millennium: with a new preface, volume III, second edition with a new preface*, 1st edn. Wiley, Chichester. https://doi.org/10.1002/9781444323436

Chhotray V, Stoker G (2009) Governance theory and practice. A cross-disciplinary approach. Palgrave Macmillan, Hampshire

Colgan JD, Keohane RO, van de Graaf T (2012) Punctuated equilibrium in the energy regime complex. Rev Int Organ 7(2):117–143. https://doi.org/10.1007/s11558-011-9130-9

DiMento JFC, Hickman AJ (2012) Environmental governance of the great seas: law and effect. Edward Elgar Publishing Limited, Cheltenham

Dyrcz C (2017) Safety of navigation in the Arctic. Sci J Polish Naval Acad 4(211):129–146. https://doi.org/10.5604/01.3001.0010.6742

Eguíluz VM et al (2016) A quantitative assessment of Arctic shipping in 2010–2014. Sci Rep 6(1). https://doi.org/10.1038/srep30682

Hajer M (1995) The politics of environmental discourse: ecological modernization and the policy process. Claredon Press, Oxford

Halpern BS et al (2008) A global map of human impact on marine ecosystems. Science 319(5865):948–952. https://doi.org/10.1126/science.1149345

Humpert M, Raspotnik A (2012) The future of Arctic shipping along the Transpolar Sea route. In: Arctic yearbook. Northern Research Forum, pp 281–307. https://arcticyearbook.com/arctic-yearbook/2012/2012-scholarly-papers/20-the-future-of-arctic-shipping-along-the-transpolar-sea-route

Jessop B (2004) Multi-level governance and multi-level Metagovernance changes in the European Union as integral moments in the transformation and reorientation of contemporary statehood. In: Multi-level Governance. Oxford University Press, pp 49–74. https://doi. org/10.1093/0199259259.003.0004

Keil K (2018) Sustainability understandings of Arctic shipping. In: Gad UP, Strandsbjerg J (eds) The politics of sustainability in the Arctic: reconfiguring identity, space, and time. Routledge, pp 34–51

Keohane RO, Victor DG (2011) The regime complex for climate change. Perspect Polit 9(1):7–23. https://doi.org/10.1017/s1537592710004068

Koivurova T (2013) Gaps in international regulatory frameworks for the Arctic Ocean. NATO Sci Peace Secur Ser C Environ Secur 135:139–155. https://doi.org/10.1007/978-94-007-4713-5_15

Liefferink D (2006a) In: Arts B, Leroy P (eds) Institutional dynamics in environmental governance, institutional dynamics in environmental governance. Springer. https://doi.org/10.1007/ 1-4020-5079-8

Liefferink D (2006b) The dynamics of policy arrangements: turning round the tetrahedron. In: Arts B, Leroy P (eds) Institutional dynamics in environmental governance. Springer, pp 45–68. https://doi.org/10.1007/1-4020-5079-8_3

Liu C-Y et al (2021) The Arctic policy and port development along the Northern Sea route: evidence from Russia's Arctic strategy. Ocean Coast Manag 201:105422. https://doi.org/10.1016/j.ocecoaman.2020.105422

Østreng W et al (2013) Shipping in Arctic waters: a comparison of the northeast, northwest and trans polar passages. Springer, Berlin/Heidelberg. https://doi.org/10.1007/978-3-642-16790-4

Pestman P (2001) In het spoor van de Betuweroute. Mobilisatie, besluitvorming en institutionalisering rond een groot infrastructureel project. Rozenberg Publishers, Amsterdam

Raustiala K, Victor D (2004) The regime complex for plant genetic resources. Int Organ 58(2):277–309. https://doi.org/10.1017/S0020818304582036

Sassen S (2006) Why cities matter. In: *Catelogue of the 10th international architecture exhibition*, pp 26–51

Sassen S (2009) Keynote address. Bordering capabilities versus Borders: implications for National Borders. Michigan J Int Law 30(567):567–597. https://doi.org/10.1057/9781137468857.0007

Sassen S (2013) When territory deborders territoriality. Territory Polit Gov 1:21–45. https://doi.org/10.1080/21622671.2013.769895

Sassen S (2015) Bordering capabilities versus borders: implications for National Borders. In: *Borderities and the politics of contemporary mobile borders*. https://doi.org/10.1057/9781137468857_2

Schlüter A et al (2019) Coastal commons as social-ecological systems. In: Hudson B, Rosenbloom J, Cole D (eds) Handbook of the study of the commons. Routledge, pp 170–187

Smits CCA, van Tatenhove JPMM, van Leeuwen J (2014) Authority in Arctic governance: changing spheres of authority in Greenlandic offshore oil and gas developments. Int Environ Agreem Polit Law Econ 14(4):329–348. https://doi.org/10.1007/s10784-014-9247-4

Smits CCA, van Leeuwen J, van Tatenhove JPM (2017) Oil and gas development in Greenland: a social license to operate, trust and legitimacy in environmental governance. Resour Policy 53:109–116. https://doi.org/10.1016/j.resourpol.2017.06.004

Tianming G et al (2021) Has the COVID-19 pandemic affected maritime connectivity? An estimation for China and the polar silk road countries. Sustainability 13(6):3521. https://doi.org/10.3390/su13063521

Van Bets LKJ, Lamers MAJ, van Tatenhove JPM (2017) Collective self-governance in a marine community: expedition cruise tourism at Svalbard. J Sustain Tour 25(11):1583–1599. https://doi.org/10.1080/09669582.2017.1291653

van Leeuwen J, van Tatenhove J (2010) The triangle of marine governance in the environmental governance of Dutch offshore platforms. Mar Policy 34(3):590–597. https://doi.org/10.1016/j.marpol.2009.11.006

van Leeuwen J, van Hoof L, van Tatenhove J (2012) Institutional ambiguity in implementing the European Union marine strategy framework directive. Mar Policy 36(3):636–643. https://doi.org/10.1016/j.marpol.2011.10.007

Van Tatenhove JPM (2013) How to turn the tide: developing legitimate marine governance arrangements at the level of the regional seas. Ocean Coast Manag 71:296–304. https://doi.org/10.1016/j.ocecoaman.2012.11.004

van Tatenhove JPM (2016) The environmental state at sea. Environ Polit 25(1):160–179. https://doi.org/10.1080/09644016.2015.1074386

van Tatenhove JPM (2017) Transboundary marine spatial planning: a reflexive marine governance experiment? J Environ Policy Plan 19(6):783–794. https://doi.org/10.1080/1523908X.2017.1292120

van Tatenhove JPM (2019) Regulatory mixes in governance arrangements in (offshore) oil production. In: Smart mixes for transboundary environmental harm. Cambridge University Press, pp 309–326. https://doi.org/10.1017/9781108653183.014

van Tatenhove J, Leroy P (2000) The institutionalisation of environmental politics. In: van Tatenhove J, Arts B, Leroy P (eds) Political modernisation and the environment. The renewal of environmental policy arrangements. Kluwer Academic Publishers, Dordracht/Boston/London, pp 17–33

Van Tatenhove J, Arts B, Leroy P (2000) In: Van Tatenhove J, Arts B, Leroy P (eds) Political modernisation and the environment. The renewal of environmental policy arrangements. Kluwer Academic Publishers, Dordracht/Boston/London

van Tatenhove JPM et al (2020) The governance of marine restoration: insights from three cases in two European seas. Restor Ecol. https://doi.org/10.1111/rec.13288

Chapter 12
Assembling the Seabed: Pan-European and Interdisciplinary Advances in Understanding Seabed Mining

Wenting Chen, Kimberley Peters, Diva Amon, Maria Baker, John Childs,
Marta Conde, Sabine Gollner, Kristin Magnussen, Aletta Mondre,
Ståle Navrud, Pradeep A. Singh, Philip Steinberg, and Klaas Willaert

Abstract This chapter deploys assemblage theory and thinking to bring together a unique set of insights on the seabed ranging from the ecological, to legal, practice to theoretical. It does so with a particular aim in mind: to *integrate* debates pertinent to understanding the frontier space of the sea floor. Whilst there are increasing calls for interdisciplinary integration in the marine sciences, combining the natural and social sciences research on the space of the seabed and its potential for mining tends to be siloed with work addressing component parts of such possible processes: ecosystem and ecosystem service aspects, legal dimensions, and geopolitical aspects, to name but a few. Whilst these contributions touch upon intersecting issues (society and environment; law and economics, and so on) they remained centered

The original version of the chapter has been revised. A correction to this chapter can be found at
https://doi.org/10.1007/978-3-031-20740-2_19

W. Chen (✉)
Norwegian Institute for Water Research, Oslo, Norway
e-mail: wenting.chen@niva.no

K. Peters
Helmholtz Institute for Functional Marine Biodiversity, Oldenburg, Germany
e-mail: kimberley.peters@hifmb.de

D. Amon
SpeSeas, D'Abadie, Trinidad and Tobago & Marine Science Institute,
University of California, Santa Barbara, CA, USA

M. Baker
Ocean and Earth Sciences, University of Southampton, Southampton, UK

J. Childs
Lancaster Environment Centre, Lancaster University, Lancaster, UK

© German Institute of Development and Sustainability (IDOS) 2023,
corrected publication 2023
S. Partelow et al. (eds.), *Ocean Governance*, MARE Publication Series 25,
https://doi.org/10.1007/978-3-031-20740-2_12

275

on particular disciplinary and scientific offerings to understanding the seabed and prospect of seabed mining. This chapter offers a thoroughly 'joined up' approach, which presents a prism through which to better understand the issues at stake in venturing to the new vertical frontiers of ocean extraction.

12.1 Introduction

Seabed mining is an extractive process, removing and retrieving resources from the seabed – the solid 'surface' that lies at the bottom of the ocean – otherwise known as the 'ocean floor' or 'sea floor'. The mining happens on the very surface layer of the seabed which can be rich in mineral deposits such as copper, nickel, aluminum, manganese, zinc, lithium and cobalt (IUCN 2018). In respect of seabed mining, there is mining which may be described simply as 'seabed mining' and this may occur at any depth. For example, explorations and exploitations off the coast of Namibia are described as 'seabed mining' and exist within the territorial sea (12 nautical miles (nm)) and in the Exclusive Economic Zone (EEZ, up to 200 nm) from the West African country's coast – but these deposits are not *deep*.[1] Other forms of seabed mining are explicitly named 'deep-sea mining' (or DSM) and this refers to "retrieving mineral deposits from the deep sea – the area of the ocean *below* 200m" (IUCN 2018, emphasis added). Spaces of possible extraction are located on and in

---

[1] Diamonds mined at around 130 m and exploration for phosphates is up to 300 m depth.

M. Conde
Centre for Social Responsibility in Mining (CSRM),
University of Queensland, Brisbane, Australia

ICTA, Autonomous University of Barcelona, Barcelona, Spain
e-mail: mcondep@gmail.com

S. Gollner
Royal Netherlands Institute for Sea Research (NIOZ), Texel, The Netherlands

K. Magnussen
Menon Economics, Oslo, Norway

A. Mondre
Institute of Political Science, Kiel University, Kiel, Germany

S. Navrud
School of Economics and Business, Norwegian University of Life Scienes (NMBU),
Ås, Norway

P. A. Singh
Institute for Advanced Sustainability Studies (IASS), Potsdam, Germany

Research Centre for European Environmental Law (FEU), University of Bremen,
Bremen, Germany

P. Steinberg
Department of Geography, Durham University, Durham, UK

K. Willaert
Faculty of Law and Criminology, Maritime Institute, Ghent University, Ghent, Belgium

the seabed in EEZs globally, as well as on and in the seabed beyond EEZs, on the continental shelf and in the 'Area' – the seabed beyond these zonal markers. This chapter is concerned with both seabed and deep-sea mining – in other words, the surface of the seabed – as a space of extraction.

The seabed has been long recognized as an ocean 'frontier' for exploration and exploitation (Zalik 2018). In an article in the *American Journal of International Law* in 1969, Louis Henkin noted the existence of an 'untapped' global extraction space under the liquid surface of the sea, stating that "a new environment of golden promise looms on the distant horizon" (Henkin 1969, 504). Some 50 years on, this 'golden promise' does indeed 'loom' large. As Matthew Taylor has recently noted, "the world's oceans are facing a 'new industrial frontier' from a fledgling deep-sea mining industry as companies line up to extract metals and minerals from some of the most important ecosystems on the planet" (Taylor 2019, n.p). Indeed, the seabed holds 'promise' because it is a lucrative space that may provide access to valuable minerals that are now more difficult to access from terrestrial mining sites, where resources are depleting (IUCN 2018). Seabed mining opens-up a new space to retrieve minerals that are often needed in the production of today's "high-tech applications such as smartphones and green technologies such as wind turbines, solar panels and electric storage batteries" (IUCN 2018). Yet, whilst there is huge economic benefits of the promise of such extraction, there is also a wide acknowledgement of the legal complexities of such activities at sea (especially in spaces beyond national jurisdiction); of the global challenges of enabling mining where it may be driven solely by profit and multinational corporations rather than local concerns; and where technologies, access and processes of extraction may impose irreversible harm to the seabed environment and ecosystems. To return to Henkin, then, there are many 'looming' issues in respect of seabed mining as it finally comes to fruition and into reality, as a new offshore industry, alongside the 'extractive' industries of fishing and the piping of oil and gas reserves.

Given this 'looming' issue, this chapter *assembles* a unique set of insights on the seabed ranging from the ecological, to societal, practice to theoretical. It does so with a particular aim in mind: to *integrate* debates pertinent to understanding the frontier space of the sea floor. Whilst there are increasing calls for interdisciplinary integration in the marine sciences, combining the natural and social sciences (see Markus et al. 2018) research on the space of the seabed and its potential for mining tends to be siloed with work addressing only component parts of such possible processes: legal dimensions (see Willaert 2020a, b), ecological aspects (see Simon-Lledó et al. 2019) societal perspectives (see Childs 2020; Zalik 2018). Whilst these contributions (and more) of course touch upon intersecting issues (society and environment; law and economics, and so on) they remained centered on particular disciplinary and scientific offerings to understanding the seabed and prospect of seabed mining. There is much value in these approaches but they can lack a more thoroughly 'joined up' approach, which presents a prism for better understanding the issues at stake in venturing to the new vertical frontiers of ocean space.

Recent work has attempted to 'join up' debates more concretely. For example, in a recent paper on traditional knowledge and seabed mining developments, Tilot et al. (2021) bring together indigenous and traditional knowledge with legal understandings, ecological insights and contemporary politics to understand management futures for mining in the Pacific. The chapter builds on such integrative approaches and features a series of linked interventions – assembling a dialogue – which highlights how researchers are grappling with this 'frontier space' – legally, socio-economically, environmentally and geopolitically (see Koschinsky et al. 2018). This chapter offers – in one piece – a conversation on the *complexities* of seabed science and management, where the anthropogenic drivers, historic developments and future climate impacts as well as approaches for such an aim differ across space, and through the lenses of different disciplinary approaches demonstrating the necessity of such 'joined-up' thinking. That said, whilst highlighting contemporary research and approaches for understanding the seabed, it does not offer a definitive answer in how we manage such rich, varied, contentious sites, but rather aims to demonstrate the richness of combining such work to encourage further interdisciplinary endeavors as the march towards sustainable seabed mineral extraction continues afoot.

To achieve this aim, this chapter unfolds in the following way. It begins with an analytic consideration of 'assemblage' – a theoretical tool used for drawing together heterogeneous parts, into a 'whole' (DeLanda 2006). This approach makes it possible to assemble a set of disparate debates, which tend to remain separate in discussions about the seabed, and can create new modes of knowing and making sense of seabed governance issues. Following this framing, the chapter then 'assembles' a series of interventions, collating and linking these into the chapter as a whole[2] to enliven an understanding of the range of actors, issues, knowledges, techniques and practices that must combine to understand seabed and deep-sea mining, past present and future. In doing so, it aims to demonstrate the potential of combining numerous voices for an integrated understanding of the impacts of the development of the new industry. The chapter ends with a conclusion of future possibilities and required knowledge for deepening our understanding of the seabed.

12.2   Assembling Knowledge: Assembling the Seabed

Assemblage thinking or 'theory' is a mode of post-structural understanding, attuned to understanding the multiplicity of the world. It aims to provide a means of making sense of how phenomena are always emerging and 'becoming' (in other words, is never 'finished' but always in the making). As such, it is attuned to the ongoing

---

[2] The term 'whole' draws from work in assemblage theory which contends that multiple, heterogeneous 'parts' cohere together to form more or less territorialised 'wholes' - a complete picture of something for us to grasp. That said, the 'whole' is always open (and ever becoming) as other parts may be inserted or other parts may drop away as the assemblage comes together and apart over time. This chapter is a snapshot of seabed mining (a 'whole') configured of different parts: the ecological, political, economic etc. In 5 years times the picture of seabed mining may look quite different as certain parts hold fast or fall away, or new parts come to play an important role.

co-constitution of given issues, rather than presenting a situation, place, or politics as static, unchanging and complete. Moreover, rather than only allowing a dominant narrative in understanding a particular place, phenomenon or issue to be revealed, assemblage thinking enables scholars to consider the multiple, heterogeneous, seemingly separate, 'parts' (human and non-human actors, influences, discourses, environments) that 'hold together', making complexity known (DeLanda 2006). Indeed, key to assemblage is that there is a 'pause' – a moment of stability – at which a phenomenon and its parts 'territorialise' for us to assess it. However, it is always acknowledged that such an assemblage is always open, and could 'deterritorialise' and change in the future as new parts are added or detracted (a new stakeholder opinion, scientific finding, or policy, for example). As Dovey states, any assemblage comes "from flows becoming…which then produce relative points of stability", only for that stability to be shaken as parts of a particular assemblage are 'unplugged' or different parts become 'plugged in'. In sum, as Venn notes, assemblage allows a focus on "the dynamic character of interrelationships between heterogeneous elements" in the case of any given phenomena (2006, 107). It thus, arguably, can enable a more detailed, careful and critical consideration of the world.

For Deleuze and Guattari, the key 'architects' of assemblage thinking, we can think of virtually anything as an 'assemblage' – be it a person, animal, home (2004, 503–4), or as DeLanda shows, a city, or even something more intangible such as an 'issue' or 'discourse' (DeLanda 2006). Indeed, under the remit of 'assemblage' thinking, assemblage is a device that can be used for understanding almost any given topic that is emergent and complex. For example, scholars have used this framework for making sense of the ongoing construction of places (cities, streets, towns); for understanding social movements and protest; environmental justice regimes (Bickerstaff and Agyeman 2009) to a mode of thinking about the Blue Economy (Winder and LeHeron 2017). Accordingly, then, as Anderson and McFarlane note, "there is no single 'correct' way to deploy the term" (2011, 124) and it may be applied in a variety of contexts. The concept itself then, is and constantly re-becoming an assemblage.

Although seemingly abstract, the theory provides a useful framework for this chapter, in collating a series of insights about seabed mining to *integrate* debates than often remain siloed. Seabed mining can be understood as an issue – one that does not exist in and of itself – but that is *assembled* of emergent and evolving 'parts' (law, local communities, material resources, technology, economic and environmental concerns and so on) that come together to define it at any given moment. With this aim in mind, the chapter now assembles a series of voices and perspectives on seabed mining. We begin by drawing on definitional work that sets out what the seabed is – as a geographical space and site of potential governance and extraction, before highlighting why it is such an emergent zone of extraction and, hand-in-hand, of possible ecological harms. Our next logical step shifts us to ecosystem service dimensions, before we highlight to how these 'parts' assemble with the legal and geopolitical terrain of seabed mining potentials. In assembling these sections into an integrated coherent 'whole' we also bring together different country perspectives from Aotearoa New Zealand to Papua New Guinea; from the seabed mining potentials in territorial waters to the deep sea (or the 'Area').[3]

12.3 Setting the Scene: Defining Who and What

Deep-sea mining (known as DSM), is currently being pursued by many industries and national governments. At the same time, it is being heatedly opposed-to by environmental and local groups who fear the unknown impacts and potential risks this activity can cause on the environment and the affect this may have on lives and livelihoods. Currently, regulations to manage DSM are being drafted by several countries as well as the International Seabed Authority (ISA), the body in charge of overseeing this process in the 'Area' – the seabed beyond national jurisdiction (see Van Dover 2011 and Wedding et al. 2015, on advances, possible impacts and regulatory frameworks related to seabed mining). However, in spite of this work, two fundamental questions remain largely underexplored. First, how, ontologically, do we understand what the seabed *is*, and secondly, who is the stakeholder to whom seabed issues relate? Who will profit and exploit, who will be impacted, who will take decisions and govern, and vitally – who is excluded?

These definitions are vital because how the seabed is defined influences governance in national and international settings and shapes regulations in innumerable ways. At the most basic level, in respect of defining the seabed, this space is regarded as either an extension of land (in which case seabed mining could be regulated by adapting terrestrial mining laws) or an area of ocean (in which case there is a greater need to consider a broader range of ecological impacts on, for instance, the water column and its users). Some governments have considered the seabed as an extension of land seeking inspiration in regulatory instruments for onshore mining activities. For example, Papua New Guinea, the country that is arguably most advanced in pursuing DSM, has issued permits based on an extension of onshore mining protocols that, for purposes of the seabed, redefine "land" as "the offshore area being the seabed underlying the territorial sea from the mean low water springs level of the sea to such depth as admits of exploration for or mining of minerals". Other countries such as Japan, Canada, and several European Union members, will likely extend existing onshore mining regulations to the seabed to allow for seabed mining in areas of national jurisdiction. As a point of contrast, New Zealand has developed and applied specific regulations on seabed mining, the Exclusive Economic Zone and the Continental Shelf Act 2012, which rather than taking land as their reference point, place seabed mining within New Zealand's overall marine

---

[3] The 'Area' refers to the zone of "seabed and ocean floor and the subsoil thereof, beyond the limits of national jurisdiction. The international seabed area represents around 50 per cent of the total area of the world's oceans" (International Seabed Authority, n.d.) It is under the jurisdiction of the International Seabed Authority or ISA.

management strategy. Building on a recognition of the divisions within New Zealand's maritime space – the territorial sea, exclusive economic zone, and outer continental shelf – New Zealand mandates that when permitting seabed mining "the [Environmental Protection Authority] must take into account the…effects that may occur in New Zealand or in the waters above or beyond the continental shelf" (EEZ 2012). This understanding implies that models for best practice might come *less* from the onshore mining and *more* from forms of marine management used in other extractive industries such as Marine Protected Areas (MPAs).[4] As we can see, then, the definition of the seabed ultimately matters how use, and governance, emerge.

Similarly, although there has been an increasing interest in, and attention to, 'stakeholders' – how they might engage in Environmental Impact Assessments and their limited participation (see Lallier and Maes 2016; Lodge et al. 2014; Jaeckal et al. 2017) – a careful analysis of *who* stakeholders are (and could be) in the first place, remains underexplored. For DSM, where people have different connections and dependencies to this contentious and inhabited space, it is crucial to develop a new understanding of who stakeholders are, and by default, who may be excluded from debates. Who is identified and recognized as having a legitimate connection or interest; who is included (or has access) in the drafting of the regulations; who is involved in decision-making if a project is to go ahead; who has had, in effect, meaningful participation in its governance? All these questions become more complex to answer when referring to activities in the ABNJ (Areas Beyond National Jurisdiction).

Turning again to New Zealand's regulations, the EEZ provided not only for a Māori Advisory Committee that can 'advise' and 'comment on' regulation changes, but also allows for the wider participation of stakeholders: 'any person' that the EPA considers to "have existing interests that may be affected by the application" can provide 'submissions' in favour or against a marine project. Other legislation, beyond New Zealand, has similar understandings of who a stakeholder 'is', but its implementation has been criticized. For instance, the Cook Islands' Seabed Minerals Policy stated that "the entire nation and its people are the "community" affected by seabed mining activities and that related decisions are best-handled with participation of all concerned citizens, at the relevant level". However, the Cook Islands Seabed Minerals Advisory Committee created for this endeavour has been criticized for a lack of representation and transparency. Accordingly, seeking inclusion of varied stakeholders does not ensure such representation manifests.

Regulations in other countries have a 'fuzzy' or incomplete understanding of what a stakeholder is. For Portugal, where seabed mineral exploration has started in the Azorean sea, the specific regulations developed in 2015 state that a "compulsory consultation" shall be carried out "of the municipalities in their respective areas of

---

[4] Although not traditionally regarded as extractive or as industries, MPAs do have these qualities, when, for example, we see them as geopolitical resources. In this context they allow sovereign states to extract security assets from the environment under the guise of conservation. Likewise, they can play into the hands of global discourses around environmental protection and extract resources from traditional and indigenous users in modes of neocolonialism.

territorial jurisdiction (...)" (Lei 54/2015 Portugal). Although stakeholders seem to be represented here through the municipalities and competent bodies, the extent of 'territorial jurisdiction' of these municipalities on marine areas and how exactly the stakeholders will be involved, is unclear. Other countries are still developing specific regulations for DSM. In the case of Namibia the new Minerals Policy draft made public in 2018 states that "the Government will ensure community participation through consultation before companies are allowed to commence metallurgical operations" however, it is unclear how 'communities' are to be defined in the context of the seabed or if 'metallurgical' (i.e. the extraction and modification of metals) applies to the seabed.

As such, at the start of any discussion of seabed or deep-sea mining, unpacking what the seabed is and who the stakeholders are, is not a purely academic or philosophical exercise, it is a *political decision* which may be influenced by various lobby groups. The political decision shapes outcomes for potential use, and governance. It is a key 'part' of understanding seabed mining. Specific definitions of the seabed are likely to influence governance in national and international settings, from what particular ministry is given lead regulatory authority, to the calculation of risk and the scope of Environmental Impact Assessments (EIA). Indeed, whilst it is vital to assemble the question of 'what' the seabed is, and 'who' it matters to, this must be held in the context of *why* it matters – its economic benefits in the short term, but the possible ecological harms in the long term. We next integrate this vital 'part' of understanding to our assemblage of seabed perspectives.

12.4   Socio-Economic Dimensions: Marine Ecosystem Services and Values of Deep-Sea Mining

The Ecosystem Services (ES) framework, linking the environment to human well-being, is important for sustainable management of the deep-sea, which could provide a quantitative basis for future practice of Marine Spatial Planning (MSP), a crucial 'part' of understanding emerging DSM regimes by recognizing various values, including economic values. Figure 12.1 shows how the ecosystem services framework can be incorporated into the various stages of MSP.

Le et al. (2017) identify, in detail, the ES that could potentially be affected by DSM in terms of polymetallic sulfide mining, ferromanganese crusts mining, polymetallic nodules mining and phosphorites mining. ES, when considering provisioning of fish catch, for example, may be affected by disrupted breeding grounds and nursery habitat, altered secondary production and trophic support, and dispersal connectivity. Pharmaceuticals and biomaterial provisioning ES will also be affected by the changes in biodiversity and metabolic activities. Regulating services will be impacted through many channels such as surface photosynthesis, chemosynthesis, carbon flux, bioturbation, bio-irrigation, aerobic methane oxidation, greenhouse gas

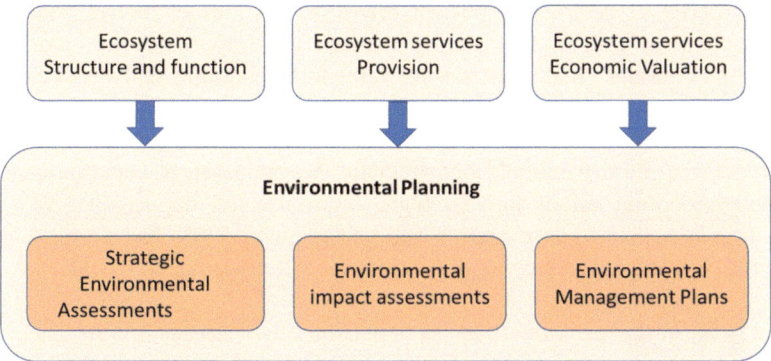

Fig. 12.1 Relationships among ecosystem services, their study and phases in environmental planning where ecosystem services can be incorporated. (Adapted from Le et al. 2017)

regulation, biological control of population and waste absorption. As noted already in this chapter, deep-sea mining will also affect the *cultural* ES that the deep-sea and seabed provides; such as the educational, aesthetic including arts, existence and stewardship values; often termed 'non-use' values.

As knowledge of deep-sea ecosystems and their dynamics is still limited, it has – to date – been difficult to connect the ecosystem function with the services they provide (Hanley et al. 2015), and it has been yet more difficult to quantify the impacts of deep-sea mining on the affected ES (Le and Sato 2013). If impacts could be quantified on provisioning services like lost fish catch, this can be valued using market prices. Impacts on regulating services can also be valued by market prices through the replacement costs approach, but it can be difficult to find a perfect substitute project that in theory could replace the loss in these ES. However, impacts on cultural ES can be very challenging to put an economic value on, as people are unfamiliar with these deep-sea, with ES and the long-term risk to these ES posed by DSM (Hanley et al. 2015). However, some environmental valuation studies have tried to address this issue.

For example, a Delphi based method was developed by Jobstvogt et al. (2014a) to communicate the ecological value of the deep-sea ecosystem. Jobstvogt et al. (2014b) conducted a Stated Preference (SP) survey in terms of a Choice Experiment (CE) to elicit households' willingness-to-pay (WTP) for creating additional MPAs in the Scottish deep-sea in order to protect them from potential destructive impacts from, for example, DSM. Aanesen et al. (2015) and Sandorf et al. (2016) conducted CE surveys of Norwegian households' WTP for extending the national MPAs for deep water, cold water corals (CWCs), including also the uncertainty of the ecological role of CWCs in their assessment. They experimented with different survey modes and different ways of presenting the ES of these unfamiliar public goods to the general public in order to improve the validity and reliability of these non-use values. These CWC valuation estimates were later included in a bioeconomic fisheries model of destructive bottom trawl versus non-destructive coastal gear (Armstrong et al. 2017); that clearly showed the importance of incorporating impacts on cultural ES and their non-use value into economic analysis of extraction of natural resources.

The uncertain linkage between the deep-sea ecosystem, ecosystem services and their benefits to humans should not preclude the inclusion of ecosystem services and their economic values into strategic environmental impact analysis (SEA), monitoring systems and ecosystem based management. DSM has strong economic motivation but impacts on ecosystem services and their values to humans should also be considered in the economic analysis in order to support a sustainable development path. This is a crucial 'part' of understanding deep-sea mining.

Ecosystem Accounting (EA), a framework proposed by UN, views nature as an asset, and aims to incorporate the environmental assets into the system of national accounts (UN 2014). EA involves not only the physical terms of ecosystem such as ecosystem extent and condition, but also the supply and use of ecosystem services, and the monetary valuation of supply and use, as well as the periodic revision of asset values based on changes in predicted future flows of ecosystem services (UN 2017). Although marine ecosystem accounting is still in an early stage, EA is a potentially valuable 'part' to consider in seabed mining assemblages as it can enhance transparency in governance, and link stocks and flows of natural resources with a broad spectrum of ecosystem services and benefit values (Chen et al. 2020). In the context of DSM, EA could provide a flexible monitoring framework (Grimsrud et al. 2018) to support ecosystem-based management as it maps the changes in ecosystem extent, condition, or physical supply and use of the ES or changes in the economic value of the ES if there is sufficient knowledge to monetize the ES. The spatial focus of EA could highlight the different geographic impacts of DSM, helping to identify management hotspots and create MPAs, if needed. However, it has to be admitted that EA faces the similar challenge as those related to quantifying ES and ES values mentioned in the section above. Yet it is still a crucial arena of research and key 'part' of the assemblage in helping to make sense of the current 'state of play' in emergent seabed mining activities.

12.5   Tackling the Legal Perspectives: Insights from Law and Policies

Given the complexity previously described, the regulation of DSM, particularly in the Area Beyond National Jurisdiction, represents a unique challenge legally. From the perspective of the United Nations Convention on the Law of the Sea (UNCLOS), the seabed has garnered great attention, and this before any real activity has taken place (an unusual feat in maritime governance, to seek to govern a still largely unpracticed mining activity). Yet with the *exploitation phase* rapidly approaching, many interests are at stake and research on legal dimensions is highly relevant, not only for academics and lawyers, but for all parties involved in exploration and exploitation of the continental shelf and the deep seabed. Moreover, future issues can be anticipated and remedied through thoughtful analysis of changes in international regulations and national legislation. By studying all the relevant legal sources, including the applicable conventions, the regulations, guidelines and standards of the International Seabed Authority, the national legislation of specific states, pertinent case law and authoritative literature, research can offer insights into a highly complex legal regime and how its implementation unfolds on the ground. Furthermore, research may fill the blanks and make a substantive contribution to the legal literature surrounding this topic by 'zooming in' on issues which have been largely neglected until now, such as the effective implementation of the status of common heritage of mankind, the interaction between the regimes of the deep seabed and the continental shelf and the differences between relevant national laws (Willaert 2020a).

Legal designation of the seabed began as early as 1970, with the Area and its mineral resources, declared as the 'common heritage of mankind' (sic). Since this point, the Area as well as seabed resources *within* national boundaries, have been the subject of regulation on the scales of national to international politics. Some research is, crucially, exploring the potential legal conflicts between DSM and the status of the seabed and its natural resources as the 'common heritage' of all (Willaert 2020b). The objective of this work is to find out if the current international legal framework and the national legislation of selected states fully respect the applicable legal principles linked to the common heritage of mankind (sic). As the research has observed, fairly quickly, the legal framework with regard to the deep seabed is not in a final state (see Hunter et al. 2018) and keeps progressing. As with any assemblage, it is in a state of 'becoming', and corrections and improvements can still be made.

But beyond this, what is at stake legally? With respect to activities in the area beyond national jurisdiction (short: the Area and high seas), mining practices are considered to comprise of the exploration and exploitation for three different types of minerals. These are polymetallic nodules, polymetallic sulphides, and cobalt-rich ferromanganese crusts. UNCLOS clearly stipulates that minerals of the Area cannot be subject to any sovereign claim by an individual state. As such, access to the resources is only possible through the regime designed by the UNCLOS and the regulations of the ISA. UNCLOS, which provides the general framework to govern deep seabed mining in the Area, confers upon the ISA the requisite mandate to actually develop all the necessary rules, regulations and procedures to administer the mineral resources of the Area. Accordingly, since its inception in 1994, the ISA has been working to this end. The ISA comprises of 168 Member States and is headquartered in Kingston, Jamaica. Member states of the ISA, all of whom are represented in the UN Assembly, meet annually. In recent years, the executive organ of the ISA, the Council, has been meeting twice a year (a sign of increasing demands for legal decision-making in respect of the seabed). It is noteworthy to mention that a number of non-Member States, most notably, the United States of America, regularly participates in the work of the Authority by attending annual sessions.

Regulations for the exploration of minerals have been in place since the year 2000, in the case of polymetallic nodules (amended in 2013), while exploration regulations for polymetallic sulphides were adopted in 2010, and the same for cobalt-rich ferromanganese crusts in 2012. Since 2014, the ISA has shifted its focus towards developing regulations to govern the exploitation of mineral resources. Instead of designing individual regulations for each type of minerals, the Member States have proceeded to develop one set of regulations that applies across the board (Willaert 2019).

One area that raises some significant questions when it comes to the exploitation of mineral resources is the harm that will be inflicted on the marine environment. Harm is another key 'part' or factor to be considered in an assemblage of understanding DSM and seabed mining per se. It is widely accepted that mining activities could cause irreparable harm on the marine environment and ecosystems at the mining site, and plumes that are generated from mining activities could spread well beyond the mining site, thereby disrupting surrounding ecosystems. In response to this, numerous Member States have called for the development of Regional Environmental Management Plans or REMPs. The prevailing view is that REMPs should be in place *before* any mining activity is permitted within a specific region. However, there is yet to be clear consensus on the actual legal force that REMPs actually connote. Like any assemblage, then, the legal and management provisions of mining remain emergent, ever in process.

One view is that REMPs are merely planning instruments that guide decision-making, while another view is that REMPs are binding instruments that instruct decision-making processes. If the latter view is adopted, this would mean that the ISA could actually reject exploitation applications on the basis that its approval would not conform with the goals and objectives of the applicable REMP. It is expected that some progress will be with respect to the legal force or effect of REMPs and the dynamics between REMPs and decision-making at the ISA. Similarly, discussions pertaining to the adoption of Standards and Guidelines that should apply to exploitation activities are also currently ongoing. The regulatory assemblage of DSM, then, is in an ever-changing and evolving state, which will be important for scholars and stakeholders to keep abreast of.

However, such instruments, and other regulative apparatus are not the only legal considerations that are a 'part' of making sense of seabed mining. Competent legal knowledge is also vital. An important area of research which arguably requires more detailed interrogation is the actual ability of the ISA to ensure the effective protection of the marine environment. From an institutional perspective, it appears that the ISA lacks the appropriate expertise in this regard. While it has an expert subsidiary body, known as the Legal and Technical Commission (LTC), a significant majority (80–90%) of the Commission's members are lawyers and geologists, who do not have environmental-related expertise. This is worrying, as the LTC is entrusted to make recommendations to the Council on environmental-related matters such as the design of appropriate regulations, the consideration of environmental impacts (including the need for emergency action), and whether or not to approve environmental monitoring and management plans submitted by contractors. This workload is substantial for three experts. However, stakeholder input enables further deep-sea ecology expertise can be injected into the process. To add further, it should be noted that it is difficult for the Council to disagree with any recommendations made to it by the LTC, as this would require two-thirds majority of Council members present and voting in most cases.

Moreover, more clarity is needed with respect to the confidentiality of data related to DSM decision making by the ISA. On the one hand, contractors insist on the need to protect proprietary interests and by extension, can withhold data obtained. On the other hand, the Law of the Sea clearly states that all environmental data should be promptly released to the ISA. Environmental information, such as baseline data, and the analysis thereof, is essential for the ISA to take necessary measures to ensure the effective protection of the marine environment. As such, there is a crucial need to clarify which information can be deemed as confidential, and which information is essential for environmental-related measures and must be disclosed. Given the current developments in international environmental law and the status of the deep seabed and its resources as common heritage of mankind, public participation, which is closely linked to the topic of transparency, is also a hot issue.

Finally, two critical 'parts' that require greater attention from a legal perspective are the financial terms of exploitation contracts, and the appropriate mechanism for benefit sharing. With regards to the former, the ISA is currently taking steps to design a suitable method to calculate how payments that emerge from mining, should be made (and who they should be made to). As concerns the latter, efforts remain at a preliminary stage and are currently, as of writing this, elusive. It is anticipated that efforts to make some progress in the appropriate benefit-sharing mechanism will intensify in coming years (altering the assemblage of DSM) and the operationalization of the Enterprise, an organ through which the ISA can develop its own mining activities, will also play a vital role in providing benefits for 'mankind' as a whole. The chapter next turns to economic dimensions, particularly ecosystems services, in greater detail.

However, it should be noted that legal research on the seabed is challenging. It is highly likely that new developments, in the form of new agreements, regulations or changes to relevant national legislation, may occur during any given research period. However, these risks can be limited by closely observing recent evolutions and anticipating such changes, thereby ensuring that the research results do not lose their relevance if these developments eventually take place. Indeed, apart from studying the *existing* legal framework, it is very useful to focus on the law-making process of the International Seabed Authority by attending and observing the annual sessions of the ISA Council in order to enhance knowledge with regard to the future regulations on DSM and the topical issues under discussion. Under this remit, legal scholars are also integrating semi-structured expert interviews informed by the findings of the desk research. By interviewing a representative of each of these involved parties, such as environmental NGOs, scientists, commercial mining operators, sponsoring states, developing states and the International Seabed Authority, the various points of contention are underlined, allowing for a better analysis of the different factors influencing stakeholder perceptions of the current legal framework and enabling a more accurate assessment of future policy changes.

12.6   Recognizing the (Geo)Political and Associated Socio-Cultural and Temporal Dimensions

Whilst this chapter has, so far, addressed the definitional, ecosystem services, economic and legal 'parts' that help us understand seabed mining, geo-politics is also crucial in this emergent industry and overlaps and converges (see Anderson 2012) with the parts introduced so far. Through an approach conceptually grounded at the interface of critical geography, political ecology and resource anthropology, research by Childs has been focusing on how the seabed has emerged as a new political *terrain* of struggle (see 2018, 2019, 2020). Moving beyond geopolitical approaches that understand the world largely in the narrow terms of interstate relations, this research instead seeks to understand the seabed as a space of politics produced by a relational congregation of socio-natural forces, considering 1) the temporalities of DSM (Childs 2018); 2) a corporate anthropology of a DSM firm and its strategies (Childs 2019); and 3) the impacts of DSM upon indigenous communities and the political potential of art to counter-narrate the seabed (Childs 2020). Shifting back to the earlier section on definitions, certainly, a vibrant, unstable and agentive seabed, that is in flux and changing, is seen as generative of DSM's evolving geo-politics.

The seabed as a geopolitical concern has emerged where, in recent years, it has been re-imagined by industry and policy makers not as an inert edge of a politically insignificant watery volume, but as the latest 'frontier' of resource extraction. Variously scripted by global capital as both a solution to global resource security and as a more sustainable alternative to the terrestrial mining industry, DSM has thus emerged as a new iteration of *spatial fix*. This 'fix' encompasses both a tendency to 'sink money into physical objects' (for example, ports, ships, deep-sea mining equipment) and a metaphorical 'addiction' to resource extraction (Brent et al. 2018: 3). In other words, for those who work DSM into the blue economy narrative, the seabed becomes a key geographical site for capital's ongoing expansion.

Yet for all the spatially centered critiques that it provokes, DSM also invites us to think about the (geo)political effects of its unique *temporalities*. As Childs argues, temporal dimensions 'may be projected forwards; DSM's target metals and minerals have been constructed both historically and currently as 'resources of the future', global finance is courted by corporate pronouncements of DSM's 'resource potential', 'waste' from the extractive process is included in predictions of environmental impact and so forth. But the temporal also engages with the geological time of deep-sea topographical formation; for example, where polymetallic sulphides form at very different speeds to polymetallic nodules, or where the status of resources can be either materially altered by physical forces such as volcanism or through discursive shifts inspired by (human) knowledge production and commodification' (Childs 2018: 2). *Time* and *space* then, are vital parts to critically consider in understanding the politics of DSM.

Yet it is also vital to consider DSM as *social*. A geopolitical approach critiques the oft-understood domain of DSM as largely asocial (its industry proponents often describe it as having 'no human impact'). To date, there have been very few ethnographic studies of those affected by or invested in the activity. Childs has sought to partially address this gap by analyzing the emotional and affectual aspects of DSM upon communities in Papua New Guinea (PNG) situated closest to the world's first commercial DSM license. Using a range of participatory methods and creative practices, including drawing, sculpture and participatory theatre, these small-island communities sought to find an alternative vocabulary for making the seabed visible to DSM actors including the corporation, the PNG state and activist groups at local, national and global scales (Childs 2020). Building on earlier sections of the chapter – research on DSM is not only scientific but understanding the assemblage of the issue involves engaging with legal analysis, interviews and as demonstrated here, more novel methodologies. Indeed, creative practices, in particular, emerge as 'submerged perspectives' that seek to 'pierce through the entanglements of power' associated with blue growth and proclamations of 'sustainable' DSM and which seek to 'differently organize the meanings of social and political life' (Gomez-Barris 2017: 11 in Childs 2020: 7). In other words, they make possible a greater understanding of geopolitics in the context of DSM.

There is also a need to understand the political possibilities enabled by the deep-sea's unique materialities, not least in terms of the ways that these prefigure the legitimizing strategies of certain human actors to mine the seabed. For example, the Deep-Sea Mining Corporation frames DSM activity in a way quite specific to the deep-sea environments in which it operates. By engaging with the matter or materiality of deep-sea mining (for example, the violence and unruliness of its associated volcanism, and the temporalities of sulphide 'chimneys'), the DSM industry is able to position itself as a more sustainable version of mining than its terrestrial equivalent (Childs 2019), geopolitically legitimizing itself. Understanding geopolitical 'positioning' then, of this multinational and complex industry and its physical materialities are an important part of any seabed mining assemblage.

Finally, connecting studies of deep-sea mining to a broader turn in the social sciences towards 'critical ocean studies' (DeLoughrey 2019) is essential to understanding its place in the politics of the Anthropocene. This means taking the geophysical processes of the seabed and deep-water column as well as a broadened cast of political actors (including spirits and deep-sea fauna) seriously in understanding how DSM's politics is wrought. Such an approach can draw upon work that has urged us to think 'with' the ocean (Steinberg and Peters 2015; Peters and Steinberg 2019) and the still hidden (post)colonial histories that it reproduces (Deloughrey 2017).

12.7   Endings, and Beginnings

This chapter has been ambitious in scope, assembling together a collection of perspectives and knowledges about seabed and more specifically DSM. Unlike other
publications on the seabed – *which have largely tended to focus solely on single
issues* – this chapter has deployed the post-structural theory of 'assemblage' – as a
mode of bringing together disparate parts, territorializing them into one coherent
whole – a multipart, complex and varied discussion of the issue of seabed mining.
Each 'part' of the chapter – on definitions, legal dimensions, ecosystems services
and values, to science communication and geopolitics – has aimed to demonstrate
that DSM cannot be understood through only one approach, but requires a *conversation* and *collaboration* across fields of knowledge and academic disciplines, and
across the many approaches of those disciplines (from quantitative modelling, to
qualitative interview data to scientific findings).

Taking an 'assemblage' approach has enabled a chapter that pays attention not to
one 'master narrative' of mining, but rather the many parts that constitute this
global, underwater development. The 'part' focused on definition, raised the vital
issue that seabed mining does not, or will not, emerge outside of how we define,
know and understand what the seabed – as a space to 'save' or a space to 'exploit'.
Intersecting closely, the 'part' on socio-economic assessment identified the need to
understand seabed ecology and measure potential harms. This must be done in situ
with understanding 'legal' and 'geopolitical' parts of the story. Together our 'prism'
has demonstrated the potential of assemblage in has enabling a perspective that
identifies often overlooked or previously unrecognized dimensions of the issue.
Indeed, assemblage theory encourages scholars to be critical in thinking through the
many parts that make a 'whole' (whether they seem immediately important or not)
which come together to form an understanding of an issue. It permits also, "an alternative account", one rich in its diversity and attention to heterogeneous elements (in
this case, law, geopolitics, science, governance, management) showing how they
interrelate. However, our assembled analysis is not complete and further questions
could be asked. Whilst the chapter touches on seabed mining in different areas – the
'Area' and within national jurisdiction, it would be beneficial to dig deeper and
explore (as one Reviewer urged us to do): 'how the assemblage of seabed mining
under national jurisdiction differs from assemblage of seabed mining beyond
national jurisdiction, if there is a difference, and if so why: which are the dominant
parameters?'

Yet also crucial to assemblage is the acknowledgement, as noted at the start of
the chapter, that any assemblage is always open, and could 'deterritorialise' and
change in future as new parts are added or detracted. This chapter has 'held together'
a set of 'parts' arising at one moment in time. In this sense, this chapter provides a
certain sort of conclusion for how to think about and understand seabed mining. But
the chapter also, with the knowledge that assemblages change and evolve – and that
seabed assemblages, in particular, are a terrain of flux (physically, legally, and

beyond) – argues that this is just the beginning. Future studies of this emergent assemblage will be necessary in the future, and we hope they may take inspiration from our approach here.

Appendix: The Deep-Ocean Stewardship Initiative

The body of scientific literature relating to environmental aspects of DSM has exploded in the past few years with examples too numerous to cite here. Of course, this is excellent news for increasing scientific knowledge of the remote areas of the deep ocean that maybe targeted for mineral extraction. It provides far more information to work with to try to gauge potential impacts for those that inhabit our deep oceans and the effects on the important services they provided to the planet and its occupants. However, this wealth of information can be overwhelming in complexity for scientists who work in this field, let alone for other stakeholders who have alternative interests in this realm (however those stakeholders may be defined, see previous part).

In trying to grapple with this wealth of information, the Deep-Ocean Stewardship Initiative (DOSI) – consisting of mainly scientists but also lawyers, policy makers, economists, conservationists and industry experts from around the globe – help to collate, disseminate and translate the current scientific literature for all. DOSI network members work to advance deep-ocean science in UN and other intergovernmental policies as well as on a national level, and translate science into digestible information at all levels. This is a key occupation for this group who unite to advise on ecosystem-based management of resource use in the deep ocean (both within and beyond national jurisdiction), including on DSM.

As the industry is rapidly approaching the transition from exploration to exploitation in our world's oceans (see above), one of the current primary focus areas is to provide independent scientific advice to the ISA and other stakeholders on DSM issues, including the development of exploitation regulations for the Area. DOSI is in a unique position to be able to collate this knowledge and deliver it directly to policy makers, neither advocating for mining or opposing it. Moreover, the initiative has been an official observer at the ISA Annual Sessions since 2016, delivering scientific side-events and interventions to highlight environmental aspects of DSM, working with country delegates and other stakeholders and producing related policy briefs (for example on climate-change considerations, the importance of biodiversity assessment and monitoring, and strategic environmental goals and objectives).

The network also actively encourages and funds the engagement of a broad spectrum of scientists, including those from developing nations, as well as early career individual, in these activities. Capacity development is an important aspect of DOSI's work, especially where there may be unequal power relations between those who seek to extract and exploit resources, and those subject to such extraction and exploitation. Between the Annual Sessions, DOSI Minerals Working Group members (which number around 175), contribute expert commentaries on the draft regulations and other ISA documents, produce peer-reviewed publications (for example, Tunnicliffe et al. 2018), reports and outreach materials, convene and attend workshops and meetings relating to environmental planning and management aspects of DSM, and have regular communications with the enormous flux of information coming from scientific papers and meetings. Proactive development and implementation of comprehensive management practices, frameworks and policies prior to the onset of commercial mining will ensure protection and preservation of the marine environment, whilst enabling the use of seabed mineral resources. This, however, requires a deep understanding of law, a vital part of the seabed mining 'assemblage'.

References

Aanesen M, Armstrong C, Czajkowski M, Falk-Petersen J, Hanley N, Navrud S (2015) Willingness to pay for unfamiliar public goods: preserving cold-water corals in Norway. Ecol Econ 112:53–67

Anderson J (2012) Relational places: the surfed wave as assemblage and convergence. Environ Plann D Soc Space 30(4):570–587

Armstrong CW, Kahui V, Vondolia GK, Aanesen M, Czajkowski M (2017) Use and non-use values in an applied bioeconomic model of fisheries and habitat connections. Mar Resour Econ 32(4):351–369

Bickerstaff K, Agyeman J (2009) Assembling justice spaces: the scalar politics of environmental justice in North-East England. Antipode 41(4):781–806

Brent Z, Barbesgaard M, Pedersen C (2018) The blue fix: unmasking the politics behind the promise of blue growth. Transnational Institute, Amsterdam

Chen W, Van Assche K, Hynes S, Bekkeby T, Christie H, Gundersen H (2020) Ecosystem accounting's potential to support coastal and marine governance. Mar Policy 112:103758. https://doi.org/10.1016/j.marpol.2019.103758

Childs J (2018) Extraction in four dimensions: time, space and the emerging geo (–) politics of deep-sea mining. Geopolitics 25:1–25

Childs J (2019) Greening the blue? Corporate strategies for legitimising deep sea mining. Polit Geogr 74:102060

Childs J (2020) Performing blue degrowth: critiquing seabed mining in Papua New Guinea through creative practice. Sustain Sci 15:117–129

DeLanda M (2006) A new philosophy of society: assemblage theory and social complexity. Continuum, London

Deleuze G, Guattari F (2004) A thousand plateaus. Athlone Press, London

DeLoughrey E (2017) Submarine futures of the Anthropocene. Comp Lit 69(1):32–44

DeLoughrey E (2019) Toward a critical ocean studies for the Anthropocene. Engl Lang Notes 57(1):21–36

Gómez-Barris M (2017) The extractive zone: social ecologies and decolonial perspectives. Duke University Press

Grimsrud K, Lindhjem H, Barton D, Navrud S (2018) Challenges to ecosystem service valuation for wealth accounting. In: Managi S, Kumar P (eds) UNEP inclusive wealth report 2018: measuring Progress toward sustainability. Routledge, New York, USA

Hanley N, Hynes S, Patterson D, Jobstvogt N (2015) Economic valuation of marine and coastal ecosystems: is it currently fit for purpose? J Ocean Coastal Econ 2(1). https://doi.org/10.15351/2373-8456.1014

Henkin L (1969) International law and "the interests:" the law of the seabed. American J Int Law 63(3):504–510

Hunter J, Singh P, Aguon J (2018) Broadening common heritage: Addressing gaps in the deep sea mining regulatory regime. Harvard Environmental Law Review (online) https://www.researchgate.net/profile/Pradeep_Singh53/publication/326518886_Broadening_Common_Heritage_Addressing_Gaps_in_the_Deep_Sea_Mining_Regulatory_Regime/links/5b51fc9b0f7e9b240ff200f1/Broadening-Common-Heritage-Addressing-Gaps-in-the-Deep-Sea-Mining-Regulatory-Regime.pdf. Accessed 21 Feb 2020

International Seabed Authority (n.d.) About the Authority (online) https://www.isa.org.jm/frequently-asked-questions-faqs. Accessed 01 May 2021

IUCN (2018) Deep Sea Mining (online) https://www.iucn.org/sites/dev/files/deep-sea_mining_issues_brief.pdf. Accessed 28 Feb 2022

Jaeckel A, Gjerde KM, Ardron JA (2017) Conserving the common heritage of humankind–options for the deep-seabed mining regime. Mar Policy 78:150–157

Jobstvogt N, Hanley N, Hynes S, Kenter JO (2014a) Twenty thousand sterling under the sea: estimating the value of protecting deep-sea biodiversity. Ecol Econ 97(January):10–19. https://doi.org/10.1016/j.ecolecon.2013.10.019

Jobstvogt N, Townsend M, Witte U, Hanley N (2014b) How can we identify and communicate the ecological value of Deep-Sea how can we identify and communicate the ecological value of Deep-Sea ecosystem services? PLoS One 9(7):1. https://doi.org/10.1371/journal.pone.0100646

Koschinsky A, Heinrich L, Boehnke K, Cohrs JC, Markus T, Shani M, Singh P, Smith Stegen K, Werner W (2018) Deep-sea mining: interdisciplinary research on potential environmental, legal, economic, and societal implications. Integr Environ Assess Manag 14(6):672–691

Lallier LE, Maes F (2016) Environmental impact assessment procedure for deep seabed mining in the area: independent expert review and public participation. Mar Policy 70:212–219

Le JT, Sato KN (2013) Ecosystem Services of the Deep Ocean. pp. 49–54. Report, ocean-climate. org. Retrieved from http://levin.ucsd.edu/people/satofiles/Le&Sato_ecosystem-services-deep-ocean_OCPScientificNotes_2016.pdf

Le JT, Levin LA, Carson RT (2017) Incorporating ecosystem services into environmental management of deep-seabed mining. Deep-Sea research part II. Top Studies Oceanography 137:486–503. https://doi.org/10.1016/j.dsr2.2016.08.007

Lodge M, Johnson D, Le Gurun G, Wengler M, Weaver P, Gunn V (2014) Seabed mining: international seabed authority environmental management plan for the clarion–Clipperton zone. A partnership approach. Mar Policy 49:66–72

Markus T, Hillebrand H, Hornidge AK, Krause G, Schlüter A (2018) Disciplinary diversity in marine sciences: the urgent case for an integration of research. ICES J Mar Sci 75(2):502–509

Peters K, Steinberg P (2019) The ocean in excess: towards a more-than-wet ontology. Dialogues Hum Geogr:1–15

Sandorf ED, Aanesen M, Navrud S (2016) Valuing unfamiliar and complex environmental goods: a comparison of valuation workshops and internet panel surveys with videos. Ecol Econ 129:50–61

Simon-Lledó E, Bett BJ, Huvenne VA, Schoening T, Benoist NM, Jones DO (2019) Ecology of a polymetallic nodule occurrence gradient: implications for deep-sea mining. Limnol Oceanogr 64(5):1883–1894

Steinberg P, Peters K (2015) Wet ontologies, fluid spaces: giving depth to volume through oceanic thinking. Environ Plan D Soc Space 33(2):247–264

Taylor M (2019) Deep-sea mining to turn oceans into new industrial frontier Available at: https://www.theguardian.com/environment/2019/jul/03/deep-sea-mining-to-turn-oceans-into-new-industrial-frontier. Accessed 25 Nov 2019

Tilot V, Willaert K, Guilloux B, Chen W, Mulalap CY, Gaulme F, Bambridge T, Peters K, Dahl A (2021) Traditional dimensions of seabed resource management in the context of Deep Sea Mining in the Pacific: learning from the socio-ecological interconnectivity between island communities and the ocean realm. Front Mar Sci 8:257

Tunnicliffe V, Metaxas A, Le J, Ramirez-Llodra E, Levin, LA. (2018) Strategic Environmental Goals and Objectives: Setting the basis for environmental regulation of deep seabed mining. Marine Policy. ISSN 0308-597X. https://doi.org/10.1016/j.marpol.2018.11.010

United Nations (2017) Technical Recommendations in support of the System of Environmental-Economic Accounting 2012. Experimental Ecosystem Accounting. Retrieved from https://seea.un.org/sites/seea.un.org/files/technical_recommendations_in_support_of_the_seea_eea_final_white_cover.pdf

United Nations, European Commission, F. and A. O. of the U., Nations, International Monetary Fund, O. for E. C. and, & Development, T. W. B. (2014) System of Environmental-Economic Accounting 2012 Central Framework. Retrieved from https://unstats.un.org/unsd/envaccounting/seeaRev/SEEA_CF_Final_en.pdf

Van Dover CL (2011) Tighten regulations on deep-sea mining. Nature 470(7332):31

Wedding LM, Reiter SM, Smith CR, Gjerde KM, Kittinger JN, Friedlander AM et al (2015) Managing mining of the deep seabed. Science 349(6244):144–145

Willaert K (2019) Assessment of the ISA Draft Exploitation Regulations, WWF. pp.1–14 (online) http://www.maritimeinstitute.ugent.be/sites/default/files/2019-10/Report%20Assessment%20ISA%20Draft%20Exploitation%20Regulations%20%28April%202019%29.pdf Accessed 21 Feb 2020

Willaert K (2020a) Diverse national legislation on deep sea mining: towards sponsoring states of convenience? Belgian Review of International Law 2020 (forthcoming; accepted for publication)

Willaert K (2020b) Effective protection of the marine environment and equitable benefit-sharing in the Area: empty promises or feasible goals? Ocean Development and International Law (forthcoming; accepted for publication)

Winder GM, Le Heron R (2017) Assembling a blue economy moment? Geographic engagement with globalizing biological-economic relations in multi-use marine environments. Dialogues Hum Geogr 7(1):3–26

Zalik A (2018) Mining the seabed, enclosing the Area: proprietary knowledge and the geopolitics of the extractive frontier beyond national jurisdiction. International Social Science Journal (Early Online) https://doi.org/10.1111/issj.12159

Chapter 13
Societal Transformations and Governance Challenges of Coastal Small-Scale Fisheries in the Northern Baltic Sea

Pekka Salmi, Milena Arias-Schreiber, and Kristina Svels

Abstract Our chapter adds a northern dimension to the discussion about the past, present and future of small-scale fisheries and their governance. For centuries, extraction of fish resources has been of utmost importance in many coastal areas of the Baltic Sea and small-scale fisheries have survived due to the robustness of the social institutions that have helped them adapt throughout periods of economic and social upheaval. Lately, the fishing livelihood has been undergoing a continuous process of contraction and concentration in terms of vessel numbers and employment. Leisure use of water areas, nature conservation and science-based governance systems have challenged fishers' access to fish resources. Especially in the northern parts of the Baltic Sea, the viability and future of coastal small-scale fisheries is severely challenged by problems caused by fish-eating animals, mainly grey seals and cormorants. We draw upon interactive governance theory to compare experiences on Finnish and Swedish small-scale fisheries governance. Our conclusion is that the present governance system is incompatible with the small-scale fisheries context, and propose creating new co-governance arrangements where small-scale fishers' interests, values and local knowledge are better integrated into a governance system.

13.1 Introduction

The Baltic Sea is located in Northern Europe and is considered the second largest brackish semi-enclosed water body in the world (Bonsdorff et al. 2015). With a coastline of approximately 1600 km in length and estimated 29 million people living within 50 km of the coast (Baltic Sea 2019), it stretches along nine countries

P. Salmi (✉) · K. Svels
Natural Resources Institute Finland, Turku, Finland
e-mail: pekka.salmi@luke.fi; kristina.svels@luke.fi

M. Arias-Schreiber
University of Gothenburg, Gothenburg, Sweden
e-mail: milena.schreiber@gu.se

© German Institute of Development and Sustainability (IDOS) 2023
S. Partelow et al. (eds.), *Ocean Governance*, MARE Publication Series 25,
https://doi.org/10.1007/978-3-031-20740-2_13

295

including Finland and Russia, the Baltic countries (Estonia, Latvia and Lithuania), Denmark, Sweden, Poland and Germany.[1] Throughout history the availability of natural resources has been of crucial importance for coastal inhabitants of the northern Baltic Sea whose coastal waters become ice-covered during the winter. Fish and seals have been of major importance dating back to the food cultures of Stone Age coastal dwellers (Pääkkönen et al. 2016). Between the 15th and seventeenth century, the Baltic Sea including its northern waters were central in the development of the important commercial late medieval fishery (Lajus et al. 2013).

Since centuries, coastal fishing has adapted to changing conditions and transformations in nature and society. During the last 70 years many coastal fishing communities along the Baltic Sea have lost their vitality and, in many locations, the remaining coastal fishers are struggling to continue with their livelihoods. At the same time, these livelihoods produce environmentally friendly healthy food and nutrition as well as maintain cultural and economic values of communities (Waldo and Loven 2019). These benefits create opportunities for coastal fishing communities to develop and self-sustain, yet these fisheries are being increasingly steered and restricted from external influences. National and supranational top-down governance systems, and market-based fisheries management, are among those factors (Hultman et al. 2018). Presently, external environmental governance is also affecting Finnish and Swedish small-scale fishers[2] allowing populations of seals and cormorants to rise which creates further challenges for small-scale fisheries.

While transformations seem to be part of the dynamics of certain social groups, at the onset of the twenty-first century the concept of societal transformations has emerged in academic, the media and policy spheres to emphasize that radical changes are required to tackle major global challenges. The concept is used also to point out that small social or technical readjustments will not be enough to for example fully implement the Sustainable Development Goals by 2030 or to confront global environmental and climate change. Societal transformations are defined as "deep and sustained, nonlinear systemic change, generally involving cultural, political, technological, economic, social and/or environmental processes" (Linner and Wibeck 2020, page 222). As it is regarded non-linear, many outcomes of the transformations might be unforeseen and not predictable. The term gained wide recognition through Karl Polanyi's classic book "The Great Transformation", despite its scarcely explicit mention in the book or its use as a synonym for secular social change (Merkel et al. 2019). Governance, understood as the formal and informal processes of collective decision-making, planning, deliberating, and capacity building by governmental, market, and civil society actors is inevitably affected by societal transformations.

In this chapter, we examine the connections between societal transformations, present challenges and future opportunities of the northern Baltic Sea small-scale fisheries. The main geographical focus is on the coastlines and archipelagos of

---

[1] For the purposes of the Helsinki Convention the "Baltic Sea Area" is defined as the Baltic Sea and the entrance to the Baltic Sea bounded by the parallel of the Skaw in the Skagerrak at 57° 44.43′N.

[2] Throughout this chapter the terms coastal fishers and small-scale fishers are used as synonyms.

Sweden and Finland, stretching from the Northern Baltic Proper area to the north (Fig. 13.1). These regions share similar ecosystem features but also face akin challenges related to fishing, livelihoods and transformations of their local owner-based management systems, where the landowners could use and manage the local fisheries in the past.

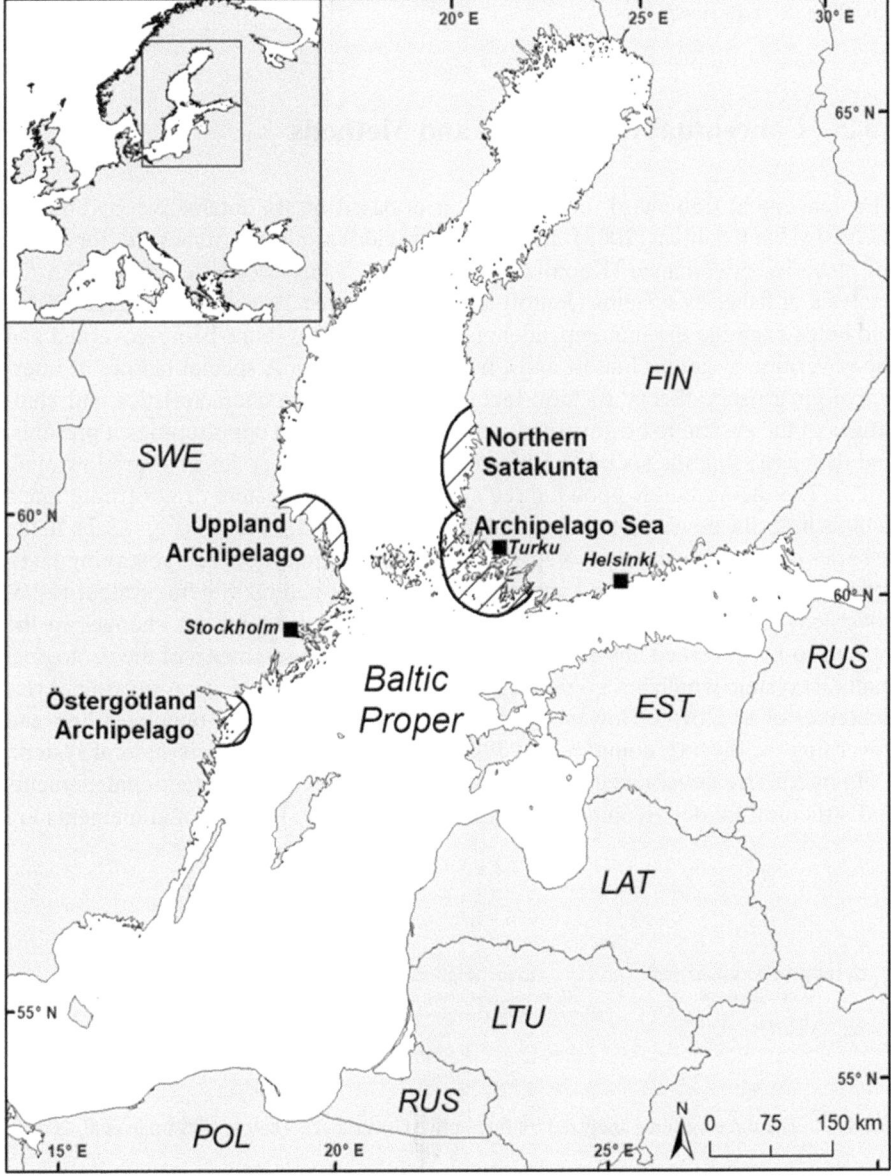

Fig. 13.1 The Baltic Sea case study areas in Sweden and Finland. (Source: Authors)

The northern Baltic Sea coastal fisheries are scantly studied, although it differs in many respects from other parts of the Baltic Sea, regarding its particular environmental conditions, historical, cultural and economic settings, and governance systems. Using interactive governance theory (Kooiman 2003), our core question is whether the current governance system fits or harmonize with the system-to-be-governed. How has this compatibility changed along with societal transformations? What have been the consequences of these changes from the perspective of small-scale fisheries?

13.2 Conceptual Framework and Methods

The conceptual framework in this chapter is based on the interactive governance theory by Jan Kooiman (2003). This theory provides a suitable framework for studying fisheries governance (Kooiman et al. 2005). It has been often applied in the analysis of fisheries systems (Jentoft and Chuenpagdee 2019; Partelow et al. 2020) and helps examine the interdependencies between the systems being governed and the governing systems (Jentoft and Chuenpagdee 2015). A special feature of interactive governance theory is "first identifying the unique characteristics and challenges of the system to be governed, i.e., the problems and opportunities it presents, and then assessing the social organizational factors of governance" (Partelow et al. 2020). The theory builds upon a three systems model: a system of governing interactions links the governing system and the system-to-be-governed (Fig. 13.2). In the fisheries context, governance is a result of the interactions between governing institutions (the governing system) and the targeted social-natural system (system-to-be-governed) (Kooiman et al. 2005). In the Baltic Sea, research on changes in the system-to-be-governed has mostly focused on regular assessments of the ecological (natural) system which has guided science-based policy and decision-making (Arias Schreiber et al. 2019). However, this scope is insufficient for understanding and governing the diverse, complex and dynamic coastal fisheries socio-natural system.

In interactive governance theory, interactions are divided in intentional elements and structural modes (Kooiman and Chuenpagdee 2005). Intentional elements are

Fig. 13.2 The three systems analyzed by Interactive Governance Theory. (Kooiman et al. 2005)

images, instruments and action. Images come in many types: visions, knowledge, facts, judgements, presuppositions, hypotheses, convictions, ends and goals. These images constitute the guiding lights as to how and why of governance (Kooiman and Bavinck 2005) including background ideas about the major problems and challenges (Jentoft et al. 2010). An image can be for example the vision that a sustainable fishery will be achieved by matching fishing effort to the size of the fish populations in a certain area. The main focus in our analysis, however, is on governance instruments, which is an intermediary element that links images to action. Instruments are not a neutral medium; their design, choice and application frequently elicit strife (Kooiman and Bavinck 2005). One's position in society determines the range of instruments available. Instruments can be for example policies, management plans, governing structures or governance actors' constellations.

Interactive governance theory distinguishes three types of structural modes: hierarchical governance, co-governance, and self-governance (Kooiman and Bavinck 2005). Hierarchical governance is the most classical of the governance modes, characteristic of the top-down interactions between a state and its citizens. This style of intervention expresses itself in policies and in law being control and steering key concepts in hierarchical governance. In co-governance, societal parties join hands with a common purpose in mind and stake their identity and autonomy in this process (ibid.). Governance theory contains numerous manifestations of co-modes, such as communicative governance, public-private partnerships, networks, regimes, and co-management. Co-governance is at the core of governance theory, as the necessity of broad participation is, for instance in the context of fisheries, seen as essential from a normative and from a practical standpoint (Kooiman and Bavinck 2005). Finally, self-governance, where actors take care of themselves outside the purview of government, is rare in the governance of modern fisheries.

Governance has an analytical but also a normative application. In fisheries and coastal governance, governing systems and the system-to-be-governed should often be compatible (Jentoft and Chuenpagdee 2009). Jentoft et al. (2010) emphasize that the governing system needs to correspond to images of the system-to-be-governed and note that when the system-to-be-governed is perceived to be simple and stable, the governing system with a top-down, centralized government pyramid may be most effective. Conversely, when the system-to-be-governed is complex and unstable, the governing image of a rose (a coalition of stakeholder groups) may be more desirable (Jentoft et al. 2010). Similarly, conflicting images among stakeholders impact the governability or the quality of governance of a fishing system (Arias Schreiber et al. 2019).

Fisheries is not managed in a vacuum (e.g., Jentoft 2000). It is important to study societal transformations, because they affect both the system-to-be-governed and the governing system. In this chapter, our point of departure is, firstly, that recent struggles of coastal fisheries in the northern Baltic Sea are largely due to societal transformations in the governing institutions towards post-productivism,[3] globalized economies of scale and environmentalisation.[4] Secondly, we hold that these transformations shape the compatibility and balance between the governance system and the system-to-be-governed, as these transformations affect both fisheries and the governance systems, including images, instruments, structures and actions. Moreover, we discuss options for striking a better balance broadening the perspectives to understand and steer the future of the northern Baltic Sea.

13.2.1 Methods

We use secondary documentary sources and draw on an overview of the literature. We also use primary material collected by the Baltic Sea Seal and Cormorant Transnational Cooperation (TNC) project (Svels et al. 2019) and the project "Fishing for solutions: community economies for a coastal rural development in Sweden". For both projects, surveys and semi-structured interviews were used for data collection. The regular participation of the authors in workshops, meetings and other fora where the situation of the small-scale fisheries is analysed and discussed and continuous dialogues with Baltic fishers during at least the last 5 years contributed also to the data collection. In the following sections we describe the empirical contexts of small-scale fishing in the Baltic Sea and changes in their governing system. By providing regional examples, the next section examines the multiple and changing conditions of small-scale fisheries in Finland (Northern Satakunta and the Archipelago Sea region) and Sweden (from Östergötland up to the Uppland archipelagos) and their relations to the governance systems. Further, two main contemporary governance challenges – the quota-based management and seals and cormorants-related conflicts – are analysed. We finish this chapter with a conclusions section where the theoretical framework is linked to our results.

---

[3] In the post-productivist setting, coastal areas are viewed as sources of individual experiences and consumption, as well as places for nature conservation (Rannikko, 2008; Rannikko and Salmi, 2018).

[4] Environmentalisation is defined as "the adoption of a generic environmental discourse by different social groups, as well as the concrete incorporation of environmental justification to legitimate institutional, political and scientific practices" (Acselrad, 2010, p. 103). The story of environmentalisation reflects an emphasis on ecological values in society and tends to disempower the primary producers and other rural people (Marsden, 2004). The change towards environmentalisation can be clearly seen in the fisheries sector (Salmi, 2009).

13.3   Transformations of Small-Scale Fishing and the Governance System in the Northern Baltic Sea

13.3.1   The Northern Baltic System-to-Be-Governed

Baltic Sea is one of the world's largest semi-enclosed bodies of brackish water (HELCOM 2021). Eutrophication, caused by nutrient pollution, is a major concern in most areas. The only coastal areas not affected by eutrophication are confined to the Gulf of Bothnia. Recently, inputs of nitrogen and especially phosphorus to the Baltic Sea have been substantially reduced. Living organisms and bottom sediments are affected by hazardous substances in all parts of the Baltic Sea (HELCOM 2021). In Finland, fish stocks targeted by coastal small-scale fishers have not been under serious threat and are recommended as environmentally friendly food (Salmi 2018; WWF 2021). The coastal areas of the northern Baltic Sea are ice covered usually for 3 or 4 months each year, which amplifies the seasonality of capture fisheries.

After WWII, the number of coastal fishers started to decline at the same time when trawlers' landings increased rapidly (Zeller et al. 2011). Numbers of active coastal fishers in the Swedish Baltic Sea peaked in 1945 (6000 fishers) and after that these numbers fell by a half towards the end of the century (Piriz 2000). Between 2008 and 2018, the number of coastal vessels in Sweden decreased from 852 to 660 while in Finland there are still 1413 active fishers left (STECF 2019). The patterns in number of coastal fishers in the Baltic Sea are part of a European trend. The small-scale fisheries fleet (vessels under 12 m of length) in the European Union decreased significantly over the first decade of the twenty-first century, from around 90 thousand vessels in 2000 to just over 70 thousand in 2010 (Lloret et al. 2018). Three countries that were most affected by this decrease, i.e., Lithuania, Poland and Sweden, all operated in the Baltic Sea and showed significant decreases (over 30%) during that period.

In 2017, the coastal fishing fleet in the Baltic Sea was estimated at 5418 vessels and equivalent to 92% of the total Baltic Sea fleet and around 9% of the total fishing EU small-scale fishing fleet (Lloret et al. 2018; STECF 2019). Almost 6500 fishers are estimated to be involved in this fishery. Landings of coastal fishing in the Baltic Sea are notoriously more diverse in comparison with large-scale fisheries. Apart from herring and cod which are heavily fished also by large-scale vessels current larger economic contribution to the Baltic Sea coastal fishers are yielded by perch, eel, and pikeperch (Fig. 13.3). Instead of cod, which is unimportant for the studied northern Baltic Sea coastal fisheries, trap net and gill net fishing for European whitefish and Baltic salmon are of importance.

The relatively high diversity in the studied fishing culture is explained by the extraordinarily and unique climatic and ecological variable conditions in the Baltic Sea, which has led to correspondingly large variations in fishing technologies and adaptation strategies of the coastal fishers (Eklund 1991).

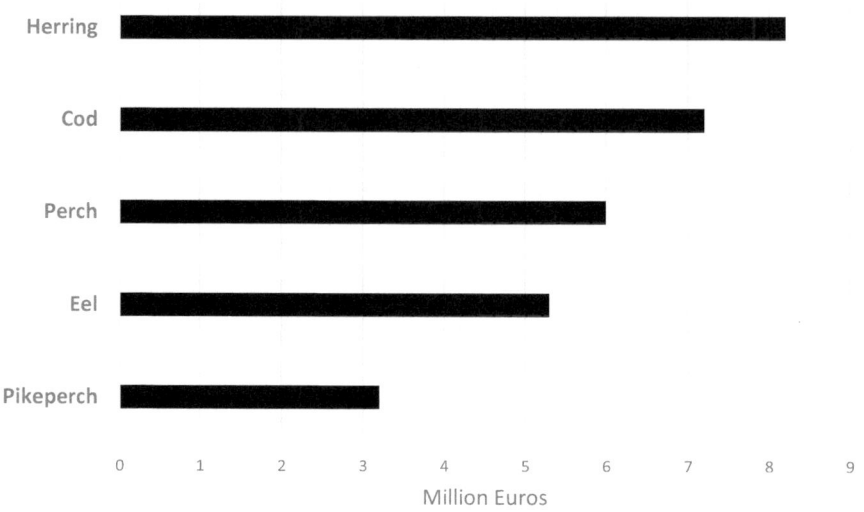

Fig. 13.3 Top 5 species landed in value by coastal fishers operating in the Baltic Sea in 2017. (Source: STEFC 2019)

Particularly in the northern Baltic Sea coast, alone the fact that fishing waters are frozen in the winter-time has led to specific demands for fishing methods. Fishing activities are seasonal also due to changes in availability of target species. Hence, coastal fishers in these areas employ a large variety of fishing gears including seine nets, gill nets, trap nets, long lines among others, and have been since centuries commonly shifting between traditional activities such as fishing, farming, forestry and shipping. Typically, the economic unit of coastal fishing is the household and fishers use small boats for accessing near-by water areas. This means that the above-mentioned statistical procedure of using vessels as the unit does not fit well with the studied northern Baltic Sea small-scale fisheries.

13.3.2   The Northern Baltic Governing System

Fisheries in the Baltic Sea have always been regulated by local rules since at least the 1500s. In these times, harbour authorities of spring-summer fishing settlements kept records of the number of fishers, controlled the landings and quality of the products, and supervised that fishers did not go fishing on weekends or before the daily morning signal, usually by the sound of a horn (Hessle 1934). These authorities were also in charge of collecting the taxes from fishing activities. Over those centuries, fisheries resources were considered inexhaustive and fishing regulations were limited to organizational matters and the collecting of taxes. Additional local fisheries rules on, for example, the division of revenues from fishing among crews, or the procedures to avoid gear collisions in fishing areas, were decided within the local fishing community.

In Sweden and Finland, the owner-based fisheries management has been important part of the governance system at the local level for centuries. Most of the coastal and inland waters are under private ownership and associated with the possession of land. In Finland this was codified in a statute in 1766 when Finland was part of Swedish territory (Eklund 1994). Later, the system became included in fisheries legislation. The landowning farmers were allocated more decision power over fishing in their water areas, and the fishing rights of the landless people – the majority of professional fishers – became weaker (Eklund 1994). Within this spatial management system, waters inside a line between the outermost islands longer than 100 meters are private according to the Swedish rule for separating private and public waters (Bruckmeier and Neuman 2005).

In the local owner-based system, fisheries management is commonly driven by a collective; a shareholders' association, which jointly represents the interests of individual owners in fishery matters. From the commercial fishers' perspective, problems in getting fishing user rights have occurred as the management rights are in the hands of other groups, like the landowners (Salmi 2012). While the local owner-based management system still exists, and water owners are according to law responsible for managing their fishing waters, new governance institutions and levels have become important players. Swedish and Finnish fisheries governance system became highly complex as a result of the coexistence of several systems: governmental fisheries management, local fisheries management and public and private ownership of coastal waters (Bruckmeier and Neuman 2005). When fisheries management became part of national and later international authority, ecological and economic concerns gained attention and little to no attention was given to the viability and sustainable development of coastal communities that depended on fisheries.

Besides local transformations, external developments in Europe and worldwide contributed to shaping the current governance of the Baltic Sea. Firstly, the threats to fish stocks from fishing became acknowledged and the "cod wars" between the UK and Iceland triggered the process which ultimately led to the UN Convention on the Law of the Sea (UNCLOS) in 1982 (see Nakamura Ch. XX in this book). As threats of overfishing and environmental degradation, in particular euthrophication, in the Baltic Sea become notorious and fisheries governance went national – and later international – the regulating and governing actors expanded accordingly. In 1980, the first Convention on the Protection of the Marine Environment of the Baltic Sea Area, known as the Helsinki Convention, was signed.

The governance of fisheries in European seas including the Baltic Sea changed substantially with the institutionalization of the European Union Common Fisheries Policy (CFP), which has served as the overarching governing framework for fisheries management in all European Union member states through its directives and regulations. Finland and Sweden became EU members in 1995. Prior to 2004 and the inclusion of Poland and the Baltic countries (Estonia, Lithuania and Latvia) into the EU, the states bordering the Baltic Sea managed fisheries issues multilaterally via *the International Baltic Sea Fisheries Commission* (IBSFC).

The Baltic Sea fisheries governance has been described lately as a complex, multi-level governance case (Burns and Stöhr 2011) where different actors from a variety of sectors interact to steer formal and informal processes to achieve collective (socially) beneficial outcomes. Over the last decades, governance in the Baltic Sea region has witnessed a process of "Europeanisation" at a significant incremental level (Gilek et al. 2016; Zeller et al. 2011) where not only fisheries but also the environmental governance system plays important roles.

In the following we describe cases about how the above-described changes in the governing structures and principles have affected the systems-to-be-governed, namely the small-scale fisheries along the northern Baltic Sea coasts.

13.4 Regional Examples of Transformations in Small-Scale Fisheries in the Northern Baltic Sea

13.4.1 Northern Satakunta Region

In the Northern Satakunta region, on the Finnish West coast, the heyday of commercial fishing was at the end of the 1910s, when a general shortage of food supplies increased the demand for Baltic herring especially during wartime (Salmi and Salmi 2009). At that time boats were progressively being equipped with motors which increased efficiency and reduced workload. However, a new fishing method, trap net fishing for Baltic herring, was adopted slowly among the professional fishers in the Northern Satakunta region. In spite of the tensions over fishing rights, contradictions between the local owners of the fishing rights and the professional fishers were not as severe in the Northern Satakunta region as it was the case in archipelago areas in South-west Finland (Salmi 2018).

After WWII the number of fishers in the region declined rapidly and traditional coastal fishing methods and operation models were reassessed and renewed in many ways (Salmi and Salmi 2010). The number of commercial fishers in the major fishing village Merikarvia decreased to less than one half in 30 years. New fishing methods were adopted; for example, stationary long line fishing for Baltic salmon in autumn months became widely appreciated as it enabled the sharing of economic risks through fishing co-operation. The episode of stationary long line fishing started a wider strategic change in the commercial fishing in the region, as the fishing moved from Baltic herring towards targeting river-spawning species: Baltic salmon and European whitefish. Moreover, targeted species were affected by the damming of rivers for the purposes of electric power production and forestry. The consequent decline was partly compensated by restocking of fish fingerlings.

Trawling for Baltic herring started in the 1950s but this capital-intensive method failed to attract the Northern Satakunta fishers, who considered the striving of economic profit as morally dubious (Salmi and Salmi 2010). At the beginning of the 1960s driftnet fishing for European whitefish was invented along with the introduction of synthetic fibres in fishing nets (Salmi 2011). Fishers were paid good prices for the large-sized European whitefish. Therefore, driftnet fisheries for European whitefish soon became an important element in the fishing culture of the Northern Satakunta coast.

Unlike today, opportunities for fishing livelihoods before the 1960s were less often directly affected by the state – whether supporting or restricting the livelihood. The major legislative issue was connected to the definition of water ownership and access rights for the fishers. Thus, the local community, alongside fishers' life modes and identities, was of major importance in responding and adapting to transformations in the society. During the last 50 years however, the role and effects of the national and international levels of governance, and especially measures for biodiversity protection, have substantially increased.

In the 1980s the Baltic salmon stocks increased in abundance partly due to restocking programs. In addition to using long lines to fish Baltic salmon, fishers started to use trap nets. No local license had previously been needed as salmon fishing was operated outside the privately owned water areas. With time, state-initiated regulations for Baltic salmon fisheries have become stricter and increased in complexity over the last 30 years. In the 1980s the salmon fishery was regulated by closed seasons and by restricting the number of fishing gears (Salmi and Salmi 1998). As a consequence of new top-down fishing policies, which were found unreasonable by fishers, a national fishery organization, the Finnish Fishermen's Association, was founded to defend their rights.

Much later, drift net fishing was banned by the EU in 2008 legitimized by the protection of the Baltic Sea harbour porpoise. This decision terminated not only open sea netting for Baltic salmon, but also the drift net fishing for Baltic herring, adopted already in 1861, and later for European whitefish, and to some extent also trout and salmon (Saiha 2009). Small-scale fishers considered the ban to be an example of centralized decision making based on insufficient knowledge, as, for instance, porpoises seldom entered Finnish coastal areas (Salmi and Mellanoura 2019).

The current situation was examined in an annual national survey that monitors changes in commercial fishing. Fishers in the municipality of Merikarvia stated that the seal-induced losses had increased in 2017 (Setälä et al. 2018) and that the effects of the seals, accompanied with the cormorants, pose the greatest threat to their fishing livelihood. In addition, the local herring and salmon fishers are affected by the launching of the transferrable quota system, which is often considered to increase bureaucracy rather than to enhance independent decision making in fishing operation. The consequences of the quota system and the seal and cormorant-induced conflicts will be in more detail studied in Sect. 13.5 below.

13.4.2 Archipelago Sea Region

The Archipelago Sea, South-west Finland, consists of unique coastal landscape with shallow water areas between thousands of islands, skerries and fragmented shorelines (Salmi 2018). Extraction of fish resources has been of utmost importance. Baltic herring was landed for centuries cooperatively during the ice cover period by winter seining methods. Tensions between the professional fishers and the land-owning peasants became evident in the 1920s, causing more severe and long-lasting contradictions within the local communities than those in the Northern Satakunta region. In the inner waters of the archipelago local peasants had started, in addition to seine nets, capturing Baltic herring by large trap nets. The breakthrough of trap nets for herring occurred after WWI. With this method, the landowners captured large quantities of herring during the spawning season in spring, which created conflicts with the full-time gill net fishers in the outer archipelago (Eklund 1991, 1994).

In 1934 the number of commercial fishers in the southern parts of the Archipelago Sea was 3447, of them 42% being full time fishers (Salmi 2018). The golden era of the archipelago fisheries lasted until the end of WWII. After the war, fisheries started to lose their position as co-providers of welfare and nutrition, and consequently the number of commercial fishers declined. Already in the 1930s the demand for salted herring started to decline as a consequence of the reduced purchasing power of the consumers after the depression and import restrictions and new preservation methods for fish products (Jónsson 2009). For centuries, live pikes, perches and other species were transported from the archipelago especially to Stockholm and Helsinki, where fish demand had grown (Eklund 1991). This wellboat shuttle continued until the 1950s when Sweden forbid the import of live fish (Soldéus 2013).

In some islands the traditional fishing peasant livelihood still continued; root crops were produced on small fields and the Baltic herring was smoked in order to gain added value. Cattle were also a part of this self-sufficient economy. Processing, transportation and marketing chains in the archipelago were underdeveloped, as the Finnish state prioritized the development of agriculture. While agriculture was generously subsidized, the archipelago and coastal fishing was being supported only in the late 1960s, once the former fishing villages were transformed to sites for agriculture, recreation and services (Eklund 1991). The adoption of industrial trawling since the 1950s moved part of the fishing activities away from the archipelago and contributed to its depopulation (Jaatinen 1961).

Still in the 1970s nearly half of the archipelago inhabitants were employed in primary production, like agriculture and fisheries (Andersson and Eklund 1999). Thereafter the importance of primary production has decreased as recreational activities, shipping and nature protection became more significant. A new fisheries innovation, rearing of rainbow trout in net cages in the sea, was introduced and partially transformed the 'capture economy' into 'fish culture'. Moreover, new post-productivistic interests, e.g., summer cottage dwellings and nature conservation, captured large archipelago and coastal water areas without prior discussions with the fisheries sector. Also the number of recreational fishers multiplied. Regarding several fish species, the annual landings by recreational fishers exceed those by the commercial fishers. Furthermore, recreational rod fishers have successfully gained fishing rights and easy access to fishing waters (Salmi 2012). Especially the launching of the province-wide lure fishing fee system in 1997 raised bitterness among water owners as the new state-organized license system narrowed local opportunities for managing local recreational fisheries – the local owners could not anymore control rod fishing in their waters. Environmental conservation has intensified in the study area since the 1970s along with initiations for establishing national and international protected areas. Commercial fisheries find the expanded spatial occupation by nature conservation interests problematic (Salmi 2018) due to the restrictions of access and overlapping protective zones (Svels 2017).

Parallel to diminishing availability and demand for the Baltic herring, the archipelago fishers have relied on other fish species like perch, pikeperch and European whitefish. Lately pikeperch stocks built up and occupied wider areas in the archipelago, and compensated for the reduced herring landings. Yet, after the turn of the millennium increased seal and cormorant populations have created the greatest threat to the viability of commercial fishing (Salmi 2015). The effects of the grey seals have also changed the fishing locations; instead of more open archipelago waters the best pikeperch catches are currently yielded in sheltered and shallow coastal fishing grounds (Saarinen 2013). Archipelago fishers have traditionally relied on local support whereas the state and authorities are more often perceived as restricting fishers' operations and independence (Salmi 2005). The ownership of waters has partly shifted to the summer cottage dwellers, not inevitably willing to grant licenses for commercial fishing. This challenges fishers who have tried to move their pikeperch fisheries towards more shallow waters. Moreover, the minimum size of pikeperch in the Archipelago Sea commercial fishing was, by law, raised to 40 cm in 2019. Fishers objected the change doubting the idea that the pikeperch stocks were endangered (Sonck-Rautio 2019).

13.4.3 From Östergötland to Uppland Archipelago

Fishing, farming and animal husbandry have been important livelihoods in the history of several archipelago areas in Östergötland, Sweden (Norr et al. 2018). The land and water are connected in use, ownership and livelihoods alike, as in the above case of the Finnish Archipelago Sea. Fishing is one element of archipelago peoples' identity and one part of the pluriactivity providing income to the permanent population. In Hållnäs on the Uppland coast, fishing culture and its specific life mode has been important for local identity as well as supporting local tourism using fishers' knowledge and stories in creating tourism products (Rådberg et al. 2018). In spite of traditions and the importance of fishers' livelihood recruitment of new fishers is challenging.

In the southern part of the Östergötland coast, in Valdemarsvik, Baltic herring and eel capturing has dominated. In the past, Baltic herring was targeted in winter and spring by seine nets near the bays and by gill nets in the outer archipelago. Later trawling was adopted and consequently the Baltic herring landings substantially increased. Eel was fished in late summer and in autumn with trap nets. Today, there are only a few commercial fishers in Valdemarsvik, while in S:t Anna's archipelago, another part of Östergötland coast, there is only one fisher left (Norr et al. 2018).

On the Swedish Baltic Sea coast recreational rod fishing has been free since 1985. The lack of charges and regulation has been criticized as it is considered to hamper management of local fish resources (Norr et al. 2018). In the Östhammar-Singö archipelago on the Uppland coast, narrowed economic opportunities for the local management organization irritated water owners, as they found unjust that one group could operate without fees and others needed pay for the funding of the fishery management. This situation is similar to local perceptions regarding the widened rights of rod fishing in the Finnish Archipelago Sea (Salmi 2002). People on the Swedish Baltic Sea coast have found that 'free rod fishing' is a consequence of general change in fisheries policies towards downplaying the local level management.

According to the Swedish archipelago fishers, fisheries policy has contributed to supporting large-scale fisheries at the expense of small-scale fisheries (Norr et al. 2018). Fishers claim they have been neglected by the society while large-scale trawling has been supported. Small-scale fishers state difficulties in following new requirements which necessitate investing, e.g., new fish processing facilities, and bureaucracy has increased. Similarly, water owners of the Östhammar-Singö archipelago were convinced that the large-scale trawlers caused damaging effects on the environment and fish stocks (Salmi 2002). Fishers in the Hållnäs region state that small-scale fishing is a sustainable way to utilize the sea resources and that the livelihood supports development of local communities (Rådberg et al. 2018).

Fishers in the Östergötland archipelago state that the price and demand for local fish is good (Norr et al. 2018). Thus, opportunities for small-scale fishing could be seen as promising, nevertheless the core problem is that the landings by the remaining small number of fishers are too small to revitalize a fishing industry. This is due to the seals and cormorants that reduce catches and discourage fishers participation. In Hållnäs, fishers process their catch for direct sale to tourists during the summer, claiming they could sell more provided they had better opportunities to decide what and where to fish as it was in the past for small-scale fishing (Rådberg et al. 2018).

The contemporary seal and cormorant-induced problems were highlighted both in the Östergötland archipelago and in Hållnäs (Norr et al. 2018; Rådberg et al. 2018). During interviews with local fishers made in 2018 it became clear that especially the seal population had exploded within 2 years. Seals are allowed to be hunted, but the hunter must be accompanied by a registered commercial fisher (Norr et al. 2018). Seal hunting in the Swedish archipelago, like in Finland, is hampered also by the EU ban of trading seal products (European Union 2009a). Fishers are displeased with the compensation payments because they are bound with turnover instead of the amount of losses. Likewise, cormorants eat a lot of fish. In Hållnäs, informants said that although protective hunting of cormorant is allowed, the amounts are too small to make a difference (Rådberg et al. 2018). Seals, cormorants and authorities are perceived as prime factors of Hållnäs fishers' problems as they feel powerless and not recognized in the decision-making process. In this area the number of commercial fishers has decreased between 1930 and 2018 from 250 to only two fishers left (Rådberg et al. 2018).

13.5 Contemporary Governance Challenges in the Northern Coasts

The studied examples of coastal small-scale fisheries in the northern Baltic Sea have shown capacity to adapt to a variety of changes. Despite this potential, recent societal transformations driven by post-productivist governance images, instruments and structures have changed, and become increasingly pivotal for the existence of this fishery. Alongside these developments, this section focuses on two core issues that challenge the northern Baltic Sea small-scale fisheries: market-based governance and natural predators-related conflicts.

13.5.1 Quota Management as a Market-Based Governance Instrument

The quota management (individual transferable quota, ITQ) was introduced as an instrument for fisheries governance widely during the twentieth century (Hultman et al. 2018). A basic idea is that individual 'ownership' of fishing quotas, combined with a free market for the selling and buying of these quotas, will foster a more economically efficient and sustainable fishing sector. These quota systems arrived in the northern Baltic Sea more recently. In Sweden, individual and transferable quotas were introduced in 2009 in the large-scale pelagic (herring and alike species) fisheries and later individual quotas with annual leasing were set up in the demersal (cod and alike species) fisheries (Hultman et al. 2018). The pelagic ITQ system was introduced for a period of 10 years (Stage et al. 2016) and was renewed for the same period in 2019, while for the demersal fisheries the individual quota system was a provisional regulation which ended in 2018 and was not renewed in 2019 (Arias Schreiber et al. 2019). Finland introduced individual quotas (more specifically: transferable fishing concessions, TFCs) in the Baltic herring and Baltic salmon fisheries in 2017.

According to Hultman et al. (2018) quota markets have been contested and opposed by large groups of fishers, and the implications have been subject to public debate in Nordic countries. This introduction of market mechanisms in the distribution of fishing quotas and rights has changed the previous economic and social base, and hence the very basic conditions for the coastal fishing sectoral development.

It is evident that this policy change has favoured volume-based large-scale fisheries (Hultman et al. 2018) in comparison to small-scale fisheries. The impacts of the ITQ system for the coastal fisheries in Sweden have not been systematically evaluated and are a matter of debate.

In contrast to the Swedish quota system, the Finnish system also covers small-scale coastal fishing. According to a survey made in 2018, many Finnish fishers were sceptical of the benefits of the newly launched system, especially for the coastal small-scale Baltic herring and Baltic salmon fisheries (Salmi et al. 2019). The system was considered complicated, bureaucratic and unsuitable for the seasonal trap net fishing for Baltic herring and Baltic salmon. These views are similar with those among the Northern Satakunta fishers (Sect. 13.4). Baltic salmon fishers, however, found a positive aspect of the quota system, as it enabled partial relieving of temporal regulation in Baltic salmon fisheries (Salmi et al. 2019).

13.5.2 Natural Predators-Related Conflicts; Governing Seals, Cormorants and Fisheries

Environmental and predators-related themes have emerged along with the post-productivist transition not only in the public discourse but also in governance institutions and practices of coastal resource utilization. The increased seal and cormorant populations ('natural predators') form the greatest contemporary conflict issue for commercial fisheries in most parts of the Baltic Sea coast and challenge the future of small-scale fisheries also in the studied Finnish and Swedish coastal locations (Sect. 13.4).

Seals' and cormorants' impacts on coastal fishing livelihood can be divided in six types: damage to fishing gear, damages in caught fish, reduction of catches by taking fish from the gear and fishing grounds, reduction of catches due to changes in fish stocks and behavior, an increase of the work load, and an increase of the operation costs of fishing (Svels et al. 2019). The seal-induced impacts are perceived as considerable or serious in nearly all the studied Baltic Sea coastal areas and the cormorant-induced impacts are usually considered less serious than those of seals (Arias Schreiber and Gillette 2021).

The multi-level and sectored governance systems have been unable to solve the conflicts, although technological measures have provided partial mitigation of the seal-induced problems (Salmi 2009; Svels et al. 2019). In general, perspectives on seal politics are steeply divided between fisheries and hunting organizations on one hand and nature conservationists and environmental administrators on the other. The former groups want to restrict the seal population and the latter like to restrict hunting and enhance conservation. Similar tensions are present in the cormorant conflict, where environmental perspectives hold more power, largely due to the cormorants' status as non-hunted species under the EU Bird Directive (European Union 2009b).

Grey seals can be hunted with certain limitations in Finland and Sweden. The motivation for hunting is, however, often low due to the EU ban for the trade of seal products (European Union 2009a). In the cormorant case derogation permits can be locally applied for harassment, shooting or oiling eggs. Fishers call for wider opportunities for hunting seals and cormorants as the primary choice of governance instrument (Svels et al. 2019. Active hunting near fishing gear may improve catches for a couple of days, nevertheless fishers stress that over time it will affect the behavior of the animals, becoming yet again more fearful towards humans. The fishers interviewed by the TNC project (Svels et al. 2019) recommend that killed animals should be considered a resource; thus, a repeal of the EU ban for trading seal products is considered to be central for this option. The opinion is divided regarding the role of monetary compensations as a governance tool for mitigating the animal-human contradictions. Development of fishing technology has been considered as a primary instrument in the mitigation of the seal-induced problems, as gear development aims to enable commercial fishers to continue their livelihood without challenging the protection of seal populations (Svels et al. 2019).

National and local management plans have been made for seals and cormorants in Sweden and Finland. The process of making the plans may support learning and finding common ground among stakeholders and interest groups. However, the final plans are typically science-based and lack recommendations for improving the institutional arrangements and efficient governance tools for conflict mitigation (Petersson et al. 2012). Discussion on inclusion of local knowledge in the conflict mitigation processes has been controversial (Bruckmeier and Höj Larsen 2008).

13.6 Conclusions: Societal Transformations and Governance

Changes and challenges of the Baltic Sea natural system including its fishing stocks has been extensively researched while social changes in the fisheries system, and in particular related to coastal small-scale fisheries, have been seldom addressed (Arias Schreiber et al. 2019). The lack of research persists despite significant transformations that have been witnessed in the governing system and the way it affects the small-scale fisheries sector. This chapter has focused on the changing govern systems and their compatibility with small-scale fisheries operated on the northern coasts and archipelagos of the Baltic Sea. The presented case studies show that a minor number of northern Baltic Sea coastal fishers have been able to adapt and continue their livelihood in current difficult circumstances. Adaptation to the present challenges, especially the quota systems and the natural predators-related conflicts, is, more than ever, in the hands of the governors, their images of the small-scale fisheries, and the functionality of the governance structures.

Societal transformation towards post-productivism, globalized economics of scale and environmentalisation with a multilevel hierarchic science-based governance, have shaped both the system-to-be-governed, namely the studied small-scale fishers and their communities, and the fisheries governance systems. Under a post-productivism context, coastal areas provide individual experiences, touristic attractions and places for nature conservation where commercial fishing is not valued. Globalized economics of scale allow for large amounts of imports of cheap seafood that tends to limit marketing opportunities of the coastal fishers. Moreover, since ecological concerns are prioritized in the current governance system, science-based hierarchical governance has turned small-scale fisheries voices and knowledge irrelevant for fisheries management. The reduced number of small-scale fishers are not well organized and their participation and values are rarely considered in the decision-making process (Arias Schreiber et al. 2019). The governance challenges are substantial as the system-to-be-governed has become more diverse, complex and dynamic along with arrival of new interests and values, such as those of the environmental movement and recreational users of the area. Many fishers have coped with the new situation by adopting new strategies that make use of the increased demand, for instance, for environmentally friendly niche products (Hultman et al. 2018). However, the narrowing scope of action, bureaucracy and consequent reduced independence poses a challenge for recruiting new generations of coastal fisheries.

Figure 13.4 connects our theoretical framework of interactive governance with the elements and systems relevant for the studied Baltic Sea small-scale fisheries. Main changes in the governing mode have been described as a shift from local owner-based governance to a hierarchic science-based fisheries and environmental governance. In the past the local community was of major importance in responding and adapting to transformations in society. During the last 50 years the role of the

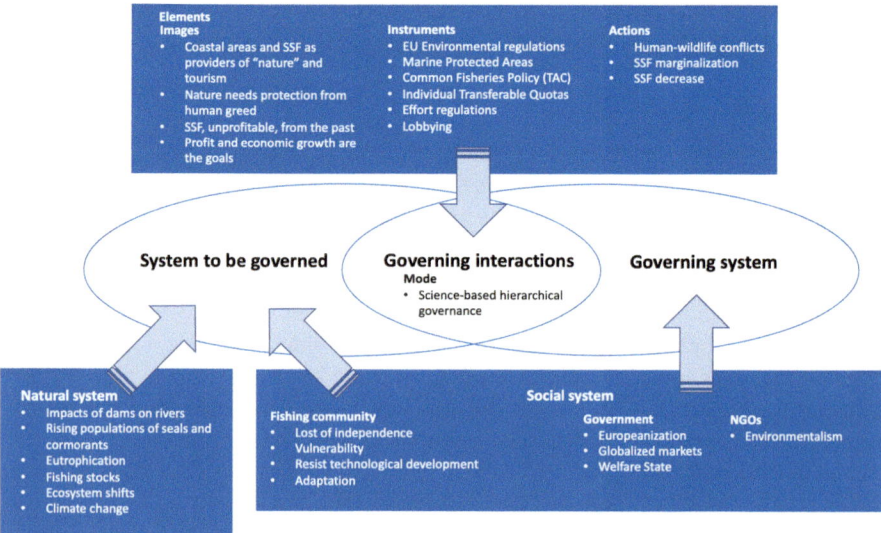

Fig. 13.4 Interactive governance framework adjusted for the analysis of the societal transformation and governance challenges of the small-scale fisheries in the Northern Baltic Sea. (Modified on the basis of Jentoft and Chuenpagdee 2009 and Kooiman et al. 2005)

national and international levels of governance has become central. At the same time the societal transformations have brought new governance principles, structures and practices that have increasingly affected coastal small-scale fisheries. As our analysis shows, neither the local level governance has been free of contradictions. Tensions regarding fishing opportunities of professional fishers vis-á-vis land owners, who possess more power in local fisheries management, have provoked obstinate conflicts in the studied archipelago fisheries until today.

Northern coastal fishers perceive that the hierarchic science-based systems enhance large-scale fisheries, and do not take the coastal fishers' interests, values and cultures properly into account. These perceptions are visible e.g., in the presented examples of the drift net ban, the natural predators-fisheries conflicts and the transferable quota system. The quota systems form a governance instrument that fits well with fisheries based on mobility, large investments, and economies of scale to achieve profitability. This is in contrast to small-scale fishers who lack financial capital, and normally avoid large investments (Arias Schreiber et al. 2019). In Finland small-scale fishers considered that the consequent distributional outcomes, were not properly taken into account when launching the quota system.

It is not far-fetched to conclude that the fisheries and environmental governance systems are incompatible with the system-to-be-governed, small-scale fisheries in the northern parts of the Baltic Sea. How can a better balance be achieved? There is an obvious need for improvements both regarding structural modes and intentional elements. The emergence of hierarchic science-based governance structure is largely behind the problems, but there is no turning back to self-governance if that ever existed. However, a narrowly science-based governance approach does not guarantee a fitting to small-scale fisheries values and interests. The third and most promising alternative is co-governance where small-scale fishers' interests, values and local knowledge are integrated into a governance system that considers economic, ecological and also social sustainability concerns. At present the EU-funded Fisheries Local Action Groups provide an opportunity for small-scale fishers also in the northern coasts of the Baltic Sea to collaborate widely and get support for applying locally relevant survival strategies (Salmi and Svels 2022). New networks and hands-on collaboration with the fishers may raise the awareness of small-scale fishers' challenges and every-day life, which is needed for improving compatibility in the governance interactions.

Governance collaboration, creating better balance and wider awareness of small-scale fisheries, could thus enhance the overall sustainability of the Baltic Sea coastal regions. Similarly, the science-based policy setting needs to rely on more disciplines than just the ones that provide knowledge on ecological and economic aspects within the complex fisheries system. Transdisciplinary research that includes the co-production of knowledge with resource users offers an alternative in this regard. As this chapter shows, interactive governance theory is well fitted for analyzing the modes and elements in the complex governing systems that affect opportunities for small-scale fisheries in the constantly changing system-to-be-governance.

References

Acselrad H (2010) The "Environmentalization" of Social Struggles: the Environmental Justice Movement in Brazil. Trans. by Jeffrey Hoff, Estudos Avançados 24(68):103–119. http://www.scielo.br/pdf/ea/v24n68/en_10.pdf. Accessed 15 Dec 2019

Andersson K, Eklund E (1999) Tradition and innovation in coastal Finland: the transformation of the Archipelago Sea region. Sociol Rural 39(3):377–393

Arias Schreiber M, Gillette M (2021) Neither fish nor fowl: navigating motivations for fisheries participation and exit in Sweden. Soc Nat Resour 34(8):1019–1037. https://doi.org/10.1080/08941920.2021.1925381

Arias Schreiber M, Linke S, Delaney AE, Jentoft S (2019) Governing the governance: small-scale fisheries in Europe with focus on the Baltic Sea. In: Chuenpagdee R, Jentoft S (eds) Small-scale fisheries governance: transdisciplinary analysis and practices. Springer, Dordrecht, pp 357–376

Baltic Sea (2019) New World Encyclopedia. Retrieved 12:11, December 1, 2019 from http://www.newworldencyclopedia.org/p/index.php?title=Baltic_Sea&oldid=1022236

Bonsdorff E, Andersson A, Elmgren R (2015) Baltic Sea ecosystem-based management under climate change: integrating social and ecological perspectives. Ambio 44(3):333–334

Bruckmeier K, Höj Larsen C (2008) Swedish coastal fisheries – from conflict mitigation to participatory management. Mar Policy 32:201–211

Bruckmeier K, Neuman E (2005) Local fisheries management at the Swedish Coast: biological and social preconditions. Ambio 34(2):91–100

Burns TR, Stöhr C (2011) Power, knowledge, and conflict in the shaping of commons governance: the case of EU Baltic fisheries. Int J Commons 5:233–258

Eklund E (1991) Kustfiskets historia i Finland. Skärgård 2:5–15

Eklund E (1994) 'Kustfiskare och kustfiske i Finland under den industriella epoken. Studier i en yrkesgrupps yttre villkor, sociala skiktning och organisation 1860–1970', SSKH Skrifter 5, Svenska social- och kommunalhögskolan vid Helsingfors universitet. Forskningsinstitutet, Helsinki

European Union (2009a) Regulation (EC) No 1007/2009 of the European Parliament and of the Council of 16 September 2009 on trade in seal products. https://op.europa.eu/en/publication-detail/-/publication/a6064cc0-48ee-4698-b380-106c6ea30e89/language-en

European Union (2009b) Directive 2009/147/EC of the European Parliament and of the Council of 30 November 2009 on the conservation of wild birds (amended from Directive 79/409/EEC in April 1979). https://eur-lex.europa.eu/legal-content/EN/TXT/?uri=CELEX:32009L0147

Gilek M, Karlsson M, Linke S, Smolarz K (2016) Environmental governance of the Baltic Sea: identifying key challenges, research topics and analytical approaches. In: Gilek M, Karlsson M, Linke S, Smolarz K (eds) Environmental governance of the Baltic Sea. Springer International Publishing, Cham, pp 1–17

HELCOM (2021) Ecosystem Health of the Baltic Sea 2003–2007. HELCOM Initial Holistic Assessment. Baltic Sea Environment Proceedings No. 122. Helsinki Commission. https://www.helcom.fi/wp-content/uploads/2019/08/BSEP122-1.pdf

Hessle C (1934) Die Schwedish Ostseefisherei. Handbuch der Fisherei Nordeuropas 8(3b):31

Hultman J, Säwe F, Salmi P, Manniche JB, Holland E, Høst J (2018) Nordic fisheries at a crossroad. TemaNord 2018(546). http://norden.diva-portal.org/smash/get/diva2:1253218/FULLTEXT01.pdf

Jaatinen S (ed) (1961) Text till Atlas över skärgårds-Finland. Helsinki, Nordenskiöld-samfundet i Finland

Jentoft S (2000) The community: a missing link of fisheries management. Mar Policy 24(1):53–59

Jentoft S, Chuenpagdee R (2009) Fisheries and coastal governance as a wicked problem. Mar Policy 33:553–560

Jentoft S, Chuenpagdee R (eds) (2015) Interactive governance for small-scale fisheries. Global reflections, MARE publication series no. 13. Springer

Jentoft S, Chuenpagdee R (2019) The quest for Transdisciplinarity in small-scale fisheries governance. In: Chuenpagdee R, Jentoft S (eds) Transdisciplinarity for small-scale fisheries governance, MARE publication series, vol 21. Springer, Cham. https://doi.org/10.1007/978-3-319-94938-3_1

Jentoft S, Chuenpagdee R, Bundy A, Mahon R (2010) Pyramids and roses: alternative images for the governance of fisheries systems. Mar Policy 34:1315–1321

Jónsson G (2009) Fishing nations in crisis: the response of the Icelandic and Norwegian fisheries to the great depression. Int J Marit Hist 21(1):127–151. https://doi.org/10.1177/084387140902100107

Kooiman J (2003) Governing as governance. SAGE Publications, London, p 249

Kooiman J, Bavinck M (2005) 'The governance perspective'. In: Kooiman J, Bavinck M, Jentoft S, Pullin R (eds) Fish for life. Interactive governance for fisheries, MARE publication series no 3. Amsterdam University Press, Amsterdam, pp 11–24

Kooiman J, Chuenpagdee R (2005) Governance and governability. In: Kooiman J, Bavinck M, Jentoft S, Pullin R (eds) Fish for life. Interactive governance for fisheries, MARE publication series no. 3. Amsterdam University Press, Amsterdam, pp 325–349

Kooiman J, Bavinck M, Jentoft S, Pullin R (eds) (2005) Fish for life. Interactive governance for fisheries, MARE publication series no. 3. Amsterdam University Press, Amsterdam, p 427

Lajus J, Kraikovski A, Lajus D (2013) Coastal fisheries in the eastern Baltic Sea (gulf of Finland) and its basin from the 15 to the early 20th centuries. PLoS One 8(10):e77059. https://doi.org/10.1371/journal.pone.0077059

Linnér B-O, Wibeck V (2020) Conceptualising variations in societal transformations towards sustainability. Environ Sci Pol 106:221–227. https://doi.org/10.1016/j.envsci.2020.01.007

Lloret J, Cowx IG, Cabral H, Castro M, Font T, Gonçalves JMS, Gordoa A, Hoefnagel E, Matić-Skoko S, Mikkelsen E, Morales-Nin B, Moutopoulos DK, Muñoz M, Dos Santos MN, Pintassilgo P, Pita C, Stergiou KI, Ünal V, Veiga P, Erzini K (2018) Small-scale coastal fisheries in European Seas are not what they were: ecological, social and economic changes. Mar Policy 98:176–186. https://doi.org/10.1016/j.marpol.2016.11.007

Marsden T (2004) The quest for ecological modernisation: re-spacing rural development and Agri-food studies. Sociol Rural 44:129–146

Merkel W Kollmorgen R, Wagener H (2019) Transformation and Transition Research: An Introduction. In The Handbook of Political, Social, and Economic Transformation. Oxford: Oxford University Press, 2019. Oxford Scholarship Online. p. 2019. https://doi.org/10.1093/oso/9780198829911.003.0001

Norr H, Olsson E, Lovén E, Rohlin L (2018) Fiskeförvaltning och mervärden I S:t Annas, Gryts och Tjusts skärgård In: Sandström, E. (ed.) Det småskaliga kustfiskets förändrade förutsättningar och mervärden. Urban Rural Rep 2018(1):21–30

Pääkkönen M, Bläuer A, Evershed RP, Asplund H (2016) Reconstructing food procurement and processing in early comb ware period through organic residues in early comb and Jäkärlä ware pottery. Fennosc Archaeol XXXIII:57–75

Partelow S, Schlüter A, Armitage D, Bavinck M, Carlisle K, Gruby R, Hornidge A-K, Le Tissier M, Pittman J, Song AM, Sousa LP, Văidianu N, Van Assche K (2020) Environmental governance theories: a review and application to coastal systems. Ecol Soc 25:19. https://doi.org/10.5751/ES-12067-250419

Petersson E, Salmi P, Parz-Gollner R (2012) The incorporation of scientific contributions and other stakeholders' views into management plans: an analysis for 'conflict' species. In: Marzano M, Carss DN (eds) Essential social, cultural and legal perspectives on cormorant-fisheries conflicts. Intercafe-project, COST, pp 58–84

Piriz L (2000) Dependence, modernisation and the coastal fisheries in Sweden. In: Symes D (ed) Fisheries Dependent Regions. Fishing News Books, Blackwell Science, pp 118–131

Rådberg B, Karlsson L, Lindberg M (2018) Det kustnära fisket i Hållnäs – utmaningar och mervärden In Sandström E (ed) Det småskaliga kustfiskets förändrade förutsättningar och mervärden. Urban Rural Rep 2018(1):31–40

Rannikko P (2008) Postproduktivismi metsässä. In: Karjalainen T, Luoma P, Reinikainen K (eds) Ympäristösosiologian virrat ja verkostot. Juhlakirja professori Timo Järvikosken 60-vuotispäivänä. Oulun yliopisto, Oulu, pp 83–95

Rannikko P, Salmi P (2018) Towards neo-Productivism? – finnish paths in the use of forest and sea. Sociol Rural 58(3):625–643

Saarinen M (2013) Saaristomeren elinkeinokalatalouden nykytila. Strategiatyöryhmä 31.3.2013. Report, p. 32. http://www.sameboat.fi/wp-content/uploads/2016/06/Saaristomeren-elinkeinokalatalouden-nykytila-2013.pdf. Accessed 28 Feb 17

Saiha M (2009) Itämeren rannalla. Valokuvaajan meriselityksiä. Porvoo, WS Bookwell

Salmi P (2002) Local fishery management and private property of coastal waters – case study Östhammar-Singö archipelago, Swedish east coast, Human Ecology Research Series, SUCOZOMA report 2002:7

Salmi P (2005) Rural Pluriactivity as a coping strategy in small-scale fisheries. Sociol Rural 45:22–36

Salmi P (2009) Rural resource use and environmentalisation: governance challenges in Finnish coastal fisheries, Maaseudun uusi aika. Finnish J Rural Res Policy Special Issue 2/2009(17):47–59

Salmi J (2011) Merikalastuksen vaiheet – jokivirran kuohuista kansainvälistyviin aaltoihin' In Hiedanpää J, Jussila I, Lehto S, Louekari S, Ruohonen J, Salmi J, Salonen T, Savola A, Ylikoski K (eds) Kokemäenjoen aalloilla ja rannoilla. Satakuntaliitto 2011, Sarja A(304), pp. 100–125

Salmi P (2012) The social in change: property rights contradictions in Finland. Maritime Studies 11(2) http://www.maritimestudiesjournal.com/content/11/1/2

Salmi P (2015) Constraints and opportunities for small-scale fishing livelihoods in a post-productivist coastal setting. Sociol Rural 55(3):258–274

Salmi P (2018) Post-productivist transformation as a challenge for small-scale fisheries: changing preconditions and adaptation strategies in the Finnish Archipelago Sea region. Reg Stud Mar Sci 21:67–73

Salmi P, Mellanoura J (2019) Finnish small-scale fisheries: marginalisation or revival? In: Pascual-Fernandez J, Pita C, Bavinck M (eds) Small-scale fisheries in Europe: status, resilience and governance. Springer, MARE Publication Series

Salmi J, Salmi P (1998) Livelihood and way of life: Finnish commercial fisheries in the Baltic Sea. In: Symes D (ed) Northern waters: management issues and practice. Fishing News Books, Blackwell Science, pp 175–183

Salmi J, Salmi P (2009) Ammattikalastajuuden synty: yhteiskunnallinen murros ja kalastajan identiteetti Pohjois-Satakunnan rannikolla, Riista- ja kalatalous. Tutkimuksia nro 7:35

Salmi J, Salmi P (2010) 'Ammattikalastuksen selviämiskamppailu – Elinkeinon kriisi ja yhteisön tuki', Riista- ja kalatalous, Selvityksiä nro 23, 22pp

Salmi P, Svels K (2022) Marginalization and reinvention of small-scale fisheries – the Finnish case study of social justice. In: Jentoft S, Chuenpagdee R, Said A, Isacs M (eds) Blue justice: small-scale fisheries in a sustainable ocean economy, Mare publication series. Springer

Salmi P, Mellanoura J, Niukko J, Saarni K, Setälä J, Virtanen J (2019) Kalastuksen toimijakohtaisen kiintiöjärjestelmän käyttöönoton vaikutusten arviointi, Natural resources and bioeconomy studies. Natural Resources Institute Finland, Helsinki. Submitted

Setälä J, Harjunpää H, Jaukkuri M, Lehtonen E, Mellanoura J, Niukko J, Keskinen T, Salmi P, Saarni K (2018) Kalastuksen olosuhdekatsaus 2017, Natural Resources Institute (Luke). https://merijakalatalous.fi/wp-content/uploads/Ammattikalastuksen-olosuhdekatsaus-2017.pdf

Soldéus LG (2013) Fiskköpare, sumpskeppare och deras seglande sumpar. Handeln med levande fisk i Stockholm och skärgårdarna, 1st edn. Malmö, SoldEko HB

Sonck-Rautio K (2019) The endangered coastal fishers in the coast of the Archipelago Sea – the environmental conflict in policy-making. Ethnologia Fennica 46:5–35

Stage J, Christiernsson A, Söderholm., P. (2016) The economics of the Swedish individual transferable quota system: experiences and policy implications. Mar Policy 66:15–20

STECF (2019) The 2019 annual economic report on the EU fishing fleet (stecf-17–12), Scientific, Technical and Economic Committee for Fisheries Luxembourg

Svels K (2017) World Heritage management and tourism development: A study of public involvement and contested ambitions in the World Heritage Kvarken Archipelago. Doctoral Thesis. Åbo Akademi University, Vaasa

Svels K, Salmi P, Mellanoura J, Niukko J (2019) 'The impacts of seals and cormorants experienced by Baltic Sea commercial fishers', natural resources and bioeconomy studies 77/2019. Natural Resources Institute Finland, Helsinki, p 50

Waldo S, Loven I (2019) Värden i svenskt yrkesfiske, Agrifoods Economics Centre. RAPPORT 2019:1

WWF (2021) WWF:n kalaopas. World wildlife fund. https://wwf.fi/kalaopas/. Accessed 8 Oct 2021

Zeller D, Rossing P, Harper S, Persson L, Booth S, Pauly D (2011) The Baltic Sea: estimates of total fisheries removals 1950–2007. Fish Res 108(2):356–363

Chapter 14
The Plastic Bag Habit and the Ocean Bali: From Banana Leaf Wrappings to Reusable Bags

Roger Spranz and Achim Schlüter

Abstract The pollution of the oceans by plastic waste is a growing threat to marine life, ecosystems, livelihoods of coastal communities and the health of human beings in general. Indonesia is the world's second largest source of marine plastic pollution. As an island state, plastic litter is regularly flushed into the sea. In this study we explore some behavioural and cultural reasons for the high consumption and pollution by plastic bags on Bali and locally adapted solutions. The data was collected from interviews and surveys with shop owners and customers, religious leaders, students, lecturers and activists during 3 years of research in the region. The analysis is structured in three parts: first, understanding the relevant concepts that inform Balinese perception of the natural environment; second, analysing the popularity and aversions among local Balinese in regard to plastic bags; third, investigating a local initiative working towards a ban of plastic bags. Based on these three parts we identified promising approaches that can effectively support local initiatives and awareness campaigns.

Keywords Environmental anthropology · Indonesia · Plastic bags · Marine pollution · Bali

R. Spranz
Making Ocean Plastic Free e.V., Bali, Indonesia

Making Ocean Plastic Free e.V., Freiburg, Germany

A. Schlüter (✉)
Leibniz Centre for Tropical Marine Research, Bremen, Germany

Jacobs University, Bremen, Germany
e-mail: achim.schlueter@leibniz-zmt.de

© German Institute of Development and Sustainability (IDOS) 2023
S. Partelow et al. (eds.), *Ocean Governance*, MARE Publication Series 25,
https://doi.org/10.1007/978-3-031-20740-2_14

319

14.1   Introduction

Plastic bags make up 9.4% of the world's coastal litter. More than a million birds, marine mammals and turtles die from ingesting plastics each year (Jeftić et al. 2009). Indonesia is the second largest source of plastic marine pollution (Jambeck et al. 2015). There is a growing number of studies suggesting that plastic particles taken up by marine life (Desforges et al. 2015) causes adverse health effects in a number of creatures ranging from nano-organisms to whales to human beings (cf. Andrady 2011; Thompson et al. 2009). Most of the Balinese people that have contributed to this research have little knowledge of this. Their understanding of nature and potential damage to it is largely based on concepts and perceptions that are different. The variety and differences are examined in this study.

The approach of the study at hand is similar to other studies contributing to the growing field of environmental anthropology. Sponsel (2007) notes how these studies extend the former focus of ecological anthropology from local to now global considerations. Whether identity related factors for example in connection to the increasing role of transnational media or migration, or whether aspects of natural phenomena, such as climate change, the relation of humans with their natural environment can better be understood by looking beyond a narrow local and a more holistic approach. The pollution of the seas by plastic bags is a phenomenon that cannot be understood without its local and global dimensions. The fact that Balinese use plastic bags is a direct result from its global connectedness. Whether the use, disposal and pollution by plastic bags are perceived as a problem depends on local and global discourses and the identities and the interpretation of their specific components.

The two authors of this study are deeply rooted in Western culture. We can be described as being WEIRD (Western, Educated, Industrialised, Rich, and Democratic) (Henrich et al. 2010), from a global perspective. Similar to us, we would claim that many Westerners going to Bali would perceive according to their perspective the "plastic problem" of Bali being all too apparent. From our perspective plastic is at the wrong place (Douglas 2003). There seem to exist clearly distinguishable cultural and behavioural characteristics and understandings of people geographically misleading described as coming from the 'West' in comparison to 'Non-Western' people (Henrich 2020). However, obviously there is the strong danger of oversimplification (cf. Dove et al. 2003) and there is a huge diversity among them. This diversity is even more pronounced, when looking at 'Non-Western' perspectives, as cultural diversity is much more abundant and less aligned. Aiming to understand what we perceive as the plastic problem of Bali, we believe it is useful to examine the so called 'Western' and "the" 'Balinese' concept of nature to better appreciate an important difference in understanding the natural environment. According to Hviding the 'Western' concept sees nature as the antithesis of culture, nature vs. man, the material as opposed to spiritual (cf. Hviding 2003). Nature in the 'West' is hence largely understood as an autonomous category with its own set of rules. This is set in contrast to the 'Non-Western' conceptualizations of the environment in which culture and nature are not separate units. Bruun and Kalland (1995) describe a moral unity of human and nature referring to Asia as a whole. The anthropomorphisation of natural objects and phenomena – the attribution of human characteristics to the latter – we often found among Balinese people is an expression of this idea. Nature becomes more understandable, accessible and manageable. Many rituals and offerings can be seen as such interactions with the aim to influence proceedings in the environmental realm. In this study we will learn how the anthropomorphisation and interconnectedness of nature with other spheres in life, such as religion, helps us to understand the meaning and perceptions of nature as it presents itself to many Balinese.

These Balinese perspectives, meaning and perceptions are due to globalisation more and more interacting with Western, exported patterns of production and consumption and the corresponding concepts of nature. A central aspect in 'Western' separation of human and environment is the elevation of humans to control and manipulate the environment. It is often that this kind of manipulation has led to environmental destruction. For example, Ramseyer et al. (2001) perceive that the rise of materialism and consumerism induced from Western into Balinese culture serves as a vehicle for attitudes favouring exploitive behaviour. Materialistic and consumptive values are increasing all over Indonesia (see e.g. Gerke 2000; Spranz et al. 2012). Further, they play an important role in the constitution of people's social status and identity (see e.g. Douglas 1976, 1997; Jackson 2005). On the other hand, Western concepts of caring and conserving of nature are exported and interacting with Balinese or other concepts in the region ever more often (Pauwelussen et al. 2017). Due to an important political debate on plastic bag use on Bali, going on while fieldwork was done, we decided to try to understand factors that influence shopping bag choice.

14.2   Methods

We selected Bali as a case study, because as a well-established, fast growing, and substantial tourism hub in South East Asia it is producing a lot of plastic waste. It is a very religious place, with high respect in God's Creation and nature. It went through a rapid transformation, challenging many values, norms and institutions in more general that require to be adapted to the new ways of living. This happened under the condition of a state with limited governability. Roger, Spranz, the first author, has collected the data for this study in between 2013 and 2016 during which he spent most of his time on Bali conducting different research projects. Data was collected using a variety of research methods ranging from qualitative interviews to quantitative surveys. The research consisted of expert interviews, semi-structured interviews, group discussions and informal talks with more than 80 informants. Participant observation was a big part in the research as well. From grocery shopping with and without plastic bags and different everyday life situations that involve interaction with the environment, waste, especially plastic bag waste, he has been able to further approach an emic perspective. He also participated at community gatherings and meetings of local initiatives concerned with reducing plastic bag waste. With the help of research assistants, we have been able to conduct surveys with another 60 informants, all of which were owners of little grocery shops.

The data has been analysed using an inductive approach based on principles of the *Grounded Theory* by Glaser and Strauss (1967). An iterative process of data collection, analysis and interpretation was ongoing throughout the research. As a result of this different categories and concepts have emerged and been constructed from this process. These categories and concepts represent hybrid points of where the 'Western' and 'Balinese' concepts meet. They are intended to serve as translational bridges.

The article aims to contextualize the research by means of thick description (Geertz 1973) using categories and concepts of environmental knowledge, attitudes and behavior as well as the motivations and adversities for plastic bag use. To describe details about the data collected, we hope to increase the understanding and at the same time validity of the interpretations provided (cf. Lincoln and Guba 1985).

14.3 Local Environmental Knowledge in Bali

In this section we present the views of nature by Balinese people in different categories: History, Education and Religion. We developed and employed these categories by departing from familiar 'Western' categories, which also represent our starting point and therewith the starting points of many interviews and informal conversations held. Within, across and in between those categories 'Balinese' concepts of nature come to the front. However, we will learn that 'Western' and 'Balinese' concepts are not mutually exclusive and people may interconnect or constitute them in parallel ways.

14.3.1 History

Less than half a century ago environmental pollution by waste was not a problem on and around the island of Bali. Buying food at the market, taking it home or to work had been done using sustainable practices. As found in several talks with Balinese – but also are still often seen – baskets carried on top of the head have served to carry larger amounts of shopping goods for a long time. These are fine for carrying unprocessed fresh vegetables or fruits. In the case of meat, fish or processed food (Tofu, Tempe[1] etc.), they were first wrapped in coconut or banana leaves. As plates a

---

[1] Fermented soybean patty.

coconut shell or wooden plate worked well. This practice has existed for many centuries, 'before the era of plastic began in Bali' (Hindu Priest).

To the Balinese Hindu the daily offerings prepared and placed to the different manifestations of God appear as important as the provision of food to themselves. This together with around 50 more ceremonies throughout the year, requiring even larger amount of offerings, has always produced a substantial amount of abandoned resources. However, the traditional content of offerings, such as flowers, fruits, rice, along with the baskets, disposed banana and coconut leaf wrappings, have been more of a fertilizer than a source of risk for people's health or the ecosystem. 'The offerings used in the ceremony (…) will degrade over time, such as leaves, coconut shell, it can decay so that the old offerings – after the completion of the ceremony – are used as organic fertilizer.' (Hindu Priest). To dispose the organic matter into rivers or burn the waste to regain space and also for fertilizing the soil showed to be rather well adapted waste management practices through time.

14.3.2   Education

As with many other areas of life in Bali, there are strong dynamics underlying people's perceptions and concepts of the natural environment. There are some indications that Balinese with higher education and those living in urban areas share views of nature similar to 'Western' concepts. The most significant difference can be seen between generations. However, there are no clear cut lines, and younger Balinese very much share traditional 'Non-Western' Balinese concepts of nature and at the same time they refer to 'Western' concepts. This is not surprising acknowledging the increasing exposure of young Balinese to different Western media and many Western tourists coming to their island. But there is another important factor changing the way young Balinese think about environmental issues including eco-systems, pollution, waste, climate change and health: Their education in schools. 'The perspective of environmental education in the curriculum 2013 is packed with the expectations that learners gain awareness and sensitivity, gain a variety of experience and a basic understanding of the environment' (Prihantoro 2015: 83).

Prihantoro (2015) shows the important role of environmental education in Indonesian school curriculum today. We learned about different environmental initiatives in collaboration with schools. These ranged from education on waste management to holistic social, cultural and environmental approaches as represented by the Adiwiyata Mandala program by the Indonesian Ministry of Environment (cf. KLH 2012). Four schools have been visited in Bali throughout 2014 and 2015. It was very clear that the students knew about many concepts part of 'Western' environmental discourses. We had discussions on recycling, degradable and non-degradable waste, waste separation, pollution of the sea and climate change. Since the students were part of specific environmental programs, on-site waste separation, as well as composting was part of their daily practices.

The Directorate General of Primary and Secondary Education included environmental education into national curriculum already in 1984 (KLH 2012). Since then the subject has grown to be a more important part of education in Indonesia (cf. Kusmawan et al. 2009). In interviews and discussions with parents of young Balinese on environmental topics such as pollution by plastic bags and waste separation, the parents often referred to their children and how their children learned about these things in schools. 'Western' concepts of nature and the environment – including the vulnerability of the ecosystem – as taught in Indonesian schools, are of increasing relevance to the way Balinese think about nature, as well as their respective behaviour.

14.3.3   Religion

While around 90% of all Indonesians are Muslims, around 90% of the Balinese are Hindu (BPS 2010). Common to the vast majority of Indonesians is the important role of religion in their lives. Perceptions, understanding and interaction with the natural environment are largely influenced by religious beliefs and practices. Hindu-Dharmaism, the Balinese form of Hinduism, has been explained in interviews emphasizing different aspects. One is the concept of *Tri Hita Karana*, the harmonic relationship in between God, society and the environment. At different points in the interviews with and talks to Balinese they returned to their God's trinity of Shiva, Brahma and Vishnu as a starting point to clarify what the environment is about.

> So there is like Shiva for wind, Brahma for fire and Vishnu for water [...]. When Shiva gets angry there will be a tornado. When Vishnu gets angry there will be a tsunami, and when Brahma gets angry there will be fire, like forest fires, also without people making the fire. It also happens when people destroy the environment, like Lapindo [Indonesian gas company].[2] That's why in Bali there is no drilling. There was a demonstration that, if they need to drill in order to produce electricity, it is better to go back to life without electricity. (Male Balinese teacher)

This quote serves as an example on how the belief of many Balinese Hindus is connected to animistic ideas of nature or rather natural 'elements', which are perceived as emotional beings. The animistic or anthropomorphic quality of nature is a central concept to their understanding of their relationship with the environment and also how to understand it. Natural disasters are angry outbreaks by nature due to misbehaviour on behalf of individuals or the society at large. 'There is no eruption of volcanoes because we make a ceremony every year' a Balinese woman said. 'Why is there always water on Bali? Even in dry season? Because we pray at the temples close to the lake.' another young man explained. Central to maintaining a healthy and balanced relationship to nature are religious practices such as ceremonies,

---

[2] He refers to a gas drilling accident in 2006 also known as the Sidoarjo mud flow. After drilling for gas an ongoing mud flow has inundated several nearby villages. Thousands of people had to be permanently evacuated. To this day mud continues to come out of the drilling hole.

including offerings and prayers to one of the many manifestations of God. Beyond purely religious activities other behaviour towards the environment can upset its spirits and provoke environmental problems or disasters in return. A teacher said: 'I believe that the problem of dryness is because people are not praying and there is no balance in between construction and trees'. Several of the people talked to were worried about a land reclamation project, which is currently planned for the south of Bali. They fear 'the sea will get angry' and strike back with a tsunami. Whether it is drilling for gas, excess construction or land reclamation, many are expecting an environmental crisis as a consequence.

Ibu Pertiwi – which translates to *Mother Earth* – is another central religious and mythological concept showing in Balinese people's perceptions, ideas and actions involving the natural environment. 'Ibu Pertiwi is the entire world. This is Ibu Pertiwi. So don't hurt her.' A male merchant in his late thirties explained. During the socialization of an environmental program in a local community, the village head also referred to the environment by talking about *Ibu Pertiwi*. A locally well-known religious leader emphasized in an interview 'We really respect the earth. We call her mother, like our mother, *Ibu Pertiwi*'.

So if the earth, with its trees, rivers, lakes and sea is so respected and often seen as holy, from a 'Western' perspective we tend to wonder, how come it is being polluted and littered so badly? Besides the lack of proper waste management services, we believe that the answer to this has to be mainly seen in the perception of many Balinese that polluting and littering is not understood as disrespectful or irreverently behaviour. Again, Balinese frequently view environmental disasters and crisis not necessarily caused by environmentally adverse behaviour, often it is a general moral misconduct. Views based on 'Western' environmentalist concepts may be on the rise but are rather rare. One of the few people who connected an environmentalist view with a Hindu-Balinese concept of nature was a shop owner who believed that littering may also provoke anger in *Ibu Pertiwi:* 'There are many problems for Ibu Pertiwi. […] Everybody put some rubbish in the river, everywhere, put it in the ocean, to Ibu Pertiwi. […] So when Ibu Pertiwi gets angry, maybe there will be an earthquake'. What most tourists on the island see as an overwhelming and disturbing problem, when they spot plastic bags and other waste at the side of the roads, in rivers, on beaches and in the sea, may not irritate Balinese residents in the same way. The waste and littering seems to not interfere with the principles guiding Balinese towards respecting nature as a sacred environment. Another revealing perspective was shared by a Hindu priest talked to about the problem of plastic bag waste. He explained:

> Actually Balinese Hindu believe that anything that can go to the market and be purchased is considered holy. For example, there are eggs after the ceremony washed and then sold to the market again, to be purchased and considered holy, because they believe in the God of the market, Dewi Melanting. Same with plastic becoming holy, with them not knowing about the plastic and its effects. So their actions do not make them realize that they suffer from their own actions. (Hindu Priest)

As the priest further argued, these beliefs and practices stem from the times when fruits, vegetables and other unprocessed food and spices were traded at the market.

All were seen as blessings from nature. The temple next to each local market makes sure everything coming to the market is blessed. What has formerly been the banana or coconut palm leaf, a blessed material from *Ibu Pertiwi*, to carry shopping goods and eventually returned to nature behind the house or in a river, has within a short time switched to being a plastic bag. The way to handle plastic and the way it is culturally seen is very much the same as any other blessed organic material originating from *Mother Earth*.

14.4   Plastic Bag Use

So far we have discussed the concepts shaping Balinese people's understanding and behaviour in relation to the environment. We also know more about what issues are perceived as problematic towards the environment or not, and the cultural reasons for that. As much as the use or non-use of plastic bags needs to be viewed in connection to its negative effects on the environment we also need to go beyond the environmental context. In this section we will explore the reasons for the frequent use of plastic bags by Balinese, and why – in the view of Balinese – plastic bags may not be a good choice.

14.4.1   Reasons for Plastic Bags Use

For several interviews on the advantages and disadvantages of plastic bag use, we instructed university and high school students to ask shoppers about their views. Analysing the data of around 20 interviews showed three major benefits to shoppers: Using plastic bags is practical, easy and cheap. These were the terms the shoppers told us, but they also serve as categories for other advantages people mentioned, the use of plastic bags brings to them. It is 'practical' to shoppers that plastic bags are foldable, hygienic, durable, reusable, multi-functional and water-proof. All of these had been mentioned more than twice to us. Related to the concept of being practical was the reason of being easy to use. 'Easy' was the second most often answer to us and in this category we also find answers that point out that plastic bags are readily available, lightweight and easy to dispose of. Somewhat of a different aspect is being highlighted in answers that state using plastic bags is modern. Using baskets or banana leaf wrappings as in the former days makes people embarrassed. To explain this further, one person gave this example to illustrate: 'When they come to the cities first, they also wear sarong.[3] But when they come a second time, they may start to get embarrassed.' Just as choosing what dress to wear, the shopping bag they use is a fashion and status statement of modernity.

---

[3] traditional lower garment worn by people on Bali and other parts of South-East Asia.

So while some appear to be taking identity and status concerns into account and change towards the new, there are others arguing that using plastic bags has just become their habit: 'It's normal'. In the survey with 60 shop owners, we also included a question of why they use plastic bags. Similar to the data above, almost two thirds argued that plastic bags are the most practical solution to carry shopping home. Due to their different role as shop owner in the shopping process, other motives were added. From shop owners' perspective providing the purchase of the customer directly in a plastic bag is good for the sales – good and cheap customer service. Plastic bags compare to other carrying devices also have the advantage that the purchase can be seen, one shop owner added.

Based on the observations of people's use of plastic bags in everyday life, it shows how widely plastic bags are being used. There are many situations well beyond the grocery shopping. Small plastic bags are being used as food packaging for the popular small crackers which are often produced by home-business and wrapped piece by piece. Parts of the many offerings which are placed at the Hindu temples every day are often wrapped in plastic bags. Often the big fruit basket offerings are also entirely wrapped in plastic bags. You find plastic bags also along the rice fields attached to poles in order to scare away birds picking the seeds. Shirts and dresses in fashion stores in some cases have each piece hanging sealed in large plastic bags. Something remarkably from a Western perspective: Whether keyboards from computers, frames of TVs, cushions on chairs, entire sofas, and many other objects in Indonesia are sealed into plastic. Among the reasons for that may certainly be the protection against dust and dirt, but in some cases the plastic represents the new, a recent purchase, something beyond the practical, more towards the status partially based on wealth, spending money, partially taking care properly, keeping it clean, being a good household.

14.4.2 Disadvantages of Plastic Bags

Many people enjoy the benefits from using plastic bags, not only in Bali and the rest of Indonesia, but in many other parts of the world. But there are also reasons for avoiding plastic bags. In the 'West' the negative environmental effects from using, producing and disposing or dumping plastic bags has created a critical attitude towards them. Do Balinese share these critical attitudes, and what are reasons to them to avoid plastic bags or choose alternatives?

Asking about disadvantages of plastic bags the most frequent answer we heard was the problem of flooding caused by plastic bags. The ditches and drainages in between roads and houses are often clogged by waste, especially plastic bags. In case of heavy rains, as during rainy season, the clogged waterway in front of your house may result in inundation of your home. Plastic bags causing flooding is therefore among the biggest concerns many Balinese people have with plastic bags. Other negative qualities of the plastic bag use, explained shop owners, are that they are expensive, but as customers ask for them so they need to provide. Almost a quarter of the 60 shop owners surveyed mentioned this. Other comments were that they break easily, get dirty and quickly smell. The problem that plastic bags are contributing to the amount of waste and causing air and soil pollution have nevertheless been mentioned by several people. During a socialization event of a local anti-plastic bag initiative, the village leader pointed out: 'Plastic bags are objects very dangerous for our lives on this earth, for all living creatures. Plastic is one of the most dangerous killers, but we do not realize, we do not know how dangerous plastic is'. And a religious leader talked to said: 'In the villages they are regretful about the plastic, because the soil is becoming less productive.' These quotes represent very strong convictions and perspectives. As local leaders they are eloquent and it is noteworthy to find this environmental awareness similar to 'Western' views. But these views are not widely spread. Although many people express concern about the waste problem, insights on the polluting mechanism and risks, such as a potential decrease of soil fertility, are not shared among the majority of people.

It is not a rare exception to find critical – on environmental consideration based – views towards plastic bags. Critical views are often expressed by the younger school and student generation, but this 'Western' perspective is not so widely spread. Far more common is an understanding shared by what may be the majority that the plastic bags contribute to the flooding problems in different parts of the island when heavy rainfall, together with clogged waterways, results in flooded homes and streets. The flooding issue is rarely part of 'Western' discourses, although there is a case of flash flood in L.A. being caused by clogged plastic bags having been discussed against the background of the plastic bag ban in California (San Jose Mercury News 2016). In India and Bangladesh, the infrastructural conditions had caused similar problems to Bali and can be seen to have been the main driving forces there leading to regulations and bans of plastic bags, again due to flooding and subsequent health concerns (Ritch et al. 2009; Gupta 2011). Despite the mentioning of negative effects from flooding, we did not come across these arguments as a specific and sole reason for those Balinese explaining their reduction of plastic bags use. Those Balinese who avoid or reduce the plastic bag use argue – if not solely along 'Western' environmental reasons (cf. Cherrier 2006) – at least in combination with those.

14.5   Plastic Bag Free

So far we have learned about the 'Western' and 'Balinese' concepts allowing the Balinese to understand and behave in a certain way towards and within the natural environment. After obtaining a better understanding of benefits and disadvantages of plastic bags to the Balinese, we will now turn towards the efforts and achievement of a local initiative to stop the use of plastic bags on Bali.

During the fieldwork Roger Spranz has been able to join the monthly meetings and participate in a number of activities by the local initiative Bye Bye Plastic Bags (BBPB). We will discuss the approach and results of their campaign in the following to learn more about effective ways of reducing plastic bag consumptions. What could be learned from BBPB's activities, is it a successful model for replication in other parts of Indonesia or beyond?

Bye Bye Plastic Bags (BBPB) is a local initiative mainly driven by teenagers between 11 and 17 years of age to make Bali plastic bag free. It was founded by two Indonesian sisters in 2013, their father Indonesian, their mother Dutch. Most of the active members are teenagers from international schools, often with at least one parent being from another country. BBPB's core activity had been the collection of one million signatures – on- and off-line, to hand over to the Governor of Bali in order to ban plastic bags. BBPB have also gained the support of a local village head who offered his village as a pilot village for being plastic bag free. To follow up this opportunity there have been a number of community meetings and village days, during which the teenagers of BBPB have distributed free reusable bags and conducted surveys with shop owners and villagers. Beyond these activities there BBPB has received large attention from local, national and international media. The teenagers have started to visit more and more local Indonesian schools to spread the word and motivate new members to join. BBPB have given presentations at large conferences, as well as at INK talks in India and TED global in London. BBPB is being supported in different ways by the Rotary Club, Jane Goodall foundation and UNORCID, which Secretary General of the UN Ban Ki-moon initiated after a visit to the school of the BBPB founders.

Mass media and nowadays social media play an increasingly decisive role. More than one and a half million viewers have seen the TED talk by BBPB; a result which would not have been possible without sharing the video on social media. The same month the TED talk has been released – February 2016 – several big supermarket and retail chains across Indonesia, introduced a fee for plastic bags. Among many other very committed groups, e.g. the activists Diet Kantong Plastik from Java, BBPB have surely contributed to this achievement.

After several attempts with the former governor of Bali, who had signed letters stating intentions to reduce plastic pollution, but who never moved forward in terms of tangible legislation, in 2018 a new governor took office. The new governor Koster quickly took action and passed a plastic ban law end of 2018, coming in effect after a grace period of 6 months. In June 2019, aside from plastic straws and Styrofoam, plastic bags were banned. From major supermarkets to smaller grocery stores the ban was effectively followed by big retail chains and franchises. However, among individual stores and especially traditional local markets no big changes were seen. One thing the ban did achieve, plastic pollution and its problematization, among society evolved from being a marginal topic to a mainstream one. While the degree of problematization may vary and among certain groups still be weak, the governor eventually speaking out about why he is issuing this ban, explaining the need for society to refuse single use plastics, has represented an important milestone regarding the use and end of using plastic bags.

After considerable progress in the economic, political and societal spheres the use of plastic bags and other plastics came recently to a renaissance due to the impact of the Covid-19 and the increase of single use plastic products and packing around the world and in Bali, too. But it is less than before and there is little doubt that BBPB's activities and activism in Bali's communities and directed at the local government in Bali has been supportive to raise awareness and introduce legislation of banning plastic bags.

The BBPB initiative is quite different compared to other more traditional community based approaches and campaigns. As such, to what degree can the approach of the BBPB initiative be a model beyond Bali, for other parts of Indonesia and other countries? There are several difficulties for replicating the approach, the ethnic and culturally differing background of the BBPB members is not 'accessible' to Indonesian children. Whether an initiative of Indonesian children – due to their traditionally more subordinating role in families and society – is possible and how well it will be perceived remains an open question. While a similar success such as in Bali is hard to identify, the fact that local BBPB campaign groups have now been organized in more than 50 locations in other regions of Indonesia and around the globe, e.g. in the US, India and Australia shows a first step and faith by other activist to give the BBPB model a chance for successfully campaigning against the use of plastic bags elsewhere. Only the next months and years can tell us more on what impact can be expected from the BBPB in a variety of contexts.

14.6   Discussion and Concluding Remarks

This article has analysed the issue of plastic bag pollution on Bali and possible solutions. We approached the phenomena by attempting to understand and explain Balinese people's perceptions and concepts of the environment. Different dimensions of the environment have been presented, as they are being constructed and formed through history, in the educational system and in the religious context. The environment is not perceived to be harmed by the waste management practices of burning and dumping waste as it is a practice that has been done throughout past centuries. What has changed is that plastic bags and other sorts of waste have been added to the picture. However, when it comes to cleaning-up and disposing waste many Balinese perceive a discarded plastic bag just like the traditional banana leaf wrapping that has helped to bring some food home from the market. This view that disposed plastic waste is not problematic to the environment is supported by religious concepts that appreciate all goods –including plastic bags – traded at the market as blessed. Hence the traditional practices continue while the material – from mainly organic to more and more non-organic and plastic – is quickly changing.

When it comes to understanding Balinese people's relationship towards nature, it is important to remember that most have specific religious ideas on 'who' the environment is, and why it 'behaves' in a certain way. This anthropomorphisation of natural objects and phenomena is a concept about the environment widely and firmly held by most Balinese. Very much like other human beings, *Ibu Pertiwi* (*Mother Earth*) can get hurt and react angry, although the reasons to get upset are dominantly political and moral failures. Frequent natural disasters, like tsunamis, earthquakes, floods and volcano eruptions are therefore seen as the angry outbreaks and consequences of the moral misconduct. Environmental wrongs in the 'Western' sense are rarely taken as source of anger for *Ibu Pertiwi*. 'Western' explanations and interpretations are however increasingly added to views of Balinese people. Among younger generations the vulnerability of the ecosystem from a scientific point of view is getting acknowledged more frequently, such as the problematisation of plastic bags and other waste pollution. Nevertheless 'Western' views are hardly dominant and only rarely part of the discourse. While there is a potential for contradicting local Balinese perspectives, 'Western' aspects are in fact often added, integrated or held in a parallel manner in complementary reference systems for understanding and interpreting the environment. This leads to what Nygren (1999) appropriately describes as 'heterogeneous knowledges'. The variety in hybridizations of environmental concepts in Balinese lives helps explain the frequent surprises or seeming contradiction one comes across on the way to a better understanding of local environmental knowledge.

In line with research by other scholars (Pasang et al. 2007; Tejalaksana 2012), the data collected shows that pollution by waste is not widely perceived as problematic. Given the current low problematisation of (plastic) pollution, future awareness campaigns must recognize, embed and connect their approach well to the respective 'Balinese' environmental concepts. As has been pointed out, *Ibu Pertiwi* gets hurt from the pollution of waste and she can get angry and strike back in form of natural disasters.

Then we narrowed the focus from the environment in general towards the use and perceptions of plastic bags in Balinese people's views. We learnt that to most people the striking negative effect from plastic bags are floods caused by the clogging of waterways. In Bangladesh, a plastic bag ban was implemented largely to reduce the negative effects from floods. These consequences are easy to understand and more relevant to current priorities in the views of many Balinese. It can therefore be very useful to raise awareness toward the negative effects of plastic bags by including and connecting to this existing problematisation of plastic bags in discourses in Bali and other flooding prone areas in Indonesia and beyond.

Problematic views of plastic bags by Balinese people can support more effective awareness campaigns, but it is just as important to understand the positive qualities and popularity of plastic bags to inform promising behavioural change approaches. The dominant reasons for using plastic bags in the view of Balinese shoppers and shop owners are very pragmatic. They are practical, easy and cheap, pointing all into the same direction for the vast majority. This reasoning is very much in line with findings from other studies looking at plastic shopping bag use, such as Hawkins (2001), who describes it as the 'easy convenience of plastic bags'. Gupta (2011) points to the 'easy availability' of plastic bags. But also the plastic bag's role for status and identity has come to the fore. Choosing a non-plastic shopping bag in the 'Western' context is often a conscious ethical and environmental decision, making people feel better about themselves (cf. Cherrier 2006). In other contexts, the contrary may hold true. Often it is the use of plastic bags that allows people to feel better, more modern. Examples for this can be seen in what Yasmeen (2013) describes as the postmodern 'plastic bag housewives' in the case of Bangkok in

Thailand. Stone (2006) points out how in the Turkish context minarets were referred to as symbols of tradition and plastic bags as symbols of modernity. Along this argument the use of plastic bags may represent a modern, a preferred attitude and identity for Balinese. When Hawkins (2001) analyses 'The object marketed for its convenience evokes a modernist asceticism and temporality (...)' he shows how both aspects – convenience and modernity – are related and play together in the choice of plastic bags. Beyond these considerations the repetitive use and mainstreaming of plastic bags lead to their normality and the habitualisation of use (Ohtomo and Ohnuma 2014). Knowing the motivations leading towards the plastic bag use habit can be helpful also in regard to creating eco-friendly alternatives to plastic bags.

In the last section of this paper we turned to finding solutions for the plastic bag problem on Bali. The analysis of the Bye Bye Plastic Bags campaign showed the power of a charismatic social initiative by teenagers in receiving the attention of local, national and international media. Jordan and van Tujil (2000) and Wright (2000) point out that the success of a campaign is crucially linked to its presence in mass media. Gritten and Kant (2007) explain the important role of local and national media in environmental campaigns and provide an example in the region of an effective campaign against an Indonesian pulp and paper company. Those young teenagers, coming from hybrid families, being very familiar with Balinese and 'Western' concepts of nature might have been able to build the required translational bridges. They might have the necessary "amphibiousness", which Pauwelussen and Verschoor (2017 #3100) describe as "the ability to move in and relate different worlds that do not add up, yet partly flow into each other" (p. 295), when they report on the important role of local people, who are familiar with both worlds, when 'Western' conservation NGOs try convince in this case Bajau people in Sulawesi to conserve coral reefs. Those translational bridges provided by Bye Bye Plastic but also other actors lead to, as has been shown e.g. by quotes the Hindu priest mentioned above that perspective get a higher 'plasticity'. A contamination, but also a fructification between the different knowledges is facilitated.

The success of campaigns such as BBPB is often difficult to assess due to its multidimensionality (Cf. Keck and Sikkink's 1998). It is therefore hardly possible to define the scope of BBPB's influence on the government's decision in regard to introducing fees on plastic bags for selected commercial sectors and areas in Indonesia. A fee on plastic bags has shown to be a very effective tool to reduce plastic bag use in many countries across the globe, for example in the UK, Germany and Ireland (New York Times 2008). However, the political and societal will to implement such policies, or marketing strategies by retailers, only recently emerging in Indonesia, has to be nourished by societal change, among others fostered by initiatives like Bye Bye Plastic Bags.

The public support and media attention for BBPB also resulted in government representatives inviting the initiative for a meeting. The governor received and listened to the teenagers' request of stopping the plastic bag pollution. As a result of the meeting the governor and environmental agency of Bali have announced to support the goal of making Bali plastic bag free within their jurisdiction and responsibilities. While there is still no legally binding document, this could be a step towards banning plastic bags, which has been the central request by BBPB. A plastic bag ban has already effectively worked in a number of countries, for example in Uganda, Kenya, and Bangladesh (Cf. Teh et al. 2014). These policies often take a long time to be applied, monitored and effectively enforced.

In the meantime, and with the insights of this article we hope to contribute to the knowledge about the perception and understanding at work that contextualize and influence the use of plastic bags. To connect with the local perceptions of nature and existing problematisations of plastic bags, as specified in this article, can inform effective approaches for awareness campaigns, local initiatives and political programs. The role of fashion, identity, and convenience related factors are crucial in people's choice and use of plastic bags. Alternatives to plastic bags will have to consider these factors in order to successfully facilitate a behavioural change. There are hence opportunities not only for environmental initiatives and NGOs, but also politicians and businesses towards creating an environment free from plastic bags.

Acknowledgement   We are grateful to the many Balinese people, who have given their time and thoughts to us, allowing to understand slightly better the puzzle of plastic pollution on Bali. One step in addressing the global marine litter problem. These insights have helped us to design experimental studies aiming to reduce plastic bag use on Bali (Spranz 2017, 2018). The financial support provided by ZMT and therewith by the Bremen and German tax payer is gratefully acknowledged.

References

Andrady AL (2011) Microplastics in the marine environment. Mar Pollut Bull 62(8):1596–1605. https://doi.org/10.1016/j.marpolbul.2011.05.030
BPS (2010) Penduduk Menurut Wilayah dan Agama yang Dianut. http://sp2010.bps.go.id/index.php/site/tabel?tid=321&wid=0. Accessed 05 June 2015
Bruun O, Kalland A (eds) (1995) Asian perceptions of nature: a critical approach. Curzon press, Richmond
Cherrier H (2006) Consumer identity and moral obligations in non-plastic bag consumption: a dialectical perspective. Int J Consum Stud 30(5):515–523. https://doi.org/10.1111/j.1470-6431.2006.00531.x
Desforges J-PW, Galbraith M, Ross PS (2015) Ingestion of microplastics by zooplankton in the Northeast Pacific Ocean. Arch Environ Contam Toxicol 1–11. https://doi.org/10.1007/s00244-015-0172-5
Douglas M (1976) Relative poverty, relative communication. In: Butler V, Halsey AH (eds) Traditions of social policy: essays in honour of violet Butler. Basil Blackwell, Oxford
Douglas M (1997) In defence of shopping. In: Falk P, Campbell C (eds) The shopping experience. Sage, London [etc.], pp 15–30
Douglas M (2003) Purity and danger: an analysis of concepts of pollution and taboo. Routledge
Dove MR, Campos MT, Mathews AS, Meitzner Yoder LJ, Rademacher A, Rhee SB, Smith DS (2003) The global mobilization of environmental concepts: re-thinking the Western/non-Western divide. In: Selin H (ed) Nature across cultures: views of nature and the environment in non-western cultures, Science across cultures. Kluwer Academic Publishers, Boston, pp 19–46
Geertz C (1973) The interpretation of cultures: selected essays. Basic Books, New York

Gerke S (2000) Global lifestyles under local conditions: the new Indonesian middle class. In: Chua BH (ed) Consumption in Asia: lifestyles and identities, The new rich in Asia series. Routledge, London/New York

Glaser B, Strauss A (1967) The discovery of grounded theory. Weidenfield & Nicolson, London

Gritten D, Kant P (2007) Assessing the impact of environmental campaigns against the activities of a pulp and paper company in Indonesia. Int For Rev 9(4):819–834. https://doi.org/10.1505/ifor.9.4.819

Gupta K (2011) Consumer responses to incentives to reduce plastic bag use: evidence from a field experiment. http://www.sandeeonline.org/uploads/documents/publication/954_PUB_WP_65_Kanupriya_Gupta.pdf

Hawkins G (2001) Plastic bags: living with rubbish. Int J Cult Stud 4(1):5–23. https://doi.org/10.1177/136787790100400101

Henrich J (2020) The WEIRDest people in the world: how the west became psychologically peculiar and particularly prosperous. Penguin UK

Henrich J, Heine SJ, Norenzayan A (2010) Beyond WEIRD: towards a broad-based behavioral science. Behav Brain Sci 33(2–3):111–135

Hviding E (2003) Both sides of the beach: knowledges of nature in Oceania. In: Selin H (ed) Nature across cultures: views of nature and the environment in non-western cultures, Science across cultures. Kluwer Academic Publishers, Boston, pp 245–275

Jackson T (2005) Motivating sustainable consumption: a review of evidence on consumer behaviour and behavioural change a report to the sustainable development research network. University of Surrey

Jambeck JR, Geyer R, Wilcox C, Siegler TR, Perryman M, Andrady A, Narayan R, Law KL (2015) Plastic waste inputs from land into the ocean. Science 347(6223):768–771. https://doi.org/10.1126/science.1260352

Jeftic L, Sheavly SB, Adler E, Meith N (2009) Marine litter: a global challenge. Regional Seas, United Nations Environment Programme, Nairobi

Jordan L, van Tuijl P (2000) Political responsibility in transnational NGO advocacy. World Development

Keck ME, Sikkink K (1998) Activists beyond borders: advocacy networks in international politics. Cornell University Press, Ithaca

KLH (2012) Informasi Mengenai Adiwiyata. http://www.menlh.go.id/informasi-mengenai-adiwiyata/

Kusmawan U, O'Toole JM, Reynolds R, Bourke S (2009) Beliefs, attitudes, intentions and locality: the impact of different teaching approaches on the ecological affinity of Indonesian secondary school students. Int Res Geogr Environ Educ 18(3)

Lincoln YS, Guba EG (1985) Naturalistic inquiry. Sage, Beverly Hills

Nygren A (1999) Local knowledge in the environment-development discourse: from dichotomies to situated knowledges. Crit Anthropol 19:267–288. Accessed 25 Jan 2016

Ohtomo S, Ohnuma S (2014) Psychological interventional approach for reduce resource consumption: reducing plastic bag usage at supermarkets. Resour Conserv Recycl 84:57–65

Pasang H, Moore GA, Sitorus G (2007) Neighbourhood-based waste management: a solution for solid waste problems in Jakarta, Indonesia. Waste Manag 27:1924–1938

Pauwelussen A, Verschoor GM (2017) Amphibious encounters: coral and people in conservation outreach in Indonesia. Engag Sci Technol Soc 3:292–314

Prihantoro CR (2015) The perspective of curriculum in Indonesia on environmental education. Int J Res Stud Educ 4(1):77–83

Ramseyer U, Tisna IGRP, Surya R (2001) Bali: living in two worlds : a critical self-portrait. Museum der Kulturen Basel, Basel. http://www.maff.go.jp/j/nousin/kaigai/inwepf/i_document/pdf/sympo_sutawan.pdf

Ritch E, Brennan C, MacLeod C (2009) Plastic bag politics: modifying consumer behaviour for sustainable development. Int J Consum Stud 33(2):168–174. https://doi.org/10.1111/j.1470-6431.2009.00749.x

San Jose Mercury News (2016) José M. Hernández and Hans JohnsonOpinion: plastic bag ban supported by science and conscience. http://www.mercurynews.com/opinion/ci_29529247/jose-m-hernandez-and-hans-johnsonopinion-plastic-bag

Sponsel LE (2007) Ecological anthropolgy. http://www.eoearth.org/view/article/151926/. Accessed 30 Nov 2015

Spranz R (2017) Reducing plastic bag use in Indonesia. PhD in Economics, Jacobs University

Spranz R, Lenger A, Goldschmidt N (2012) The relation between institutional and cultural factors in economic development: the case of Indonesia. J Inst Econ 8(04):459–488. https://doi.org/10.1017/S1744137412000124

Spranz R, Schlüter A, Vollan B (2018) Morals, money or the master: the adoption of eco-friendly reusable bags. Mar Policy. https://doi.org/10.1016/j.marpol.2018.01.029

Stone L (2006) Minarets and plastic bags: the social and global relations of Orhan Pamuk. Turk Stud 7(2):191–201. https://doi.org/10.1080/14683840600714608

Teh J, Taljanovic MS, Monu J (2014) International skeletal society outreach 2013: Rwanda. Skelet Radiol 43(5):563–565. https://doi.org/10.1007/s00256-014-1822-9

Tejalaksana A (2012) Water management pollution policy in Indonesia. http://www.wepa-db.net/pdf/0712forum/paper26.pdf. Accessed 20 Jan 2016

Thompson RC, Moore CJ, Vom Saal FS, Swan SH (2009) Plastics, the environment and human health: current consensus and future trends. Philos Trans R Soc Lond Ser B Biol Sci 364(1526):2153–2166. https://doi.org/10.1098/rstb.2009.0053

Wright BG (2000) Environmental NGOs and the dolphin-tuna case. Environ Polit 9(4):82–103. https://doi.org/10.1080/09644010008414552

Yasmeen G (2013) "Plastic-bag housewives" and postmodern restaurants? Public and private in Bangkok's foodscape. Urban Geogr 17(6):526–544. https://doi.org/10.2747/0272-3638.17.6.526

Chapter 15
Futuring 'Nusantara': Detangling Indonesia's Modernist Archipelagic Imaginaries

Hendricus A. Simarmata, Irina Rafliana, Johannes Herbeck, and Rapti Siriwardane-de Zoysa

Abstract Archipelagic identities have long patterned Indonesian historic imaginaries, collective memory, and its postcolonial modernist narratives on nation-building. This chapter examines and puts into conversation two distinct and interrelated concepts undergirding archipelagic thinking – 'Nusantara' and the lesser studied 'Tanah Air' – against speculative visions of Indonesia's developmental trajectories. These concepts intersect with Indonesia's aspirational vision as a maritime nation that is to take its place within a regional and globalist paradigm of ocean-centric economic growth. Inspired by critical ocean studies and by drawing on narrative analysis, we begin by considering the paradoxes within Indonesia's contemporary blue economy growth visions in relation to its older land-based biases in planning and nation-building. In critically engaging with Indonesia's own oceanic turn towards a blue growth orthodoxy, we consider three aspects of its futuring trajectory, namely industrialization, infrastructural development, and its recent choice of relocating its administrative capital to east Kalimantan. While engaging with paradigmatic land-locked biases and political path dependencies that unwittingly entrench 'Java-centric' development, we illustrate how Indonesia's distinct archipelagic thinking has co-evolved in recent history, and with what cultural resonance for its nation-building vision in the decades to come.

H. A. Simarmata (✉)
Universitas Indonesia (UI), Indonesian Association of Urban and Regional Planners (IAP), Jakarta, Indonesia
e-mail: hendricus.andy@ui.ac.id

I. Rafliana
University of Bonn and the German Development Institute (DIE), Bonn, Germany

J. Herbeck
Sustainability Research Center (artec), University of Bremen, Bremen, Germany

R. Siriwardane-de Zoysa
Leibniz Centre for Tropical Marine Research, Bremen, Germany

© German Institute of Development and Sustainability (IDOS) 2023
S. Partelow et al. (eds.), *Ocean Governance*, MARE Publication Series 25,
https://doi.org/10.1007/978-3-031-20740-2_15

337

Keywords *Nusantara* · Archipelagic imaginaries · Speculative ecologies · Indonesia

15.1 Introduction

This chapter analyses core interrelated trajectories of discourses and policies around Indonesia's recent paradigmatic 'oceanic' transition. We do this by tracing the different ideas and concepts around Wawasan Nusantara, looking at the political relevance as well as translation attempts and ideological interpretations of the concept from a historical perspective. Against this background, we will then analyse Indonesia's more recent ideas of a maritime nation as a future engine of development and address their translation in three different areas of policy – industrialisation, science and research, and urbanization.[1] We also analyze the potentiality for strengthening Nusantara-based ideas with the broadly discussed plan to relocate capital functions to east Kalimantan as potential game changer and as a trigger of development transformation. Thus, the chapter contributes to discussions on ocean governance, in particularly the nexus of maritime policy and regional socio-economic development in the Indonesian context. Furthermore, it will also elaborate the relevance of the notion of regional ocean governance (UNEP 2016) and Ocean Economy Agenda (OECD 2016) to Indonesian development.

From early childhood, Indonesians are raised with the awareness that they are part of a vast archipelagic community that has historically been shaped by fishing and seafaring. The folk song "Nenek Moyangku Seorang Pelaut", composed by Ibu Sud in 1940, is a song learnt by most elementary students at primary education stage to "introduce them to the glory of their ancestors, who have been described as brave sailors who explored and sailed the vast ocean" (Iswatiningsih and Fauzan 2021, p. 216). As by the time Indonesia was still under Dutch colonial rule, the lyrics of the song were also meant to encourage and inspire Indonesia's young people to fight and defend the Indonesian seas (Titiek Suliyati 2012).

In 1982, the once critical Indonesian musician and now cultural icon, Iwan Fals, also composed a song that bemoaned a changing ocean from the 1960s when he was

---

[1] The authors applied qualitative research, using used secondary data retrieved from books, academic journals, news, as well as grey literature publication types (e.g., government policy paper, regulation, project reports, etc.). In addition, information has also been generated from the reflective practices of the lead author as urban planning practitioner. This experiential knowledge has been used to enrich the information in this chapter. Content analysis has been used for probing the information from secondary data. Also, this chapter serves as a speculative analysis of different imaginaries of Wawasan Nusantara. The research scope explored in this chapter will be divided into three different areas of ocean-related policies – industrialisation, science and research, and urbanization. Referring to the industrialization aspect, we analysed the industrial policies that can influence the ocean economy. For science and research, we particularly discussed on the research for disaster management. For urbanization, we analysed the national urban strategy, including the capital relocation as game changer of regional development.

child, right up until the 1980s, when he discovered how his surrounding ocean was being progressively polluted. Both these songs are examples of a vast corpus of popular works of art inspired by oceanic thinking and imagination. Some of them can be seen as a reaction to *devide et impera* politics of Dutch colonial rule that instrumentalized and reinforced the multiple island identities of the archipelago. In contrast, late colonial and post-Independence Indonesian politicians tried to establish the idea of the ocean as an integrator, not separator by enlivening the vernacular concept of 'Nusantara' and its more applied aspect 'Wawasan Nusantara'.

(Wawasan) Nusantara is an all-encompassing notion and a fundamental concept for Indonesian geopolitics, to address ideology, politics, economy, sociocultural, security opportunities and challenges (Situmorang (2013). Etymologically, the Ministry of Research, Technology and Higher Education of Indonesia (2016) explains the word as a combination of two words, i.e., "Nusa" and "Antara" (pp. 212–213). In Sanskrit, "Nusa" means island or archipelago, while "Antara" can be interpreted as the sea, across or outside. In Latin, it suggests that "Nusa" comes from the word "Nesos", which means peninsula or nation, while "Antara" might have the equivalent meaning with the word "in" and "terra" taken to mean "between" or "within a group." In English, "Nusa" and "Antara" may bear similar resonance with the words "nation" and "inter", respectively. Based on that explanation, the Ministry notes that the merging of the words "Nusa" and "Antara" into the word "Nusantara" could be interpreted as an archipelago between seas, or nations connected by the sea.

Lestari (2018) argues that the word "Nusantara" was first recorded in medieval Javanese literature (around the twelfth to the sixteenth centuries) to describe the concept of a state adopted by a kingdom called "Majapahit" (pp. 57–58). In the context of Majapahit Kingdom, she describes that the State has been divided into three regional categories, i.e., "Negara Agung"; "Mancanegara"; and "Nusantara". "Negara Agung" was the area around the royal capital where the king reigned. "Mancanegara" were areas on the island of Java and its surrounding border areas whose folk culture was similar to "Negara Agung". "Nusantara" was an area outside the influence of Javanese culture but was still claimed as an area subject to the "Majapahit" Kingdom, where the ruler must pay tribute. Santoso et al. (2020) posits that the medieval Javanese literary book *Kitab Pararaton* indirectly mentions that a prime minister (locally noted as "Patih") of the Majapahit Kingdom called Gajah Mada declared an oath to the kingdom dignitaries, known as "Sumpah Palapa" or "amukti palapa" (p. 46). In analyzing the oath, it was stated that the majesticity of Gajah Mada would not come to pass until "Nusantara" was unified under the rule of the Majapahit Empire. A translated excerpt of Gajah Mada's oath is as follows:

> *Sira Gajah Mada patih Amangkubhumi tan ayun amukti palapa, sira Gajah Mada: "Lamun huwus kalah nusantara isun amukti palapa, lamun kalah ring Gurun, ring Seran, Tanjung Pura, ring Haru, ring Pahang, Dompo, ring Bali, Sunda, Palembang, Tumasik, samana isun amukti palapa."*
>
> [He *Gajah Mada Patih Amangkubumi* (prime minister) did not want to break his fast. He *Gajah Mada*: "If I have subdued the entire archipelago under *Majapahit* rule, I will (only) break my fast. If I defeat *Gurun, Seram, Tanjung Pura, Haru, Pahang, Dompo, Bali, Sunda,*

Palembang, Tumasik, that's how I (will) break the fast".] (cited and translated from lestari 2018, p. 57)

As a counter-discursive maritime narrative, the concept was further popularized by Ki Hajar Dewantara, a national leader on education in early twentieth century. He revived the term Nusantara to contest and replace the colonial imaginary of the Dutch East Indies. During and after independence, Nusantara – as a vernacular term – gained increasing attention and support, and was complemented by the more politically practicable concept of *Wawasan Nusantara*. Wawasan Nusantara entails a mindset or paradigm of Indonesian nationalism that, until today, carries the ideological purchase in strategizing Indonesia's environment, reconfiguring imagination(s) of unified nationhood, while moving towards regional integration for achieving nationally determined goals. However, it was the Djuanda Declaration 13 December 1957 that set the historical moment of the re-creationing Wawasan Nusantara, as a political doctrine and vision, aimed at reclaiming the once *Territoriale Zee Maritiem Kringen Ordinatie* boundaries set by the Dutch colonials.[2] In general, Djuanda's declaration states that Indonesia adheres to the principles of the Archipelagic State so that the seas between islands are also the territory of the Republic of Indonesia and are not free areas (Ernawati 2015). Although initially contested geopolitically (e.g., by the USA and Britain), the conception of Wawasan Nusantara (from an archipelagic perspective) as reflected in the Juanda Declaration, was legally strengthened in Indonesia under Law No. 4/Prp 1960 concerning Indonesian Waters (Kusumaatmadja 2001 cited in Nurhidayati 2021, p. 45)

Since then, the sea territories of Indonesia were enlarged, and this was globally acknowledged during the Geneva's Maritime Law Conference in 1978. Furthermore, the concept of Indonesia as an Archipelagic State was internationally recognized after the United Nations Convention on the Law of the Sea (UNCLOS) ratified on December 10, 1982, and Indonesia further ratified it under Law No. 17 of 1985[3] (Shalihah 2016). Shortly after, the concept of Wawasan Nusantara was formally accepted by UNCLOS. Since Indonesia's post-Soeharto Reformation Era, Wawasan Nusantara was emplaced within the long-term development planning agenda (Law 17/2007). From its earliest definitions in national regulations, Wawasan Nusantara as a doctrine aimed at reshaping collective political identity and meanings of nationhood: *"the perspective and attitude of Indonesian people in understanding oneself*

---

[2] The striving effort was continued and succeeded by Juanda Declaration to replace 1939 Ordonantie in December 13, 1957. Based on this Juanda Declaration, the government of Indonesia proposed the concept to the international forum and was accepted by UN Convention on the Law of the Sea (UNCLOS) in 1982. At national level, this declaration was further legalized in 1960 through UU 4/Prp/1960. It regulates the Indonesian ocean to cover the inland ocean, the sea border line on 12 miles, and the inland ocean sea lane located next to the baseline (Rimbakita 2021). Law 4/60 later was updated by the law 6/1996 concerning Indonesian Ocean. This law regulates the shipping, transit, communication, and information rights on the international sea lane that crossed Indonesian region.

[3] Law of Republic Indonesia No.17 of 1985 concerning the Enactment United Nations Convention on the Law of the Sea.

and his/her diverse environment and with strategic values, by forefronting the unity of the nation through living as a community and as a nation to reach the national goals." (Indonesian Archipelagic Vision 2013).

The translation of Wawasan Nusantara into development discourse has been highly dynamic and depended on the priorities of development regimes over time. After UNCLOS 1982, there was no specific national direction to explore or articulate maritime issues, except through a conventionally-oriented maritime focus on national security, transportation, and oil and gas mining. During the Orde Baru regime,[4] Indonesia achieved major successes in rice production through socioeconomic change concentrated primarily on Java Island, and a shift towards a more agriculture-dominated development path focusing on rice production and its export. Since 1998, during the Reformation Era, the focus on marine resources has been continuously expanded. President Abdurrahman Wahid (1999–2001) established the Ministry of Ocean Exploration that has been extended in 2000, now renamed as the Ministry of Ocean and Fisheries.[5] Under President Joko Widodo (2014–present), the vision to develop the nation from the outer islands has been advanced, and the idea of Indonesia as a global maritime axis has been articulated in political discourse. Widodo further established the Coordinating Ministry of Maritime Affairs[6] to facilitate the cross-sectoral development issues coordination. This dynamic translation demonstrates that there is no blueprint on how Wawasan Nusantara is actually filled with life in the different phases of Indonesia's shift from an agriculture to a maritime state.

15.2   *Wawasan Nusantara*, Past and Present

Out of the many transcultural vernacular concepts emerging from oceanic human history, is the beguiling and oft-romanticized notion of 'Nusantara', suggesting a shared identity that spans the Indo-Malay Archipelago, geographically the largest of its kind, until the farthest reaches of the Southeast Asian region. The concept itself appears as an empty signifier in Laclau's (2017) sense of the term – easily filled with different meanings and used as a shorthand reference by historians and other scholars to allude to 'maritime' Southeast Asia. In popular history, imaginaries interpret the idea of Nusantara as long journey in human civilization and modern nation-building across archipelagic Indonesia and beyond. Its various narratives were

---

[4] The New Order (in Indonesia Language called Orde Baru, abbreviated Orba) is used to characterize the second Indonesian President Suharto regime (1966–1998). This term has replaced his predecessor regime, the first Indonesian President Sukarno (known as "Old Order," or Orde Lama).

[5] Referring to the official website of the Ministry of Maritime Affairs and Fisheries (https://kkp.go.id/page/6-sejarah), President Abdurrahman Wahid, with Presidential Decree No. 355/M of 1999 dated 26 October 1999, formed the Department of Marine Exploration.

[6] See also: http://www.pubinfo.id/instansi-544-kemenko-bidang-kemaritiman%2D%2Dkementerian-koordinator-bidang-kemaritiman.html

historically documented and retold by the Kutai and Srivijaya Kingdoms, and long after, by the Majapahit Empire (1293–1527) that took control of maritime trade routes across the archipelago. Ancestral identities were closely associated to seafaring and the ability to cross the ocean. Trade activity in the high seas at that time was largely patterned by the growing historic demand for spices and other rare commodities. Precolonial and colonial cities as trading hubs grew across Indonesia, mostly along archipelagic coastlines. Oceanic connectivity, thus remained the prerogative in securing local and regional political influence over the course of its pre- and colonial history, thus lending privileged value to certain sites that became centers of power, trade, and socio-cultural exchange.

Besides Batavia in early nineteenth century, there were several kingdoms (for example Paser) that grew because of inter-insular trading. Known for its gold mining and forest resources from East Kalimantan, Paser was famously known for the Sadurangas Kingdom that existed in the fifteenth century. Mining and timber trading led to the development of riverine and sea transportation networks over time, further expanding its reach during Dutch colonialism. Traditional ships and vessels such as the Pinishi and the Pencalang from Siak Riau were manufactured, with each island possessing its own 'brand' of ship. Pre-colonial coastal settlements also evolved through water sensitive design and engineering coupled with traditional architectural forms, such as Bajau houses Sulawesi and the Gadang dwellings known in southern coastal Sumatra. Those examples show that Indonesian port-cities were places in which archipelagic sea – based cultures and urbanization met and thrived.

Yet, a curious paradox remains – the seemingly conceptual disconnect between archipelagic thinking and hubbing on the one, and the theorization of contemporary coastal life on the other. In the context of Java and the region on the whole, Southeast Asia has been conceptually divided between its 'mainland' and 'insular', the littoral and the hinterlands. By extension, socio-economic activity too was imaginatively separated in terms of the agrarian and its maritime other – constituting seafaring and trade, naval enterprise, commercial fishing, etc. While the watery bodies materially divided, they also constituted shared cultural borderlands of diverse kinds of exchange over the course of history. Might *Nusantara* then, also serve as both metaphor and method with which to counter land-locked paradigms in container thinking, and seeing spaces as capsular territories, and as mere transport surfaces and routeways? In a contemporary sense, the oceanic – replete with its fluid expansiveness, depths, and seafloor voluminalities– have long been critically theorized as economic and bio-frontiers, while shoreline beaches and oceans themselves have lent themselves as neoliberal playgrounds and sites for late-capitalist resource extraction. A vastly different and historically flecked metaphor appears in the case for Nusantara – the revisioning and melding of ocean as an economic engine and that of postcolonial nation-building. Drawing on recent concepts from the social and cultural sciences on the study of land-sea continuums (Steinberg 2001; Steinberg and Peters 2015), the territorialisation of seas and coasts (Campling and Colás 2018; Foley and Mather 2019), and amphibious lifeworlds (Krause 2017) we are particularly interested in how the discursive shift of an essentially cultural

(Nusantara) into an ideological concept (Wawasan Nusantara) is reflected through actual changes in the negotiation of policies.

But what does Nusantara stand to signify as a historically-embedded and as a culturally 'lived' concept? How much of a shadowy backdrop has the sea and maritime life been relegated to in the postcolonial Indonesian imaginary? Despite its older origin, Nusantara as an imaginary took root arguably in the early twentieth century when Ernest Douewes Dekker (or Setiabudi), the Indo-Dutch politician and nationalist offered up Nusantara as a name for the independent country of Indonesia, far removed from its former echoes to the Dutch East Indies (Evers 2016).

Since, the Wawasan Nusantara doctrine has been consistently introduced across many layers of public education, from formal schools to official public officer trainings, as well as military training. The idea however leans more towards securing the risk of the Indonesian state's multiple disintegrations and calls for island(ed) autonomy. The doctrine is embedded in the Parliamentary Decision (Ketetapan MPR) in 1999, which stated that Wawasan Nuasantara is a way to view and comprehend oneself and the society that aims at the unity of the nation. There are in fact two explicit objectives of the doctrine, both internally (inward for Indonesians) and externally, which is rather reflecting the role of the State in the global connectedness and connectivity. The external role however is rather addressing the importance to maintain peace and stability, social justice and equal respect on each sovereignty. In other words, it was not part of the doctrine to view Nusantara beyond the State. Together with the concept of 'Tanah Air', Nusantara resonates a deep meaning of homeland or motherland that comes together as a unified entity of a nation. Tanah Air is rooted in the Malay language, and consists of two fundamental elements of the archipelago; *tanah* (soil or land) and *air* (water, pronounced *ah-yer*), meaning that both elements were inseparably intertwined. The concept of 'maritime' explicitly blends the two, of which sea, rivers and lands construct the culture and the way of living of people dwelling in the "motherland" Indonesia.

The integrated concept of land-water at a micro level and *Wawasan Nusantara* at the macro level has been used to reimagine "maritime" mindset of a built infrastructural environment. The scope of both the concepts taken in concert, bear far-reaching implications, from influencing how people build their houses in adapting to wetland environments while connecting spaces through various water-based transportation options. The imperative is then that urban development in Indonesia thus should not only be dominated by land-based development, but should harmonize spaces and places that crosscut both land and water. As time goes by, these archipelagic imaginaries were expected to organically co-evolve with distinct community-based coastal lifeworlds. Littoral sensibilities were further expected to influence hinterland-based communities that re-settled in coastal areas that offered economic opportunities due to the agglomeration of industries, including port facilities and warehouses. Therefore, the Indonesian government initially coined the term "Wawasan Nusantara" drawing from interpretations of antiquity and its precolonial narratives, mining what was seen as the strategic geopolitical and geostrategic needs of Indonesia, while being further 'translated' to appeal to national ideology and the *realpolitik* of communal identity-making, defence and security.

Yet the puzzle that remained was that Javanese culture was irremovably more agrarian-oriented and land-based in its outlook towards the hinterland, marked by lush terrain and volcanicscapes. Still, Indonesia has continued to shift its narrative from being that of an 'agricultural' to a 'maritime' nation in the past decades. The next section tries to unpick the very rationales and narrative shifts witnessed in materially reimaging new built environments – as urban coast transforms and transposes into the urban riverine. Even more importantly, we try to sketch how in political discourses, a transition has taken place to strengthen the maritime elements of a new economic strategy.

15.3   Indonesia's Maritime Vision as Future Economic Engine

If we go from Aceh to Lampung through the east coast corridor, and continue on to northern coastal Java, we can identify that most of the major cities in Indonesia are located in both corridors. More than 50 % of the population resides in coastal areas. However, for the last 30 years, most of the development throughout Indonesia draws on the Javanese developmental experience as a reference point. The expansion of paddy fields and rice-led agriculture through Soeharto's transmigration policy in 1980s had profoundly transformed original food, nutritional and dietary habits of the other islands, creating a high dependency on rice (cf. Manning 1987). These changes were also witnessed in state provided housing development programs that favoured hinterland-based housing designs. These architectural forms proved markedly different from those of Marind Anim, Bajau, and Maluku homes often built on stilts, many offer to be taken as examples of 'amphibious' construction, favouring tidal flows and other forms of coastal flooding (Manurung 2014; Wahyudi et al. 2022).

Numerous policy-oriented scholars, Delima et al. (2019), Karim (2019), Son Diamar (2021), Jompa (2022) have normatively suggested that Indonesia should re-define or strengthen its political identity as a 'maritime nation', while drawing on its plans of relocating its capital city as an intrinsic part of a momentum to reach these ambitions. Alamsyah (2010) harks back to the time in which Indonesia had an active role in UNCLOS 1982, thereby settling questions of sovereignty spanning surface and seabed marine space in relation to its oceanic natural resources. Further arguments pointed towards the development potential of marine resources and a re-centring of maritime culture, formerly downplayed during Indonesia's first post-Independence development trajectory. There were also further calls in exploring how the interface between the sea and its coastal cities can be developed. For example, Baumeister (2020) conceptualizes sea cities as an urban adaptive tactic in attending to the urgency of sea level rise. Arguments have been made favouring the development of amphibious and aquatic floating infrastructures, to mining opportunities afforded by an oceanic-driven 'blue economy' (Alamsyah 2004). Yet, the choice of location in moving the administrative arm of Jakarta City arguably hinders further possibilities for 'sea city' development and that of an ocean economy, given its emplacement in east Kalimantan. Among different reasons, the bad experiences of coastal risks management in Jakarta and the future prediction of Jakarta Bay have been raised (VOI 2021) why to select the location that far from the sea.

In close connection with the described political-ideological definitions of a maritime identity for Indonesia, concrete efforts have been made in recent years to establish a maritime development strategy for the economic growth of the country. With the term "maritime fulcrum", there were efforts to infuse a maritime vision to long-term development planning by creating a so-called 'Navigation towards National Maritim' or Haluan Maritim Nasional (HMN), initiated by the Coordinating Ministry for Maritime Affairs, which engaged several ministries under their coordination. It is expected that the HMN will provide Indonesia with a 2045 Maritime Vision as a world's maritime fulcrum.[7] Within this Indonesian Maritime Vision 2045, Indonesia is looking to navigate its development as the World's Maritime Fulcrum which includes core sectors of maritime development: maritime transportation, maritime tourism, fisheries and mining. Yet considering urban development, the *State of Indonesian Cities* 2017 (Kementerian PUPR 2018) recounts 21 metropolitan spaces demarcated by law. Of these, 18 of them are littoral; nine metropolitan areas alone are located in the Island of Java. Yet, the national economy is still highly dependent on Java Island. The dense concentration of population and infrastructure still remains very "Javanese-centric" in its outlook, resulting in veritable socio-political and cultural biases, especially in reducing the regional disparity.

In the last decade, initial development policies to integrate the many islands of Nusantara were, for example, demonstrated by Masterplan of Acceleration and Expansion of Indonesia's Economic Development (MP3EI in Bahasa Indonesia acronym). This masterplan was legalized under the Presidential Regulation No. 1/2011 and followed the basic idea to connect the major island through economic corridors. The corridors build upon the road or railway network and serve the industrial zones, agribusiness, tourism destination, and cities. Based on MP3EI, it is expected that this corridor can work to increase job employment and investments. However, the connectivity between islands has not yet developed as planned. Multimodal interaction between sea-land transportation still remains to occur at high logistic costs.

Under the Joko Widodo Presidency, the idea to integrate Nusantara came from the opposite direction – from outside Java. Joko Widodo aimed to initiate and support economic development from the outer islands, state-border areas, and underdeveloped regions. He argued that those regions required strong intervention from the government compared to the existing metropolitan or strong economic areas that can be facilitated through cooperation between government and private sectors. In the last 5 years, there were many infrastructural projects that developed in those

---

[7] https://maritim.go.id/wujudkan-indonesia-emas-tahun-2045-kemenko-marves-terus/ accessed 23 November 2021.

regions, such as harbours, airports, road networks, and water irrigation. In the second term, he also sharpened the vision of a Global Maritime Axis (PMD or Poros Maritim Dunia in Bahasa Indonesia acronym) that he instigated in his first phase in office commencing with the development of a sea toll.[8] Subsequently, he signed the Presidential Regulation No. 16/2017 concerning an Indonesian Ocean Policy, including a corresponding action plan. Opportunities to capitalize on the geopolitical and geo-strategic location of Indonesia are also mentioned as the reason why PMD should be implemented. But in the initial steps of its implementation, there were challenges in balancing the volume of trading from the western to eastern parts of Indonesia. We argue that the huge gap in logistical demand proved to be one of the biggest challenges. Again, Nusantara connectivity still appears to be mired by economic gaps and old land-locked path dependencies.

The main difference – and dissonance – between MP3EI and PMD appears with respect to connectivity (Fig. 15.1). The impression that is shown from both pictures is easily contrasted against the arrows that denote movement and connectivity. MP3EI provides policy directions that increase land connectivity, promote defensive-oriented strategies, and offer inter-city linkages and other development nodes or hubs. On the other hand, the PMD promotes policy directions related to sea connectivity, proactive 'offensive-oriented' strategies while emphasizing port-city development. MP3EI thus focuses on the mainland, while the PMD attempts to develop sea-land connectivity. The latter is therefore closer to achieving the vision of Wawasan Nusantara and is more suggestive of a holistic translation from the top-brass of Indonesian leadership with respect to a more acute understanding of Indonesia's archipelagic geo-strategic and geo-political position. Taking this line of reasoning, national development planning is crucial in entrenching opportunities that at the same time minimize potential challenges by establishing the inter-connectivity of land and sea as a response for enacting an offensive-oriented 'game plan' in the region.

Furthermore, the strategy to develop Indonesia as maritime fulcrum is also supported by the claimed transition from an agricultural state to a global maritime country. In his speech on October 24, 2016, President Joko Widodo stated that:

> We must work harder to reclaim Indonesia as a maritime country. Oceans, straits, and bays are our future civilization. We have ignored for them too long. It is the time to give back, so *Jalesveva Jayamahe* (trans: ocean is our glory), as was our ancestors' motto in the past, that returns as an echo. (Joko Widodo, translated by Author, 20 October 2014).

---

[8] The global maritime axis is Jokowi's vision that was initially established in the first period of presidency. It consists of five pillars: (i) developing Indonesian maritime culture, (ii) establishing marine sovereignty through fisheries industries and fishermen as the main stakeholder, (iii) developing marine infrastructures and connectivity through sea-toll, sea-port, logistic and vessels, and marine tourism, (iv) marine defence, and (v) maritime diplomacy. On the last pillar, the Government of Indonesia (GoI) has been establishing the Archipelagic and Island States Forum (AIS Forum) to establish the inter-countries cooperation in climate change adaptation and mitigation, blue economy, marine debris, and ocean governance. In the period 2018–2020, the GoI through Coordinating Ministry of Maritime and Investment held several ministerial-level meetings.

Fig. 15.1 Masterplan of Economic Development 2011 (MP3EI in Bahasa Indonesia Acronym) (above) and Global Maritime Axis 2014 (PMD in Bahasa Indonesia Acronym) (bottom) (Source: MP3EI Report, 2011 and Bappenas/PT.Pelindo in Indonesia.go.id)

As mentioned, the first phase witnessed numerous infrastructure developments for increasing the regional connectivity and competitiveness of small islands groupings.. It was anticipated that the economic impacts of those infrastructures would not be short-lived.[9] Yet, statistics show that the Gross Domestic Regional Production (GDRP) of Java Island still dominates the national GDRP. If we look at the population share compared to 2014, the proportion of Java was still 56.9% of the national population. GDRP only slightly decreased from 56.7% (2016) to 56.2% (2020). In the industrial sectors, the number of mid- and large-scale industries is still Java-dominated as well (Fig. 15.2). Only 10–15% is shared with Kalimantan and other

---

[9] As reported by the President Office in 2018, and as documented in Kompas Online (20/10/2018), the sea infrastructures were 27 new ports, including eight that are still on progress, ten ferry-ports, five ferry boats and three motor boats. The air infrastructures were ten new airports and 408 improved airports. The land infrastructures were new road 3432 km, toll road 947 km, bridges 39,8 km, and suspension bridges 134 units. Those infrastructures are located mostly outside of Java Islands.

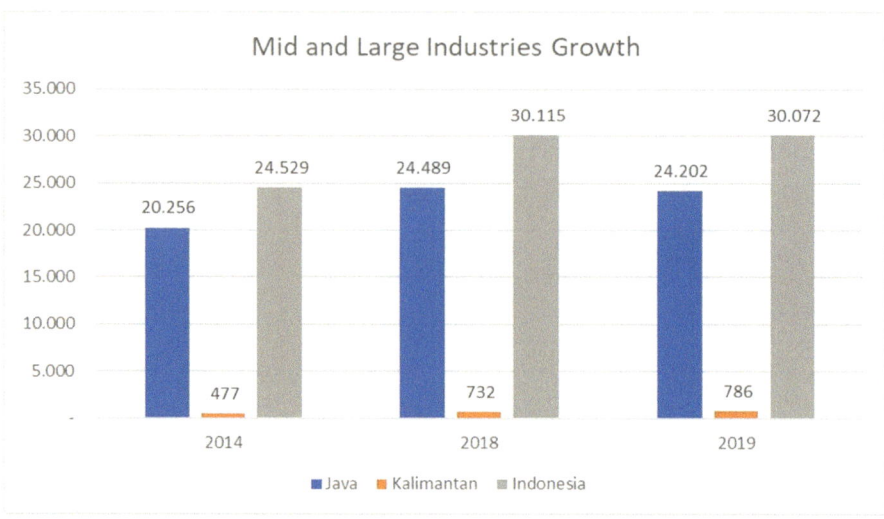

Fig. 15.2 Mid and Large Industries Growth in Java and Kalimantan Island. (Source: BPS Indonesia, 2014, 2018, 2019)

regions. It shows that the industrial transformation to outside of Java is needed, especially to Kalimantan where the new capital will be located.

However, when looking at the growth of special economic zones (SEZ), there are hints that some of the strategic decisions yield mixed results. In the first period (2014–2019), the total SEZ in 2018 was 12 areas, with only one in Java Island and the rest in the other islands. However, in the second period (2019–now), the number of SEZs is growing to 19 areas and six of them are in Java Islands. In the last three years, the number of SEZs in Java increased by five areas, while there were only three outside of Java. Therefore, it is apparently the strategy of economic de-concentration to include areas outside Java that still remains a challenge. It is obvious that the business sectors still consider Java Island as the most suitable area for investment.

Figure 15.3 below shows that the international shipping or sea-trading lane that had been established since UNCLOS 1982. The purple line denotes an international sea lane of Indonesia as the potential trigger to develop global port cities in the future. The figure indicates that from a transportation policy angle, numerous cities along the coast (marked with red dots) are strengthened to connect with the international shipping lanes (marked by purple lines), thus showing the relationality between the marine economy (primarily through shipping), and its concomitant port cities, especially the lane between Kalimantan and Sulawesi Island (Makassar Strait) where the new capital located. It should be a opportunity to develop a maritime urban corridor.

Both figures underline that the economic policies towards an ocean-centric development have been mainly implemented by developing industries and SEZ, and various sea infrastructures outside of Java. The competitiveness through

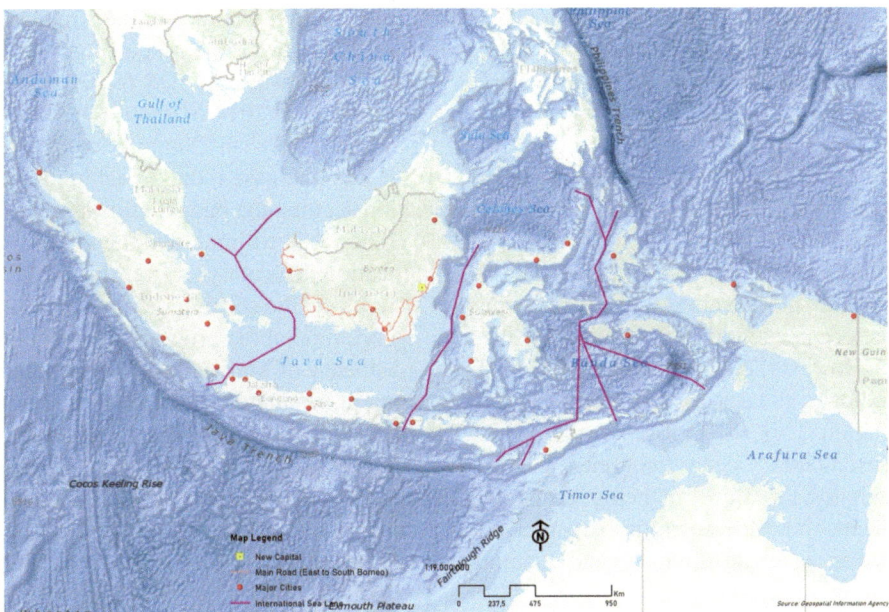

Fig. 15.3 International Sea Lanes of Archipelagic Indonesia (Source: Author, 2022, modified from various sources)

infrastructures thereby remains the key transformational strategy. However, the economic indicators show the slow response of the market to capture the immense infrastructural potential outside Java, and a maritime vision that reduces regional disparity.

We have seen that there are first signs of improvement in many regions outside of Java. Yet, the basic regional economy structure of land-based enterprise (e.g., plantations and agriculture) and agro-based industries have not changed. However, in the cultural dimension, the Ministry of Education and Culture has enlivened the history of Nusantara's former spice route by connecting old port cities. In the *Jalur rempah-rempah Nusantara* website, the program demonstrates the effort to trace back old routes and to reorient future routes of *rempah-rempah*. Port cities serve as hubs and centers of collection and distribution, and emerge as key actors.

In his second term, under the Coordinating Ministry of Maritime and Investment (MMI), Mr. Luhut Panjaitan advanced and continued policies towards becoming 'maritime nation'. The MMI kicked off the preparation by making the national document titled '2045 National Maritime Direction' (Haluan Maritim Nasional 2045 in *Bahasa Indonesia*) on September 21, 2021 in Jakarta (Kemenkomarves 2021a, b). The vision is to develop Indonesia as the centre of the largest maritime-based 'civilization' in the world. It is expected that this document will be mainstreamed into the national long-term development planning (RPJPN in Bahasa Indonesia acronym). The RPJPN 2025–2045 will be launched by 2025.

Again, the maritime vision in this second term faces challenges in the structure of national budget expenditure. In the period of 2014–2019, the range of government budget for Ministry of Ocean and Fisheries (MOF) was only IDR 6–10 trillion or USD 70–130 million, compared to total expenditure in 2014 of about IDR 1800 trillion (Ophelia 2019). According to the Chairman of Ocean Commission at the time, the budget is still insufficient: By 2019, it was decreased, only reaching IDR 5 trillion or USD 60 million. In sea-transportation sectors, the budget allocation has been increasing slightly, if compared to 2016 (IDR 11.24 trillion), and then in 2019 (IDR 12.84 trillion). The increasing budget allocation does not yet cover the marine tourism, marine energy, and other industries. Therefore, the shifting focus to capture the ocean economy opportunity is still taking more time at the programming level.

However, the presence of the Coordinating Ministry of Maritime and Investment indicates the seriousness of the central government in pursuing maritime transformation. It is shown by the active presence of Indonesia in High Level Panel for Sustainable Ocean Economy and the cooperation with UNDP in making the Indonesia's Blue Financing Strategic Document (Kemenkomarves 2021a, b). It means that blue economy is perceived as a concept to use the potential economic resources in an effective and sustainable way, avoiding pollution activities and overexploitation.

15.4 Translations of a "Maritime Indonesia" into Science Policy

Predictably, the outbreak of the novel coronavirus pandemic in 2020 contributed to a professed slow-down in debates around the maritime fulcrum and the implementation of Wawasan Nusantara into politically actionable steps. Still, we argue that Nusantara and the other ideological notions of developing a shared "maritime" Indonesian identity and development model continues to impact goal-setting agendas of other sectors and policy fields.

One policy area that has continually captured political imagination is that of national science research and science policy. With regard to maritime research activities, Indonesia's role continues to be marginal compared to the output of other G20 countries. Augy Syahailatua, Director for the Research Center for Oceanography at BRIN[10] (National Research and Innovation Agency) shared his opinion in Kompas News,[11] arguing that although national maritime research emerged 115 years ago,

---

[10] The research center for Oceanography was once under LIPI (Indonesian Institute of Sciences), now refurbished to a new super agency that host all research functions in all ministries in Indonesia and merged all research related agencies including LIPI into BRIN, commencing 1st September 2021.

[11] https://www.kompas.com/sains/read/2020/08/29/173400523/riset-kelautan-di-indonesia-maju--tapi-tertinggal?page=all (Indonesian Maritime Research – Progressing but Left Behind? 29 August 2020, downloaded 16 November 2021).

the value added of maritime research in Indonesia had yet to deliver significant impacts on the development progress of the nation, mostly due to the low number of marine-focused scientists, and the lack of concerted research work in the country of which 95% of its territories comprise waters and sea. He noted that merely 1000 of academia/university lecturers in 108 Fisheries and Maritime faculties, and only 500 marine researchers, were found to be scattered across different research organizations. The Oceanography Research Center (then still under LIPI, see footnote), had developed the *2020–2035 Foresight of Maritime Research*, but it appeared that the foresight paper was rather looking inward on how the research institute could potentially contribute to the vision of the Maritime Fulcrum or a Blue Indonesian vision – much in line with BRIN's overall research strategies.

The *Foresight* document reviewed the topics of potential future research, past research efforts, and the gaps between the two. Nevertheless, the investments in marine research are merely 0.25% of the national Gross Domestic Products (GDP) in 2017 (Ariana et al. 2017). It is relatively small proportion, compared to for example Norway, as one of the leading countries in maritime industry, with a share of 2.5% of the national GDP in 2017 (Koilo 2021). The foresight also incorporated the analysis on the Indonesian Relative Competitive Advantages (RCA) of marine research. Several topics have moderately high RCA such as marine biodiversity, coastal ecosystem research and governance. While on the other hand, research conducted by LIPI related with marine biotechnology, oceanography, ocean warming, ocean safety and deep-sea research are relatively sparse. These are typically the domains of marine research that require substantial and long term investments of research, particularly in funding. Although the foresight aimed at achieving a transformative process and becoming a new research frontier (between 2031 and 2035) through strengthening industrialized oceanography, the assumed roles are again, relatively inward looking, with the absence of ideas on expanding leadership beyond Indonesia, attending only to the imagined Maritime Fulcrum. It would be prudent to then challenge the scientific foundation of maritime governance, and how it was translated to the *Wawasan Nusantara* or the views of Nusantara doctrine in everyday life. And the expansion of knowledge and shifting of boundaries are also not supported, as argued by Syahailatua, by strong knowledge and scientific productions and investments in the maritime domain (LIPI 2017).

Also, the policies around disaster risk management could be a potential alley into strengthening the maritime focus of Indonesia. The strategic role of Indonesia in ASEAN's disaster response policies might change the course of the vision, learning how the region is prone to external dependencies after a disaster occurred, for example in the 2004 Indian Ocean tsunami that hit Indonesia, Malaysia and posed threats to Singapore or the 2008 Nargis Cyclone in Myanmar (Gottlieb 2019). The establishment of AHA Center (ASEAN Humanitarian Assistances) was in fact politically driven by the Indonesian key actors and accepted through the ASEAN conventions as part of the campaign of One ASEAN identity, including one joint response for disasters. Indonesia placed its geopolitical interests in the initiative and offered to host the secretariat and the Operation Center of AHA in Jakarta, where mobilizations of resources are managed, by agreements from all ASEAN member states (Luu and Rafliana 2021).

Similarly, in the climate change negotiation, the government of Indonesia is convinced that coastal mangrove and marine ecosystems are natural assets that can contribute to the future economic development through carbon markets. The enabling environment for this mechanism has opened through Presidential Regulation No. 98 Year 2021 concerning carbon economic value to achieve NDC. The regulation states that the climate change mitigation action on blue carbon is managed by the Ministry of Ocean and Fisheries. According to Alongi et al. (2016), "Indonesia's seagrasses and mangroves conservatively account for 3.4 Pg C, roughly 17 % of the world's blue carbon reservoir". The government also argues that there is a slowing down of deforestation in the last decade. Still, the transition of energy policy to renewable and clean energy to reduce carbon emission is needed to be accelerated. In the NDC, the mix of primary energy will be composed by 34% coal, 25% gas, 8% oil, and 33% of other non-renewable energy sources by 2050. Therefore, not only green development but the blue economy is also considered as the direction of regional economy. At the COP 26 in Glasgow, President Joko Widodo informed that Indonesia was honored to circulate a joint statement with leaders of the Archipelago and Small Island State (AIS) forum to advance maritime cooperation and climate action at UNFCCC (Setkab 2021).

As a final area of policy intervention, we will turn our attention to the relocation of the capital of Indonesia. Since the maritime vision has been brought for the last seven years but the implementation faces challenges, the government needs a game changer that can shift the focus from Java primacy to Indonesian archipelagic region. The capital relocation could have potentially offered such a transformative moment.

15.5 From Delta to Forest: De-littoralising a New 'Dry' Jakarta

Announced in 2019, the motivation to relocate the administrative functions of Jakarta City to Kalimantan Island stands to be read as a strategically oriented move favouring the archipelago. Indeed, it was proposed to name the new capital 'Nusantara'. It is hoped that its relocation reduces the state's dependency on Java, while at the same time attending to the metropolitan's socio-ecological problems such as overcrowding, flooding and subsidence (Bappenas 2019). However, an enduring question is whether moving the capital can be regarded as part of the implementation of Indonesia's maritime vision, since the selected location rests in a commercially grown forest area three hours from Balikpapan, the nearest coastal city. While we think that visions of metropolitan relocation might not seem new or novel in Indonesia's historic political discourse, narratives that feature (hinterland) East Kalimantan as a central geostrategic location to archipelagic sea-lanes, and its close proximity to the Straits of Malacca and the South China Sea, offer a new vantage point in which to study changing futures of coastal urbanities, and questions of uneven development and urbanization (Batubara et al. 2018).

Yet, at the same time, contesting vernacular and at times traditionalist notions of 'restoring' archipelagic living within urbanized settings, otherwise believed to have been lost, continues to remain the mainstay of regional planning mantras. We therefore ask, what promises and perils have been discursively articulated by the 'new' relocated Jakarta City in its hinterland spaces? In particular, while 'futuring' a new cityscape, what kinds of urban natures are envisioned? Furthermore, does its new material, socio-cultural and political positioning from 'delta-to-forest' mean muting, recalibrating and/or enlivening other notions and enactments of *Nusantara*, as understood in Indonesia's hegemonic political narrative(s)? In particular, which visions and narratives of a futuristic 'dry' city are privileged, and how do these different visions of the future intersect and at times contradict, in determining a new urban imaginary for Bornean Jakarta?

At the beginning of the most recent attempt to relocate the capital function, the Ministry of Development Planning (Bappenas) conducted location studies. At the first stage, Kalimantan Island was chosen because the island is not located in the ring of fire, although experiencing forest fires. According to the BNPB (2021), Kalimantan has a lower risk index. Still, as the mean temperature in the region is projected to increase by 1.2 °C by 2050, the potential risks are still uncertain. At the second stage, the study pointed to three potential areas: Central, South, and East Kalimantan provinces. Based on personal experiences in various public seminars, it is clear that government then aimed at selecting which of the provinces has an adequate space of around 200,000 ha, low environmental risks, clearer and cleaner land status, no social conflicts, is safe and secure, and is next to the nearest cities for ease of mobility. Yet, if we examine the selection criteria, the maritime vision was not explicitly included as major considering factor for the site selection of the new capital. The result then showed eastern Kalimantan, particularly in the area of Penajam Paser and Samboja districts. The area faces the international sea lane (ALKI II) and the Makassar Strait, but the capital land is not coastal but a forested area. Unfortunately, the new capital city does not represent a decentralized maritime connectivity to areas outside Java Island. Considering that the new capital city will adopt a forest city concept (Mutaqin et al. 2021), it may show that the maritime vision would also not be considered in the capital planning phase.

The proposed site can be reached through a three-hour trip overland from Balikpapan city, and lies on a commercially grown forest space, while being surrounded by tropical rain forests, some of which are conserved natural reserves such as Bukit Soeharto, and the protected Forest Wein River. Given these geo-spatial characteristics, the capital development will clearly face limitations to growth due to protected forest areas and high-value ecosystem barriers. In addition, the physical condition of the intended metropolitan land problematically remains hilly. Strategically, the state promotes the branded vision of a new Forest City, while its urban development still offers to be prudently cautionary, given its high cost of redevelopment. The same notion of Forest City is also developed by the private sectors in Johor Baru where a reclaimed area was used to develop forestry urban area with thousands of trees planted. The new Indonesian capital city is planned to take only 35% of forest area for the urban area (Bappenas 2020).

Drawing on the recent discourse around the capital relocation, the Forest City concept aims to develop an eco-city with modern facilities combined with minimal disaster risks. The capital must not be affected by flooding and forest-fire hazing. The location of the capital city amid the industrial forest is expected to eradicate the environmental risks faced by former Jakarta.. It is apparent that the government aims to develop a 'dry' capital to be more resilient and sustainable in the future. Yet, it leads to the absence of maritime conception in the capital relocation. Still, the waterscape in the design of the new city, particularly in relation to its built façade and landscaping of the envisioned presidential palace communicates an interpretation of 'Tanah Air.' It also connotes the abandoning of colonial symbols in building structures, features that are undetachable from the current palaces in Bogor and Jakarta (Fig. 15.4).

The absence of maritime narratives in the discourse of capital land selection has no clear explanation in recent date. However, it is clear that this branding of a 'forest city' has minimized the primacy of the maritime narrative in the future administrative capital, ironically named Nusantara. The selection of Penajam Paser Utara and Kutai Kertanegara, that actually located in the sea-lane (ALKI-2), is a weak argument to convince the public that the capital movement is key for a shift in momentum from an agriculture to a maritime country, particularly as governmental centers are to be relocated in upland Penajam Paser Utara. The central government only supported the idea of moving the development epicentre, shifting away from a Java-centric to an Indonesia-centric paradigm since Penajam is still centrally located geographically. It is hoped that there will be a new impetus of local and international transportation to frequently visit the hinterland capital.

The second reason explaining the absence of maritime narratives owes much to the placement of the new capital within a land-locked region, with no coastline or seaport. It barely embodies the faintest identity of a maritime city, as previously anticipated in plans of capital relocation. Instead, the site selection and the Forest

Fig. 15.4 The "Forest City" Image Illustration of the new capital (Source: IKN.go.id, 2022)

City concept illustrates a material and symbolic epistemic shift from that of a delta city to a hinterland forest city.

If maritime narratives were to be embodied in the vision and site selection of its capital, it may have produced a different result. As argued by UNEP (2019), the 'ocean city' may serve as a useful a concept to promote an integrated approach in planning and urban development that strengthens the linkage between nature and urbanity, climate resilience, and coastal ecosystems. It enlivens the coast and the sea as convivial living places, bringing together human habitation with natural systems.

Nevertheless, it is worth contemplating on maritime historian Adrian Lapian's work (2006 in Utomo and Sholiha 2019) on the Tidung ethnic clan groups inhabiting East Kalimantan. Theirs was a terra-aqueous culture and lifeworld patterned by the riverine and the deltaic. Lapian's argument extends to the contemporary, for it is more likely that new kinds of interconnectedness on the one hand, may lessen the division between hinterland forest and the sea, through means of transportation and digitised communication networks and technologies. On the other hand, the compounding of new socio-ecological risks also usher novel complexities that create new cycles of precarity and uncertainty that the newly rebuilt city ought to be preparing for.

In sum, placing the agenda of capital relocation within the broader matrix of Wawasan Nusantara, thereby accelerating momentum towards a concerted maritime vision would have seemed a rational decision for Indonesian planning. As Fig. 15.5 shows, this vision of Nusantara portends an imaginary of a united archipelagic region. The blue lines indicate the continental boundaries that delineate islands from inland seas, while the red marks territorial sea lanes and the dotted lines indicating Indonesia's Exclusive Economic Zone. Interestingly, the green-white line indicates the Nusantara Pendulum that connects west-central-eastern regions of the vast archipelago, pointing to the unique configuration of Nusantara as it is imaginatively mapped while missing a central feature of its construction – the capital city. Arguably then, there seems to be little 'maritime spirit' in the selection and the planning of the new city, while its main rationale appears to be discursively framed in terms of minimizing environmental risks and increasing land-based connectivity in Kalimantan.

Besides the limited connection to marine-related development components, the selected site is also surrounded by Javanese transmigration villages.[12] Rural livelihoods in the area are rooted in agricultural and forestry production. Several fisheries villages are located in the delta along Balikpapan Bay, but are not economically nor culturally connected to the proposed site for city relocation. Although the new capital will be primarily occupied by urban dwellers from Jakarta, it is likely that the current conditions of the 'new' urban social economy in Kalimantan would remain far distanced from the rhythms, sensibilities and lifeworlds of previously existing maritime-based settlements, at least in its first evolutionary phase (Fig. 15.6).

---

[12] https://nasional.tempo.co/read/1241908/ibu-kota-pindah-penajam-paser-utara-mayoritas-dihuni-orang-jawa

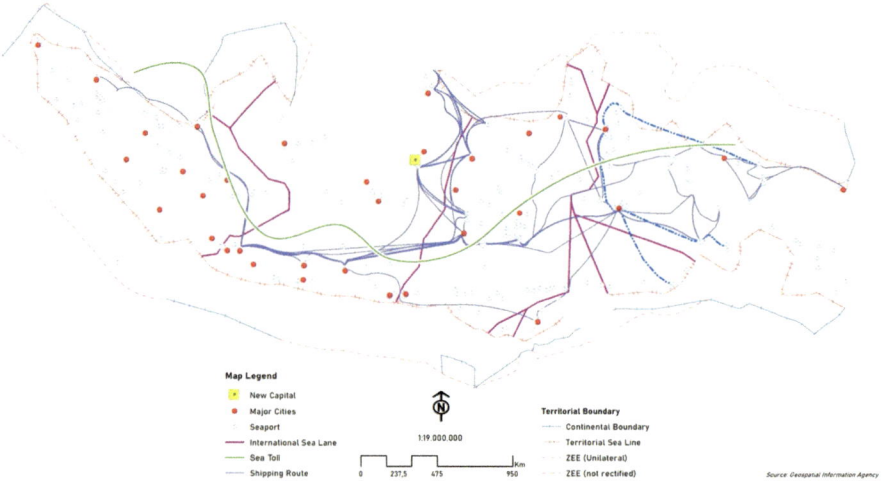

Fig. 15.5 The geostrategic location of Indonesia. The green line shows the Pendulum Nusantara, purple line shows the international sea lane, and the blue ones are the existing shipping routes. (Source: Authors, modified from various sources)

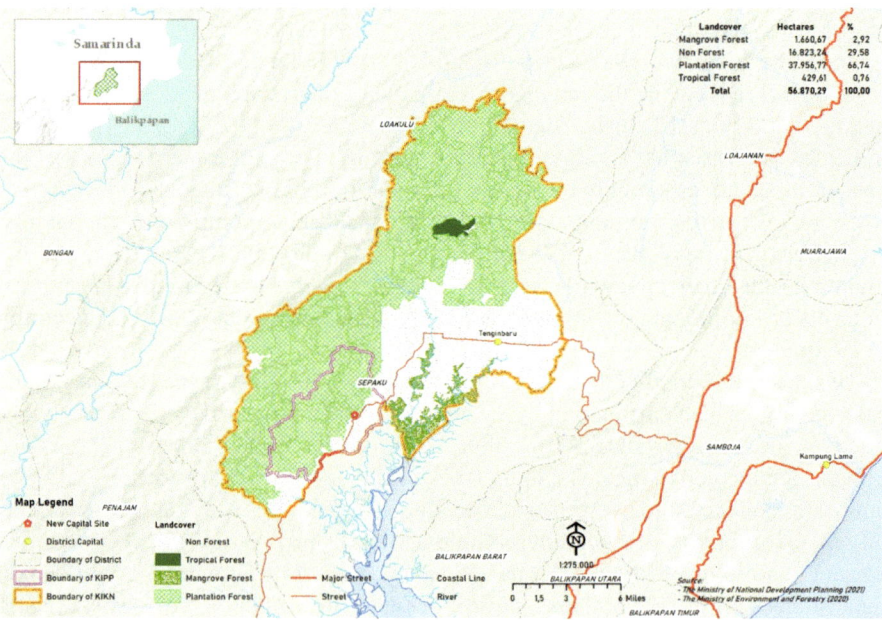

Fig. 15.6 The location of new capital. Pink-lined polygon shows the boundary of capital district, the yellow-lined one shows the capital area. While the green cross-boxes show the plantation forest, the white dots area show the non-forest area, and the green yellow area show the mangrove. (Source: Authors, modified from Ministry of National Development Planning 2021 and Ministry of Environment and Forestry 2020)

Nevertheless, the moving of the Indonesian capital city to 'dry' Kalimantan could be seen as a setback from the idea of cementing the nation as a maritime fulcrum. On the other hand, the maritime fulcrum itself is an outlook beyond the Indonesian archipelago, particularly beyond Java, in the sense of a contemporary form of Nusantara in-the-making. To a certain extent, the placemaking of a new capital city in Kalimantan then is in fact supportive of such a notion, as it approaches a closer proximity to the ASEAN neighbours and also the strategic role of the South China Sea, compared to Jakarta or Java.

15.6 *Quo vadis*: Futuring Wawasan Nusantara

In which direction archipelagic thinking can evolve in future in Indonesia, at a time in which its planners strive to co-develop new futurities for Nusantara? Archipelagic identities have long patterned Indonesian historic imaginaries, collective memory, and its postcolonial modernist narratives on nation-building. This chapter serves as a speculative analysis of different imaginaries of Wawasan Nusantara. Notwithstanding, 'Nusantara' appears to be read as a historically complex vernacular concept. Yet we focus on its enduring resonance and purchase as a political doctrine and vision by asking how its discursive meanings speak to Indonesia's recent 'oceanic' policy turn in envisioning the entire archipelago's strategic emplacement as a global maritime fulcrum.

This analysis of three sectoral domains – industrial development, science policy, and urbanization – shows how Wawasan Nusantara continues to remain as an inward-looking doctrinal discourse, while being more conservatively instrumental than socio-economically strategic. At the very least, Nusantara – as a unifying archipelagic vision – attempts to hold sway as a consolidative dream of nationhood that runs counter to narratives of political disintegration and island autonomy. Yet, it barely extends and translates into policies that attempt to bridge the material and conceptual distance between hinterland and coastal dynamics, including its landlocked Java-centrism that has flecked much of Indonesia's post-Independence politics.

Furthermore, Indonesia's articulations of its maritime fulcrum, encased in an all-too-familiar vocabulary of blue growth and the blue economy remains limp, lacking in concerted policy action and equal measures of institutional will and capacity. The fragmented silos under which Indonesia's 'maritime' sectors stand to be questioned. Although the relocation of its capital offers to be read as a missed opportunity in some regards, it may still integrate the grand vision of Wawasan Nustantara, especially by integrating small island clusters within its matrix of generating socio-economic and political value through untapped sea-borne connectivities. This calls for experimental practices for inclusively integrating existing maritime cultures and lifeworlds into existing policy plans and blueprints on fisheries, other marine industries, and the future of Indonesia's blue economy entailing disparate aspects from coastal tourism to urban development.

In the context of urbanization, the missing conception of maritime vision in the capital relocation show that the meaning of *Wawasan Nusantara* only occurred at the macro level, but was not translated at its micro level, when for example, decision-making on site selection for the relocation of its capital relocation and a new urban vision. The formation of an archipelago state relies on the very existence of coastal cities alongside the sea that they can utilize marine resources. In the context of ocean governance, ocean policy plays an important role to enable coastal cities to become maritime cities. The Indonesian urban development strategy should be mainstreamed by the maritime economy and the future blue economy. In this case, Indonesia's capital city relocation plan is part of an urban strategy that is supposed to be developed with a maritime spirit.

To empower Indonesian ocean governance, we suggest that first, ocean policies need to gradually change from insular connectivity (internal mobility) to more expansive transregionally-oriented connectivity (also serving international sea lanes). Second, the urban infrastructure also needs to support maritime sector development. Third, ocean economic orientation should shift, from initially depending solely on an extractive marine resource economy. Fourth, the institutional roles of Indonesian ministries and agencies in supporting Indonesian ocean economic policy should be more strategically integrated also in ways that synergize across scales with local city governments.

Acknowledgments We thank Nala Hutasoit, Soli Iman Santoso, Dea Amelia Yusrina, and Dhian Ivanna at the Research Center of Urban and Regional Studies, Center of Strategic and Global Studies at Universitas Indonesia for their invaluable assistance in data collection and compilation. This work has been generously funded by the German Science Foundation's 'SPP 1889 Regional Sea Level Change and Society' funding line, within the broad scope of the BlueUrban project (blueurban.org) and in concert with its regional partner Universitas Indonesia.

References

Alamsyah AT (2004) Sealand regionism – new paradigm for water based settlement development of the Jakarta metropolitan area. Urban Planning Overseas

Alamsyah AT (2010) Indonesia as an archipelagic state. Course Materials of Urban Development in tropical archipelagic region. Urban Studies Postgraduate Program. In: Universitas Indonesia, Jakarta

Alongi DM, Murdiyarso D, Fourqurean JW et al (2016) Indonesia's blue carbon: a globally significant and vulnerable sink for seagrass and mangrove carbon. Wetlands Ecol Magaz 24:3–13. https://doi.org/10.1007/s11273-015-9446-y

Ariana L et al (2017) Foresight Riset Kelautan 2020-2035 (Foresight of maritime research 2020-2035). Pusat Penelitian Oseanografi LIPI, Jakarta

Batubara B, Kooy M, Zwarteveen M (2018) Uneven urbanisation: connecting flows of water to flows of labour and capital through Jakarta's flood infrastructure. Antipode 50(5):1186–1205

Campling L, Colás A (2018) Capitalism and the sea: sovereignty, territory and appropriation in the global ocean. Environ Plan D Soc Space 36(4):776–794

Delima PS, Wirutomo P, Moersidik SS, Alamsyah TA (2019) Developing social resilience and building a culture of nationalism in the City of Batam, Indonesia. Asian Soc Sci 15(6):19–29

Evers HD (2016) Nusantara: history of a concept. J Malays Branch R Asiatic Soc 89(1):3–14

Foley P, Mather C (2019) Ocean grabbing, terraqueous territoriality and social development. Territory Polit Govern 7(3):297–315. https://doi.org/10.1080/21622671.2018.1442245

Gottlieb G (2019) 10 years after, Cyclone Nargis still holds lesons for Myanmar. The Conversation. http://theconversation.com/10-years-after-cyclone-nargis-still-holds-lessons-for-myanmar-95039. Accessed 15 Mar 22

Koilo V (2021) Evaluation of R&D activities in the maritime industry: managing sustainability transitions through business model. Prob Perspect Manag 19(3):230–246. https://doi.org/10.21511/ppm.19(3).2021.20

Krause F (2017) Towards an amphibious anthropology of delta life. Human Ecol 45(3):403–408

Laclau E (2017) Why do empty signifiers matter in politics? In: Deconstruction. Routledge, pp 405–413

Luu TTG, Rafliana I (2021) The AHA trajectory: breaking dependence through information and knowledge transfer for disaster risk management in Southeast Asia. Interdisciplinary Term Paper – ZEF-BIGS-DR

Manning C (1987) Public policy, rice production and income distribution: a review of Indonesia's rice self-sufficiency program. Southeast Asian J Soc Sci:66–82

OECD (2016) The ocean economy in 2030. OECD Publishing, Paris. https://doi.org/10.1787/9789264251724-en

Setkab (2021) COP 26: President Joko Widodo says Indonesia is committed to tackling climate change. Available online at https://setkab.go.id/en/cop26-president-Joko Widodo-says-indonesia-is-committed-to-tackling-climate-change/

Situmorang F (2013) 'Wawasan Nusantara' vs UNCLOS. Jakarta Post online. Available online at https://www.thejakartapost.com/news/2013/01/29/wawasan-nusantara-vs-unclos.html

Steinberg PE (2001) The social construction of the ocean, vol 78. Cambridge University Press, Cambridge

Steinberg P, Peters K (2015) Wet ontologies, fluid spaces: giving depth to volume through oceanic thinking. Environ Plan D Soc Space *33*(2):247–264

UNEP (2016) Regional oceans governance making regional seas programmes, regional fishery bodies and large marine ecosystem mechanisms work better together

UNEP (2019) Ocean cities: regional policy guide for delivering resilience solution in Pacific Island Settlement. Available online at https://www.unescap.org/sites/default/files/Ocean%20Cities%20Policy%20Guide_300519.pdf

Voice of Indonesia (VOI) (2021) Joe Biden predicts Jakarta sinking in 10 years, can mangroves be prevented? Retrieved from https://voi.id/en/bernas/72396/joe-biden-predicts-jakarta-sinking-in-10-years-can-mangroves-be-prevented

Wahyudi SI, Adi HP, Lekerkerk J, Jansen J, Boogaard FC (2022) Expectation of floating building in Java Indonesia, case study in Semarang City. In: WCFS2020. Springer, Singapore, pp 491–504

Indonesian Language Weblinks and References

BPS (Biro Pusat Statistik) (2020) Statistik Sumber Daya Laut Pesisir 2020 (The Statistics of Coastal Resources in 2020). Digital Media: https://nasional.tempo.co/read/1241908/ibu-kota-pindah-penajam-paser-utara-mayoritas-dihuni-orang-jawa

Badan Pusat Statistik. Capaian Kemantapan Jalan Nasional (km), 2012–2014. Diakses pada 22 Okytober 2021, dari https://www.bps.go.id/ide/17/1197/2/capaian-kemantapan-jalan-nasional.html

Badan Pusat Statistik. Jumlah Pelabuhan PenyeberideIndonesia (Pelabuhan), 2014–2016. Diakses pada 22 Oktober 2021, dari http://ideps.go.id/indicator/17/1208/1/-jumlah-ide-penyeberangan-di-indonesia.html

Badan Pusat Statistik. Jumlah Penduduk Hasil Proyeksi Menurut Provinsi dan Jenis Kelamin (Ribu Jiwa), 2015–2017. Diakses pada 22 Oktober 2021, dari https://www.bps.go.id/indicator/12/1886/2/jumlah-penduduk-hasil-proyeksi-menurut-provinsi-dan-jenis-kelamin.html

Badan Pusat StideJumlah Penduduk Menurut Provinsi di Indonesia (ribu), 2014–2018. Diakses pada 22 Oktober 2021, dari https://jatim.bps.go.id/statictablide/10/1715/jumlah-penduduk-menurut-provinsi-di-indonesia-ribu-2014-2018.html

Badan Pusat Statistik. Panjang Jalan di Pulau Jawa Tahun 2020. Diakses pada 22 Oktober 2021, dari https://databoks.katadata.co.id/datapublish/2021/08/31/panjang-jalan-di-pulau-jawa-104-ribu-km-pada-2020ide

Badan Pusat Statistik. Persentase Penduduk Daerah Perkotaan menurut Provinsi, 2010-2035. Diakses pada 22 Oktober 2021, dari https://www.bps.go.id/statictable/2014/02/18/1276/persentase-penduduk-daerah-perkotaan-menurut-provinsi-2010-2035.html

Badan Pusat Statistik. Produk Domestik Regional Bruto (Milyar Rupiah). 2018-2020. Diakses pada 22 Oktober 2021, dari https://www.bps.go.id/indicator/52/286/1/-seri-2010-produk-domestik-regional-bruto-.html

Badan Pusat Statistik. Profil Pelabuhan Perikanan 2016. Diakses pada 8 November 2021, dari https://www.bps.go.id/publication/2017/12/27/2017000000000000102373/profil-pelabuhan-perikanan-2016.html

Badan Pusat Statistik. Statistik Industri. Jumlah Industri Sedang dan Besar. Diakses pada 22 Oktober 2021, dari https://www.bps.go.id/subject/9/industri-besar-dan-sedang.html#subjekViewTab4

Badan Pusat Statistik. Statistik Pelabuhan Perikanan 2018. Diakses pada 8 November 2021, dari https://www.bps.go.id/publication/2019/10/22/008497e443749da7787717a8/statistik-pelabuhan-perikanan-2018.html

Badan Pusat Statistik. Statistik Pelabuhan Perikanan 2019. Diakses pada 8 November 2021, dari https://www.bps.go.id/publication/2020/11/27/fa482d8bc9274f46d6967f54/statistikiden-perikanan-2019.html

BNPB (2021) Laporan Akhir Kajian Risiko Bencana Kawasan Ibukota Negara (Final report disaster risks for the new capital city)

BPS (Biro Pusat Statistik) Kalimantan Timur (2021) Kalimantan Timur dalam Angka (Kalimantan Timur province in figures)

Dewan Nasional Kawasan Ekonomi Khusus Republik Indonesia. Laporan Tahunan Dewan Nasional Kawasan Ekonomi Khusus Tahun 2018. Diakses pada 22 Oktober 2021, dari https://kek.go.id/assets/images/ride9/LAPORAN-AKHIR-TAHUN-2018.pdf

Dewan Nasional Kawasan Ekonomi Khusus Republik Indonesia. Laporan Tahunan Dewan Nasional Kawasan Ekonomi Khusus Tahun 2019. Diakses pada 22 Oktober 2021, dari https://kek.go.id/asseide/report/2020/LAPORAN-AKHIR-TAHUN-2019.pdf

Dewan Nasional Kawasan Ekonomi Khusus Republik Indonesia. Peraturan Pemerintah Tentang Kawasan Ekonomi Khusus Tahun 2014-2016. Diakses pada 22 Oktober 2021, dari https://kek.go.id/peraturan-pemerintah

Dewan Nasional Kawasan Ekonomi Khusus Republik Indonesia. Peta Sebaran KEK. Diakses pada 22 Oktober 2021, dari https://kek.go.id/peta-sebaran-kek

Ernawati E (2015) Implementasi Deklarasi Djuanda dalam Perbatasan Perairan Lautan Indonesia. In: Seminar Nasional Multi Disiplin Ilmu Unisbank. Stikubank University. Retrieved from https://media.neliti.com/media/publications/171776-ID-implementasi-deklarasi-djuanda-dalam-per.pdf

Fahriyad (2013) Ada 14 bendungan groundbreaking akhir tahun ini. Diakses pada 8 November 2021, dari https://nasional.kontan.co.id/news/ada-14-bendungan-groundbreaking-akhir-tahun-ini

Hutapea E (2019) Pembangunan 29 Bendungan Kelar Akhir 2019. Diakses pada 8 November 2021, dari https://properti.kompas.com/read/2019/03/21/070000721/pembangunan-29-bendungan-kelar-akhir-2019

Iswatiningsih D, Fauzan F (2021) Semiotika Kebudayaan Masyarakat Indonesia Pada Syair Lagu. Satwika: Kajian Ilmu Budaya dan Perubahan Sosial 5(2):214–228. https://doi.org/10.22219/satwika.v5i2.18073

Jompa J (2022) Pembangunan IKN momentum bangkitkan Maritim Indonesia. Diakses Januari 2022, dari https://www.jawapos.com/ibu-kota-baru/13/01/2022/pembangunan-ikn-momentum-bangkitkan-maritim-indonesia/

Karim M (2019) Menuju Ibukota Maritim. Diakses pada Januari 2022, dari https://news.detik.com/kolom/d-4682404/menuju-ibu-kota-negara-maritim

Kementerian Keuangan Republik Indonesia (2017) 24 Bendungan Dianggarkan Rp2,38 Triliun. Diakses pada 8 November 2021, dari https://www.djkn.kemenkeu.go.id/berita_media/baca/12497/24-Bendungan-Dianggarkan-Rp238-Triliun.html

Kementerian Keuangan. Buku II Nota Keuangan Beserta Anggaran Pendapatan dan Belanja Negara Tahun Anggaran 2018. Diakses pada 22 Oktober 2021, dari www.kemenkeu.go.id/media/6665/nota-keuangan-apbn-2018-rev.pdf

Kementerian Keuangan. Buku II Nota Keuangan Beserta Anggaran Pendapatan dan Belanja Negara Tahun Anggaran 2019. Diakses pada 22 Oktober 2021, dari https://www.kemenkeu.go.id/media/11212/nota-keuangan-beserta-apbn-ta-2019.pdf. pada 22 Oktober 2021

Kementerian Keuangan. Buku II Nota Keuangan Beserta Anggaran Pendapatan dan Belanja Negara Tahun Anggaran 2020. Diakses pada 22 Oktober 2021, dari https://www.kemenkeu.go.id/media/14041/nota-keuangan-beserta-apbn-ta-2020.pdf

Kementerian Keuangan. Nota Keuangan dan Anggaran Pendapatan dan Belanja Negara Tahun Anggaran 2008. Diakses pada 22 Oktober 2021, dari www.kemenkeu.go.id/media/6612/apbn-2008.pdf.

Kementerian Keuangan. Nota Keuangan dan Anggaran Pendapatan dan Belanja Negara Tahun Anggaran 2009. Diakses pada 22 Oktober 2021, dari www.kemenkeu.go.id/media/6613/apbn-2009.pdf

Kementerian Keuangan. Nota Keuangan dan Anggaran Pendapatan dan Belanja Negara Tahun Anggaran 2014. Diakses melalui pada 22 Oktober 2021, dari www.kemenkeu.go.id/sites/default/files/nk%20dan%20apbn%202014%20full_0.pdf

Kementerian Keuangan. Nota Keuangan dan UU Nomor 14 Tahun 2006 Tentang Perubahan Atas UU Nomor 13 Tahun 2005 Tentang Anggaran Pendapatan Dan Belanja Negara Tahun Anggaran 2006. Diakses pada 22 Oktober 2021, dari www.kemenkeu.go.id/media/6610/apbn-2006.pdf

Kementerian Keuangan. Nota Keuangan dan UU Nomor 28 Tahun 2003 Tentang Anggaran Pendapatan Belanja Negara Tahun Anggaran 2004. Diakses pada 22 Oktober 2021, dari melalui. www.kemenkeu.go.id/media/6608/apbn-2004.pdf

Kementerian Keuangan. UUAPBN. Diakses pada 22 Oktober 2021, dari www.kemenkeu.go.id/uuapbn

Kemenkomarves (2021a) Dorong Ekonomi Biru dan Pembangunan Berkelanjutan di Indonesia, Pemerintah Susun Rancangan Dokumen Strategis Biru. Pers Release, diakses Februari 2022 dari https://maritim.go.id/dorong-ekonomi-biru-pembangunan-berkelanjutan-indonesia-pemerintah-susun/

Kemenkomarves (2021b) Peringati HMN ke-57, Presiden Jokowi: Identitas Indonesia Sebagai Bangsa Maritim Bukan Hanya Melalui Jargon Tapi Perlu Kerja Nyata. Available online at https://maritim.go.id/peringati-hmn-ke-57-presiden-jokowi-identitas-indonesia/

Kementerian PUPR. Dirjen Sumber Daya Air. Pagu Anggaran Tahun 2020 Sebesar Rp 120,21 Triliun : Kementerian PUPR Lanjutkan Penyelesaian Bendungan dan Bangun 20 Ribu Hektar Irigasi. Diakses pada 8 November 2021, dari https://sda.pu.go.id/balai/bwssumatera1/article/pagu-anggaran-tahun-2020-sebesar-rp-12021-triliun-kementerian-pupr-lanjutkan-penyelesaian-bendungan-dan-bangun-20-ribu-hektar-irigasi

Kementerian PUPR. Dokumen Proyek Strategis Tahun 2018 : Peta Pembangunan Infrastruktur Strategis Tahun 2018 Bidang Sumber Daya Air. Diakses pada 8 November 2021, dari http://ebudgeting.pu.go.id/dokumen/dokproyekstrategis.pdf

Kementerian PUPR. Informasi Statistik Infrastruktur PUPR Tahun 2020. Diakses pada 22 Oktober 2021, dari https://data.pu.go.id/sites/default/files/Informasi%20Statistik%20Infrastruktur%20PUPR%20Tahun%202020.pdf

Kementerian PUPR. Rencana Strategis PUPR Tahun 2015–2019. Panjang Jalan Nasional. Diakses pada 22 Oktober 2021, dari https://pu.go.id/assets/media/1666676046Renstra-2015-2019.pdf

Kementerian PUPR. Sebaran Infrastruktur PUPR. Diakses pada 22 Oktober 2021, dari https://sigi.pu.go.id/ast/

Kementerian PUPR. Sekjen Pusat Data dan Pusat Informasi (PUSDATIN). Buku Induk Statistik Tahun 2016. Diakses pada 22 Oktober 2021, dari https://setjen.pu.go.id/source/File%20pdf/Buku%20Induk%20Statistik/Buku%20Induk%20Statistik%20Tahun%202016.pdf

Kementerian PUPR. Sekjen Pusat Pengolahan Data (PUSDATA). Buku Informasi Statistik Infrastruktur Pekerjaan Umum Tahun 2014. Diakses pada 22 Oktober 2021, dari https://setjen.pu.go.id/source/File%20pdf/Buku%20Induk%20Statistik/Buku%20Induk%20Statistik%20Tahun%202014.pdf

Kementerian PUPR (2018) The State of Indonesian cities 2017. Buku Laporan Akhir Badan Pengembangan Infrastruktur Wilayah, Januari 2018

Kementerian Riset, Teknologi, dan Pendidikan Tinggi Republik Indonesia, Direktorat Jenderal Pembelajaran dan Kemahasiswaan. 2016. Buku Ajar Mata Kuliah Wajib Umum. Pendidikan Kewarganegaraan. ISBN 978-602-6470-02-7. http://luk.tsipil.ugm.ac.id/atur/mkwu/9-PendidikanKewarganegaraan.pdf. Accessed February 2022

Kontan.co.id (2021) Peneliti ITB: Bukan Jakarta yang berpotensi tenggelam pertama, tapi wilayah lain. Retrieved from https://regional.kontan.co.id/news/peneliti-itb-bukan-jakarta-yang-berpotensi-tenggelam-pertama-tapi-wilayah-lain?page=all

Lestari B (2018) Hubungan Internasional Kuno Indonesia (Konsep Perdagangan Sistem Barter di Selat Malaka dan Pemberian Nama Nusantara – Indonesia). Jurnal Majalah Ekonomi 14(1). Retrieved from http://jurnal.unipasby.ac.id/index.php/majalah_ekonomi/article/view/1373/1199

LIPI (2017) Foresight Riset Kelautan 2020-2035. Pusat Oceanografi LIPI, Jakarta

Mutaqin DJ, Muslim MB, Rahayu NH (2021) Analisis Konsep Forest City dalam Rencana Pembangunan Ibu Kota Negara. Bappenas Working Papers 4(1):13–29

Nasution DD (2019) PUPR: Jalan Nasional Terbangun 3.387 Kilometer. Diakses pada 22 Oktober 2021, dari https://republika.co.id/berita/ekonomi/korporasi/19/02/21/pn9gsu370-pupr-jalan-nasional-terbangun-3387-kilometer

Nurhidayati N (2021) Dari Deklarasi Djuanda ke Wawasan Nusantara: Peranan Mochtar Kusumaatmadja dalam Mencapai Kedaulatan Wilayah Laut Indonesia, 1957-1982. Jurnal Kajian Sejarah dan Pendidikan Sejarah (J Hist Stud Hist Educ) Susurgalur 9(1):37–54. Retrieved from https://www.journals.mindamas.com/index.php/susurgalur/article/view/1429/1215

Manurung P (2014) Arsitektur Berkelanjutan, Belajar Dari Kearifan Arsitektur Nusantara. Simp. Nas. RAPI XIII-2014 FT UMS, pp A-75–A-81

Ophelia S (2019) Anggaran untuk sektor kelautan dan perikanan belum maksimal. Available online at https://ekonomi.bisnis.com/read/20191204/99/1177774/anggaran-untuk-sektor-kelautan-dan-perikanan-belum-maksimal

Pusat Statistik. Persentase Penduduk Daerah Perkotaan Indonesia (2010-2035). Diakses pada 22 Oktober 2021, dari httpsideks.katadata.co.id/datapublish/2021/08/18/sebanyak-567-penduduk-indonesia-tinggal-di-perkotaan-pada-2020

Rimbakita N. Pengertian, Sejarah, Wilayah, dan Wawasan Nusantara. Rimbakita.com website. Available online at https://rimbakita.com/nusantara/

Santoso D, Firmaningsih A, Setyowati DN (2020) Sejarah Peristiwa Sumpah Palapa Dalam Kitab Pararaton. SULUK: Jurnal Bahasa, Sastra, dan Budaya 2(1):44–51. Retrieved from http://jurnalfahum.uinsby.ac.id/index.php/Suluk/article/view/303/176

Setiawati TW, Mardjo M, Paksi TFM (2019) Politik Hukum Pertanian Indonesia Dalam Menghadapi Tantangan Global. Jurnal Hukum Ius Quia Iustum 26(3):585–608

Shalihah F (2016) Perlindungan Hukum Terhadap Kedaulatan Wilayah Negara Republik Indonesia Menurut Konsep Negara Kepulauan Dalam United Nation Convention On The Law Of The Sea (UNCLOS) 1982. In: Prosiding Seminar Nasional Perbatasan, Kemaritiman. Retrieved from http://repository.uir.ac.id/490/1/2.%20snpk%20umrah%20full.pdf

Sinaga O (2014) Pidato Joko Widodo: Sudah Lama Kita Memunggungi Laut. 2014. Diakses pada 20 Oktober 2021, dari https://nasional.tempo.co/read/615707/pidato-JokoWidodo-sudah-lama-kita-memunggungi-laut/full&view=ok

Susilowati D, Qur'ani HB (2021) Analisis Puisi Tanah Air Karya Muhammad Yamin dengan Pendekatan Struktural. Jurnal LIterasi 5(1). Accessed 18 Sept 2021

Utomo IN, Sholihah F (2019) Dari Hilir ke Hulu: Perkembangan Sejarah Maritim Indonesia dan Permasalahannya. In: Seminar nasional dan Temu Alumni HMPS 2019. FIS UNY Yogyakarta. Accessed 6 Sept 2021

Zulfikar F (2021) 82 Nama Pelabuhan di Indonesia, Terbentang dari Sumatera hingga Papua. Diakses pada 22 Oktober 2021, dari https://www.detik.com/edu/detikpedia/d-5679505/82-nama-pelabuhan-di-indonesia-terbentang-dari-sumatera-hingga-papua

Chapter 16
Market Initiatives of Small-Scale Fisheries in the Mediterranean: Innovation in Support of Sustainable Blue Economy

Jerneja Penca and Alicia Said

Abstract The study of traditional marine stakeholders, such as small-scale fishers in the Mediterranean, represents a site of a changing seascape. This is characterized by impeding factors of the past but also a possibility for improved future trajectories. Small-scale fisheries (SSF) have played a crucial socio-economic role in the Mediterranean for decades, and they continue to comprise over 80% of the fishing fleets and provide direct and indirect economic contributions to coastal communities. Their contribution to blue economy has so far been described as low, but this is largely due to a narrow conception both of benefits to be drawn from the development of maritime sectors (which have focused strongly on economic growth) and types of innovation that are capable of supporting the transition to sustainability (which have overlooked social innovation). This chapter outlines the multi-scale contributions of the small-scale fisheries and presents innovative approaches of the sector towards the markets, both of which support the inclusion of SSF in the blue economy sector. The chapter focuses on key instances of recently developed initiatives by the SSF across the Mediterranean with impacts on the supply chain and the marketing of their products. We argue that these market interventions contribute to the ultimate governance objectives, and challenge the conception of SSF as a non-innovative sector. We propose that a richer engagement with the blue economy paradigm supports the perception of the SSF as a prospective sector, to match the promotion of aquaculture among others.

J. Penca (✉)
Euro-Mediterranean University, Piran, Slovenia
e-mail: jerneja.penca@emuni.si

A. Said
Department of Fisheries and Aquaculture, Luqa, Malta

© German Institute of Development and Sustainability (IDOS) 2023
S. Partelow et al. (eds.), *Ocean Governance*, MARE Publication Series 25,
https://doi.org/10.1007/978-3-031-20740-2_16

16.1 Introduction

Blue economy has emerged as a policy notion to refer to the use of seas and oceans as the Rio+20 summit in 2012 reaffirmed the commitment to a sustainable future of the planet at the highest level.[1] The policy notion has effectively raised the profile of the marine and maritime space in global, regional and sub-regional contexts to an unprecedented level, but it has not been not without contestation. The development of "blue" visions of futures has been particularly problematic for taking place without the participation of, and careful attention to the needs of coastal communities that depend on and live within these stretches of space (Barbesgaard 2018); for lacking clarity of terms and supporting competing discourses (Silver et al. 2015; Keen et al. 2018; Penca 2019a) and for promoting wrong targets (Hadjimichael 2018; TBTI 2019). The notion of blue economy has managed to direct the political and public discourse and political action particularly to those sectors that bring new opportunities for investment and hold a potential for future development, such as marine renewable energy, coastal and cruise tourism, maritime transport, marine biotechnology and aquaculture. In policy reports, fisheries and particularly small-scale fisheries were for a long time not perceived as a prospective blue economy sector.

Small-scale fisheries (SSFs) constitute one of the sectors that are impacted by the dual nature of the policy notion of blue economy as both an opportunity and a threat for their empowerment. Small-scale fishers across the world have been key players in the marine socio-economic realm providing direct contributions to coastal communities in terms of local economies, nutrition and their identity, as well as indirectly to tourism. Various studies have highlighted the invisibility of SSFs in the blue economy discussions. For instance, SSF were not mentioned in any of the EU's documents related to blue economy (Stobberup et al. 2017). Arguments have thus been made for SSF to secure their space in the marine realm globally (Cohen et al. 2019), with some going as far as replacing the rhetoric in the institutionalization of policies governing the marine space with new concepts, such as 'blue justice' (TBTI 2019), 'blue commons' (Standing 2019) and 'blue degrowth' (Hadjimichael 2018). Despite being key, and probably the pioneer users of the sea, SSFs remain the missing sector in the discussions surrounding the vision of blue economy and how it ought to shape the future of the marine resource use.

Convincing appeals have been made to consider SSF within the promulgation of policies, owing to the fact that the features of SSF are much more compatible with a blue economy and sustainable fisheries than industrial fisheries (Pauly 2018; Said and MacMillan 2020). The case for including the SSF sector in blue economy,

---

[1] Early policy documents and scholarly literature used the term 'blue growth' alongside 'blue economy', but this has gradually become fully replaced by 'blue economy', In 2021 the EU, an early advocate of the term 'blue growth' settled for 'sustainable blue economy' (EC 2021). This chapter intentionally avoids the discussion over the meaning of each and implied preference for one over the other. Instead, it uses the notion of blue economy as a policy paradigm that has a policy and strategic, but no legal, nature.

rather than pushing it outside, is rooted in their positive social impacts of the enhanced economic wealth and the avoidance of environmental risks (Cohen et al. 2019). According to this view, SSF can effectively be considered as contributing to blue economy insofar as they contribute, on a sectoral level, to achieving some other sustainability targets, such as biodiversity conservation, reduction of poverty, gender equality and climate resilience. SSF are aligned with the wider transformation required to accomplish the Sustainable Development Goals (SDGs) and sustainability (Said and Chuenpagdee 2019). One aspect from within the sector of SSF that has so far not yet been put to the focus as contributing to blue economy, is the SSFs dynamic and innovative adjustment to markets as a form of social innovation and building of resilience.

In this chapter, we argue that in addition to SSF's close alignment with sustainability, SSF have recently demonstrated a level of innovation in using markets, providing new prospects and jobs. We highlight the rise of tangible actions in different EU and some non-EU Mediterranean countries, related to organization of the supply chain of SSF as a means of overcoming the multiple structural challenges faced by the SSF. By taking into account these innovative activities in support of sustainability, we argue, the sector is well placed to be acknowledged as a driver of the sustainability transition.

We focus on the Mediterranean, but research has shown that the innovative marketing and selling activities by SSF are not unique to this region; their presence as means of resistance to mainstream monopolized markets has been observed in other parts of the world with relatively strong institutions (Stoll et al. 2015; Witter and Stoll 2017; Penca 2019b; Prosperia et al. 2019; Duggan et al. 2020). The Mediterranean initiatives have generated interest for the variety of activities, taking place against an alarming state of Mediterranean fisheries as well as the strong tradition of fishing and seafood consumption (Penca et al. 2021; Gómez and Maynou 2021). In this chapter, we focus on some instances of innovative marketing initiatives from across the Mediterranean, as gathered through a mapping exercise. We, present these as specific tangible opportunities for and by SSF in the context of the policy paradigm of blue economy. While these market activities have been overlooked in the reports of the further potential of the blue economy by the policymakers, they firmly position SSF within the ambit of a sustainable, job-generating and innovative economy of the future. As such, the described market initiatives defy the negative outlooks for SSF in the past decades and hold promise in the context of the future policy opportunities. We consider these market innovations underpinning community economies as driving forces for the recognition of the SSF within the core of future maritime strategies.

The chapter is structured as follows. By way of background, in Sect. 16.2, the chapter briefly outlines the history of the SSF's struggle for participation in the governance of the seas, characterised by the lack of voice by SSF in both decisions over the use of the sea and those impacting the markets of fishing resources, due to which the SSF have been pushed to the periphery. This overview of the past allows us to appreciate the recent signals of a more proactive intervention by SSF in markets and their supply chains. These are described in Sect. 16.3, which aims at highlighting the diversity of market initiatives, rather than their comprehensive overview. Section 16.4 highlights value of the SSF market activities first in the context of resilience and innovation, as two values underpinning blue economy, and secondly, as supportive of the policy impulses that are shaping the future of ocean governance to argue for their greater support than received so far.

16.2 Struggling to Be 'There': Historical Invisibility of SSF in Policies and Markets

Systemic marginalization of SSF has impacted on the specific resilience strategies. Although fishing in the Mediterranean was set off through the effort of small vessels with their passive gear, this sector became increasingly lost with the industrialized global development of the fishing sector. Driven by public policies aimed at economic efficiency, the growth of the large-scale sector came at a cost for the SSF and their role in the production of fish catches, ultimately impacting on their relative invisibility in the markets. While this is a global trend, it is particularly visible in the Mediterranean. Here, SSF fleet comprises 80% of the fleet and SSF account for 74% of employment in fisheries, but lands only 20% of the total landings (FAO 2018), making SSF thus unable to be the main players in the seafood markets. While facing market competition by both the large-scale and aquaculture productions, SSF – in the Mediterranean as much as elsewhere – have also been unable to differentiate their products and make them more visible. To a large extent this is contingent on the policies that made no effort to treat SSF as any different, or worthy of special measures and approaches. This section provides a brief overview of the unfavourable situation for SSF at two levels: (inter)governmental policies and strategies on the one hand, and the recognition on the markets on the other. Jointly, these seemingly independent spheres reflect the 'blindness' of the policy-makers to the significance and distinctiveness of the needs and complexities of small-scale fisheries.

16.2.1 The Policy Context

Over most of the history of fisheries management, and predominantly commencing in the post-war period, public policies (national and international) have been favouring large-scale fleets and not paying much attention to the small-scale sector (Chuenpagdee and Jentoft 2018). Through much of the twentieth century, globally SSF struggled to be included in the decisions about management approaches, funding and access to resource (vis-à-vis larger fleet, but also other users of the sea) (Griffiths et al. 2007; Carpenter et al. 2016). The emergence of the rift between the SSF and industrial fishing can be related to the governments' perception that trawlers are associated with 'efficient fishing', and a subsequent heavy support offered to them. Large-scale fleet became heavily supported by subsidies leading to overinvestment and overfishing (Schuhbauer et al. 2017; Jacquet and Pauly 2009). This led to a rapid growth in Mediterranean fishing fleet and a proliferation of trawlers by mid-twentieth century that enabled many more vessels to fish further offshore and in deeper waters (Pauly, 2018). In comparison with the much more dispersed small-scale sector, the large-scale fleet is considered as easier to monitor, negotiate with and extract data from. In addition, fishery policies and management systems were built on data that only large fisheries were required to provide (Kolding et al. 2014). In turn, also research has largely focused on industrial fishing (Smith and Bassurto 2019).

Marginalization and resilience characterise the Mediterranean SSF fleet. Here, SSF have been historically very important both in terms of social contribution, catches and economic value, but have undergone a serious decline (Guyader 2008). Nevertheless, SSF still account for the greatest part of the fleet (circa 80%) in the region and more than half of the total workers employed in the sector, albeit with great variety across the region (FAO 2019).[2] As we see, multiple interrelated drivers of the structural support for a certain type of fishing and policy sequence set a long-lasting focus on industrial fishing, and a concomitant disregard towards the needs and challenges of SSF. Apart from the access to the resources, SSF have been heavily affected by other disenabling factors of environmental and governance nature. Competition for space from the spread of aquaculture, marine tourism, marine protected areas and maritime transport; threatened material base due to expanding pollution from land and sea, overexploitation and unsustainable fishing practices; hazardous and uncertain working conditions of the fishers, irregularity and seasonality of their income and low returns to their fishing are all factors that inhibit the progress of SSF (FAO 2019). As a result of these challenges, the SSF is largely unattractive to the young generation and is indeed not being rejuvenated.

The international policy framework has only recently given a new hope for the SSF. This came in the form of the adoption of Voluntary Guidelines on Small Scale Fisheries in 2014 and the SDGs in 2015, with a specific target on SSFs. On the Mediterranean regional level, the General Fisheries Commission for the Mediterranean is set to strengthen and support sustainable small-scale fisheries in the Mediterranean region through a regional plan of action (RPOA), which was signed in 2018. This aims at setting the scene for better management of small-scale fisheries in the next ten years and beyond. Also, the EU seeks to make progress towards more socially, environmentally and economically sustainable fish stocks, and better integration of SSF, as stated in the objectives of the Common Fisheries Policy (CFP) (Regulation 1380/2013). Indeed, the criticism of an overly large EU fishing fleet, harmful subsidies and a lack of focus on ecosystem management have been the drivers of the reforms of the CFP (1992, 2002 and 2013). The most recent CFP reform contains the EU's declaratory statement in support of SSF ("the CFP

---

[2] In the EU alone, the SSF represent 80% of the fleet, and provide for around 60% of jobs but only 23% of landings (EC 2019).

should contribute to increased productivity, to a fair standard of living for the fisheries sector including small-scale fisheries") (Art 2.5(f) of Regulation 1390/2013). However, in practice, this still remains to result in any practical impact on SSF and to challenge the status quo (Said et al. 2020).

There are however some challenges to the full consolidation of a more positive policy towards SSF. In the EU, the SSF have not been brought into its blue economy discussions, previously dubbed as 'blue growth' (EC 2012). It has been argued that fisheries had not been considered because over 80% of the assessed stocks are overexploited and thus growth would simply exacerbate the worrying situation of the stocks (Da-Rocha et al. 2019). The EU's Blue Growth Strategy, which focused on economic benefits, rather than social and environmental aspects, highlighted the following priority sectors with a high potential for job creation and research and development: aquaculture, coastal tourism, marine biotechnology, ocean energy and seabed mining (EC 2012). The favouring of these sectors has implied a loss of attention with regards to some other prospective sectors, such as SSF (Said and Macmillan 2020). Also the subsequent strategy by the EU, issued in 2021 titled Transforming the EU's Blue Economy for a Sustainable Future and announcing a more holistic approach to the blue economy does not single out the SSF sector as in need of specific attention from within the fisheries and sustainable food systems (EC 2021). Reference as: European Commisssion (EC) Communication on a new approach for a sustainable blue economy in the EU: Transforming the EU's Blue Economy for a Sustainable Future. COM/2021/240 final. 17.5.2021.

16.2.2   The Market Context

Indirectly, markets have borne the impact of public policies that have, through laws, regulations and market interventions, "mainly focused on increasing productivity and facilitating the development of capital-intensive fisheries with larger and more productive vessels" (Pascual et al. 2019). They have done little to offset the heavy burden of globalised seafood markets onto local markets using traditional market systems (Gomez and Maynou 2021). A complex interplay of factors contributes to a situation of the fisheries markets supporting neither socio-economic well-being of the fishers nor the environmental sustainability. The intricate relationship between various pressures has been detailed in in-depth studies (Ertör et al. 2020; Penca et al. 2021; Gomez and Maynou 2021). In essence, small scale fishers suffer from heavy pressures imposed by globalised value chains, lack of transparency and traceability of these value chains, rigid consumer demand, and poor entrepreneurial attitude of most of fishers. In most parts, SSFs are deeply entrenched into the existing models to ensure everyday survival, which prevents them from transforming the existing socially exploitative and environmentally unsustainable marketing patterns.

Part and parcel of the predominant governance paradigm that pushes SSFs out of, rather than into centre stage, is the lack of systematic measures for ensuring visibility of SSF's products in the markets and for ensuring organisational aspects of their value chains. From a perspective of a market participant, the SSF product cannot be formally distinguished from the catch of the industrial fleets and aquaculture, and at best it can be distinguished informally. When placed on the market, SSF-sourced species sit next to catches sourced by other fleets, and their distinction can only be deciphered by specific knowledge that a consumer could potentially hold.

The current markets for seafood are characterised by the overall blurring between the industrial, farmed, imported and even illegal products. The products by SSF are hardly explicitly distinguished from products heavily implicated in international trade. Symptomatic of the globalised markets is the wide presence of farmed Norwegian salmon or imported tuna across coastal towns of the Mediterranean, while the products caught in these markets are often traded somewhere else to receive a better price. With the powerful marketing that salmon receives, including regular presence and campaigns about its health benefits as well as due to the ease of its preparation, salmon has become a pervasive species replacing the traditional fish catches of local communities. Ironically, such campaigns are also able to water down the actual environmental and health concerns including those associated with viruses and eutrophication (Taranger et al. 2015). A significant challenge lies in ensuring transparency of the products, enforcement and consumers' awareness. Catch of the same species that enter the market from abroad is equalled to the domestic catch without consumers necessarily noticing it. An example which the authors came across whilst conducting research is that of swordfish from the Pacific sold as local in the Mediterranean regions, as well as common sea bream from Oman.

The mandatory product labelling rules, to the extent they are even required in different countries beyond the EU, do not mandate a sufficient differentiation either (Penca 2020). The EU's legal framework for labelling of seafood, for instance, had the ambition of providing a high level of protection to the consumer (EP 2011). However, while the EU regulations[3] require the statement of the fishing gear that was used and the origin of the product, it does not communicate the information in a way that allows the consumer to gain information about the exact provenance and freshness of the product (Penca 2020), to which also consumers have expressed complaints (Eurobarometer 2018). Thus, the EU's rules do not allow the consumer to infer whether a product was fished by a small-scale fisher, or instead an industrial fishing boat. In addition, the labelling system further suffers from a very low transparency of the supply chains and low compliance, where products are mislabelled or the mandatory labelling is missing altogether (Helyar et al. 2014; Esposito and Meloni 2017). The role of mandatory labelling is thus marginal in better communication about the product to the consumer.

One possibility for distinguishing the products in the seafood markets was private certification. Building on the experience from the forestry sector, seafood certification developed during the 1990 and proliferated in the form of various certifications schemes (such as Friend of the sea, Dolphin-free tuna, or the largest of

---

[3] European Commission (n/a), A pocket guide to the EU's new fish and aquaculture consumer labels, available at https://ec.europa.eu/fisheries/sites/fisheries/files/docs/body/eu-new-fish-and-aquaculture-consumer-labels-pocket-guide_en.pdf

all, the Marine Stewardship Council -MSC) (Gulbrandsen 2009; Auld and Cashore 2013). However, certification schemes developed as a seeming solution to failed state governance and sluggish pace of change, primarily to the issue of overfishing. Conceptually, they do not respond to the demands of SSFs (Ponte 2012; Hadjimichael and Hegland 2016; Penca 2019b). They focus primarily on environmental attributes of the products and the environmental context of the catch. While in some cases those environmental indicators overlap with the SSF product, this is not always the case. While the currently dominant seafood certification scheme on the market, MSC, is not designed against the SSF, in practice SSF and fisheries from developing countries of the Global South find considerable obstacles to attain a label, mostly due to its costs to the participating fishers (Duggan and Kochen 2016). The certification schemes' focus on single-species is another significant challenge in the Mediterranean context, as Mediterranean small-scale fisheries mostly target mixed fisheries. The fact that certification process is performed in relation to the species in a particular fishery, rather than the fisher or community, is fundamentally at odds with the nature of SSF. The very low number of MSC-certified fisheries in the Mediterranean reflect the tension between the design of MSC (or any other certification scheme) and the needs of the Mediterranean SSF. Indeed, the MSC has itself acknowledged the difficulties of engaging the SSF and their relative under-performance, and as a consequence has devoted special attention to facilitating the SSF in the pre-certification phase (MSC 2019).

A considerably more meaningful response to the non-distinctiveness in the market have been various actions of collectivisation and cooperation between SSF. Triggered by the consumers' emergent interest for sustainably sourced seafood (McClenachan et al. 2016) and by the governance indication of the need to progress on improving the access of SSF to markets (SDG14b), a number of dissimilar activities relating to branding, marketing and retailing of SSF products have been observed in various parts of the world and have been dubbed alternative seafood marketing programmes (Witter and Stoll 2017; Duggan et al. 2020; Gomez and Maynou 2021) and market empowerment tactics (Penca 2019b). In these, SSF have started innovating in supply chains and the marketing of their products, as well as in cooperating within themselves more closely, with the view of gaining a stronger position in the market. Individually and collectively, these activities are believed to have brought about and made visible the benefits to SSF by increasing profit-taking, consumers choice and building a stronger community identity (Stoll et al. 2015; Duggan et al. 2020), as well as contributed to the empowering of the SSF as a stakeholder in policy-making (Penca 2019b).

16.3   Mediterranean SSF Innovations in Markets

The Mediterranean towns and regions have been part of the trend of the rise of novel approaches to marketing and selling the product by SSF. In this section, we identify such activities as tools by SSF to counter the past negative policy trends affecting

SSF, including the impact of export markets on the prices of their local produce, their inability to access quotas, and other challenges that have watered down their resilience, such as competition from recreational fisheries (Said et al. 2018) and reduced fishing grounds due to coastal development (Said et al. 2017). In this section, we offer a brief account of the various strategies that have been set off to differentiate the SSF seafood from the rest of the market, seek to retain the value in the SSF sector or add value to their product. In line with the prior studies on SSF marketing initiatives (Verhaegen and Van Huylenbroeck 2001; Kitts and Edwards 2003; Devaux et al. 2009; Barham and Chitemi 2009; Kaganzi et al. 2009; Foley and McCay 2014; Pascual-Fernández et al. 2019; Penca 2019a; Duggan et al. 2020), our account highlights the range of activities across the Mediterranean region, set off by the small-scale fishers. We prioritise the breadth of the initiatives over a more granular analysis of an individual case, precisely to highlight the diversity of the activities in the movement and its uncoordinated materialisation. We give but some examples of various types of such initiatives, rather than a comprehensive account of them.

To begin with, the Mediterranean small-scale fishers are increasingly becoming engaged in awareness-raising and promotional activities that seek to showcase the quality of SSF products and highlight the specificities of their catch. These are probably the most widespread of the approaches to alternative marketing and empowerment. The ultimate objective of these activities is widening the SSF markets and valorisation of SSF products. Awareness-raising can be done through compiling and distributing consumer information about the value of SSF, including by concrete consumer guides as to what fish to eat in a certain area and which not. QuickFish Guide by Fish4tomorrow NGO in Malta provides an example of surveying commonly purchased species, evaluating their sustainability and providing a recommendation on their purchase. A more proactive approach to promotion and awareness raising is typically run in form of food shows, festivals, classes and similar gastronomic events that introduce new types of SSF products or facilitate their preparation, and thus contribute to their popularity. These activities can be implemented either on an ongoing basis by a local community or association of fishermen, but are often kicked-off by a publicly-funded project. A few examples of such campaigns are the Cephs & Chefs project working on promoting the use of cephalopods (squid, octopus, cuttlefish) in the Atlantic area; the summer festival called Barche aperte ("Open boats") run in a coastal town of Caorle in the Veneto region, Italy that welcomes people to fishing boats and allows them to purchase their fish directly from fishermen; the Mediterranean Culinary Academy in Malta that trains chefs in preparation of local seafood according to old and forgotten traditions; or open-air cooking shows by an association Pescados con Arte in Cartagena, Spain.

Another popular approach to improving the position of small-scale fishers is the setting-up of short supply chains. The idea in these activities is to either improve the distribution or the valorisation of local catch by SSF by involving as few middlemen as possible. Short supply chains encompass direct sales at local markets or individual stands that are characteristic of many towns across the Mediterranean. Some initiatives involve the use of agreements between SSF and restaurants or hotels on the purchase of the 'catch of the day'. For example, some high-end hotels in Istria,

Croatia, have secured the purchase of the locally caught Norway lobster. In Slovenia, as in some other countries, fishers are allowed to sell seafood directly from the boat up to a certain amount – in this case up to 50 kg daily. Direct sales by the fishers reduce the middleman costs, securing higher profit margins from their catches. But short supply chains do not necessarily result in less kilometres travelled. Fresh seafood can be sent to where the expected value is higher. Quite often, urban centres provide a better selling point because of the higher purchasing power. Thus, seafood caught by small-scale fishers from the Gulf of Lion in France is sent over 700 km away to Paris to supply the high-end restaurants with quality fish.

Efforts in ensuing short supply chains have in recent years also made use of technology to expand their customer base, improve logistics and valorise the product. In many cases, the use of technology allows the fishers to reach new customers, for instance younger and more urban generations that would otherwise have chosen a competing product (farmed, imported or processed). Some online markets specialise in SSF products, typically informing the consumers of the daily catch of the artisanal fisher through an online platform or communication system, sometimes even before the landing. The project Fresh Fish Alert taps into the Sicilian market and scales up the practice of direct communication. In many other places, the predominant form being used to reach the customers is a simple text message with an indication of the catch of the day.

A special type of a short supply chain are fish boxes (fish baskets). Originating from North America, in recent years they have made their way to the Southern parts of Europe. They operate in a number of countries, for instance in Italy under the name FishBox, in Portugal as Cabaz do Peixe or in Gökova Bay in Turkey. This model is particularly appropriate for taking into account the Mediterranean characteristics of a large variety of species and a high degree of unpredictability of the catch. In a typical arrangement, the consumer agrees to receiving a certain weight, rather than the exact type of seafood and thus accepts an element of surprise as to what to receive. The fish basket system usually involves communal distribution at an agreed place (e.g. a public square, school etc.), but a personalised delivery at the customers' door has also been observed.

A distinct method of branding of the SSF product is the creation of individual labels testifying that products have been caught by SSF. SSF labels are mostly of very recent origin and many of them are only developing their recognition. One of the better-known ones is a label Golion which marks the products that have been caught by SSF in the Gulf of Lion, France and mostly sold to restaurants in Paris and some other big cities. At the national level, the idea of a nation-wide French label for SSF products has been considered in detail (Petit Peche) but not yet applied. There is considerable interest by consumers for a similar national label to be implemented also in other countries (Zander and Feucht 2018: 40). In Tunisia, Association Blue Club Artisanal has put in place a system, whereby artisanal Tunisian fishermen are providing identifiable, traceable and quality-controlled SSF products to certified restaurants, including to Sicily, Italy. Such a collaboration across national jurisdictions is a very rare example of the transboundary initiative.

Despite the close proximity of borders in the Mediterranean and straddling fish stocks as well as cross-border fish trade, the initiatives have been mostly targeting the customers of the same region or country at most. There had been very little consideration of transnational efforts towards the joint objective. The only form of cooperation in SSF product branding involving multiple countries can be found in the efforts led by the Slow Food Foundation, through its Slow Fish arm. Slow Fish operates a network of SSF across the world and is gaining popularity in the Mediterranean. Its activities include a logo that small-scale fishers, who are concurring to the principles of the movement, can use on their processed (not fresh) products. The Slow Fish movement subscribes to the slogan of "good, clean and fair fish" that demonstrates a commitment to the values of social benefit, low environmental impact and quality product. Slow Food also labels restaurants that subscribe to this vision and offer SSF products. As such, it represents a rare instance of an international scaling-up of the SSF efforts.

The most structural response to the fragile socio-economic situation facing SSF is the strengthening of SSFs organisation in order to address the fishery marketing processes alongside fishery management processes. Activities can take the form of establishing cooperatives to establishing Producer organization (PO), as a body with a particular capacity to lead to strengthening the participating SSF representation as well as their organizational and governance capacity, established under EU fishery laws (Cazalet and O'Riordan 2020). Two examples of strengthened SSF's representation are the establishment of the PO of artisanal fishers in Lonja de Conil from Spain and a proposal for a PO submitted by the Golion Economic Interest Group from France.

16.4 Significance of SSF Initiatives: Resilience and Innovation

Initiated and run either by small-scale fishers or other stakeholders on their behalf, SSF marketing activities have been borne to resist the status quo for SSF. A clear feature that emerges from an overview of initiatives is that all the initiatives are local or, at best, regional in nature. Interestingly, the initiatives do not show any ambition of expanding geographically, responding instead to highly local-specific functioning of the markets. While being limited in scope and character; the initiatives aspire to build bridges among themselves only informally at best, and mostly do not reference one another or coordinate among them. Their resistance to globalised and largely similar value chains seems to be demonstrated precisely by insisting on their limited, local outreach. Two sets of qualities related to SSF emerge from observing the initiatives: resilience and innovation on the one hand, and their contribution to sustainability.

16.4.1   Recognizing Resilience and Innovation

The initiatives were triggered by different reasons and pursued different tactics. Some sought to secure a market for their high-quality product and established a delivery to upscale markets, even if these were located further away. Others focus on a local distribution of common and less-valued species. In so far as they have responded to different needs, also their impact on the fishers and the consumers varies. Yet, what they jointly demonstrate is the underlying rationale of the SSF attempting to differentiate their product in the seafood markets through branding, marketing or retailing. The development of distinct channels of recognition or sale is a tool for resistance to market competition from large-scale fishery products, foreign imports, and aquaculture products. With the regulatory frame on labelling as well as distribution not conducive to differentiating the SSF products (consumers in the value chain were not able to trace the product or to differentiate between large and small scale fishers), small-scale fishers were subjected to competition in globalised trade conditions. The initiatives deployed improvements in infrastructure (ice carrying, distribution logistics), marketing (use of apps, development of new channels of sale) or organisation (clustering the SSF fishers to jointly present their product) in order to position SSF as suppliers of a distinct, high-quality product.

By recognizing the role that small-scale fishers have had in developing bottom-up marketing strategies within different contexts, a clearer link emerges between the small-scale fishers' ability to valorise their products and their contribution to the blue economy strategy at the Mediterranean level. The initiatives defeat the portrayal of the fishers as passive actors, lacking the ability to change the way livelihoods are earned. Instead, they speak of a proactive attitude in resisting the challenges related to resource access in regulated environments (in the EU, these are even highly regulated) and to change the course of action, by improving the visibility of SSF products, their marketing or the organisation of the value chain. In this respect, the initiatives demonstrate once again an inherent resilience of SSF and their long-term viability (Nayak and Berkes 2019).

These efforts should be recognised as social innovation, a segment of innovation as a central concept of the notion of blue economy, alongside jobs (Bluemed 2018; OECD 2019; EC 2021). Innovation in this case is not about instituting new kinds of material production (or technological breakthroughs in key technologies) – these are difficult to influence as the fish is fixed natural resource. Instead, innovation related to SSF lies in tackling challenges (of the organisation of the market, changing consumer preferences etc.) with new immaterial approaches to them, such as IT selling tools, marketing strategies or preparing the product in a way to better correspond to consumers' needs. SSFs marketing approaches often combine the agentic factor (entrepreneurial attitude) with structuralist factors (aiming to change the fishers' collectives and ways in which they organise) (Cajaiba-Santana 2014). By coming closer to the consumer's interests and enhancing sustainability, they accrue the value to society as a whole rather than only its own sector (Phills et al. 2008). Social innovation has been largely overlooked by the key policy documents on blue economy that have focussed on science and technology to propel economic growth. The dominant framing of innovation as technological change is restrictive to maritime sectors with inherent but unexplored capacities to challenge systemic structures for social progress. As a consequence, these are pushed out from policy attention and priorities. However, in the case of SSF, social innovation contributes to the establishment of a customer base, widening of the outreach and optimisation of the supply chain, and ultimately a better profitability and viability of the sector. In turn, improving profitability in the markets can set off a positive upward spiral of the formalization of the fisher profession in the sector, their improved welfare and finally also improved fisheries management (FAO 2019). The value of market initiatives in triggering the dynamic change should be recognised.

16.4.2 Transitioning to Sustainability

Not only have the market initiatives by SSF provided a way for SSF to sidestep the legislative and policy obstacles that they have been facing (from struggles to ensure fishing opportunities to damages from unfair competition by subsidized industrial fishing), they are a promising tool to transform governance structures because they embody a synergistic sustainability practice at the intersection of environmental, economic and social objectives.

The sustainability-centred approach offers a considerably richer perception of SSF and their role in the blue economy. It highlights that growth in fisheries can be constituted in terms other than growth in landed catches, such as an increased value of the same volume of fish, accomplished for instance through improved marketing and retailing. It draws attention to the social implications of those catches on fishers and coastal communities. Moreover, it brings to light that SSF and their market initiatives are capable of a number of positive synergies with other maritime activities (Stobberup et al. 2017).

Indeed, such a holistic consideration of SSFs is more in line with the FAO interpretation of blue economy, which is inclusive of SSF and has been used to boost messages of sustainability and to further promote the FAO Code of Conduct for Responsible Fisheries (adopted in 1995) as well as other relevant instruments, including the Voluntary Guidelines on Small Scale Fisheries (FAO 2014). The same view is found in the vision of ocean-based sustainable economy, put forth by the Ministerial declaration of the Union for the Mediterranean (UfM), which recognises the importance of building a more cohesive and sustainable Euro-Mediterranean region. The understanding of blue economy in the Euro-Mediterranean has the ambition to have a positive distributional effect and reduce disparities (Penca 2019a) and explicitly recognises the role of fisheries as a sector and the work done by the General Fisheries Commission for the Mediterranean in that regard (UfM 2015). Finally, a consultation process at the national levels in 9 EU and non-EU member countries that looked at ways to develop a healthier, productive, resilient, better known and valued Mediterranean Sea has confirmed that SSF have a role to play in that vision (Bluemed 2018).

The SSF market initiatives are consistent with key policy documents to which the states in the Mediterranean have committed. The first among these is the EU's Common Fisheries Policy (Art 1) and the Mediterranean ten-year RPOA, signed in 2018 by the contracting parties of the GFCM encompassing both EU and non-EU countries. The RPOA calls for increased focus on the key elements for SSFs value chain. In that context, RPOA encourages new ventures for SSF products, such as the creation of cooperatives, producer organizations or other organizations. The plan also calls for improvements in the profitability and viability of SSF in so far as these are environmentally sustainable, through increased quality and traceability. These provisions of the RPOA support the implementation of a system that differentiates SSF products at the market level and promotes their visibility to consumers. This would include enhancing the link between harvesters and consumers through direct sales and awareness-raising campaigns of SSFs catches, especially of an underutilized nature. Finally, the RPOA also promotes the role of certification and branded labels to increase the visibility of small-scale fisheries catches.

Another key document is the Voluntary Guidelines for Securing Sustainable Small-Scale Fisheries in the Context of Food Security and Poverty Eradication, adopted in 2014, by the FAO Committee on Fisheries. The SSF Guidelines have specific segments on the role of the markets for SSF and means through which market access can be enhanced for SSF. For example, the SSF Guidelines state that countries "should foster, provide and enable investments in appropriate infrastructures, organizational structures and capacity development to support the small-scale fisheries post-harvest subsector in producing good quality and safe fish and fishery products, for both export and domestic markets, in a responsible and sustainable manner". This speaks of the need of increased investment in capacity building, also to strengthen the organizational capacity of fishing communities to become better involved in the marketing of their products. In the context of the Mediterranean, this could imply capacity-building for fishing communities to establish marketing plans and strategies, or if no organizations are in place, to establish groups or associations to improve the role that SSF play in the markets.

The overarching roadmap to take into account sustainability as a more long-term, overarching and legal norm (Bosselman 2017) are Sustainable Development Goals, adopted in 2015. Having been adopted by the United Nations General Assembly, the SDGs enjoy a high level of international consensus and commitment. SDGs contain specific targets on SSF (SDG 14.b), which aims at "provid[ing] access for small-scale artisanal fishers to marine resources and markets". In this respect, progress is planned with regards to "application of effective legal/regulatory/policy/institutional framework, which recognizes and protects access rights for small-scale fisheries." Moreover, the empowerment of the SSF sector contributes to a number of other SDGs apart from the ocean-focused one (14), notably the eradication of hunger and malnutrition, and increase of small-scale producers and sustainable food production (Targets 2.1–2.4.), promotion of sustainable tourism that creates jobs and promotes local culture and products (Target 8.9), promotion of gender equality

(Targets 5A and 5C) (Said and Chuenpagdee 2019). In the era of increased policy convergence, where only the causes that are embedded within (a few) broader goals can survive, SSF activities through markets need to be recognised for the contributions they are making to other policy agendas.

Finally, SSF initiatives are conducive to the direction of the required sustainability transition in the domain of food systems. As food systems are subjected to pressures from increased population as well as more urbanised population, a greater focus on localising food production, and re-establishing the linkages between urban centres and their rural surroundings has been suggested (Jennings et al. 2015; SAPEA 2020). A closer re-connection between urban and rural areas, through more effective provision of ecosystem services, shorter food supply chains and encouraging regional food businesses, are believed to deliver more sustainable ecologic and socio-economic returns (Jennings et al. 2015) and are experiencing a particular renaissance after the Covid-19 pandemic (Vittuari et al. 2021). Additionally, they are sought after by the rise of conscientious consumers (Fiorino et al. 2018) and their preference for non-market values, including to creating a conscious food governance (Witter and Stoll 2017). In that context, SSF market initiatives contribute a concrete practice to demonstrating that transformative change is possible and that it can emerge from bottom-up.

16.5   Conclusion

Reflecting the reality of many other parts of the world, SSFs of the Mediterranean demonstrate the benefits that are daily drawn from marine resources, and play the role as their custodians. However, small-scale fishers of the Mediterranean have continually struggled over a number of issues of ocean governance, including disadvantaged access to fish resources, competition for space and a diminishing material base. Additionally, their products have been systematically under-appreciated by the consumers. The rise of the policy rhetoric of blue growth provided yet another episode in that struggle, as it has downplayed the contribution of SSF in future visions of the seas.

In this chapter, we deployed the rhetoric of blue economy not as inhibiting the visibility and support to the SSF practices and interests, but as enabling the value of innovative practices by SSF and their resilience. We offered to acknowledge the existing endogenous capacities by small-scale fishers to innovate through the organisation of small-scale fishers, the operation of their value chains and marketing of their products as concrete means to valorise their products, create jobs and improve socio-economic growth, while having a low impact on the environment. These values are in line with the key motivations underlying the blue economy notion, understood originally as "the final frontier for humanity and its quest for sustainable development" (n/a 2014) and supporting sustainable development and cohesion of the Mediterranean (UfM 2015). As such, we suggest to consider the blue economy policy paradigm, especially when it is embedded in the overarching goal of sustainability, as conducive to the needed policy transformations and a more prosperous future of the SSF.

The conceptual re-framing of the blue economy from a threat on SSF to an opportunity allows us to depart from the complaints over the lack of attention of existing policy priorities on the SSF and instead highlight concrete actions and activities by SSF that are both innovative and sustainable. By recognising the recent SSF market initiatives as valuable, we can also identify the points of entry for adequate governance interventions. For such initiatives to scale-up and multiply, active engagement is needed in form of stimulating networking, learning and capacity building. Targeted policy measures (public campaigns for promotion and valorisation of localised, artisanal and small-scale production, tax incentivisation, increase in funding dedicated to starting-up and scaling-up), can also be beneficial for the development of these positive SSF activities (Penca et al. 2021), alongside an acceleration of the policy process dedicated to improving access to resource of the SSF (Said et al. 2020). Finally, better coordination among the existing initiatives regarding marketing innovation and capitalization on various, unconnected efforts would result in a stronger voice of this traditional sea-centred activity and its role in the future of oceans.

References

Auld G, Cashore B (2013) Mixed signals: NGO campaigns and Non-state Market Driven (NSMD) governance in an export-oriented country. Can Public Policy 39(Supplement 2)

Barbesgaard M (2018) Blue growth: savior or ocean grabbing? J Peasant Stud 45(1):130–149. https://doi.org/10.1080/03066150.2017.1377186

Barham J, Chitemi C (2009) Collective action initiatives to improve marketing performance: lessons from farmer groups in Tanzania. Food Policy 34(1):53–59

Bluemed (2018) Strategic and Innovation Research Agenda (SRIA), Updated version 2018. Available at http://www.bluemed-initiative.eu/wp-content/uploads/2017/09/BLUEMED-SRIA_Update_final.pdf

Bosselman K (2017) The principle of sustainability: transforming law and governance, 2nd edn. Routledge, London

Cajaiba-Santana G (2014) Social innovation: moving the field forward. A conceptual framework. Technol Forecast Soc Chang 82:42–51

Carpenter G, Kleinjans R, Villasante S, O'Leary BC (2016) Landing the blame: the influence of EU Member States on quota setting. Mar Policy 64:9–15. ISSN 0308-597X. https://doi.org/10.1016/j.marpol.2015.11.001

Cazalet B, O'Riordan B (2020) The pros and cons of creating Producer Organisations (PO) for Mediterranean small-scale fishers. Published by Low Impact Fishers of Europe

Chuenpagdee R, Jentoft S (2018) Transforming the governance of small-scale fisheries. Maritime Stud. https://doi.org/10.1007/s40152-018-0087-7

Cohen P et al (2019) Securing a just space for small-scale fisheries in the blue economy. Front Mar Sci. https://doi.org/10.3389/fmars.2019.00171

Da-Rocha J-M, Guillen J, Prellezo R (2019) (Blue) Growth accounting in small-scale European Union fleets. Mar Policy 100:200–206. https://doi.org/10.1016/j.marpol.2018.11.036

Devaux A, Horton D, Velasco C, Thiele G, López G, Bernet T, Reinoso I, Ordinola M (2009) Collective action for market chain innovation in the Andes. Food Policy 34(1):31–38

Duggan DE, Kochen M (2016) Small in scale but big in potential: opportunities and challenges for fisheries certification of Indonesian small-scale tuna fisheries. Mar Policy 67(May):30–39

Duggan GL, Jarre A, Murray G (2020) Alternative seafood marketing in a small-scale fishery: barriers and opportunities in South Africa's Southern Cape Commercial Linefishery. Maritime Stud 19:193–205. https://doi.org/10.1007/s40152-020-00175-1

Ertör I, Brent Z, Gallar D, Josse T (2020) Situating small-scale fisheries in the global struggle for agroecology and food sovereignty. Transnational Institute. Available at https://www.tni.org/en/small-scale-fisheries

Esposito G, Meloni D (2017) A case-study on compliance to the EU new requirements for the labelling of fisheries and aquaculture products reveals difficulties in implementing Regulation (EU) n. 1379/2013 in some large-scale retail stores in Sardinia (Italy). Reg Stud Mar Sci 9:56–61. https://doi.org/10.1016/j.rsma.2016.11.007

Eurobarometer Special 475 (2018) EU consumer habits regarding fishery and aquaculture products, (D.-G. f. C. European Commission, Trans.)

European Commission (EC) (2012) Blue Growth opportunities for marine and maritime sustainable growth. COM/2012/0494 final. https://ec.europa.eu/maritimeaffairs/sites/maritimeaffairs/files/docs/publications/blue-growth_en.pdf. Accessed 11 Dec 2019

European Commission (EC) (2019) Directorate-General for Maritime Affairs and Fisheries. The EU blue economy report 2019. Publications Office. Available at https://data.europa.eu/doi/10.2771/437478

European Commission (EC) (2021) Communication from the Commission on a new approach for a sustainable blue economy in the EU Transforming the EU's Blue Economy for a Sustainable FutureCOM(2021) 240 final

European Commission (EC). Specificities of the Mediterranean Sea. Available at https://ec.europa.eu/fisheries/cfp/mediterranean/specificities_en

European Parliament (EP) (2011) Regulation (EU) No 1169/2011 of the European Parliament and of the Council of 25 October 2011 on the provision of food information to consumers

Fiorino GM, Garino C, Arlorio M, Logrieco FA, Losito I, Monaci L (2018) Overview on untargeted methods to combat food frauds: a focus on fishery products. J Food Qual 3:1–13. https://doi.org/10.1155/2018/1581746

Foley P, McCay B (2014) Certifying the commons: ecocertification, privatization, and collective action. Ecol Soc 19(2)

Food and Agriculture Organization (FAO) (2014) Voluntary Guidelines for Securing Sustainable Small-Scale Fisheries in the Context of Food Security and Poverty Eradication. FAO Doc. COFI/2014/Inf.10, 9–13 June 2014. Available at http://www.fao.org/3/i4356en/I4356EN.pdf

Food and Agriculture Organization (FAO) (2018) The State of World Fisheries and Aquaculture 2018 – meeting the sustainable development goals. Rome. Retrieved from http://www.fao.org/3/i9540en/I9540EN.pdf

Food and Agriculture Organization (FAO) (2019) Social protection for small-scale fisheries in the Mediterranean region – a review. Rome. Available at http://www.fao.org/3/ca4711en/ca4711en.pdf

Gómez S, Maynou F (2021) Alternative seafood marketing systems foster transformative processes in Mediterranean fisheries. Mar Policy 127(2021):104432. https://doi.org/10.1016/j.marpol.2021.104432

Griffiths RC et al (2007) Is there a future for artisanal fisheries in the western Mediterranean? FAO-COPEMED, Rome

Gulbrandsen LH (2009) The emergence and effectiveness of the Marine Stewardship Council. Mar Policy 33(4):654–660

Guyader O (coord.) (2008) Small-scale coastal fisheries in Europe. Final Report of the project. European Commission FISH/2005/10

Hadjimichael M (2018) A call for a blue degrowth: unravelling the European Union's fisheries and maritime policies. Mar Policy 94:158–164. https://doi.org/10.1016/j.marpol.2018.05.007

Hadjimichael M, Hegland TJ (2016) Really sustainable? Inherent risks of eco-labeling in fisheries. Fish Res 174:129–135

Helyar SJ et al (2014) Fish product mislabelling: failings of traceability in the production chain and implications for illegal, unreported and unregulated (IUU) fishing. PLoS One 9:e98691. https://doi.org/10.1371/journal.pone.0098691

Jacquet J, Pauly D (2009) Funding priorities: big barriers to small-scale fisheries. Conserv Biol 22(4):832–835

Jennings S, Cottee J, Curtis T, Miller S (2015) Food in an urbanised world: the role of city region food systems. Urban Agric Magaz 29:6–7

Kaganzi E, Ferris S, Barham J, Abenakyo A, Sanginga P, Njuki J (2009) Sustaining linkages to high value markets through collective action in Uganda. Food Policy 34(1):23–30

Keen MR, Schwarz A-M, Wini-Simeon L (2018) Towards defining the Blue Economy: practical lessons from Pacific Ocean governance. Marine Policy (February):333–341

Kitts AW, Edwards SF (2003) Cooperatives in US fisheries: realizing the potential of the fishermen's collective marketing act. Mar Policy 27(5):357–366

Kolding J, Béné C, Bavinck M (2014) Small-scale fisheries: importance, vulnerability and deficient knowledge. In: Garcia SM, Rice J, Charles A (eds) Governance of marine fisheries and biodiversity conservation. Wiley, New York, pp 317–331

Marine Stewardship Council (MSC) (2019) Making waves: small-scale fisheries achieving sustainability with the MSC. Available at https://www.msc.org/docs/default-source/default-document-library/what-we-are-doing/msc-small-scale-fisheries-report-2019.pdf

McClenachan L, Dissanayake ST, Chen X (2016) Fair trade fish: consumer support for broader seafood sustainability. Fish Fish 17(3):825–838. https://doi.org/10.1111/faf.12148

Nayak PK, Berkes F (2019) Interplay between local and global: change processes and small-scale fisheries. In: Chuenpagdee R, Jentoft S (eds) Transdisciplinarity for small-scale fisheries governance, MARE publication series, vol 21. Springer, Cham

OECD (2019) Rethinking innovation for a sustainable ocean economy. OECD Publishing, Paris. Available at https://doi.org/10.1787/9789264311053-en

Pascual-Fernández JJ, Pita C, Josupeit H, Said A, Garcia Rodrigues J (2019) Markets, distribution and value chains in small-scale fisheries: a special focus on Europe. In: Transdisciplinarity for small-scale fisheries governance. Springer, Cham, pp 141–162

Pauly D (2018) A vision for marine fisheries in a global blue economy. Mar Policy 87(Jan):371–374

Penca J (2019a) Blue economy in the Euro-Mediterranean: implications of the policy paradigm. Int J Euro-Medit Stud 12(1):69–92

Penca J (2019b) Transnational localism: empowerment through standard setting in small-scale fisheries. Transnation Environ Law 8(1):143–165

Penca J (2020) 2020. Frontiers in Marine Science, Mainstreaming sustainable consumption of seafood through enhanced mandatory food labelling. https://doi.org/10.3389/fmars.2020.598682

Penca J, Said A, Cavallé M et al (2021) Sustainable small-scale fisheries markets in the Mediterranean: weaknesses and opportunities. Maritime Stud 20:141–155. https://doi.org/10.1007/s40152-021-00222-5

Phills J, Deiglmeier K, Miller D (2008) Rediscovering social innovation'. (Fall) Stanford Soc Innov Rev

Ponte S (2012) The Marine Stewardship Council (MSC) and the making of a market for "sustainable fish". J Agrar Chang 12(2–3):300–315

Prosperia P, Kirwanb J, Mayeb D, Bartolinia F, Vergaminia D, Brunoria G (2019) Adaptation strategies of small-scale fisheries within changing market and regulatory conditions in the EU. Marit Policy 100:316–323. https://doi.org/10.1016/j.marpol.2018.12.006

Regional Plan of Action for Small-scale Fisheries in the Mediterranean and the Black Sea (RPOA-SSF) (2018) 26 December 2018

Regulation (EU) No 1380/2013 of the European Parliament and of the Council of 11 December 2013 on the Common Fisheries Policy, amending Council Regulations (EC) No 1954/2003 and (EC) No 1224/2009 and repealing Council Regulations (EC) No 2371/2002 and (EC) No 639/2004 and Council Decision 2004/585/EC. OJ L L 354/22, 28.12.2013, pp 22–61

Said A, Chuenpagdee R (2019) Aligning the sustainable development goals to the small-scale fisheries guidelines: a case for EU governance. Mar Policy 107. https://doi.org/10.1016/j.marpol.2019.103599

Said A, MacMillan D (2020) 'Re-grabbing' marine resources: a blue degrowth agenda for the resurgence of small-scale fisheries in Malta. Sustain Sci 15(1):91–102. https://doi.org/10.1007/s11625-019-00769-7

Said A, MacMillan D, Schembri M, Tzanopoulos J (2017) Fishing in a congested sea: what do marine protected areas imply for the future of the Maltese artisanal fleet? App Geo 87:245–255. https://doi.org/10.1016/j.apgeog.2017.08.013

Said A, MacMillan D, Campbell B (2018) Crossroads at sea: escalating conflict in a marine protected area in Malta. Estuar Coast Shelf Sci 208:52–60. https://doi.org/10.1016/j.ecss.2018.04.019

Said A, Pascual-Fernández J, Iglésias Amorim V, Højrup Autzen M, Hegland TJ, Pita C, Ferretti J, Penca J (2020) Small-scale fisheries access to fishing opportunities in the European Union: is the common fisheries policy the right step to SDG14b? Mar Policy 118. https://doi.org/10.1016/j.marpol.2020.104009

SAPEA, Science Advice for Policy by European Academies (2020) A sustainable food system for the European Union. SAPEA, Berlin. Available at https://doi.org/10.26356/sustainablefood

Schuhbauer A, Chuenpagdee R, Cheung WWL, Greer K, Sumaila UR (2017) How subsidies affect the economic viability of small-scale fisheries. Mar Policy 82(March):114–121. https://doi.org/10.1016/j.marpol.2017.05.013

Silver J et al (2015) Blue economy and competing discourses in international oceans governance. J Environ Dev 24(2):135–160

Smith H, Basurto X (2019) Defining small-scale fisheries and examining the role of science in shaping perceptions of who and what counts: a systematic review. Front Mar Sci, 7 May 2019. https://doi.org/10.3389/fmars.2019.00236

Standing A (2019) From blue growth to the "blue commons". Coalition for Fair Fisheries Arrangements. Available at https://cape-cffa.squarespace.com/en-blog/2019/3/4/from-blue-growth-to-blue-commons

Stobberup K, Dolores Garza Gil M, Stirnemann-Relot A, Rigaud A, Franceschelli N, Blomeyer R (2017) Research for PECH Committee – small-scale fisheries and "Blue Growth" in the EU. European Parliament, Policy Department for Structural and Cohesion Policies, Brussels. Available at https://www.europarl.europa.eu/RegData/etudes/STUD/2017/573450/IPOL_STU(2017)573450_EN

Stoll JS, Dubik BA, Campbell LM (2015) Local seafood: rethinking the direct marketing paradigm. Ecol Soc 20(2):40. https://doi.org/10.5751/ES-07686-200240

Sustainable Development Goals Knowledge Platform. https://sustainabledevelopment.un.org/sdg14

Taranger GL, Karlsen Ø, Bannister RJ, Glover KA, Husa V, Karlsbakk E, Kvamme BO, Boxaspen KK, Bjørn PA, Finstad B, Madhun AS, Morton HC, Svåsan T (2015) Risk assessment of the environmental impact of Norwegian Atlantic salmon farming. ICES J Mar Sci 72(3):997–1021. https://doi.org/10.1093/icesjms/fsu132

Union for the Mediterranean (UfM) (2015) Ministerial conference on blue economy. Declaration. 17 November 2015, Brussels

Verhaegen I, Van Huylenbroeck G (2001) Costs and benefits for farmers participating in innovative marketing channels for quality food products. J Rural Stud 17(4):443–456

Vittuari M et al (2021) Envisioning the future of European food systems: approaches and research priorities after COVID-19. Front Sustain Food Syst 5. https://doi.org/10.3389/fsufs.2021.642787

Witter A, Stoll J (2017) Participation and resistance: alternative seafood marketing in a neoliberal era. Mar Policy 80:130–131

Zander K, Feucht Y (2018) Consumers' willingness to pay for sustainable seafood made in Europe. J Int Food Agribusiness Mark 30(3):251–275. https://doi.org/10.1080/08974438.2017.1413611

Chapter 17
Towards Just and Sustainable Blue Futures: Small-Scale Fisher Movements and Food Sovereignty

Irmak Ertör and Pinar Ertör-Akyazi

Abstract Oceans and seas have been vital food sources for both coastal and terrestrial communities for thousands of years. Traditionally, the main actors were small-scale fishers adopting more ecologically-benign fishing practices either for their own subsistence or small-scale commercial use and livelihood. Members of small-scale fishing communities frequently combine other socioeconomic activities such as small-scale agriculture and animal husbandry with their fishing activity as well. Thus, they usually have broader and different understandings and narratives regarding their relations and interdependency with the fish and the seas compared with industrial capture fisheries targeting the most profitable commercial fish species using more destructive gears and high technological capacities. In this chapter, we aim to shed light on their past and present—as well as highlight their existence as a rather neglected and marginalized social group, their political agency and their global movement for food sovereignty in order to uncover their social, political and ecological roles for the future of oceans, coastal communities, and the society in general. Our research methodology relies on participant observation and action methods based on 3 years of continuous work with small-scale fishing cooperatives in Turkey, Spain and Europe, as well as following and collaborating with the WFFP (World Forum of Fisher People) members both in Europe and globally. We conducted more than 80 interviews with key actors from fisheries sector including policy makers, NGOs, members of fishing cooperatives, and fisheries and marine scientists that inform this investigation. We claim that even though small-scale fishing communities are usually neglected actors of the 'present' in most mainstream marine policies, narratives and agendas such as the Blue Economy, their 'presence' in ocean governance is of utmost importance and their future existence needs to be ensured for an ecologically, socially and economically just ocean governance.

I. Ertör (✉)
The Ataturk Institute for Modern Turkish History, Boğaziçi University, Istanbul, Turkey
e-mail: irmak.ertor@boun.edu.tr

P. Ertör-Akyazi
Institute of Environmental Sciences, Boğaziçi University, Istanbul, Turkey

© German Institute of Development and Sustainability (IDOS) 2023
S. Partelow et al. (eds.), *Ocean Governance*, MARE Publication Series 25,
https://doi.org/10.1007/978-3-031-20740-2_17

17.1   Introduction

Oceans and seas have been vital food sources for both coastal and terrestrial communities for thousands of years. Traditionally, the main actors were small-scale fishers (SSFs) adopting more ecologically benign fishing practices either for their own subsistence or small-scale commercial use and livelihood. Yet, both subsistence and commercial use can co-exist within the same community or fishing cooperative. Members of small-scale fishing communities frequently combine other socioeconomic activities such as small-scale agriculture and animal husbandry with their fishing activity. They crucially depend on marine ecosystems for their livelihoods, as opposed to the industrial capture fisheries targeting the most profitable commercial fish species using more destructive gears and high technological capacities.

Small-scale fisheries are defined differently depending on the national and legislative context. The Food and Agriculture Organization, for instance, uses the terms "small-scale" and "artisanal fisheries" interchangeably and define them as "traditional fisheries involving fishing households (as opposed to commercial companies), using relatively small amount of capital and energy, relatively small fishing vessels (if any), making short fishing trips, close to shore, mainly for local consumption".[1] Frequently, though, vessels smaller than 12 meters are identified as small-scale by national fisheries policies. A recent FAO report (FAO 2020) indicated that in 2018, 82% of all motorized fishing vessels in the world were smaller than 12 meters, identifying them as "small-scale" vessels. Their diversity with respect to species caught, harvesting technology used, institutional characteristics, and other social and economic relations make small-scale fisheries a quite dynamic sector, which can adapt relatively easily to changing ecological and social conditions. Scientific studies therefore indicate that a broader range of social, economic and ecological relations such as gender relations, value chains and the ways of interacting with the marine ecosystems should be used to complement the analysis of SSFs (Schuhbauer and Sumaila 2016).

In this chapter, we aim to shed light on the past and present of small-scale fishers—by exploring their local and global organizations/initiatives and their role in the global movement for food justice and food sovereignty (Sinha 2012; Levkoe et al. 2017; Mills 2018). This scrutiny enables us to uncover their social, political and ecological roles for a more just and sustainable future of the oceans, coastal communities, and society in general. Our research methodology relies on participant observation and action methods based on 3 years of collaborative work with a range of groups striving for agroecology and food sovereignty in fisheries. These include small-scale fishing cooperatives in the Istanbul region, Turkey—a member

---

[1] http://www.fao.org/faolex/glossary/en/

Table 17.1 Interviewed actors

Interviewed actors	Country
Small-scale fishing representatives	Turkey, Spain, Mauritania, Kenya, Indonesia, Thailand, India, Ecuador, Honduras
Environmental NGOs	Europe, Spain, Turkey
Policy makers	Europe, Spain, Turkey

of the World Forum of Fisher People (WFFP)[2] since 2017—, initiatives working on agroecology and food sovereignty in small-scale fisheries in Spain and Europe, as well as other European, regional and global WFFP members. We conducted more than 80 interviews with key actors most of whom were WFFP members from different countries such as Turkey, Spain, Mauritania, Kenya, Indonesia, Thailand, India, Ecuador and Honduras as well as with policy makers, NGOs, and fisheries and marine scientists, all of which inform this investigation. A table summarizing our interviews is provided below (Table 17.1).

As a result, we claim that even though small-scale fishing communities are usually neglected actors of the 'present' of marine policies in narratives and agendas such as Blue Economy and Blue Growth (European Commission 2012; African Union 2015), their presence in ocean governance is of utmost importance and their future existence needs to be ensured for an ecologically, socially and economically just ocean governance. This way, we aim to strengthen the voice of SSFs as marginalized actors of ocean governance as well as contribute to the political debates around food security, food sovereignty and fisheries governance.

The next section analyses the past and present of SSFs by uncovering their significant role in food production, sustainable use of the seas and oceans, and employment, as well as their political marginalization in ocean governance. Section 17.3 explores justice claims of SSFs by focusing on their struggles and social movements. Finally, the last section calls for just blue futures, where SSFs are the main actors of Community Supported Fisheries models and local food provisioning and are part of food sovereignty movement.

17.2 Contribution of SSFs to Food Security and Local Livelihoods and Their Political Marginalization

Providing about half of global fish catches, and around two thirds of fish captures destined for direct human consumption worldwide, small-scale fishers have always played a vital role for local livelihoods and food security (FAO 2015). Moreover, considering that 90 percent of capture fishers and fish workers are employed in

---

[2] WFFP is a global small-scale fisher organization and a social movement established to protect the rights of small-scale fishers and fish workers against various privatization and dispossession attempts in the seas and oceans (Pinkerton and Davis 2015).

small-scale fisheries, it is quite difficult to grasp how SSFs have been neglected and marginalized in policy-making for the last decades. This disregard for SSFs might be related to the perceived importance of the technologically more advanced industrial fishing activities especially since the 1960s. Their large-scale operations have been considered more efficient and suitable for the capitalistic mode of production, supported by considerable levels of perverse subsidies[3] leading to over-capacity and over-fishing, while SSFs were predominantly perceived as inefficient, and even backward (Knudsen 2009; Pinkerton 2015). Governments often prioritize industrial fishing activities as a source of employment and economic profits, however, small-scale fisheries provide more jobs than the combined employment generated by industrial fishers, oil and gas industries, tourism and shipping (Smith and Basurto 2019).

A second reason for the relatively little attention paid to SSFs so far seems to go hand in hand with the difficulty of precisely defining small-scale fishing and collecting statistical data about SSFs' activity, as they constitute a very diverse subsector of fisheries, often characterized differently depending on the national context (Smith and Basurto 2019). As a result of this diversity, small-scale fishing activities have often gone unreported and did not receive government support. Instead, especially with the rise of neoliberalism beginning from the 1980s, SSFs have increasingly been dispossessed of their fishing grounds via enclosures, establishment of marine protected areas, and market-based policy instruments such as individual transferable quotas (ITQs) (Mansfield 2004; Pinkerton and Davis 2015). Even though marine protected areas can benefit small-scale fishers in case they are designed in consultation with them, in practice, their implementation may lead to exclusion from their traditional fishing grounds as well as decision making in general (Segi 2014; Mallin et al. 2019). Small-scale fishers in countries adopting ITQs (such as Denmark and South Africa) were also negatively affected as ITQs gave rise to the creation of overcapitalization and large-scale industrial fishing operations in these regions (Barbesgaard 2018). The recent wave of Blue Growth ideas prioritizing extractivist[4] activities such as seabed mining, tourism, intensive aquaculture, offshore energy projects and biotechnology for the sake of continued economic growth will likely exacerbate such dispossession processes (Hadjimichael 2018).

---

[3] Globally, fishing subsidies amounted to USD 35 billion in 2018, of which USD 22 billion has been spent for enhancing capacity. China, EU, USA, Republic of Korea and Japan were the biggest subsidy-providers (Sumaila et al. 2019). About 90% of these harmful capacity-enhancing subsidies went to industrial fishers, increasing the economic vulnerability of small-scale fishers (Schuhbauer et al. 2017).

[4] Originally proposed for non-renewable resources, "extractivism" implies extraction of natural resources in huge quantities, which are sold/exported often unprocessed. However, extractivism also applies to renewable resources such as marine fish catches, since current industrial fishing practices undermine the regenerative capacity of marine resources, rendering them increasingly "non-renewable" (Acosta 2013).

Currently, scientific studies as well as policy attention to small-scale fishing seems to be rising (Smith and Basurto 2019), as the significant contribution of SSFs to employment, food security, poverty alleviation, and rural development becomes clearer. It is estimated that 22–34 million fishers are employed in the primary sector of small-scale fishing, and accounting for indirect employment in processing and trade related with small-scale fishing activities this figure climbs up to around 100 million individuals (FAO 2021; Teh and Sumaila 2013). However, increased attention to SSFs is not only needed due to their substantial contribution to employment and food security worldwide. In other words, SSFs are not only indispensable for subsistence or providing livelihoods for a large number of fisher peoples, but I they also embody social and cultural values, a particular way of life and identity, and contribute substantially to the well-being of coastal communities, as they are "firmly rooted in local communities, traditions and values" (FAO 2015, p. v). These values require more "visibility, recognition and enhancement" given the current political and economic marginalization and vulnerability of small-scale fishers (FAO 2015, p. ix).

Even though small-scale fishers are usually neglected—or ignored—by policy-makers or the investors of mega projects aiming at a high level of capital circulation and accumulation, in fact, they are key social actors for social and ecological justice. As opposed to most industrial fishing activities, SSFs usually adopt more sustainable fishing practices: they frequently use passive gear to catch fish, and their total annual fuel consumption as well as consumption per tonne of fish landed are lower (Pauly 2007, 2018). They are also characterized by relatively less bycatch and discards, and therefore have lower impact on habitats (Lloret et al. 2018). Still, the actual amount of fish caught by small-scale fishers is largely unknown, as they are often under-reported by FAO member countries. Catch reconstruction studies led by Daniel Pauly and his colleagues within the Sea Around Us Project, for instance, try to quantify the actual level of small-scale fishers' catches in order to come up with a proper statistic to evaluate small-scale fishers' ecological and social impacts (Pauly and Zeller 2016).

In contrast to industrial fishers using high-tech equipment to catch more and reach deeper, small-scale fishers have a biophysical view of the marine space, accumulated over hundreds of years through close observation of the nature with which they interact. Moreover, SSFs can often utilize their traditional ecological knowledge to respond to local ecological uncertainties in line with the recent adaptive management approaches (Berkes et al. 2000). This knowledge and continuous close interaction with marine ecosystems is invaluable for the protection of certain marine species, and for the identification of early warnings about changing ecological conditions in the seas and oceans. This is crucial, for instance, for adaptation to climate change and biodiversity conservation as fish is commonly viewed as a living being and as food, rather than a commodity among small-scale fishers (Ertör-Akyazi 2020; Levkoe et al. 2017).

17.2.1 Marginalization of SSF Due to the Economic and Political Privileges of Industrial Fishing

SSFs often come into competition and sometimes severe conflicts with industrial fishers, as the latter move to marine spaces that SSF people have been traditionally using for hundreds of years. In fact, the "ever-expanding enterprise" (Pauly 2018, p. 371) of industrial fishers, relying on heavy fossil fuel use and government subsidies to continue operations, increased their catches considerably since the 1960s, and global catches of marine fisheries peaked at about 93 million tons in 1996. Currently, about 34% of global marine fish stocks are unsustainably fished (FAO 2020). This global crisis in the marine capture fisheries emerged mainly as a result of "subsidy-driven over-capitalization" (Pauly et al. 1998, p. 860) of industrial fishers, and is visible not only in the decreasing level of landings, but also in the characteristics of fish caught. Especially in the Northern Hemisphere, the species that are caught changed drastically from larger piscivorous fishes to smaller planktivorous fishes and invertebrate species (ibid). This process of "fishing down marine food webs" (ibid) calls for a reconsideration of growth in marine capture fisheries, especially the fishing activities of the industrial fleet of the Global North, as it is already leading to environmental and social crises.[5]

Proposals to overcome these crises include rebuilding of fish stocks by abolishing subsidies to industrial fishers, preventing illegal, unregulated, and unreported fishing, and establishing marine protected areas. Pauly (2018), for instance, argues that marine fishing activities should be limited to the EEZs of countries only. This would allow the stocks to rebuild in high seas by reducing the large-scale, ever-expanding oligopolized activities of the industrial fleet of the Global North. Accordingly, if complemented by privileged access rights to small-scale coastal fishing communities, this would lead to a more equitable distribution of catches and improve environmental and social justice.

In fact, industrial fishing activities require more and more previously untouched marine spaces which they can fully exploit, after which they move to the next one. One such new space is currently African seas. IUCN (International Union for Conservation of Nature) recently warned that many marine species (such as *Maderian sardine*) are close to extinction due to illegal- and over-fishing in West and Central African Seas, as a result of which food security for local communities is in danger in the region.[6] Large-scale fleets of the EU countries have been fishing in these seas since 2006 via Fisheries Partnership Agreements (around 130 vessels mostly from Spain, Italy, Portugal, France, and Greece). According to a report by Greenpeace, these operations usually involve over-fishing, reduce catches of local fishing communities, and threaten local food security, while providing little benefits

---

[5] Marketing discards of non-commercial small planktivorous species via Blue Growth initiatives may further exacerbate this process in the future.

[6] https://www.iucn.org/es/node/27721

to the citizens of the African countries (Obaidullah and Osinga 2010).[7] Illegal, unreported and unregulated fishing is very common for both Chinese and European industrial fishers in the region (Belhabib et al. 2015) leading to conflicts between local SSF and foreign industrial fleets.

Furthermore, industrial fishing activities not rarely entail labor and human rights violations such as physical abuse, debt bondage, child labor, slavery, human trafficking, and even murder (Teh et al. 2019; Tickler et al. 2018). Fishing operations in high seas isolate fish workers, and monitoring of abusive relations is more difficult there.[8] Increasing demand for seafood and accelerating international seafood trade (FAO 2020) imply that consumers all over the world may end up eating fish caught via slavery-like practices and that these practices might become even more common if necessary measures are not taken and the businesses continue to act only in a profit-oriented fashion.

A broader human-rights based approach for ocean governance and especially for SSF communities shall encompass social justice principles such as access to and democratic control over marine resources, participation in decision making, territorial, indigenous and gender rights, right to food and right to livelihood (FAO 2015; Teh et al. 2019). SSFs are strongly embedded in larger social, economic and ecological systems in which they operate. Therefore, specific attention needs to be paid to their diverse ways of supporting food security, poverty alleviation, and social cohesion in their communities.

This section attempted to demonstrate why SSFs have traditionally been important actors for food security and provision of local livelihoods in coastal areas, as well as for the sustenance of marine ecosystems. Comparison to and competition with industrial fishers have historically led to the marginalization of SSF people in ocean governance. However, we claim that SSFs are indispensable actors to be considered in the governance of past, present and future of the seas and oceans especially for two reasons. First, small-scale fishers still "feed the world"—as peasants and small-scale farmers do on land.[9] In fact, a recent report of the UN Food and Agriculture Organization confirmed that "small-scale fisheries contribute about half of global fish catches. When considering catches destined for direct human consumption, the share contributed by the small-scale fisheries increases to two-thirds" (FAO 2015, p. ix). Second, they are mobilized social and political actors who

---

[7] The most recent FAO report acknowledges that 43% of Eastern Central Atlantic fish stocks are at biologically unsustainable levels (Food and Agriculture Organization 2018).

[8] Oceans are prone to human rights violations not only in the industrial fishing, but also in sectors such as marine transportation, offshore energy projects and shipbreaking, as enforcement and policing of international laws are very difficult in the oceans. See for instance the Ocean Foundation's webpage on "Human Rights and the Ocean" at https://oceanfdn.org/human-rights-and-the-ocean/

[9] See the report of GRAIN (2014), where they argue that small-scale farmers feed the world with less than a quarter of all farmland: https://www.grain.org/article/entries/4929-hungry-for-land-small-farmers-feed-the-world-with-less-than-a-quarter-of-all-farmland

organize even under very marginalized and difficult conditions, as discussed in Sects. 17.3 and 17.4 below (for a detailed analysis of fisher movements see Mills 2018; Sinha 2012).

17.3   Local and Global SSF Movements for Just Blue Presents and Futures

Historically, local SSF communities have self-organized in diverse ways. Some have traditionally organized in fishing cooperatives (Baticados et al. 1998; Berkes 1986; Pomeroy and Berkes 1997), while others established their own local norms and fishing rules through a range of self- and co-management mechanisms (Basurto et al. 2013; McCay et al. 2014). As opposed to corporatist structures and power relations of industrial fishing sector, most SSF communities strive for just socio-ecological governance mechanisms with their autonomous structures and social and ecological diversities. Against this background, this section explores the main justice demands of global and regional fisher movements in the context of fisheries justice and food sovereignty.

17.3.1   Social and Ecological Claims for Fisheries Justice

Recently, especially in the last two decades, local and regional SSF initiatives converged increasingly to a global social movement (Sinha 2012; Mills 2018). This global SSF movement has its roots in the first official assembly of World Forum of Fish Harvesters and Fish Workers (WFF)[10] in 1997 in New Delhi. After the regional division of WFF in 2000, the World Forum of Fisher Peoples (WFFP)[11] emerged as another global organization of SSF communities (Sinha 2012). Currently, both are allied and consist of SSF and fish worker and harvester representatives from about 50 countries all over the world. These two organizations were founded mostly as a response to the global fisheries policies that favor industrial fisheries and neglect the concerns, needs, and political agencies of SSF communities and cooperatives (Mills 2018; Levkoe et al. 2017). For more than two decades, they have been self-organizing to protect the rights of fisher people and fish workers and harvesters against a range of privatization and neoliberalization attempts through global fisheries policies leading to 'ocean grabbing' (Pinkerton and Davis 2015; Barbesgaard 2018; Mallin et al. 2019). They also resist a range of mega projects on their fishing grounds, e.g. construction of big harbors and airports, large marine conservation areas displacing

---

[10] See their Facebook page with the abbreviation "wff.fisher".

[11] See WFFP's webpage: https://worldfishers.org

local communities, industrial fish farms, as well as energy projects, most of which are leading to their dispossession and further marginalization and criminalization (Nayak and Berkes 2010; Ditty and Rezende 2014; Maharaj 2017).

SSF communities first of all demand 'fisheries justice', defined as "collective struggles for inclusion, equal rights, and the democratisation of access, ownership, and control of natural resources and fishing territories" (Mills 2018, p. 1278). They especially spotlight the wide range of injustices and inequalities between industrial and small-scale fisheries as well as expanding intensive fish farms restricting and displacing SSF activities (Pinkerton 2015; Ertör and Ortega-Cerdà 2015). They also highlight the injustices of the global food regime and express its link to climate justice (Mills 2018). In fact, SSFs are one of the social groups to be affected the most by climate change (e.g. by changing coastal ecological conditions in the presence and abundance of marine species), even though their use of fossil fuels and contribution to climate change are much lower compared to industrial fishers. These demands have made them part of broader movements such as the food sovereignty movement as well as the climate justice movement (Mills 2018; Levkoe et al. 2017).

Second, SSF movements organized within WFF and WFFP demand discontinuation of extractive industries and mega projects in their regions as well as globally (see the EJAtlas[12] for fisheries conflicts). These extractive industries with highly negative environmental and social impacts range from sand mining and seabed mining—especially promoted with the current Blue Growth strategies (for a critical discussion, see the Blue Degrowth framework: Ertör and Hadjimichael 2020)—to oil, gas and other mineral extraction from the seas, among others. Additionally, mega projects affecting fisher people include new massive airports or airport cities called 'aerotropolis' as in the case of Bulacan Aerotropolis in the Philippines[13] or in Yogyakarta[14] and Makassar[15] in Indonesia, luxury touristic residences—both projects are usually placed in small island states—, as well as big harbors and 'port cities' such as the Colombo Port City in Sri Lanka.[16] These mega projects are constructed for commercial purposes and create often conflicts not only with SSF people, but also with local farmers, trade unions, NGOs or Environmental Justice Organizations (EJOs), students, and other social movements (for a broader analysis of environmental conflicts and environmental defenders including fisherfolks, see Scheidel et al. 2020).

Third, they demand putting an end to 'ocean and coastal grabbing'—or 'resource grabbing' including freshwater areas—in a broader sense (TNI 2014; Barbesgaard 2018). This is because SSFs often envision themselves linked with each other as a

---

[12] See the global Environmental Justice Atlas (EJAtlas) which maps and analyzes the environmental conflicts including fisher people's conflicts: https://ejatlas.org

[13] See the for the conflict in Bulacan, where fisher people have been mobilized social actors with other allies against the 'aerotropolis' project: https://ejatlas.org/conflict/bulacan-aerotropolis

[14] https://ejatlas.org/conflict/international-airport-on-the-kulon-progo-coast-indonesia

[15] https://ejatlas.org/conflict/reclamation-project-makassar-indonesia

[16] https://ejatlas.org/conflict/fisherwomens-mobilization-against-the-port-city-sri-lanka

global social struggle, and their local fights converge against several grabbing attempts of capitalist projects, which lead to their dispossession and marginalization. For instance, large-scale marine protected areas established in the name of conservation of fishing resources are frequently enclosing the traditional fishing grounds, dislocating local people and affecting their livelihood in an adverse way through 'ocean-control grabbing' (Mallin et al. 2019) as well as through diverse forms of 'blue grabbing', in which "marine conservation results in the appropriation of marine resources and coastal land from previous custodians by more powerful actors, such as state and tourist operators" (Hill 2017, p. 97).

Finally, they claim their rights to capturing fish, right to food, human rights and tenure rights in their territories as well as recognition as relevant political actors of ocean governance—both for its present and future. In fact, SSF movements have been very active in the drafting of the "Voluntary Guidelines for Securing Sustainable Small-Scale Fisheries in the Context of Food Security and Poverty Eradication" (FAO 2015), which can be considered as one of the global fisher movements' recent achievements. This guideline is the first strong call for the recognition of the values and contributions of SSF at the international level. The document has been prepared as a result of tremendous efforts on the side of civil society supporting the rights of SSF and have been endorsed by more than hundred member states. Even though the guidelines are only voluntary, its strong reference to human rights gives the advantage of linking them to nationally and internationally enforceable laws such as the Universal Declaration of Human Rights (Jentoft 2014).[17] As mentioned in this document, these guidelines "support responsible fisheries and sustainable social and economic development for the benefit of current and future generations, with an emphasis on small-scale fishers and fish workers and related activities and including vulnerable and marginalized people, promoting a human rights based approach" (FAO 2015, p. ix). Human rights in the context of these guidelines include civil, political, economic, social, and cultural rights as well as the right for the fisher livelihoods and their empowerment.

It has to be noted that while SSF movements in some instances decided to engage with these international institutions such as FAO processes, at other instances, they saw the need to protest some international governance meetings, whenever they felt that stark inequalities embedded in such political spaces would not grant them an equal participation and capability to influence the discussions. There is also awareness within SSF movements that the mere acceptance of the Voluntary Guidelines by several nation states is not sufficient for their demands to be met. Rather, they insist on their actual implementation in each policy circle they join, to open up a broader political space for their needs and rightful demands as well as for their official recognition and protection.

---

[17] This is in stark contrast to rights-based approaches associated with the establishment of property rights and privatization in fisheries. Rights-based approaches advocate the assignment of fishing rights to individuals and/or communities to ensure economic efficiency and prevent overfishing. However, these processes can lead to the so-called "ocean grabbing", dispossessing and excluding SSF in the name of resource conservation (Pictou 2018).

An important dimension of their global movement-building is organization at different scales and in non-centralized ways. There is a special emphasis, for instance, to have one female and one male representative from each region in WFFP. Together with these representatives, the General Secretary of WFFP is elected in each General Assembly, organized usually every 3 years. Moreover, the sub-assemblies open up a political space for consolidation of groups that otherwise could have remained mere minority groups in the entire assembly. These sub-assemblies consist of women, indigenous peoples, young fishers, freshwater fishers, among others. This organizational structure enables them to empower different groups of fishers, who form part of the movement but have their own political voices and social, economic, and political needs.

17.3.2 Food Sovereignty: A Unifying Concept for SSF Movements with Other Small-Scale Food Producers

One of the central concepts SSF movements have been engaging with in their struggles while reclaiming their rights and positioning themselves as food providers has been the 'food sovereignty' approach (TNI 2020; WFFP 2017). Food sovereignty was first defined by La Vía Campesina in 1996, and the term has since then been transformed to become more bottom-up through direct political action of social movements (for a discussion on the etymology of food sovereignty, see Patel 2009). Currently, the most common definition of the food sovereignty is the one announced in Nyéléni Declaration (2007), which was the result of the Nyéléni Forum in Mali with the participation of more than 500 practitioners from about 80 countries. Even though it sounds at first glance similar, the term goes far beyond food security.[18] Food sovereignty emphasizes the right to food from a bottom-up perspective and bases its principles on people's relations to food and on their decisions on how to produce, distribute and consume food. It has emerged from peasants', pastoralists', beekeepers', and fisher peoples' movements and their alliances for a socially and ecologically just and sustainable food regime. The following definition was adopted by the Nyéléni Forum in 2007 (The Nyéléni 2007 International Steering Committee, p. 9):

> Food sovereignty is the right of peoples to healthy and culturally appropriate food produced through ecologically sound and sustainable methods, and their right to define their own food and agriculture systems. It puts the aspirations and needs of those who produce, distribute and consume food at the heart of food systems and policies rather than the demands

---

[18] Food security is defined as the following: "Food security exists when all people, at all times, have physical and economic access to sufficient safe and nutritious food that meets their dietary needs and food preferences for an active and healthy life in the World Food Summit in 1996. Following that, four dimensions of food security have been identified: (i) physical availability of food, (ii) economic and physical access to food, (iii) food utilization, (iv) stability of the other three dimensions over time (FAO 2008).

of markets and corporations. It defends the interests and inclusion of the next generation. It offers a strategy to resist and dismantle the current corporate trade and food regime, and directions for food, farming, pastoral and fisheries systems determined by local producers and users. Food sovereignty prioritises local and national economies and markets and empowers peasant and family farmer-driven agriculture, artisanal fishing, pastoralist-led grazing, and food production, distribution and consumption based on environmental, social and economic sustainability. Food sovereignty promotes transparent trade that guarantees just incomes to all peoples as well as the rights of consumers to control their food and nutrition. It ensures that the rights to use and manage lands, territories, waters, seeds, livestock and biodiversity are in the hands of those of us who produce food. Food sovereignty implies new social relations free of oppression and inequality between men and women, peoples, racial groups, social and economic classes and generations.

Accordingly, food sovereignty is based on the following six pillars: Food sovereignty (i) focuses on food for people; (ii) values food providers; (iii) localises food systems; (iv) puts control locally; (v) builds knowledge and skills; and (vi) works with nature. The movements defining food sovereignty claim that these principles are "interlinked and inseparable". As such, implementation requires all of them to be applied in practice (Nyéléni Declaration 2007).

Having participated in the Nyéléni Forum for Food Sovereignty in 2007 with other allies such as "the urban poor, women, Indigenous Peoples, peasants, pastoralists and other constituencies" (WFFP 2017, p. 2), global fisher movements started to discuss the relevance of the concept for their regional and international struggles and to use it as an umbrella concept for solidarity and alliance among distinct social movements. Even though SSF movements around the world do not always call their struggle in their locality a 'food sovereignty movement', all of these six pillars are usually relevant for them. More concrete forms of these discussions based on their local context have already been part of their struggles (TNI 2020). While they reclaim their rights as small-scale fishers, they feel the urgent need to make alliances with other social actors and movements striving for just food regimes as well as to focus on their specific fishing practices in their own regions.

Therefore, based on continuous debates with other small-scale food producers within the food sovereignty movement, SSF people have put an effort to conceptualize how these six pillars manifest themselves in small-scale fisheries production and movement. The report "Agroecology and Food Sovereignty in Small-Scale Fisheries" (WFFP 2017) is one of the main attempts to demonstrate in which ways agroecology and food sovereignty discussions are relevant, visible and unifying for the case of small-scale fishers. The recent literature usually indicates that these concepts are still understudied both in the academic literature and on practitioners' and social movements' side (Levkoe et al. 2017; Mills 2018). However, there are ongoing attempts discussing its relevance for the struggles of small-scale fisher communities to weave stronger ties with broader solidarity networks mobilized around food sovereignty (TNI 2020; Ertör-Akyazi 2020).

17.4 Alternatives for Just Blue Futures

It has been argued that resistance movements can open up space for experimentation with new alternatives as well as for the politicization and further mobilization of the existing ones (Pelenc et al. 2019). Similarly, global fisher movements with their local nodes are struggling for their rights and resisting ocean grabbing and blue growth projects on the ground, as well as constructing their own alternatives in terms of fishing practices, supply chains and consumption in their localities by establishing alliances and networks.

Localized food systems had been very common in the past of small-scale fishing communities, when, for instance, family members of fishers were doing agriculture and markets were more local. The presence of fishing cooperatives in the supply chains can also localize the food system by eliminating intermediaries, facilitating direct sales from small-scale fishers to consumers and supporting their members socially and economically (Ertör-Akyazi 2020). However, the transformations in ocean governance and global markets in the last decades led to the emergence and expansion of agrifood and seafood businesses involving heavily privatized production and consumption of food and fish (Mansfield 2004). As a result, the accumulation of economic and political power at a few hands led to the marginalization of small-scale fishers. However, the SSF movement focuses on food for people and struggles for a food system that values small-scale food providers and localizes food systems. Therefore, there is a need for redefining the food system and local production and consumption models as an alternative to industrial fishing and globalized value chains in fisheries and for reclaiming the rights of SSF people in order to develop viable alternative models. An example of recent discussions on such alternatives is examined below.

17.4.1 Community Supported Fisheries

One of the key alternative models is the Community Supported Fisheries (CSF). CSF has already been adopted in many parts of the world by SSF communities, and further expansion may serve a blue just future of ocean governance, subject to the continuation of the political will and mobilization of SSF communities. In its essence, it is similar to the Community Supported Agriculture (CSA) model in that it brings small-scale fishers in direct contact with consumers, who often pre-pay a fish box consisting of SSF harvests in their region (Brinson et al. 2011; McClenachan et al. 2014; Campbell et al. 2014). As such, it is a form of directly marketing seafood from "deck to dish" (TNI 2020).[19] In practice, there are diverse examples of the CSF model. Studies focusing on the CSF networks in North America have flourished since 2007 identifying their similarities and differences in terms of philosophy, structure, operations, and outcomes (Bolton et al. 2016). While their shared focus is direct marketing of seafood from fisher to consumer with a shortened supply chain and locally sourced seafood, they often differ in terms of organizational and ideological structures.

Based on local production and consumption and a deeper understanding of the needs of SSFs and consumers, most CSF models incorporate the above-mentioned six pillars of food sovereignty, i.e. they (i) focus on food for people; (ii) value food providers; (iii) localise food systems; (iv) put control locally; (v) build knowledge and skills; and (vi) work with nature. Moreover, some of them established a network of alliances such as the Local Catch Network[20] and Fish Locally Collaborative, and help producers and consumers get to know each other more closely to ensure a more just food system. Therefore, we argue that CSF models can be a driving force for the empowerment of SSF communities and give them a broader political and socioeconomic space in seeking for socially and ecologically more just and sustainable futures of ocean governance.

To a certain degree, this model also exists in European coastal cities. However, CSF members in Europe live in more marginalized conditions, in contrast to CSAs, which have examples of more established networks including younger, well-educated members active in the food sovereignty movement. Therefore, in many places where there are already well-organized CSA groups present, such as those in Spain and France, the CSFs still experience difficulties of communication and organization (for some examples, see the initiatives such as "stewardship fish" promoted by the Fundació Submon,[21] the local sale initiative of the small-scale fishing cooperative of Sitges as well as the recent CSF mapping of PleineMer in France).[22] Further, the initiatives and movements of SSF people are less visible and have difficulties to reach consumers and civil society directly, explain their sustainable fishing methods to consumers, reclaim their fishing tradition, culture and fisher identities, and demand mechanisms for local production and consumption. The farm to fork strategy proposed recently by the EU Green Deal may have the

---

[19] While "farm to fork" is a widely used term for agricultural activities and now also refers to fishing activities within the EU Green Deal, we prefer using "deck to dish" as this term is more directly related with fishing etymologically and used by civil society initiatives recently (TNI 2020; URGENCI 2019). Blue Growth agenda and discourses initially ignored the presence of small-scale fisheries, but EU Green Deal tries to incorporate small-scale producers via their farm to fork strategy.

[20] See the webpage of Local Catch Network consisting of more than 450 initiatives: https://local-catch.org

[21] http://www.submon.org/en/once-again-peix-custodia-comes-back-to-the-fish-markets-of-barcelona/

[22] https://associationpleinemer.com/les-community-supported-fisheries/

potential to support such direct marketing initiatives if implemented with meaningful participation of small-scale fishers in Europe.[23]

In the Global South, though, similar mechanisms have been more common until very recently even though they do not use the same terminology for their localized food systems. However, examples from Mauritania, Senegal or Indonesia show that the intervention of industrial fishing fleet from different countries led both to the grabbing of local fishing sources and weakened the local food production and consumption systems that are crucial for communities (Ertör-Akyazi 2020; DuBois and Zografos 2012; TNI 2020). Therefore, both existing and new CSF models need to be protected and promoted for just blue futures in which SSFs can raise their voices and can participate equally as other actors of food system.

17.5 Concluding Remarks

This chapter focused on small-scale fishers, as crucial social actors of the past and present of seas and oceans. By scrutinizing their role in coastal communities and food sovereignty, we argued that small-scale fishers and their organizations are key for just and sustainable blue futures. With this purpose, we first highlighted the structural inequalities and injustices leading to the marginalization of small-scale fisher communities around the world and then indicated models seeking for socio-environmental justice such as food sovereignty movements and community supported fisheries examples. In order to achieve just blue futures for ocean governance, however, there is the need for transforming broader political-economic systems. The analysis above establishes the food axis of such future social and economic alternatives. In this chapter, we have therefore emphasized that rather than neglecting small-scale fishers, global marine policies and politics need to put them and their needs and demands to the center, not only for a stakeholder-consultation process, but for co-developing the politics regarding how to use marine commons.

References

Acosta A (2013) Extractivism and neoextractivism: two sides of the same curse. In: Lang M, Fernando L, Buxton N (eds) Beyond development: alternative visions from Latin America. Transnational Institute, Amsterdam, pp 61–86

African Union (2015) Towards harnessing the potential of the blue economy to achieve Agenda 2063 and the post-2015 development agenda. AU Headquarters, Addis Ababa, Ethiopia. Available at: https://au.int/sites/default/files/newsevents/conceptnotes/27474-cn-concept_note_eng_0.pdf

Barbesgaard M (2018) Blue growth: savior or ocean grabbing? J Peasant Stud 45(1):130–149. https://doi.org/10.1080/03066150.2017.1377186

---

[23] https://ec.europa.eu/food/horizontal-topics/farm-fork-strategy_en

Basurto X, Gelcich S, Ostrom E (2013) The social–ecological system framework as a knowledge classificatory system for benthic small-scale fisheries. Glob Environ Chang 23(6):1366–1380

Baticados DB, Agbayani RF, Gentoral FE (1998) Fishing cooperatives in Capiz, central Philippines: their importance in managing fishery resources. Fish Res 34:137–149. https://doi.org/10.1016/S0165-7836(97)00090-8

Belhabib D, Sumaila UR, Lam VWY, Zeller D, Le Billon P, Kane EA, Pauly D (2015) Euros vs. Yuan: comparing european and chinese fishing access in West Africa. PLoS One 10:e0118351. https://doi.org/10.1371/journal.pone.0118351

Berkes F (1986) Local-level management and the commons problem: a comparative study of Turkish coastal fisheries. Mar Policy 10:215–229. https://doi.org/10.1016/0308-597X(86)90054-0

Berkes F, Colding J, Folke C (2000) Rediscovery of traditional ecological knowledge as adaptive management. Adapt Ecol 10:1251–1262

Bolton AE, Dubik BA, Stoll JS, Basurto X (2016) Describing the diversity of community supported fishery programs in North America. Mar Policy 66:21–29. https://doi.org/10.1016/j.marpol.2016.01.007

Brinson A, Lee MY, Rountree B (2011) Direct marketing strategies: the rise of community supported fishery programs. Mar Policy 35(4):542–548

Campbell LM, Boucquey N, Stoll J, Coppola H, Smith MD (2014) From vegetable box to seafood cooler: applying the community-supported agriculture model to fisheries. Soc Nat Resour 27(1):88–106

Ditty JM, Rezende CE (2014) Unjust and unsustainable: a case study of the Açu port industrial complex. Mar Policy 45:82–88. https://doi.org/10.1016/j.marpol.2013.11.018

DuBois C, Zografos C (2012) Conflicts at sea between artisanal and industrial fishers: intersectoral interactions and dispute resolution in Senegal. Mar Policy 36(6):1211–1220. https://doi.org/10.1016/j.marpol.2012.03.007

Ertör I, Hadjimichael M (2020) Blue degrowth and the politics of the sea: rethinking the blue economy. Sustain Sci 15:1–10. https://doi.org/10.1007/s11625-019-00772-y

Ertör I, Ortega-Cerdà M (2015) Political lessons from early warnings: Marine finfish aquaculture conflicts in Europe. Mar Policy 51:202–210. https://doi.org/10.1016/j.marpol.2014.07.018

Ertör-Akyazi P (2020) Contesting growth in marine capture fisheries: the case of small-scale fishing cooperatives in Istanbul. Sustain Sci 15:45–62. https://doi.org/10.1007/s11625-019-00748-y

European Commission (2012) Blue growth: opportunities for marine and maritime sustainable growth. Publications Office of the European Union, Luxembourg. Available at: https://blackseablueeconomy.eu/sites/default/files/blue_growth_opportunities_for_marine_and_maritime_sustainable_growth.pdf

FAO (2015) Voluntary guidelines for securing sustainable small-scale fisheries, Voluntary Guidelines for Securing Sustainable Small-Scale Fisheries in the Context of Food Security and Poverty Eradication

FAO (2020) The state of World fisheries and aquaculture 2020: sustainability in action. FAO, Rome

FAO (2021) Sustainable small-scale fisheries [WWW document]. http://www.fao.org/policy-support/policy-themes/sustainable-small-scale-fisheries/en/

Food and Agriculture Organization (2018) The State of World Fisheries and aquaculture 2018 – meeting the sustainable development goals. FAO, Rome

Hadjimichael M (2018) A call for a blue degrowth: unravelling the European Union's fisheries and maritime policies. Mar Policy 94:158–164. https://doi.org/10.1016/j.marpol.2018.05.007

Hill A (2017) Blue grabbing: reviewing marine conservation in Redang Island Marine Park, Malaysia. Geoforum 79:97–100. https://doi.org/10.1016/j.geoforum.2016.12.019

Jentoft S (2014) Walking the talk: implementing the international voluntary guidelines for securing sustainable small-scale fisheries. Marit Stud 13:1–15. https://doi.org/10.1186/s40152-014-0016-3

Knudsen S (2009) Fishers and scientists in modern Turkey: The management of natural resources, knowledge and identity on the eastern Black Sea coast. Berghahn Books

Levkoe CZ, Lowitt K, Nelson C (2017) "Fish as food": exploring a food sovereignty approach to small-scale fisheries. Mar Policy 85:65–70. https://doi.org/10.1016/j.marpol.2017.08.018

Lloret J, Cowx IG, Cabral H, Castro M, Font T, Gonçalves JMS, Gordoa A, Hoefnagel E, Matić-Skoko S, Mikkelsen E, Morales-Nin B, Moutopoulos DK, Muñoz M, dos Santos MN, Pintassilgo P, Pita C, Stergiou KI, Ünal V, Veiga P, Erzini K (2018) Small-scale coastal fisheries in European Seas are not what they were: ecological, social and economic changes. Mar Policy 98:176–186. https://doi.org/10.1016/j.marpol.2016.11.007

Maharaj B (2017) Contesting displacement and the struggle for survival: the case of subsistence fisher folk in Durban, South Africa. Local Econ 32(7):744–762. https://doi.org/10.1177/0269094217734330

Mallin MAF, Stolz DC, Thompson BS, Barbesgaard M (2019) In oceans we trust: conservation, philanthropy, and the political economy of the Phoenix Islands Protected Area. Mar Policy 107:103421

Mansfield B (2004) Rules of privatization: contradictions in neoliberal regulation of North Pacific fisheries. Ann Assoc Am Geogr 94:565–584. https://doi.org/10.1111/j.1467-8306.2004.00414.x

McCay BJ, Micheli F, Ponce-Díaz G, Murray G, Shester G, Ramirez-Sanchez S, Weisman W (2014) Cooperatives, concessions, and co-management on the Pacific coast of Mexico. Mar Policy 44:49–59

McClenachan L, Neal BP, Al-Abdulrazzak D, Witkin T, Fisher K, Kittinger JN (2014) Do community supported fisheries (CSFs) improve sustainability? Fish Res 157:62–69

Mills EN (2018) Implicating 'fisheries justice' movements in food and climate politics. Third World Q 39(7):1270–1289. https://doi.org/10.1080/01436597.2017.1416288

Nayak PK, Berkes F (2010) Whose marginalisation? Politics around environmental injustices in India's Chilika lagoon. Local Environ 15(6):553–567. https://doi.org/10.1080/13549839.2010.487527

Obaidullah, F., Osinga, Y., 2010. How Africa is feeding Europe: EU (over)fishing in Africa

Patel R (2009) Food sovereignty. J Peasant Stud 36(3):663–706. https://doi.org/10.1080/03066150903143079

Pauly D (2007) Small but mighty. Conserv Mag 8:25

Pauly D (2018) A vision for marine fisheries in a global blue economy. Mar Policy 87:371–374. https://doi.org/10.1016/j.marpol.2017.11.010

Pauly D, Zeller D (2016) Catch reconstructions reveal that global marine fisheries catches are higher than reported and declining. Nat Commun 7:1–9. https://doi.org/10.1038/ncomms10244

Pauly D, Christensen V, Dalsgaard J, Froese R, Torres FJ (1998) Fishing down marine food webs: it is far more pervasive than we thought. Science 279(5352):860–863. https://doi.org/10.1126/science.279.5352.860

Pelenc J, Wallenborn G, Milanesi J, Sébastien L, Vastenaekels J, Lajarthe F, Ballet J, Cervera-marzal M, Carimentrand A, Merveille N, Frère B (2019) Alternative and resistance movements: the two faces of sustainability. Ecol Econ 159:373–378. https://doi.org/10.1016/j.ecolecon.2019.01.013

Pictou S (2018) The origins and politics, campaigns and demands by the international fisher peoples' movement: an Indigenous perspective. Third World Q 39:1411–1420. https://doi.org/10.1080/01436597.2017.1368384

Pinkerton E (2015) The role of moral economy in two British Columbia fisheries: confronting neoliberal policies. Mar Policy 61:410–419. https://doi.org/10.1016/j.marpol.2015.04.009

Pinkerton E, Davis R (2015) Neoliberalism and the politics of enclosure in North American small-scale fisheries. Mar Policy 61:303–312. https://doi.org/10.1016/j.marpol.2015.03.025

Pomeroy RS, Berkes F (1997) Two to tango: the role of government in fisheries co-management. Mar Policy 21:465–480. https://doi.org/10.1016/S0308-597X(97)00017-1

Scheidel A, Del Bene D, Liu J, Navas G, Mingorria S, Demaria F, Avila S, Roy B, Ertör I, Temper L, Martinez-Alier J (2020) Environmental conflicts and defenders: a global overview. Glob Environ Change 63:102104. https://doi.org/10.1016/j.gloenvcha.2020.102104

Schuhbauer A, Sumaila UR (2016) Economic viability and small-scale fisheries – a review. Ecol Econ 124:69–75. https://doi.org/10.1016/j.ecolecon.2016.01.018

Schuhbauer A, Chuenpagdee R, Cheung WWL, Greer K, Sumaila UR (2017) How subsidies affect the economic viability of small-scale fisheries. Mar Policy 82:114–121. https://doi.org/10.1016/j.marpol.2017.05.013

Segi S (2014) Protecting or pilfering? Neoliberal conservationist marine protected areas in the experience of coastal Granada, the Philippines. Hum Ecol 42(4):565–575

Sinha S (2012) Transnationality and the Indian Fishworkers' movement, 1960s-2000. J Agrar Chang 12(2–3):364–389. https://doi.org/10.1111/j.1471-0366.2011.00349.x

Smith H, Basurto X (2019) Defining small-scale fisheries and examining the role of science in shaping perceptions of who and what counts: a systematic review. Front Mar Sci 6. https://doi.org/10.3389/fmars.2019.236

Sumaila UR, Ebrahim N, Schuhbauer A, Skerritt D, Li Y, Kim HS, Mallory TG, Lam VWL, Pauly D (2019) Updated estimates and analysis of global fisheries subsidies. Mar Policy 109:103695. https://doi.org/10.1016/j.marpol.2019.103695

Teh LCL, Sumaila UR (2013) Contribution of marine fisheries to worldwide employment. Fish Fish 14:77–88. https://doi.org/10.1111/j.1467-2979.2011.00450.x

Teh LCL, Caddell R, Allison EH, Finkbeiner EM, Kittinger JN, Nakamura K, Ota Y (2019) The role of human rights in implementing socially responsible seafood. PLoS One 14:1–21. https://doi.org/10.1371/journal.pone.0210241

Tickler D, Meeuwig JJ, Bryant K, David F, Forrest JAH, Gordon E, Larsen JJ, Oh B, Pauly D, Sumaila UR, Zeller D (2018) Modern slavery and the race to fish. Nat Commun 9:4643. https://doi.org/10.1038/s41467-018-07118-9

TNI (2014) The global ocean grab: a primer. Transnational Institute, Amsterdam. Available at: https://www.tni.org/files/download/the_global_ocean_grab.pdf

TNI (2020) Situating small-scale fisheries in the global struggle for agroecology and food sovereignty. Transnational Institute, Amsterdam. Available at: https://www.tni.org/en/small-scale-fisheries

URGENCI (2019) Deck to dish: increasing the visibility and the resilience of the community supported fisheries movement. Available at https://urgenci.net/deck-to-dish-increasing-the-visibility-and-the-resilience-of-the-community-supported-fisheries-movement/

WFFP (2017) Agroecology and food sovereignty in small-scale fisheries. WFFP, Indonesia. Available at: https://worldfishers.org/wp-content/uploads/2017/09/WFFP.Food_.Sov_.web_.pdf

Chapter 18
Ocean Acidification as a Governance Challenge in the Mediterranean Sea: Impacts from Aquaculture and Fisheries

Nina Bednarsek, Bleuenn Guilloux, Donata Melaku Canu, Charles Galdies, Roberta Guerra, Simona Simoncelli, Richard A. Feely, Greg Pelletier, Blaženka Gašparović, Jelena Godrijan, Alenka Malej, Cosimo Solidoro, Valentina Turk, and Serena Zunino

Abstract Despite the progress in the international and regional governance efforts at the level of climate change, ocean acidification (OA) remains a global problem with profoundly negative environmental, social, and economical consequences. This requires extensive mitigation and adaptation effective strategies that are hindered by current shortcomings of governance. This multidisciplinary chapter investigates the risks of ocean acidification (OA) for aquaculture and fisheries in the Mediterranean Sea and its sub-basins and the role of regional adaptive governance to tackle the problem. The identified risks are based on the biological sensitivities of the most important aquaculture species and biogenic habitats and their exposure to the current and future predicted (2100) RCP 8.5 conditions. To link OA exposure and biological sensitivity, we produced spatially resolved and depth-related pH and

N. Bednarsek (✉)
Marine Biology Station Piran, National Institute of Biology, Ljubljana, Slovenia

Cooperative Institute for Marine Resources Studies, Oregon State University,
Hatfield, Oregon
e-mail: nina.bednarsek@gmail.com

B. Guilloux
European Institute for Marine Studies, Laboratory for Law and Economics of the Sea,
Plouzane, Brittany, France

D. M. Canu · S. Zunino
Istituto Nazionale di Oceanografia e di Geofisica Sperimentale, Sgonico, Italy

C. Galdies
Institute of Earth Systems, University of Malta, Msida, Malta

R. Guerra
Department of Physics and Astronomy, University of Bologna, Bologna, Italy

Centro Interdipartimentale di Ricerca per le Scienze Ambientali (CIRSA-UNIBO), University
of Bologna, Bologna, Italy

© German Institute of Development and Sustainability (IDOS) 2023
S. Partelow et al. (eds.), *Ocean Governance*, MARE Publication Series 25,
https://doi.org/10.1007/978-3-031-20740-2_18

aragonite saturation state exposure maps and overlaid these with the existing aqua-culture industry in the coastal waters of the Mediterranean basin to demonstrate potential risk for the aquaculture in the future. We also identified fisheries' vulner-ability through the indirect effects of OA on highly sensitive biogenic habitats that serve as nursery and spawning areas, showing that some of the biogenic habitats are already affected locally under existing OA conditions and will be more severely impacted across the entire Mediterranean basin under 2100 scenarios. This provided a regional vulnerability assessment of OA hotspots, risks and gaps that created the baseline for discussing the importance of adaptive governance and recommenda-tions for future OA mitigation/adaptation strategies. By understanding the risks under future OA scenarios and reinforcing the adaptability of the governance sys-tem at the science-policy interface, best informed, "situated" management response capability can be optimised to sustain ecosystem services.

18.1 Introduction

In the era defined as the Anthropocene (Crutzen 2002), global oceans have already been profoundly altered by humans. Increasing levels of human made greenhouse gas emissions of which 25% has been absorbed by the oceans (Sabine et al. 2004; Le Quéré et al. 2018; Bindoff et al. 2019; Licker et al. 2019). This has led to an increase of surface seawater acidity of approximately 30% (Doney et al. 2009). This process, referred to as "Ocean Acidification" (OA), is a complex global phenome-non that is among the nine planetary boundaries identified by Rockström et al. (2009). Although global, OA has "situated" effects on regional and marginal seas such as the Mediterranean Sea and its sub-basins. These effects impact the regional ecosystems, biodiversity, and ecosystem services, including aquaculture, fisheries and food security (Barbier 2017; Bindoff et al. (2019), with significant implications

S. Simoncelli
Istituto Nazionale di Geofisica e Vulcanologia, Sezione di Bologna, Bologna, Italy

R. A. Feely
NOAA Pacific Marine Environmental Laboratory, Seattle, WA, USA

G. Pelletier
Washington State Department of Ecology, United States (Independent Researcher), Bellingham, WA, USA

B. Gašparović · J. Godrijan
Division for Marine and Environmental Research, Ruđer Bošković Institute, Zagreb, Croatia

A. Malej · V. Turk
Marine Biology Station Piran, National Institute of Biology, Piran, Slovenia

C. Solidoro
National Institute of Oceanography and Applied Geophysics (OGS), Trieste, Italy

International Centre for Theoretical Physic (ICTP), Trieste, Italy on the overall Mediterranean socio-ecological system (Hassoun et al. 2022; Zunino et al. 2021). OA introduces additional complexity in governance efforts because the source of the problem occurs at a different time and space scales than the locus of the affected people (Long 2009) and ecosystems.

Thousands of years of history have shaped the governance of the semi-enclosed Mediterranean Sea. From tensions and conflicts between coastal populations over access to the sea, its resources, or adjacent lands, to enhanced cooperation because of a shared cultural and natural heritage. In terms of marine biodiversity, the Mediterranean is considered a hotspot of marine biodiversity as it hosts high percentages of endemic species (Coll et al. 2012). However, areas of high diversity overlap with areas where there is high potential for cumulative threats such as unsustainable development and climate change linked events (including temperature rise, marine heat waves, OA, deoxygenation, eutrophication, freshening, overfishing, chemical pollution, and habitat destruction) (Giorgi, 2006; Shaltout and Omstedt, 2014; Lazzari et al. 2014; Hilmi et al. 2015; Goyet et al. 2016; Cramer et al. 2018; Tuel and Eltahir 2020). Multiple climatic drivers are many and well documented. They are dynamic and interactive, resulting in negative cumulative impacts on the pelagic and benthic habitats that support the structural and functional biodiversity and productivity, as well as ecosystem services, like aquaculture and fisheries (Zunino et al. 2021). OA is therefore an additional challenge for the sustainability of Mediterranean socio-ecological systems and their governance. While OA as a stressor, temporally and spatially interacts with many drivers, there are spatial and temporal windows where OA is a main driver of the biological responses. Late fall or winter for example is characterized by low pH conditions, something which coincides with the shellfish spawning period. This requires a more in depth understanding of the OA exposure even as a single stressor.

Due to higher temperature and low local buffering capacity, the Mediterranean Sea is particularly sensitive to CO_2 increases in the atmosphere as they are quickly absorbed in the surface waters and rapidly transported to deeper water by overturning circulation (Cossarini et al. 2015, Melaku Canu et al. 2015; Simoncelli et al. 2018; Pinardi et al. 2019; Jiang et al. 2019; Cai et al. 2011, 2017; Hassoun et al. 2019; Wimart-Rousseau et al. 2021) as this results in lower pH and carbonate saturation states (Álvarez et al. 2014; Hassoun et al. 2015). Despite substantial spatial and temporal OA variability, fingerprinting of OA conditions show that the anthropogenic signal is already detectable across the Mediterranean Sea; in its Western basin for example with a decrease of 0.0028 pH unit/year (Luchetta et al. 2010; Kapsenberg et al. 2017; Ingrosso et al. 2017) and the Eastern Basin with a decline of 0.0021 pH unit/year (Hassoun et al. 2019). Surface pH is projected to decrease by about 0.24 and 0.46 pH units according to the 2100 IPCC SRES scenarios (Goyet et al. 2016), which is consistent with the global average (Geri et al. 2014; Goyet et al. 2016; Kapsenberg et al. 2017), but some parts of the Mediterranean might be exceeding the projected global change (Hassoun et al. 2015). The aim of this chapter is to provide a regional vulnerability assessment of OA hotspots, risks and gaps as a knowledge basis for the development of adaptive management options to combat OA and sustain ecosystem services in the Mediterranean Sea.

18.1.1 The Risks for Ecosystem Services in the Mediterranean Sea

OA impacts affect the integrity and functionality of ecosystems with cascading effect on the provisioning of ecosystem services, such as aquaculture and fisheries, and consequences to local economies (Barange et al. 2014; Gattuso et al. 2015). Such impacts are related to either direct, species-related negative effects on foundation or economically important species (Gaylord et al. 2015), or to the indirect effects related to the alteration of biogenic habitats, loss of biodiversity, changes in the availability of habitats, and general trophic web alterations (Sunday et al. 2016, Zunino et al. 2019, Zunino et al. 2021). Aquaculture and fisheries contribute to the recreational, provisioning and cultural services of marine ecosystems (Millennium Ecosystem Assessment 2005) and with fish aquaculture production projected to increase by 112% over the Mediterranean basin (Piante and Ody, 2015). The shellfish aquaculture is dominated by the production of the Mediterranean mussel (*Mytilus galloprovincialis*), followed by the Japanese carpet shell (*Ruditapes philippinarum*), whilst there is a limited production of other species (Massa et al. 2017). Fishery landings in the Mediterranean are dominated by small pelagics (herrings, sardines, anchovies; FAO 2018), with the Western Mediterranean subregion having the highest fisheries' capture, and closely followed by the Adriatic and the Central and Eastern Mediterranean sub-basins (FAO 2018).

Mediterranean mariculture is economically relevant to the sea-bordering states as 75% of this industry relies on the health of marine habitats in which climate change and OA are determinants of their productive efficiency (Rodrigues et al. 2015; Gazeau et al. 2014). Although wide variations exist in the sensitivities of different shellfish species of bivalves to OA (Range et al. 2014), they in general show negative impacts in growth and development, reduced calcification, and immunological and physiological alterations (Lemasson et al. 2017; Franke and Clemmesen 2011).The Mediterranean mussel (*M. galloprovincialis*) is the most extensively studied species with demonstrated sensitivity to OA, with the early warning response to reduced shell integrity being compromised at pH value of 7.7 (Michaelidis et al. 2005; Range et al. 2012; Bressan et al. 2014). The assessment of OA impacts on Mediterranean fisheries is complex because of species-specific sensitivities (Hilmi et al. 2014). In general, fishes seem to be physiologically more resilient to OA than bivalves (Michaelidis et al. 2007; Réveillac et al. 2015).

Major OA effects on fisheries might be related to the indirect impact on the variety of biogenic habitats that provide a unique environment and physical structure, and constitute hotspots for fisheries and species richness. Examples include seagrass beds (*Posidonia oceanica*), shellfish beds, variety of corals, crustose coralline algae, bryozoans, and vermetid reefs, all of which represent ecosystem engineers that provide a nursery habitat for fish and modify the substrata, contributing to numerous ecosystem functions and services (Fletcher and Breitling 2012; Milazzo et al. 2014). Biogenic habitats in the Mediterranean are already listed as endangered or vulnerable (Beal et al. 2016) with decreasing pH causing increased corrosion, skeletal loss and reduced calcareous algal cover (Martin and Gattuso 2009; Martin et al. 2008; Lombardi et al. 2011, Milazzo et al. 2014). Biogenic habitat of the red coral showed a significant decrease in skeletal calcification and polyp activity at reduced pH (Cerrano et al. 2013; Bramanti et al. 2013). Biogenic reefs have been historically subjected to a dramatic decline and are predicted to further decline due to climate change, diseases, and non-native species interaction (Rosa et al. 2012; Ingrosso et al. 2018; Badreddine et al. 2019a, b; Milazzo et al. 2019; Ragazzola et al. 2021).

18.1.2   OA Side-by-Side Governance

OA is an emerging governance challenge with a complex, uncertain, ever widening and transboundary nature. This nature as well as the potential ecological, economic as well as societal repercussions call for a need for tailored-specific OA solutions and mitigation strategies (Galdies et al. 2020; Tiller et al. 2019). It was only during the first Monaco Declaration on OA (Monaco, October 2008), when OA has been discussed in the political arena as a "parallel and interactive phenomenon" (Osborn et al. 2017). It has also been included as a separate target in SDG 14 (Life below Water) with Target 3 calling on States to minimize and address the impacts of OA. Nevertheless, there has been limited success in addressing OA leading scholars describing it as a "governance failure" (Jagers et al. 2019), or a "governance gap" (IPCC 2019). This is either due to the absence of relevant institutions or mechanisms of coordination, or inadequate mandates of existing organisations and mechanisms. This is also visible at the Mediterranean level where existing institutions, despite a certain robustness, adaptability, and governability, have not addressed the issue of OA in a concrete, specific, and binding manner.

At the supranational level, three overlapping and interacting governance frameworks are of potential relevance for regulating OA on the Mediterranean scale.[1] The first and more advanced framework is that of the European Union (EU).[2] However, EU-wide actions remain "incomprehensible and uncoordinated" (Galdies et al. 2020). Depending on political choices, economic activities or environmental components, the fight against OA may fall under various policies, including the marine and coastal policy, the nature and biodiversity policy, the water policy, the common fisheries and aquaculture policies or the climate policies. European decision-makers

---

[1] The governance of the OA problem also relies on the twenty-one riparian States, through their own domestic laws, institutions and policy processes, as well as their participation in regional or sectoral agreements and organisations such as the European Union. For an in-depth study of OA governance management at EU member States' level, see Galdies et al. 2020.

[2] The EU has exclusive competence in the conservation of marine biological resources under the Common Fisheries Policy (CFP) (Art. 3 TFEU), while its competence is shared with the Member States for fisheries (except conservation of marine biological resources), environment or research (Art. 4 TFEU).

have not yet discussed or even understood how these policies, under the responsibility of different Directorates General of the European Commission, could minimize and address the impacts of OA in an integrated, or at least effective manner. No minimum standards have been specified so far, such as binding targets (pH) for achieving good environmental status under the Marine Strategy Framework Directive.

The second relevant governance framework is the 1975/1995 Barcelona Convention for the Protection of the Marine Environment and the Coastal Region of the Mediterranean (in force 1976/2004) and its six Protocols administered by the UNEP (and to which the 21 riparian States are all contracting parties). Similar to the Convention on Biological Diversity and other regional or sectoral agreements such as OSPAR, the Barcelona Convention direct mitigation mandate is limited (*sensu* Herr et al. 2014) and relates to climate change rather than OA. Most of the information on OA circulated among its bureaucracy is scientific data.

The third relevant governance framework for OA at the Mediterranean scale is the management and conservation of fisheries under the mandate of the General Fisheries Commission for the Mediterranean (GFCM-FAO) and the International Commission for the Conservation of Atlantic Tunas (ICCAT). None of these Regional Fishery Management Organisations (RFMOs) have included OA in their scientific strategies (Herr et al. 2014). Such organisations remain guided by political-economic concerns and conservative in their approach to emerging scientific issues. Moreover, no reference to OA is made in the most recent reports of their fishery science bodies, the GFCM-FAO Scientific Advisory Committee on Fisheries (SAC) and the ICCAT Standing Committee on Research and Statistics (SCRS). Only the Scientific Advisory Committee on Aquaculture (CAQ) advised the GFMC-FAO members in its last report of 2017 to "incorporate aquaculture, climate change and ocean acidification issues in the system of indicators to monitor the sustainable development of aquaculture".[3]

In the light of its actors, structure, and processes, a "side-by-side governance" model is sustained in the issue area of OA on the Mediterranean scale (as on a global level), where the loci of action are so widely dispersed, unrelated, and situation-specific that neither the relevant governmental officials nor their transnational non-governmental counterparts can usefully resort to mass mobilisation. Instead, they must rely on interactive and multiple flows of influence (Rosenau 2000).[4] A side-by-side governance model is not necessarily an obstacle to solving the OA problem but could instead serve as a basis for adaptive governance. For that, Galdies et al. (2020) recommends a continued assessment in understanding the nature of the risks posed by OA in local and regional waters. The OA-related vulnerability risk assessment, which combines chemical exposure with the biological sensitivity (*sensu* Bednaršek

---

[3] General Fisheries Commission for the Mediterranean, Report of the tenth session of the Scientific Advisory Committee on Aquaculture, Izmir, Turkey, 27–29 March 2017, Doc. FIAP/R1206(Bi), p. 8.

[4] Public awareness remains low (Buckley et al. 2017; IPCC 2019; Tiller et al. 2019), OA being still often confused in its discernible reality with climate change.

et al. 2021), helps identify OA spatial-temporal hotspots, acknowledges the gaps and suggests appropriate adaptation strategies to provide desired results that can ultimately favour economic and social factors.

With this aim, the chapter presents a comprehensive OA vulnerability risk assessment related to the ecosystem services over the broad Mediterranean Sea basin. By linking regional OA exposure with the relevant biotic thresholds, we produced OA exposure maps to delineate the spatial-vertical-temporal OA hotspots for the aquaculture and the fisheries under current and future (2100) predicted conditions (Sect. 3). We used this vulnerability assessment to discuss the adaptive governance and institutional framework and focused on providing recommendations needed to deal efficiently with OA risks and policy-management responses for the sustainability of ecosystem services (Sect. 4) (*sensu* Ziveri et al. 2017: Osborn et al. 2017; Jagers et al. 2019; Galdies et al. 2020; Galdies et al. 2021).

18.2 Methods

In an in-depth review, Tiller et al. (2019) recognized the lack of information and thus related consequences of OA on the economically important species. As such, we have focused our investigation of OA-related sensitivity on the most economically important species, *Mytilus galloprovincialis*, with the most studied OA impact (Zunino et al. 2017), showing negative sublethal response related to calcification at pH = 7.7. This value represents an early warning threshold for OA impacts.

The risks associated with fisheries in the Mediterranean are related to two factors. First, the direct, specific fish *sensitivity*, and second, the *sensitivity* and the *exposure* of the biogenic habitats which are important nursery and foraging habitats for a variety of fish in the coastal habitats (Zunino et al. 2021).We focus only on a variety of surface to near-surface biogenic habitats that show high pH sensitivity, on average at pH = 7.8, which is a value that can induce sublethal responses related to the reduced growth, calcification, and increased dissolution of many coastal biogenic habitat builders, and is lethal to red coral (Cerrano et al. 2013). To describe current and future levels of *sensitivity*, species representative of aquaculture activities with well-defined sensitive pathways occurring at specific OA thresholds can be used. Such information on thresholds was overlaid with current and future pH exposure maps to understand spatial patterns of OA risks.

A comparison of carbonate chemistry parameters (pH and aragonite saturation state, Ω_{ar}) distributions in summertime (July–September) 2019 and 2100 has been conducted across three different depths, i.e. 26, 51, and 250 m. The three depth levels were chosen to represent the habitats that various pelagic and benthic species inhabit and form respective ecosystems. Present day estimates (summer 2019) of seawater pH, dissolved inorganic carbon, chlorophyll a, salinity, and temperature have been derived from the European Copernicus Marine Environment Monitoring

Service (CMEMS; Bolzon et al. 2020; Clementi et al. 2019) at 1/24° of horizontal resolution.[5]

The projected pH values under RCP 8.5 scenario for 2100, are derived from the analysis made with the coupled physical-biogeochemical model OGSTM-BFM, 1/8° horizontal resolution (Solidoro et al. 2022). The average sea surface temperature (SST) and sea surface salinity (SSS) are projected to increase 3 °C and 0.5 respectively by 2100. Projected changes of pH in the Mediterranean basin are −0.34 pH units in the first 500 m, decreasing to −0.20 and −0.15 pH units in the 500–1000 m 1000–5000 m depth layers respectively. These values agree with the values (around 0.352 pH unit, range 0.242–0.462) from Adloff et al. (2015), Goyet et al. (2016), Richon et al. (2019), and the one projected by Feely et al. (2009) for the North Atlantic and North Pacific. The aragonite saturation state (Ω_{ar}) has been estimated from total alkalinity (μmol/kg), DIC and pH, using CO2SYS (Lewis and Wallace 1998). The 2100 projections of DIC and other carbonate chemistry variables were calculated with CO2SYS using the pH, temperature, alkalinity and salinity of 2019 CMEMS datasets corrected by the basin average projected changes of pH in agreement with Solidoro et al. 2022 (other variables have kept constant). The projected increase in temperature and salinity resulted in a small increase in Ω_{ar} of about 0.04 units.

18.3   Results

18.3.1   pH and Ω_{ar} Distributions in 2019 and 2100

There are notable spatial and vertical pH gradients across the Mediterranean basin under the current (2019) and expected OA conditions in 2100 (Fig. 18.1). A significant finding is the occurrence of lower pH values at 26 m depth, when compared to the 51 m or even the 250 m depth. This could be due to the highest temperature values in shallow waters presented in Fig. 18.2. The northern part of the Western Mediterranean, mid-Adriatic, and Aegean side generally have the highest pH values, ranging between pH of 8.0 and 8.13. The southern part of the Mediterranean basin is characterized by the lower pH values (pH range: of 8.0–8.05), with the coastal regions along the African coast, the Gulf of Gabes, the Libyan Sea (Central Mediterranean) and the coastal Levantine Sub-basin (South-Eastern Mediterranean) having the lowest pH values (about 7.96–7.99). Low pH values in offshore waters appear mainly in the Algerian-Balearic Sub-basin (Western Mediterranean), Southern Ionian Sub-basin, and the Western side of the Adriatic Sub-basin. In these same regions the coastal trend of low pH surface water is extended towards the offshore waters. Significant north-south pH differences are still observable at the depth of 51 m, examples including the Sea of Crete and in the Gulf of Iskenderun. In

---

[5] https://resources.marine.copernicus.eu/product-detail/MEDSEA_MULTIYEAR_PHY_006_004/INFORMATION

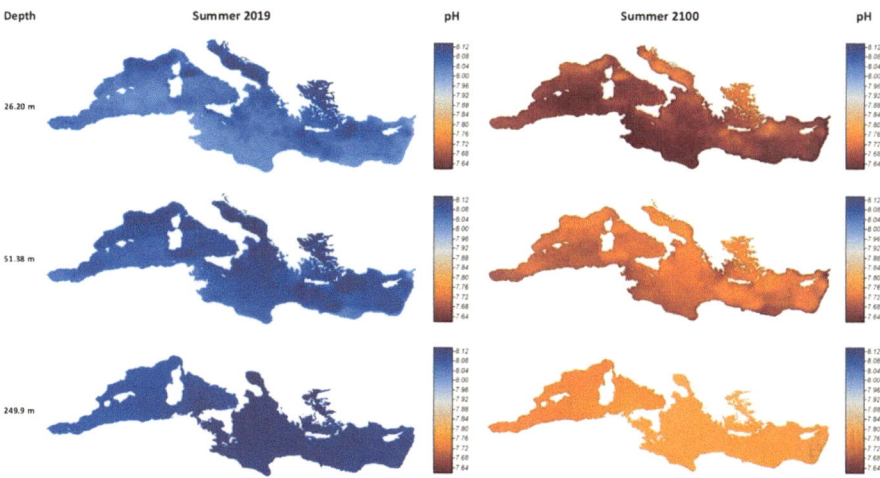

Fig. 18.1 Spatial distribution of pH for (**a**) July–September of 2019 (Bolzon et al. 2020) and (**b**) 2100 in the Mediterranean basin at three specific depths

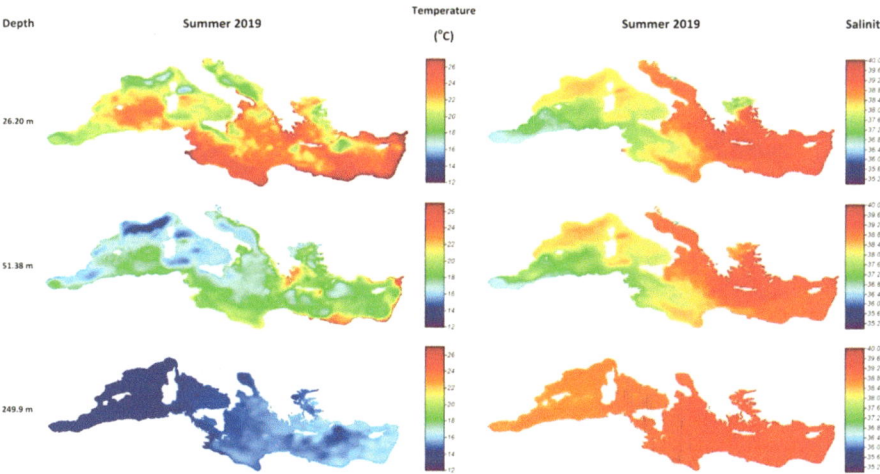

Fig. 18.2 Spatial distribution of (**a**) seawater temperature (°C) and (**b**) salinity for summer 2019 in the Mediterranean basin. (Clementi et al. 2019) at three specific depths

contrast, at 250 m depth, major pH gradients are more prominent in the west- east direction, with the Western basin having slightly lower pH values (range: 8.00–8.05) compared to the Eastern basin (range: 8.05–8.13), with the north-south pH gradient no longer apparent at 250 m. The pH projections for 2100 demonstrate the same spatial patterns as under the current (2019) conditions, with the lowest pH values in the upper 26 m projected for the Ionian Sub-basin and the eastern part of the Tyrrhenian Sub-basin. The projected frequency histogram of pH conditions for 2100 in the entire Mediterranean Sea ranges from 7.64 to 7.74 (Fig. 18.3). The 2100

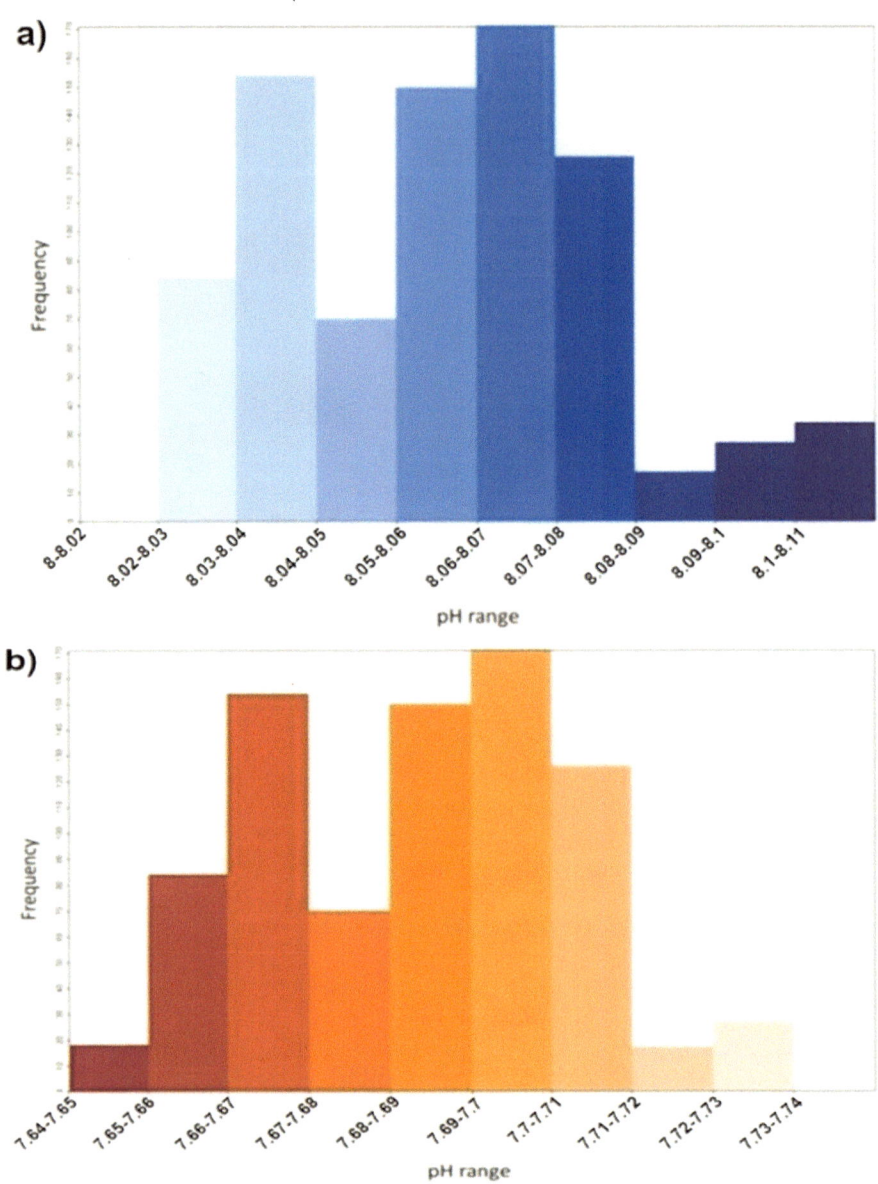

Fig. 18.3 Frequency histogram showing the pH range found within aquaculture shellfish installations in the Mediterranean (N = 335) for summer 2019 (left, blue) (Bolzon et al. 2020) vs. 2100 (right, orange) at depth of 26 m. Discrete pH classes (based on an offset pH value of −0.36 on 2019 values; Table 1) for 2019 are plotted against the number of shellfish installations

pH distribution at 50 m depth shows more uniform values across the entire sea, representing an important difference from present-day conditions. Based on the 2100 projections, it appears that the regions with currently high pH values will acidify faster than the regions with lower values. At 250 m depth, the observed pH gradient is greatly reduced, with pH values slightly higher in the Eastern basin.

The Ω_{ar} horizontal distributions for summer 2019 are roughly inversely correlated with pH, while the projected 2100 patterns generally look more uniform (Fig. 18.4). An important difference in Ω_{ar} compared to pH is that Ω_{ar} values are the highest at 26 m and consistently decrease at 51 and 250 m depths. There is a characteristic west-east Ω_{ar} gradient at 26 m depth, with the Western having lower values ($2.4 < \Omega_{ar} < 3.2$) compared to the Eastern basin ($3.4 < \Omega_{ar} < 4.3$). The regions with the highest current Ω_{ar} values are in the Aegean, coastal Levantine, Adriatic, and coastal central Mediterranean sub-basins. Although there are lower Ω_{ar} values at 51 and 250 m depths, the west-east gradient in Ω_{ar} still exists, but it is not as apparent. The 2100 projections show reduced Ω_{ar} gradient at all depth levels. Near the surface, Ω_{ar} values decrease more in the Eastern than in the Western basin, while at 250 m, the Western basin contains lower Ω_{ar} values closer to the saturation state ($1.3 < \Omega_{ar} < 1.7$) compared to the range of Ω_{ar} values ranging from 1.7 to 1.9 in the Eastern basin.

18.3.2 Assessment of the Risks Related to the Aquaculture

Here, we define *vulnerable* regions as areas associated with increased risk due to ocean acidification; specifically, the combination of OA exposure and the species sensitivity (*sensu* Bednaršek et al. 2021). Estimation of *exposure* was applied to the

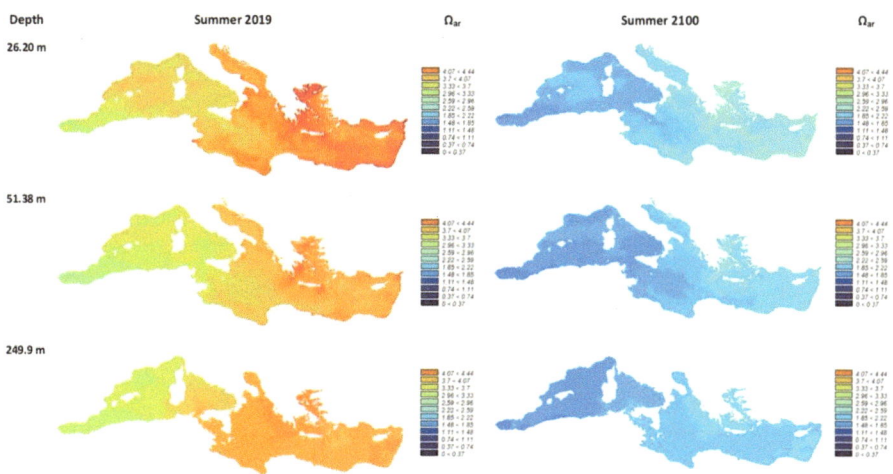

Fig. 18.4 Spatial distribution of aragonite saturation state (Ω_{ar}) for (**a**) summer 2019 and (**b**) estimated 2100 in the Mediterranean basin at three specific depths (26, 51, and 250 m)

regions of intense aquaculture activities occurring in coastal areas from the surface to 20 m depth. Spatial location of aquaculture sites used for this study were derived from the European Marine Observation and Data Network (EMODnet, Human Activities 2020; Fig. 18.5), which represents the most comprehensive database connected to the network of organisations committed to provide accurate biogeographic information and tightly coupled to the EU's integrated maritime policy. Spatial mapping of these sites shows installations located mainly along the Mediterranean shores of France, Italy, and Greece. However, EMODnet states that the absence of other locational information does not mean the absence of such installations, only indicating that the information on the aquaculture is very fragmented, especially in the southern shores of the Mediterranean.

The areas with dominant aquaculture activities are currently in the regions with the highest summertime pH range (8–8.1) in the upper 20 m. Of these, the lowest baseline pH surface conditions occur in the western part of the Ionian Sub-basin, as well as the eastern part of the Tyrrhenian Sub-basin. Under current pH conditions, exposure in all of the aquaculture-intense regions is well above the thresholds associated with physiological impairments (pH = 7.7), based on which we conclude that there is no current risk for the aquaculture. Conversely, the future (2100) RCP 8.5 summertime scenario projects a considerable decline in suitable conditions and, thus, increased exposure to conditions below the pH thresholds across all of the regions occupied by aquaculture installations. With an average projected pH value of 7.65, exposure will be consistently below the thresholds, likely impairing

Fig. 18.5 Regions with intensive aquaculture sites, identified by the European Marine Observation and Data Network, from which 2019 and 2100 pH projections were derived in this study

essential pathways, such as calcification, growth, acid-base regulation, etc. Based on this assumed increase in exposure (Fig. 18.1) and associated sensitivity, the risk for the aquaculture industry is expected to significantly increase under the 2100 scenarios, although this level of exposure would still not induce an increased mortality linked to the substantially lower pH values.

18.3.3 Indirect Risk of OA on the Fisheries in the Mediterranean Sea

In this study, we indirectly evaluate fisheries risks to OA through the *exposure* of the biogenic habitats serving as an essential fisheries habitat. The outputs were specifically related to the near-surface vertical habitats in coastal regions for the upper surface 10–20 m depth for the entire Mediterranean Sea. The lowest summer pH surface conditions, down to pH = 7.96, occur in the eastern part of the Tyrrhenian Sub-basin, the entire Eastern Mediterranean coast around the Levantine Sub-basin, the central North African coast, the southwestern Mediterranean coast, and the northern Adriatic Sub-basin, with the sole exception of the Aegean Sub-basin (Fig. 18.1). These conditions are very near the pH threshold of 7.8–7.9 that can induce negative impacts of the biogenic builders, exposing these habitats to below the thresholds fairly frequently in the summertime even under present-day conditions, especially when the duration of the exposure is the longest. Given that the 50 and 250 m pH values are higher indicating that the risks for this deeper depth biogenic habitats are lower.

Based on the future pH projections, the risk for the biogenic habitats is expected to substantially increase under future (2100) RCP 8.5 scenarios where summertime conditions below the threshold of pH = 7.8 are uniformly present over the entire Mediterranean Sea at the 10–20 m depth, with no sub-basin spared. Even the deeper depth builders at 50 m will be exposed to conditions below these thresholds, indicating that not only spatial but also vertical fisheries habitats will be compromised.

18.4 Shaping Adaptive Management Options to Combat OA in the Mediterranean Sea

Lessons from the past, including but not limited to "the aquaculture failure" (Barton et al. 2012), as well as the literature, show that OA is an emerging, complex challenge requiring polycentric and feasible nature-based solutions at the climate-OA-biodiversity nexus (*sensu* Turley et al. 2010; Galaz et al. 2012; Billé et al. 2013; Gattuso et al. 2015, 2018). Although OA is global by essence, its consequences fall asymmetrically upon respective ecosystems and the coastal and local communities that depend on them (Osborn et al. 2017). Therefore, there is indeed a need to develop context-dependent management strategies to minimise ecosystem damage caused by the continued release of CO_2 (Albright et al. 2016). Earlier in the text, we assessed the risks related to the aquaculture and fisheries under current and future conditions in the Mediterranean Sea. Such broader risk characterisation over the wider Mediterranean basin is useful in understanding regional OA hotspots and the ecosystem service sensitivity and establishes when and where OA mitigation or adaptation measures are needed. Such approach also identifies an explicit link with the adaptive governance by highlighting the core role of continued and context-based scientific assessment in decision-making.

There are currently no direct and structured regulation of OA, either internationally or for the Mediterranean Sea. The long-lasting socio-ecological effects of OA are embedded in a highly diffuse and complex regulatory and institutional framework, without any topic-related institution, mechanism, or mandate to combat it, across geographical and temporal scales, or "scale-independently" (Galdies et al. 2020; Billé et al. 2013). The difficulty in communicating with both decision-makers and stakeholders contributes to a regulatory gap in developing targeted regulatory frameworks and setting statutory pH thresholds. Barriers to generating data and information also explain why interactions between regimes pertaining to OA (for instance, climate, biodiversity, and ocean regimes) are primarily cognitive and epistemological, rather than normative and institutional.[6] Other non-OA specific impediments exist, such as mismatching institutional and legal frameworks, geopolitical tensions between riparian states, the lack or unequal distribution of resources or capacity, the lack of will and commitments of actors or pathway dependency (*sensu* Berkowitz 2020).

Several state and non-state actors have worked on the issue in recent years, mostly through informal or mixed initiatives aimed at creating awareness around the problem, synthesising and disseminating scientific information, and influencing negotiations at high-level forums.[7] Substantial outcomes, in terms of governmental action to implement concrete policies and policy tools with the capacity to generate behavioural change necessary to prevent OA, are still largely unseen (Galaz et al. 2012; Osborn et al. 2017; Ziveri et al. 2017; Jagers et al. 2019). The governance of the Mediterranean remains somehow ill-adapted to OA, without any strong dedicated or integrated policies, clear institutional mandates, or legally binding objectives, thus resulting in slow and uncoordinated progress. This issue is compounded by a linear, deterministic and state-centric approach that is still reflected in international law and governance and is unsuited to the necessary systemic vision of OA as a boundary challenge (*sensu* Capra and Mattei 2015; Rockström et al. 2009).

In light of the projected impacts of OA on marine habitats and ecosystems, as well as the fisheries and aquaculture provisioning services and other ecosystem regulating and supporting services, "best possible futures" regarding OA in the

[6] For more information, see Guilloux (2020).
[7] Cicin-Sain et al. (2019, p. 20); Ocean alliance to combat ocean acidification: https://www.oaalliance.org/; Global Ocean Acidification Observing Network (GOA-ON) and its Ocean Acidification Mediterranean Hub (OA Med-Hub); Ocean Acidification International Coordinating Centre of the IAEA (International Atomic Energy Agency).

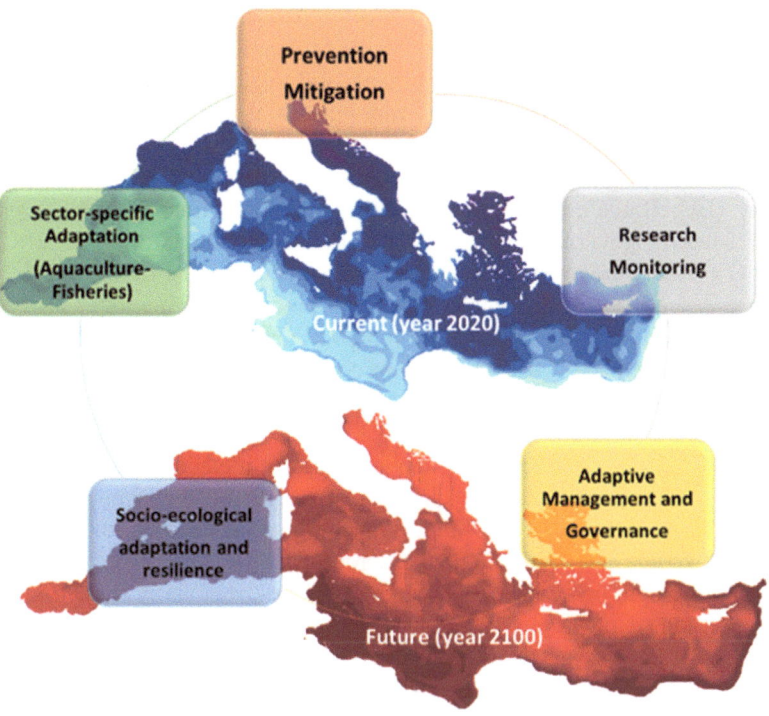

Fig. 18.6 Linkage of socio-ecological system to transition between current (2020) and future (2100) pH conditions

Mediterranean Sea and its sub-basins requires a bundle of interconnected strategies (Fig. 18.6), overall related to the central framework of an *Adaptive Management and Governance*.[8] Such framework includes the *Adaptation and Resilience* (improvement of the capacity of adaptation and resilience of the changing Mediterranean socio-ecological system to OA impacts); V*isionary research and Monitoring* (transcending disciplinary as well as political and legal divides and projecting to spatial and temporal scales adapted to socio-ecological issues); and finally *Prevention* (in the form of rapid and complementary global-scale reductions in greenhouse gases (GHGs), especially CO_2, and local reduction in anthropogenic sources of OA).

---

[8] Adaptive governance is a system of environmental governance with the potential to mediate the complexity and uncertainty inherent in socio-ecological systems (Dietz et al. 2003; Walker et al. 2004; Folke et al. 2005; Folke 2006), whereas adaptive management implies acknowledging uncertainty (and complexity) through prudent decision making rather than a static "answer" (or failure to act) by continuously gathering and integrating appropriate ecological, social, and economic information and knowledge with the goal of adaptive improvement (sensu Walters 1986, 1997; Costanza et al. 1998; Gunderson 1999; Lee 1999; Dietz et al. 2003; Chaffin et al. 2014; Monaco Ocean Acidification Action Plan, Priorities (5), 2018, p. 2).

Besides linking knowledge and decision-making through data collection and monitoring, meaningful and effective coordination and cooperation between states and non-state actors, capacity development, and exploiting governance and legal opportunities seems to be the most relevant options contributing to adaptive governance to combat OA in the Mediterranean (Sharma-Wallace et al. 2018). The side-by-side model of governance and institutional polycentricity, redundancy, and diversity (EU, Barcelona, and regional fisheries institutions) can be a premise towards adaptive governance (sensu Dietz et al. 2003; Galaz et al. 2012; Chaffin et al. 2014).Adaptive governance cannot be reduced to a list of specific prescriptions, but instead rests on situated "pattern of practises" (Brunner et al. 2005) of state and non-state actors towards common goals and behaviours in addressing OA impacts and related risks. It culminates in coordination at the bioregional scale, a scale at which the governance structure best fits ecological functions (Olsson et al. 2007; Huitema et al. 2009; Termeer et al. 2010; Cosens 2013; Chaffin et al. 2014) but has yet to find a common socio-anthropological meaning.[9] Matching regional and local governance scales is a challenge for tailored management actions to address OA. Given that EU and Mediterranean institutions (e.g., the coordinating unit for the Mediterranean Action Plan- Secretariat to the Barcelona Convention and its Protocols) might be more robust than global ones, regionally framed multilevel governance may well be an appropriate response to global challenges like OA (sensu Patt 2010; Selin and VanDeveer 2009, 2011). To avoid national policy gaps created by the flexibility of fisheries and aquaculture policies in the Mediterranean (Said et al. 2018), Nationally Determined Contributions (NDCs) and National Adaptation Plans (NAPs) have so far neglected OA (Gallo et al. 2017). Fisheries (Hidalgo et al. 2018), or aquaculture might be the most adapted tools to address OA in an integrated manner (Galdies et al. 2020). For example, scenario analysis, modelling the effects of emission reductions (Steinacher et al. 2013) should be used to assess global and local impacts of OA of several climate policy responses (IPCC 2019).

Supporting adaptive management and governance to combat OA does not necessarily require the creation of new legal and institutional settings. It depends foremost on improving the adaptive capacity of existing EU and regional policies and laws in changing or even shifting socio-ecological conditions and problems, incorporating incrementally varied actors (regional organisations and non-state actors) and processes for amending and decision-making, without creating instability or rigidity.[10] First, this involves identifying, legal constraints to an adaptive OA management in the Mediterranean and remedy to them: e.g., the "dilution" of SDG 14.3 in sector-based legal and institutional frameworks, politically-economically

---

[9] Ideally, the scale of adaptive governance will also have to be targeted to the social and ecological nature of the OA problem, as well as to societal goals like SDG 13 or healthy marine ecosystems, through sufficient response flexibility within and between existing political boundaries (Cosens 2010; Termeer et al. 2010; Chaffin et al. 2014).

[10] For more information about the role of Law in adaptive governance, see Ruhl (1997), Cosens et al. (2017), Gosnell et al. (2017), and Reinke de Buitrago and Schneider (2020).

oriented decisions rather than science-based ones, inadequate mandates, jurisdictional overlaps, narrow and prescriptive rules, potentially conflicting principles and norms leading to socio-environmental problem shifting.

Second, existing adaptive processes should also be strengthened to minimise risks and build resilience. OA should thus be factored into adaptive fisheries and aquaculture management plans (Lacoue-Labarthe et al. 2016), whose implementation varies throughout the Mediterranean between EU Member States and non-EU Member States, and, among EU Member States in the application of the reformed CFP and associated Strategic Guidelines on Aquaculture (Corner et al. 2020). It could be factored through pH-associated regime changes, thresholds, or tipping-points (Hughes et al. 2013; Goyet et al. 2016; Good et al. 2018). Regarding aquaculture, there is a strong need to support governance to establish aquaculture activities within a coordinated spatial planning process under an ecosystem approach, based on integrated coastal zone management principles and the establishment of allocated zones for aquaculture (General Fisheries Commission for the Mediterranean–FAO 2018; Corner et al. 2020). The conducted vulnerability study for the aquaculture (obtained for *Mytilus galloprovincialis*) also offer a baseline to help managers and stakeholders assess the reliability and feasibility of aquaculture in a changing sea that can generate undetected and underestimated impacts on the aquaculture sector (*sensu* Martinez et al. 2018). Third, iterative, transdisciplinary, and reflexive learning and decision-making such as marine spatial planning must also be encouraged, allowing decision-makers and stakeholders likely to be affected by OA to communicate their specific needs to researchers and decision-makers, or even participate in the decision-making process. Finally, "more attention should be paid to the potential disconnects between what science tells us is necessary for a healthy ecological system, what society wants from that ecosystem, and perhaps more importantly, what is politically feasible" (Chaffin et al. 2014; see also, Wyborn 2015).

18.4.1 Fostering the Adaptation and Resilience of the Mediterranean Socio-Ecological System to OA

At their infancy, general and sector-specific adaptation actions are mostly directed at restoring or protecting the production of services and goods harmed by OA ("supply-side oriented") (Ziveri et al. 2017). They focus on treating the localised symptoms of OA where institutions are already in place by adjusting the socio-ecological system (Jagers et al. 2019). These actions or strategies can be applied in an anticipatory (resisting change) and responsive (abating or recovering from change) ways (Billé et al. 2013; Gattuso et al. 2015). They mostly rely on sector-specific adaptation management options, especially in aquaculture and fisheries, such as relocating fisheries and aquaculture activities, and protecting food supply at local, state, and sub-regional levels.

Other targeted and interconnected bundles of solutions can be considered:

- Revisit estimation of OA into monitoring and ecosystem assessments, including how fishing grounds may impact or be impacted by OA. This means including indirect effects on the biogenic habitats that provide spawning and nursery grounds, for which little information is currently available;
- Determine the regions with more suitable future marine pH conditions and strategically invest in aquaculture operations in those regions, select resistant strains more tolerant to OA through selective alternative breeding techniques (Cooley and Doney 2009; Albright et al. 2016; Ziveri et al. 2017), and develop transportable or seasonal aquaculture infrastructure;
- Substitute and diversify fish and aquaculture species when permanently or temporarily possible by compensating fishermen and fish farmers, providing transitional support, and stimulating innovation to accelerate emergence of alternatives and technical replacement solutions (Jagers et al. 2019);
- Sustain food security by cutting or eliminating subsidies in order to reduce incentives to overexploit marine living resources, publicly investing in activities that include protecting the marine environment or supporting fishermen, where decoupling the economy from fishing is the extreme option (Ziveri et al. 2017);
- Develop adaptive and climate-friendly fisheries management so as to be able to incorporate new and emerging knowledge quickly or regularly at the level of RFMOs and the EU (*sensu* Rayfuse 2012; Herr et al. 2014); Modernise the CFP through, e.g., adapting quotas and management systems to OA; Invite RFMOs to promote stronger calls for emission reduction commitments (Herr et al. 2014).

With a view of integrated environmental, coastal and marine policies, the capacities of adaptation to and resilience under OA conditions for the Mediterranean socio-ecological system need to be activated by, *inter alia*:

- Fostering social resilience, which implies addressing the socio-economic impacts of OA by, e.g., breaking down financial, informational, cognitive, social, and cultural barriers; path dependency; compensating vulnerable coastal communities and helping them adapt to their new circumstances by stimulating education and investment; raising public awareness to change people's preferences and habits; linking adaptive environmental management to social transformative needs; etc.
- Supporting ecosystem resilience by:

  - Incorporating OA and other climate-related thresholds, goals and strategies (e.g., nature-based solutions including ecosystem-based approaches) and goals into marine spatial planning, or marine protected areas (MPAs) to conserve or restore species and ecosystems vulnerable to OA. The benefits of such actions are manifold, i.e., they address many stressors simultaneously, act within single jurisdictions or regions, and minimise transaction costs (Billé et al. 2013);
  - Treating fresh- and coastal waters (e.g., for high-value aquaculture by exploiting shellfish production) to support healthy marine waters (Ziveri et al. 2017);

- Restoring degraded "blue carbon" ecosystems (e.g., seagrass meadows), as well as other ecosystems (e.g., estuaries inhabited by oysters), with the understanding that it may slow long-term changes at the local level but can also exacerbate short-term variability (Sabine 2018);
- Embedding networks of MPAs, and combining OA exposure with biodiversity and ecosystem functioning, including connectivity, producing a multilayered, holistic conceptual space that will be instrumental for future management and protection of such networks and of marine environments in general.

- Paving the way for legal and institutional resilience (*Á propos* institutional resilience, Rayfuse 2012; or comprehensive legal approaches, Galdies et al. 2020)

18.4.2 Visionary Research and Monitoring of OA Impacts on the Mediterranean Sea

OA is among the key research questions for biodiversity sustainability and conservation. There is a persistent need for cyclical, interdisciplinary, and long-term OA information sharing and learning to address the complex, uncertain, and technical nature of this ever-widening, transboundary phenomenon (Fig. 18.6). In general, most research on OA has been conducted in the natural sciences on understanding its ecological and biogeochemical implications.[11] Against this background, it is essential to encourage interdisciplinary research on OA in the social, economic, political, and legal sciences, and between social, natural sciences and humanities. Analysing the coherency and conflicts between national, regional Mediterranean and international legal and administrative systems is of critical importance for social (including legal) sciences research with regard to OA (Jagers et al. 2019). A lack of scientific understanding of ecological interdependencies makes it more difficult to detect, avoid, and solve potential legal conflicts (Wolfrum and Matz-Lück 2003) and governance issues. The pace of decision-making partly depends on whether the type or abundance of information being offered from the bottom-up matches what is being sought from the top down (Cooley et al. 2015).

---

[11] See the *European Mediterranean Sea Acidification in a Changing Climate* (MedSeA) project's key documents available at: http://medsea-project.eu/outreach/key_documents/ [Accessed May 22, 2020]. See also the *European Project on Ocean Acidification* (EPOCA) funded by the European Commission under Framework Programme 7 from 2008 to 2012. For more information, see European Project on Ocean Acidification | EPOCA Project | FP7 | CORDIS | European Commission. Available at: https://cordis.europa.eu/project/id/211384 [Accessed May 22, 2020]. OA is also amongst the indicators of the Environmental Europe Agency, designed to answer key policy questions and to support all phases of environmental policy making. For more information, see Ocean acidification European Environment Agency. Available at: https://www.eea.europa.eu/data-and-maps/indicators/ocean-acidification-2 [Accessed May 22, 2020].

For instance, there is a lack of empirical and context-specific knowledge evaluating the resilience and adaptive capacity of legal systems to climate, biodiversity, and ocean changes. There is also a need for the provision and the communication of reliable information (long-term data, datasets, models, standards, and observational networks) aligned with the level of understanding of local stakeholders and policy makers, and their expectations and with pre-existing priorities related to climate and regional development (Cooley et al. 2015). Such alignment is paramount in order to secure the scale of investment, to develop forecasting capabilities (Monaco Ocean Acidification Action Plan, Priority No 6), and to move forward transdisciplinary and precautionary marine spatial planning (Fig. 18.6). Ideally, relevant socio-anthropological and governance information and knowledge should be integrated to guide the way best available science on OA is produced and used in decision-making. The following are suggested for an improved evidence-based decision-making:

Support of integrated *in situ* scientific information supply, monitoring and modelling to enable better-informed decision-making:

- Continuous, local carbonate chemistry monitoring (of at least two carbonate system parameters); Characterising seasonal patterns; Monitoring and understanding processes related to coastal pH variability;
- Continue researching economically important species and scale up to the ecosystem level, and the potential synergism of acidification with other climate change relevant stressors, including warming and marine heat waves, stratification, and eutrophication.
- Devise long-term experimental studies to understand adaptation as well as acclimation (Monaco Ocean Acidification Action Plan, Priority No 2).
- Strengthen regionally and financially incentivise a coordinated and institutionalised network of monitoring stations through the OA Mediterranean hub[12] of the Global Ocean Acidification-Observing Network (GOA-ON), to map the vulnerability of coastal areas to OA and to extend monitoring to near-shore systems relevant to management jurisdictions;

Definition of interdisciplinary research priorities and frameworks (*sensu* Albright et al. 2016) such as:

- Direct and indirect effects of OA on spawning and nursery habitats and their consequences on fisheries;
- Identify critical habitats sites (*sensu* Ziveri et al. 2017) based on the most ecologically critical and sensitive bases; as well as heavily under-studied Mediterranean areas, such as the South and South-East realms that might be negatively exposed to OA and would benefit from strong research coordination, building capacity and OA-related policy and management;

---

[12] For more information, see http://www.goa-on.org/regional_hubs/mediterranean/about/introduction.php [Accessed May 22, 2020].

- High-resolution physical-chemical observations and regional downscaled model outputs should be used to provide more accurate and spatially explicit OA exposure.
- A governance structure that enables continued support for the further development and use of downscaled models and significant enhancement of coastal observation processes will help understand local and regional processes, their timing, and extent that can negatively impact ecosystem services, and result in improved management-policy actions.
- Set up initiatives in each EU coastal State to assess the threat of OA to ecosystem health and human livelihoods and to evaluate strategies to mitigate local drivers (*sensu* Strong et al. 2014);
- Identify and financially support interdisciplinary research related to the socio-ecological impacts of OA or evaluating the adaptive capacity and the resilience of regional governance systems, to slow onset and abrupt environmental changes within the next research and innovation framework programme (Horizon Europe).

18.4.3   Preventing OA by Mitigating Climatic and Anthropogenic Sources

The only comprehensive solution to prevent further OA is to rapidly and drastically reduce global anthropogenic emissions of CO_2 and other GHGs uptake by the ocean. Multi-level policy tools targeting CO_2 emissions and beyond, behavioural changes, exist such as the EU climate action and the European Green Deal.[13] They are not designed to address OA specifically but can be marginally adapted to do so (Billé et al. 2013; Jagers et al. 2019) and to embrace the sustainable development and biodiversity post 2020 goals. In addition, where local and national economies rely heavily upon carbonate-dependent ecosystem services like in the Mediterranean Countries, reducing local acidifying stressors could produce results both faster and in a more politically feasible manner than at the global level (Billé et al. 2013; Gattuso et al. 2018). Such reductions could include non-atmospheric local or site-specific stressors, such as nutrient pollution (nitrogen, phosphorus) and runoff from acidic fertilisers used in agriculture (Doney 2010; Kelly et al. 2011; Carstensen et al. 2020; Duarte et al. 2020).To this end, a spatially explicit biological vulnerability assessment (as conducted in here), and the inclusion of different stakeholders of relative importance to the different causes related to OA is necessary to maximise the utility of smaller-scale policy recommendations (Billé et al. 2013; Galdies et al. 2020).

---

[13] For more information, see Cabuzel (2019).

18.5 Critical Considerations

The biogenic habitats that are uniform coastal features across the Mediterranean basin are and will be more severely impacted in the future, especially in the less developed areas around the Mediterranean coast. Currently, despite being on the list of endangered or vulnerable habitats, there are no specific conservation actions in place for them. Because of such an invaluable habitat role and their sensitivity, marine conservation planning should be considered with regional OA impact in mind. While we address regional ecological risks related to the ecosystem services, we only partially address socio-economic risks, more comprehensively described in Hilmi et al. (2014). The subsistence of human communities depends heavily on marine resources, especially from the less developed Mediterranean nations, where fishing provides a greater contribution to GDP and supports higher levels of employment. Since the majority of fishers fall into the small-scale artisanal category, this makes them more dependent on coastal inshore waters and, thus, vulnerable to the local ecosystem conditions. Concurrently, our risk estimates point towards the coastal ecosystems, such as fisheries-supportive habitats, that will be mostly impacted by OA in the very near future, thus exposing already vulnerable fishing communities even further. This is in contrast to the aquaculture industry stationed in the developed part of the Mediterranean Sea that is, through proactive OA risk management, less vulnerable than artisanal fishers. Still, large adaptation in aquaculture practises and related costs will be needed to sustain the expected increases in productivity that has quadrupled in the last 30 years (Hilmi et al. 2014).

Regional governance on a pan-European and Mediterranean scale has potential to support the development of 'localised' management strategies to minimise the exposure and risks incurred by coastal and marine ecosystems and dependent local communities to the continued release of CO_2 and subsequent OA. The future governability of OA (*sensu* Gattuso et al. 2018) will depend on the robustness, adaptability, and the quality of polycentric (interagency and institutional) coordination (*sensu* Galaz et al. 2012) amongst regional organisations and other non-state actors, while remaining inexorably limited by the global nature of the OA boundary challenge that requires Earth scale-fitted solutions. The most urgent one is the drastic reduction in GHGs/CO_2 emissions (Harrould-Kolieb and Herr 2012; Kim 2012; Herr et al. 2014; Stephens 2015). Despite existing adaptive management tools (e.g., MPAs) and processes (e.g., integrated coastal zone management or marine spatial planning), evaluations of the planning, decision-making, implementation, monitoring, and reporting related to OA are scarce. More extensive learning and experiments are necessary to foster the adaptive capacity of Mediterranean socio-ecological system to abrupt and slow-onset climate and non-climatic changes, to scrutinise potentially maladaptive incentives, mechanisms, and investments, to resort to mass mobilisation to increase OA awareness and, to acknowledge legal and institutional barriers.

Acknowledgements NB and BG share the first co-authorship as they have equally contributed to the paper, but to different sections. This article is based upon work from COST Action CA15217 – Ocean Governance for Sustainability - challenges, options and the role of science, supported by COST (European Cooperation in Science and Technology). COST (European Cooperation in Science and Technology www.cost.eu) is a funding agency for research and innovation networks. These Actions help connect research initiatives across Europe and enable scientists to grow their ideas by sharing them with their peers. This boosts their research, career and innovation. Nina Bednaršek acknowledges support from the Slovene Research Agency (ARRS *'Biomarkers of subcellular stress in the Northern Adriatic under global environmental change'*, project # 01 J12468). Roberta Guerra was funded by Short-Term Scientific Mission ATlaNTES Grant (COST-STSM-CA15217-40699) within the COST Action OCEANGOV. NB attended Piran workshop with COST Action travel support. RF was supported by the Pacific Marine Environmental Laboratory and the NOAA Ocean Acidification Programme. Contribution number 5108 from the NOAA Pacific Marine Environmental Laboratory. In memoriam T. Bednaršek.

References

Adloff F, Somot S, Sevault F et al (2015) Mediterranean Sea response to climate change in an ensemble of twenty first century scenarios. Clim Dyn 45:2775–2802

Albright R, Caldeira L, Hosfelt J et al (2016) Reversal of ocean acidification enhances net coral reef calcification. Nature 531:362–365. https://doi.org/10.1038/nature17155

Álvarez M, Sanleón-Bartolomé H, Tanhua T et al (2014) The CO_2 system in the Mediterranean Sea: a basin wide perspective. Ocean Sci 10(1):69–92. https://doi.org/10.5194/os-10-69-2014

Badreddine A, Milazzo M, Abboud-Abi Saab M, Bitar G, Mangialajo L (2019a) Threatened biogenic formations of the Mediterranean: current status and assessment of the vermetid reefs along the Lebanese coastline (Levant basin). Ocean Coast Manage 169:137–146. https://doi.org/10.1016/j.ocecoaman.2018.12.019

Barange M, Merino G, Blanchard JL et al (2014) Impacts of climate change on marine ecosystem production in societies dependent on fisheries. Nat Clim Chang 4(3):211–216. https://doi.org/10.1038/nclimate2119

Barbier ED (2017) Marine ecosystem services. Curr Biol 27(11):R507–R510. https://doi.org/10.1016/j.cub.2017.03.020

Barton A, Hales B, Waldbusser GG et al (2012) The pacific oyster, crassostrea gigas, shows negative correlation to naturally elevated carbon dioxide levels: implications for near-term ocean acidification effects. Limnol Oceanogr 57(3):698–710. https://doi.org/10.4319/lo.2012.57.3.0698

Badreddine A, Milazzo M, Abboud-Abi Saab M, Bitar G, Mangialajo L (2019b) Threatened biogenic formations of the Mediterranean: current status and assessment of the vermetid reefs along the Lebanese coastline (Levant basin). Ocean Coast Manage 169:1371–1146

Beal S, Borg J Calix M et al (2016) European red list of habitats Part 1. http://dx.publications.europa.eu/10.2779/032638. https://doi.org/10.1016/j.ocecoaman.2018.12.019

Bednaršek N, Naish KA, Feely RA, Hauri C, Kimoto K, Hermann AJ, Michel C, Niemi A, Pilcher D (2021) Integrated assessment of ocean acidification risks to Pteropods in the northern high latitudes: regional comparison of exposure, sensitivity and adaptive capacity. Front Marine Sci, p 1282

Berkowitz H (2020) Participatory governance for the development of the blue bioeconomy in the Mediterranean region. [Research Report] PANORAMED. https://hal.archives-ouvertes.fr/hal-02555685/

Billé R, Kelly R, Biastoch A et al (2013) Taking action against ocean acidification: a review of management and policy options. Environ Manag 52:761–779. https://doi.org/10.1007/s00267-013-0132-7

Bindoff NL, Cheung WWL, Kairo JG, Arístegui J, Guinder VA, Hallberg R, Hilmi N, Jiao N, Karim MS, Levin L, O'Donoghue S, Purca Cuicapusa SR, Rinkevich B, Suga T, Tagliabue A, Williamson P (2019) Changing Ocean, Marine Ecosystems, and Dependent Communities. In: Pörtner H-O, Roberts DC, Masson-Delmotte V, Zhai P, Tignor M, Poloczanska E, Mintenbeck K, Alegría A, Nicolai M, Okem A, Petzold J, Rama B, Weyer NM (eds) IPCC Special Report on the Ocean and Cryosphere in a Changing Climate

Bolzon G, Cossarini G, Lazzari P, et al (2020) Mediterranean Sea biogeochemical analysis and forecast (CMEMS MED-Biogeochemistry 2018-Present). Copernicus Monitoring Environment Marine Service (CMEMS). 10.25423/CMCC/ MEDSEA_ANALYSIS_FORECAST_BIO_006_014_MEDBFM3

Bramanti L, Movilla J, Guron M et al (2013) Detrimental effects of ocean acidification on the economically important Mediterranean red coral (*Corallium Rubrum*). Glob Chang Biol 19(6):1897–1908. https://doi.org/10.1111/gcb.12171

Bressan M, Chinellato A, Munari M et al (2014) Does seawater acidification affect survival, growth and shell integrity in bivalve juveniles? Mar Environ Res 99:136–148. https://doi. org/10.1016/j.marenvres.2014.04.009

Brunner RD, Steelman TA, Coe-Juell L et al (2005) Adaptive governance: integrating science, policy and decision making. Columbia University Press, New York

Buckley PJ, Pinnegar JK, Painting SJ et al (2017) Ten thousand voices on marine climate change in Europe: different perceptions among demographic groups and nationalities. Front Mar Sci 4:206. https://doi.org/10.3389/fmars.2017.00206

Cabuzel, T. (2019) EU climate action and the European Green Deal.. Climate Action – European Commission. https://ec.europa.eu/clima/policies/eu-climate-action_en

Cai W-J, Hu X, Huang W-J et al (2011) Acidification of subsurface coastal waters enhanced by eutrophication. Nat Geosci 4(11):766–770. https://doi.org/10.1038/ngeo1297

Cai W-J, Huang W-J, Luther GW III et al (2017) Redox reactions and weak buffering capacity lead to acidification in the Chesapeake Bay. Nat Commun 8(1):369. https://doi.org/10.1038/ s41467-017-00417-7

Capra F, Mattei U (2015) The ecology of law: toward a legal system in tune with nature and community. Berrett-Koehler Publishers, Oakland, CA. isbn:978-1-62656-207-3

Carstensen J, Conley DJ, Almroth-Rosell E et al (2020) Factors regulating the coastal nutrient filter in the Baltic Sea. Ambio 49(6):1194–1210. https://doi.org/10.1007/s13280-019-01282-y

Cerrano C, Cardini U, Bianchelli S et al (2013) Red coral extinction risk enhanced by ocean acidification. Sci Rep 3:1457. https://doi.org/10.1038/srep01457

Chaffin BC, Gosnell H, Cosens BA (2014) A decade of adaptive governance scholarship: synthesis and future directions. Ecol Soc 19(3):56. https://doi.org/10.5751/ES-06824-190356

Cicin-Sain B, Barbiere J, Cunha TP, et al (2019) Assessing progress on ocean and climate action 2019: a report of the Roadmap to Oceans and Climate Action (ROCA) Initiative, p 59. https:// roca-initiative.com/oceans-action-day-at-cop25/

Clementi E, Pistoia J, Escudier R et al (2019) Mediterranean Sea analysis and forecast (CMEMS MED-Currents, EAS5 system) [Data set]. Copernicus Monitoring Environment Marine Service (CMEMS)

Coll M, Piroddi C, Albouy C et al (2012) The Mediterranean Sea under siege: spatial overlap between marine biodiversity, cumulative threats and marine reserves. Glob Ecol Biogeogr 21(4):465–480

Cooley SR, Doney SC (2009) Anticipating ocean acidification's economic consequences for commercial fisheries. Environ Res Lett 4:1–8

Cooley SR, Jewett EB, Reichert J et al (2015) Getting ocean acidification on decision makers' to-do lists: dissecting the process through case studies. Oceanography 28:198–211

Corner RA, Aguilar-Manjarrez J, Massa F, Fezzardi D (2020) Multi-stakeholder perspectives on spatial planning processes for mariculture in the Mediterranean and Black Sea. Rev Aquac 12(1):347–364. https://doi.org/10.1111/raq.12321

Cosens B (2010) Transboundary river governance in the face of uncertainty: resilience theory and the Columbia River Treaty. J Land Resour Environ Law 30(2):229–265

Cosens BA (2013) Legitimacy, adaptation, and resilience in ecosystem management. Ecol Soc 18(1):3. https://doi.org/10.5751/ES-05093-180103

Cosens BA, Craig RK, Hirsch SL et al (2017) The role of law in adaptive governance. Ecol Soc 22(1):30. https://doi.org/10.5751/ES-08731-220130

Cossarini G, Lazzari P, Solidoro C (2015) Spatiotemporal variability of alkalinity in the Mediterranean Sea. Biogeosciences 12:1647–1658. https://doi.org/10.5194/bg-12-1647-2015

Costanza R, Andrade F, Antunes P et al (1998) Principles for sustainable governance of the Oceans. Science 281(5374):198–199. https://doi.org/10.1126/science.281.5374.198

Cramer W, Guiot J, Fader M et al (2018) Climate change and interconnected risks to sustainable development in the Mediterranean. Nat Clim Chang 8(11):972–980. https://doi.org/10.1038/s41558-018-0299-2

Crutzen PJ (2002) The "Anthropocene". J Phys IV (Proceedings) 12(10). EDP Sciences

Dietz T, Ostrom E, Stern PC (2003) The struggle to govern the commons. Science 302:1907–1912. https://doi.org/10.1126/science.1091015

Doney SC, Fabry VJ, Feely RA, Kleypas JA (2009) Ocean acidification: the other CO_2 problem. Annu Rev Mar Sci 1:169–192. https://doi.org/10.1146/annurev.marine.010908.163834

Doney SC (2010) The growing human footprint on coastal and open-ocean biogeochemistry. Science 328(5985):1512–1516. https://doi.org/10.1126/science.1185198

Duarte CM, Agusti S, Barbier E et al (2020) Rebuilding marine life. Nature 580:39–51. https://doi.org/10.1038/s41586-020-2146-7

EMODnet Human Activities (2020) Available: https://www.emodnet-humanactivities.eu/view-data.php

FAO (2018) The state of Mediterranean and Black Sea fisheries. General Fisheries Commission for the Mediterranean, Rome, p 172

Feely RA, Doney SC, Cooley SR (2009) Present conditions and future changes in a high -CO_2 world. Oceanography 22(4):36–47. https://doi.org/10.5670/oceanog.2009.95

Fletcher R, Breitling J (2012) Market mechanism or subsidy in disguise? Governing payment for environmental services in Costa Rica. Geoforum 43(3):402–411

Folke C (2006) Resilience: the emergence of a perspective for social-ecological systems analyses. Glob Environ Chang 16:253–267. https://doi.org/10.1016/j.gloenvcha.2006.04.002

Folke C, Hahn T, Olsson P, Norberg J (2005) Adaptive governance of social-ecological systems. Annu Rev Environ Resour 30:441–473. https://doi.org/10.1146/annurev.energy.30.050504.144511

Franke A, Clemmesen C (2011) Effect of ocean acidification on early life stages of Atlantic herring (*Clupea harengus* L.). Biogeosciences 8:3697–3707

Galaz V, Crona B, Österblom H et al (2012) Polycentric systems and interacting planetary boundaries — emerging governance of climate change–ocean acidification–marine biodiversity. Ecol Econ 81:21–32. https://doi.org/10.1016/j.ecolecon.2011.11.012

Galdies C, Tiller R, Martinez Romera B (2021) Global Ocean governance and ocean acidification. In: Martinho, F. (Section Editor) Encyclopedia of the UN sustainable development goals. Life below water. Springer. https://doi.org/10.1007/978-3-319-71064-8

Galdies C, Bellerby R, Canu D et al (2020) European policies and legislation targeting ocean acidification in European waters - current state. Marine Policy 118:103947. https://doi.org/10.1016/j.marpol.2020.103947

Gallo N, Victor D, Levin L (2017) Ocean commitments under the Paris agreement. Nat Clim Chang 7:833–838. https://doi.org/10.1038/nclimate3422

Gattuso J-P, Magnan A, Billé R et al (2015) Contrasting futures for ocean and society from different anthropogenic CO_2 emissions scenarios. Science 349:6243. https://doi.org/10.1126/science.aac4722

Gattuso J-P, Magnan AK, Bopp L et al (2018) Ocean solutions to address climate change and its effects on marine ecosystems. Front Mar Sci 5:337. https://doi.org/10.3389/fmars.2018.00337

Gaylord B, Kroeker K, Sunday JM et al (2015) Ocean acidification through the lens of ecological theory. Ecology 96(1):3–15. https://doi.org/10.1890/14-0802.1

Gazeau F, Alliouane S, Bock C et al (2014) Impacts of ocean acidification and warming on Mediterranean mussel (*Mytilus galloprovincialis*). Front Marine Sci 1:1–12

Geri P, El Yacoubi S, Goyet C (2014) Forecast of sea surface acidification in the northwestern Mediterranean Sea. J Comput Environ Sci 2014:1–7. https://doi.org/10.1155/2014/201819

Giorgi F (2006) Climate change hot-spots. Geophys Res Lett 33(8):L08707. https://doi.org/10.1029/2006GL025734

Goyet C, Hassoun AER, Gemayel E, Touratier F, Abboud-Abu Saab M, Guglielmi V (2016) Thermodynamic Forecasts of the Mediterranean Sea Acidification. Mediterranean marine Science 17/2, 2016:508–518

Good P, Bamber J, Halladay J et al (2018) Recent progress in understanding climate thresholds: ice sheets, the Atlantic meridional overturning circulation, tropical forests and responses to ocean acidification. Prog Phys Geogr 42(1):24–60. https://doi.org/10.1177/0309133317751843

Gosnell H, Chaffin BC, Ruhl JB et al (2017) Transforming (perceived) rigidity in environmental law through adaptive governance: a case of Endangered Species Act implementation. Ecol Soc 22(4):42. https://doi.org/10.5751/ES-09887-220442

Guilloux BG (2020) Ocean and climate regime interactions. Ocean Yearbook 34:43–88. https://doi.org/10.1163/9789004426214_004

Gunderson L (1999) Resilience, flexibility, and adaptive management – antidotes for spurious certitude? Conserv Ecol 3(1):7. http://www.consecol.org/vol3/iss1/art7/

Harrould-Kolieb ER, Herr D (2012) Climate change and ocean acidification: synergies and opportunities within the UNFCCC. Clim Pol 12:378–389. https://doi.org/10.1080/1469306 2.2012.620788

Hassoun AER, Gemayel E, Krasakopoulou E et al (2015) Acidification of the Mediterranean Sea from anthropogenic carbon penetration. Deep-Sea Res I Oceanogr Res Pap 102:1–15. https://doi.org/10.1016/j.dsr.2015.04.005

Hassoun AER, Fakhri M, Abboud-Abi Saab M et al (2019) The carbonate system of the Eastern-most Mediterranean Sea, Levantine Sub-basin: Variations and drivers. Deep-Sea Res II Top Stud Oceanogr 164:54–73. https://www.sciencedirect.com/science/article/abs/pii/S0967064518301802

Hassoun AER, Bantelman A, Canu D, Comeau S, Galdies C, Gattuso J-P, Giani M, Grelaud M, Hendriks IE, Ibello V, Idrissi M, Krasakopoulou E, Shaltout N, Solidoro C, Swarzenski PW, Ziveri P (2022) Ocean acidification research in the Mediterranean Sea: status, trends and next steps. Front Mar Sci 9:892670. https://doi.org/10.3389/fmars.2022.892670

Herr D, Isensee K, Harrould-Kolieb E, Turley C (2014) Ocean acidification: international policy and governance options. IUCN, Gland

Hidalgo M, Mihneva V, Vasconcellos M, Bernal M (2018) Chapter 7: Climate change impacts, vulnerabilities and adaptations: Mediterranean Sea and the Black Sea marine fisheries. In: Barange M, Bahri T, Beveridge MCM et al (eds) Impacts of climate change on fisheries and aquaculture: synthesis of current knowledge, adaptation and mitigation options. FAO Fisheries and Aquaculture Technical Paper 627. http://www.fao.org/3/i9705en/i9705en.pdf

Hilmi N, Allemand D, Cinar M et al (2014) Exposure of Mediterranean countries to ocean acidification. Water 6:1719–1744

Hilmi N, Allemand D, Kavanagh C et al (eds) (2015) Bridging the gap between ocean acidification impacts and economic valuation: regional impacts of ocean acidification on fisheries and aquaculture. International Union for Conservation of Nature. https://doi.org/10.2305/IUCN.CH.2015.03.en

Hughes TP, Carpenter S, Rockström J et al (2013) Multiscale regime shifts and planetary boundaries. Trends Ecol Evol 28:389–395

Huitema D, Mostert E, Egas W et al (2009) Adaptive water governance: assessing the institutional prescriptions of adaptive (co-) management from a governance perspective and defining a research agenda. Ecol Soc 14(1):26. http://www.ecologyandsociety.org/vol14/iss1/art26/

Ingrosso G, Abbiati M, Badalamenti F et al (2018) Mediterranean bioconstructions along the Italian Coast. In: Advances in marine biology. Academic Press, pp 61–136

Ingrosso G, Bensi M, Cardin V, Giani M (2017) Anthropogenic CO_2 in a dense water formation area of the Mediterranean Sea. Deep-Sea Res I Oceanogr Res Pap 123:118–128

IPCC (2019) IPCC Special Report on the Ocean and Cryosphere in a Changing Climate [H.-O. Pörtner, D.C. Roberts, V. Masson-Delmotte, P. Zhai, M. Tignor, E. Poloczanska, K. Mintenbeck, A. Alegría, M. Nicolai, A. Okem, J. Petzold, B. Rama, N.M. Weyer (eds)

Jagers SC, Matti S, Crépin A et al (2019) Societal causes of, and responses to, ocean acidification. Ambio 48:816–830. https://doi.org/10.1007/s13280-018-1103-2

Jiang L-Q, Carter BR, Feely RA et al (2019) Surface ocean PH and buffer capacity: Past, present and future. Sci Rep 9(1):1–11. https://doi.org/10.1038/s41598-019-55039-4

Kapsenberg L, Alliouane S, Gazeau F et al (2017) Coastal ocean acidification and increasing total alkalinity in the Northwestern Mediterranean Sea. Ocean Sci 13(3):411–426. https://doi.org/10.5194/os-13-411-2017

Kelly RP, Foley MM, Fisher WS et al (2011) Mitigating local causes of ocean acidification with existing laws. Science 332(6033):1036–1037

Kim RE (2012) Is a new multilateral environmental agreement on ocean acidification necessary? Review of European Community and International Environmental Law 21:243–258

Lacoue-Labarthe T, Nunes PALD, Ziveri P et al (2016) Impacts of ocean acidification in a warming Mediterranean Sea: an overview. Reg Stud Mar Sci 5:1–11. https://doi.org/10.1016/j.rsma.2015.12.005

Lazzari P, Mattia G, Solidoro C et al (2014) The impacts of climate change and environmental management policies on the trophic regimes in the Mediterranean Sea: scenario analyses. J Mar Syst 135:137–149. https://doi.org/10.1016/j.jmarsys.2013.06.005

Le Quéré C, Andrew RM, Friedlingstein P et al (2018) Global carbon budget 2018. Preprint. Earth System Science Data. https://doi.org/10.5194/essd-10-405-2018

Lee KN (1999) Appraising adaptive management. Conserv Ecol 3(2):3. http://www.ecologyandsociety/vol3/iss2/art3/

Lemasson AJ, Fletcher S, Hall-Spencer JM, Knights AM (2017) Linking the biological impacts of ocean acidification on oysters to changes in ecosystem services: a review. J Exp Mar Biol Ecol 492:49–62

Lewis E, Wallace DWR (1998) Program developed for CO2 systems calculations, ORNL/CDIAC-105, Carbon Dioxide Inf. Anal. Cent., Oak Ridge Natl. Lab., Oak Ridge, Tenn. Available at ftp://cdiac.ornl.gov/pub/co2sys

Licker R, Ekwurzel B, Doney SC, Cooley, et al. (2019) Attributing ocean acidification to major carbon producers. Environ Res Lett 14(12):124060. https://doi.org/10.1088/1748-9326/ab5abc

Lombardi C, Rodolfo-Metalpa R, Cocito S et al (2011) Structural and geochemical alterations in the Mg calcite Bryozoan *Myriapora truncata* under elevated seawater pCO_2 simulating ocean acidification. Mar Ecol 32(2):211–221. https://doi.org/10.1111/j.1439-0485.2010.00426.x

Long J (2009) From warranted to valuable belief: local government, climate change, and giving up the pickup to save Bangladesh. Nat Resour J 49(3/4):743–800

Luchetta A, Cantoni C, Catalano G (2010) New observations of CO_2-induced acidification in the Northern Adriatic Sea over the last quarter century. Chem Ecol 26:1–17. https://doi.org/10.1080/02757541003627688

Martin S, Gattuso J-P (2009) Response of Mediterranean coralline algae to ocean acidification and elevated temperature. Glob Chang Biol 15:2089–2100. https://doi.org/10.1111/j.1365-2486.2009.01874.x

Martin S, Rodolfo-Metalpa R, Ransome E et al (2008) Effects of naturally acidified seawater on seagrass calcareous epibionts. Biol Lett 4(6):689–692. https://doi.org/10.1098/rsbl.2008.0412

Martinez M, Mangano MC, Maricchiolo G et al (2018) Measuring the effects of temperature rise on Mediterranean shellfish aquaculture. Ecol Indic 88:71–78. https://doi.org/10.1016/j.ecolind.2018.01.002

Massa F, Onofri L, Fezzardi D (2017) Aquaculture in the Mediterranean and the Black Sea: a Blue Growth perspective. In: Nunes PALD, Svensson LE, Markandya A (eds) Handbook on the economics and management of sustainable oceans. Edward Elgar Publishing, Cheltenham, pp 93–123

Melaku Canu D, Ghermandi A, Nunes PALD et al (2015) Estimating the value of carbon sequestration ecosystem services in the Mediterranean Sea: an ecological economics approach. Glob Environ Chang 32:87–95. https://doi.org/10.1016/j.gloenvcha.2015.02.008

Michaelidis B, Ouzounis C, Paleras A, Pörtner HO (2005) Effects of long-term moderate hypercapnia on acid–base balance and growth rate in marine mussels *Mytilus galloprovincialis*. Mar Ecol Prog Ser 293:109–118

Michaelidis B, Spring A, Pörtner HO (2007) Effects of long-term acclimation to environmental hypercapnia on extracellular acid–base status and metabolic capacity in Mediterranean fish *Sparus aurata*. Mar Biol 150(6):1417–1429. https://doi.org/10.1007/s00227-006-0436-8

Milazzo M, Rodolfo-Metalpa R, Bin San Chan V et al (2014) Ocean acidification impairs vermetid reef recruitment. Sci Rep 4(1):1–7. https://doi.org/10.1038/srep04189

Milazzo M, Alessi C, Quattrocchi F et al (2019) Biogenic habitat shifts under long-term ocean acidification show nonlinear community responses and unbalanced functions of associated invertebrates. Sci Total Environ 667:41–48. https://doi.org/10.1016/j.scitotenv.2019.02.391

Millennium Ecosystem Assessment (2005) Ecosystems and human well-being. Island Press, Washington, D.C.

Olsson P, Folke C, Galaz V et al (2007) Enhancing the fit through adaptive co-management: creating and maintaining bridging functions for matching scales in the Kristianstads Vattenrike Biosphere Reserve, Sweden. Ecol Soc 12(1):28. http://www.ecologyandsociety.org/vol12/iss1/art28/

Osborn D, Dupont S, Hansson L, Metian M (2017) Ocean acidification: impacts and governance. In: Nunes PALD, Svensson LE, Markandya A (eds) Handbook on the economics and management of sustainable oceans. Edward Elgar Publishing and UNEP, Cheltenham, pp 396–415

Patt AG (2010) Effective regional energy governance—not global environmental governance—is what we need right now for climate change. Glob Environ Chang 20:33–35. https://doi.org/10.1016/j.gloenvcha.2009.09.006

Piante C, Ody D (2015) Blue growth in the Mediterranean Sea: the challenge of good environmental status. WWF report 2015. MedTrends Project, France, 189 pp

Pinardi N, Cessi P, Borile F, Wolfe CLP (2019) The Mediterranean sea overturning circulation. J Phys Oceanogr 49(7):1699–1721. https://doi.org/10.1175/JPO-D-18-0254.1

Ragazzola F, Marchini A, Adani, et al. (2021) An intertidal life: combined effects of acidification and winter heatwaves on a coralline alga (Ellisolandia elongata) and its associated invertebrate community. Mar Environ Res 169:105342. https://doi.org/10.1016/j.marenvres.2021.105342

Range P, Piló D, Ben-Hamadou R et al (2012) Seawater acidification by CO_2 in a coastal lagoon environment: effects on life history traits of juvenile mussels *Mytilus galloprovincialis*. J Exp Mar Biol Ecol 424–425:89–98. https://doi.org/10.1016/j.jembe.2012.05.010

Range P, Chícharo MA, Ben-Hamadou R et al (2014) Impacts of CO_2-induced seawater acidification on coastal Mediterranean bivalves and interactions with other climatic stressors. Reg Environ Change 14:19–30

Rayfuse R (2012) Climate change and the law of the sea. In: Rayfuse R, Scott SV (eds) International law in the era of climate change. Edward Elgar Publishing, Cheltenham, pp 159–161. ISBN: 978 1 84980 030 3

Reinke de Buitrago S, Schneider P (2020) Ocean governance and hybridity: dynamics in the Arctic, the Indian Ocean, and the Mediterranean Sea. Global Governance: A Review of Multilateralism and International Organizations 26(1):154–175. https://doi.org/10.1163/19426720-02601004

Réveillac E, Lacoue-Labarthe T, Oberhänsli F et al (2015) Ocean acidification reshapes the otolith-body allometry of growth in juvenile sea bream. J Exp Mar Biol Ecol 463:87–94. https://doi.org/10.1016/j.jembe.2014.11.007

Richon C, Dutay J-C, Bopp L et al (2019) Biogeochemical response of the Mediterranean Sea to the transient SRES-A2 climate change scenario. Biogeosciences 16:135–165. https://doi.org/10.5194/bg-16-135-2019

Rockström J, Steffen W, Noone K et al (2009) Planetary boundaries: exploring the safe operating space for humanity. Ecol Soc 14(2):32. https://doi.org/10.5751/ES-03180-140232

Rodrigues LC, Van Den Bergh JCJM, Massa F et al (2015) Sensitivity of Mediterranean bivalve mollusc aquaculture to climate change, ocean acidification, and other environmental pressures: findings from a producer survey. J Shellfish Res 34(3):1161–1176. https://doi.org/10.2983/035.034.0341

Rosa R, Marques A, Nunes ML (2012) Impact of climate change in Mediterranean aquaculture. Rev Aquac 4:163–177. https://doi.org/10.1111/j.1753-5131.2012.01071.x

Rosenau J (2000) The governance of fragmegration: Neither a world republic nor a global system, prepared for presentation at the Congress of the International Political Science Association, Quebec City, 1–5 August 2000. http://aura.u-pec.fr/regimen/_fich/_pdf/pub_002.pdf

Ruhl JB (1997) Thinking of environmental law as a complex adaptive system: how to clean up the environment by making a mess of environmental law. Houston Law Review 34(4):933. https://scholarship.law.vanderbilt.edu/faculty-publications/526

Sabine CL (2018) Good news and bad news of Blue Carbon. Proc Natl Acad Sci U S A 115(15):3745–3746. https://doi.org/10.1073/pnas.1803546115

Sabine CL, Feely RA, Gruber N et al (2004) The oceanic sink for anthropogenic CO_2. Science 305(5682):367–371. https://doi.org/10.1126/science.1097403

Said A, Tzanopoulos J, MacMillan D (2018) The contested commons: the failure of EU fisheries policy and governance in the Mediterranean and the crisis enveloping the small-scale fisheries of Malta. Front Mar Sci 5:300. https://doi.org/10.3389/fmars.2018.00300

Selin H, VanDeveer SD (eds) (2009) Changing Climates in North American Politics: institutions, policymaking, and multilevel governance, 1st edn. Cambridge, MA, The MIT Press

Selin H, VanDeveer SD (2011) US climate change politics and policymaking. WIREs Climate Change 2(1):121–127. https://doi.org/10.1002/wcc.94

Shaltout M, Omstedt A (2014) Recent sea surface temperature trends and future scenarios for the Mediterranean Sea. Oceanologia 56(3):411–443. https://doi.org/10.5697/oc.56-3.411

Sharma-Wallace L, Velarde S, Wreford A (2018) Adaptive governance good practice: show me the evidence! J Environ Manag 222:174–184. https://doi.org/10.1016/j.jenvman.2018.05.067

Simoncelli S, Pinardi N, Fratianni C et al (2018) Water mass formation processes in the Mediterranean Sea over the past 30 years. In: Copernicus marine service ocean state report, issue 2. J Oper Oceanogr 11(S1):s13–s16

Solidoro C, Cossarini G, Lazzari P, Galli G, Bolzon G, Somot S, Salon S (2022) Modeling carbon budgets and acidification in the mediterranean sea ecosystem under contemporary and future climate. Front Mar Sci 8. https://doi.org/10.3389/fmars.2021.781522

Steinacher M, Joos F, Stocker TF (2013) Allowable carbon emissions lowered by multiple climate targets. Nature 499:197–201. https://doi.org/10.1038/nature12269

Stephens T (2015) Ocean acidification. Chapter 26. In: Rayfuse R (ed) Research handbook on international marine environmental law. Edward Elgar Publishing, Cheltenham: 978 1 78811 057 0

Strong AL, Kroeker KJ, Teneva LT et al (2014) Ocean acidification 2.0: managing our changing coastal ocean chemistry. Bioscience 64(7):581–592. https://doi.org/10.1093/biosci/biu072

Sunday JM, Fabricius K, Kroeker K et al (2016) Ocean acidification can mediate biodiversity shifts by changing biogenic habitat. Nat Clim Chang 7:81–85. https://doi.org/10.1038/nclimate3161

Termeer CJAM, Dewulf A, van Lieshout M (2010) Disentangling scale approaches in governance research: comparing monocentric, multilevel, and adaptive governance. Ecol Soc 15(4):29. http://www.ecologyandsociety.org/vol15/iss4/art29/

Tiller R, Arenas F, Galdies C et al (2019) Who cares about ocean acidification in the Plasticene? Ocean Coast Manag 174:170–180

Tuel A, Eltahir EAB (2020) Why is the Mediterranean a climate change hotspot? J Clim 33(14):5829–5843. https://doi.org/10.1175/JCLI-D-19-0910.1

Turley C, Eby M, Ridgwell AJ et al (2010) The societal challenge of ocean acidification. Mar Pollut Bull 60(6):787–792

Walker B, Holling CS, Carpenter SR, Kinzig A (2004) Resilience, adaptability and transformability in social-ecological systems. Ecol Soc 9(2):5. http://www.ecologyandsociety.org/vol9/iss2/art5

Walters CJ (1986) Adaptive management of renewable resources. Macmillan, New York

Walters CJ (1997) Challenges in adaptive management of riparian and coastal ecosystems. Conserv Ecol 1(2):1. http://www.consecol.org/vol1/iss2/art1/

Wimart-Rousseau C, Wagener T, Álvarez et al (2021) Seasonal and interannual variability of the CO2 system in the eastern Mediterranean Sea: a case study in the North Western Levantine Basin. Front Mar Sci 8:649246. https://doi.org/10.3389/fmars.2021.649246

Wolfrum R, Matz-Lück N (2003) Conflicts in international environmental law. Springer, Berlin, p 213

Wyborn C (2015) Co-productive governance: a relational framework for adaptive governance. Glob Environ Chang 30:56–67. https://doi.org/10.1016/j.gloenvcha.2014.10.009

Ziveri P, Delpiazzo E, Bosello F et al (2017) Adaptation policies and strategies as a response to ocean acidification and warming in the Mediterranean Sea. Chapter 16. In: Nunes PALD, Svensson LE, Markandya A (eds) Handbook on the economics and management for sustainable oceans. Edward Elgar Publishing, Cheltenham

Zunino S, Canu DM, Bandelj V, Solidoro C (2017) Effects of ocean acidification on benthic organisms in the Mediterranean Sea under realistic climatic scenarios: a meta-analysis. Reg Stud Mar Sci 10:86–96. https://doi.org/10.1016/j.rsma.2016.12.011

Zunino S, Canu DM, Zupo V, Solidoro C (2019) Direct and indirect impacts of marine acidification on the ecosystem services provided by coralligenous reefs and seagrass systems. Glob Ecol Conserv 18:e00625. https://doi.org/10.1016/j.gecco.2019.e00625

Zunino S, Libralato S, Canu D, Prato G, Solidoro C (2021) Impact of ocean acidification on ecosystem functioning and services in habitat-forming species and marine ecosystems. Ecosystems. https://doi.org/10.1007/s10021-021-00601-3

Correction to: Assembling the Seabed: Pan-European and Interdisciplinary Advances in Understanding Seabed Mining

Wenting Chen, Kimberley Peters, Diva Amon, Maria Baker, John Childs, Marta Conde, Sabine Gollner, Kristin Magnussen, Aletta Mondre, Ståle Navrud, Pradeep A. Singh, Philip Steinberg, and Klaas Willaert

Correction to:
Chapter 12 in: S. Partelow et al. (eds.),
Ocean Governance, **MARE Publication Series 25,**
https://doi.org/10.1007/978-3-031-20740-2_12

The last name of the author Marta Conde was incorrectly published as Conte and without her affiliation. This has been corrected. The correct name and affiliation are as follows:

Marta Conde
Centre for Social Responsibility in Mining (CSRM), University of Queensland, Brisbane, Australia
ICTA, Autonomous University of Barcelona, Barcelona, Spain
mcondep@gmail.com

The affiliation of the author Maria Baker was incorrectly published including "John Childs-Lancaster Environment Centre, Lancaster University, Lancaster, UK". This has been corrected. The correct affiliation is as follows:

Maria Baker
Ocean and Earth Sciences, University of Southampton, Southampton, UK

The affiliation of the author John Childs was published incorrectly as "JHU-UPF Public Policy Centre, Universitat Pompeu Fabra (UPF), Barcelona, Spain". The correct affiliation is as follows:

J. Childs
Lancaster Environment Centre, Lancaster University, Lancaster, UK

The updated original version for this chapter can be found at
https://doi.org/10.1007/978-3-031-20740-2_12

Afterword

How to Engage Going Forward: Focal Governance Arenas for Students, Researchers and Policy-Makers of Ocean Governance

Stefan Partelow, Maria Hadjimichael, and Anna-Katharina Hornidge

The need for engaged individuals to continue developing expertise on ocean governance topics – whether in science, policy, industry or public policy – will only become more essential in the coming years and decades. As we begin the United Nations Decade of Ocean Science for Sustainable Development (2021–2030) and observe the rapid expansion of blue growth and economy strategies, we encourage readers to approach engagement with our oceans with a sense of responsibility but also optimism. For millennia human societies have wondered what is beyond the horizon and under the surface. Standing on the shore has shaped narratives of bewilderment and opportunity alike, providing both solice and hope while accepting that over 71% of the earth's surface was out of our hands and unknown. However, today, the ability to actively govern our oceans, know it in different ways and shape ocean futures is well within our means, and perhaps a moral imperative for action, to help steer away from harmful exploitation and injustice, towards conscious stewardship that co-supports environmental integrity and the advancement of human well-being.

Many professional careers and civil society activities contribute to shaping the ocean governance landscape of people, places, processes and institutions. Concerted efforts are needed to engage across the spectrum from individual behaviour change to systemic issues, and the countless diverse communities and activities in between who depend on healthy ocean environments and economies. The oceans are, arguably, our collective responsibility, and we encourage interested students, researchers, policy-makers and other professionals to continue active engagement. As many trends suggest, new and innovative ways to become involved or continue engaging will emerge as the ocean economy, knowledge base and public interest continue to expand.

This book has presented a select collection of analyses on important ocean governance issues with the aim to convey two key messages. First, that our oceans and their sustainability challenges are pluralistic, both in how problems manifest and influence society and the environment, but also in the types of knowledge and action needed to find constructive governance pathways forward. Second, to provide a

© German Institute of Development and Sustainability (IDOS) 2023

S. Partelow et al. (eds.), *Ocean Governance*, MARE Publication Series 25,
https://doi.org/10.1007/978-3-031-20740-2

series of articles on specific topics that lead to further inquiry and interest for those pursuing professional development in the relevant areas and beyond. Nonetheless, synthesizing the main points, from our perspective as editors, may provide further condensed insight into often complex topics.

In this Afterword, we synthesize the focal arenas that students, researchers and policy-makers at all levels can reflect on to guide their continued work on ocean governance. Focal arenas are detailed, but generally applicable areas of interest, which can be topics, themes, methods or contextual aspects linked to ocean governance. The focal arenas are presented to foster guidance and coordination across actors and activities, perhaps across those with differing agendas, acting as anchors for finding common ground and joint paths forward. The focal arenas are by no means definitive or gathered through rigorous methodology, but rather reflective, malleable and intended to foster continued discussions arising from this book and related topics. They are in many ways the take-away messages for each of the book's target audience groups.

Cross-Cutting Focal Arenas

Many of the ways to engage with ocean governance are applicable across sectors, contexts and places. While the reflections made below are reflective of the editors of this volume, they can be reflected on as generally applicable focal of ocean governance relevant across topics, perspectives or professional capacity. We refer to these focal arenas as cross-cutting, which regardless of professional focus or interest, can be used to guide continued action and engagement.

• Knowledge co-production through multi-actor and cross-sector partnerships
• Value and include diverse knowledge systems
• Innovative partnership models and cooperation strategies
• Support transparency, justice and equality while pursuing best practices
• Invest in knowledge, capacity and technology development and transfer programs
• Prioritize goal setting through inclusive and deliberative participation
• Acknowledge trade-offs and find solutions through deliberation
• Support transparency, communication and traceability
• Proactive climate mitigation and adaptation strategies
• Support environmental stewardship

Focal Arenas for Researchers

For researchers, focus will centre on collaborative knowledge practices and integration of knowledge systems that can foster pluralistic understandings, co-creation and co-learning. Building a toolbox with different methods and concepts, and competency to understand the tools of others and use those tools, will be essential. Ocean governance is also in need to context specific theories that do not rely heavily on the empirical evidence of historical terrestrial based research, but rather new empirical insights from ocean governance contexts. Transparency is a key theme for researchers, both within and between science communities, and how science is communicated beyond.

- Pursue inter- and trans-disciplinarity knowledge co-production
- Problem-driven and solution-oriented research in tangent with fundamental research
- Examine the role and use of diverse knowledge systems
- Examine opportunities, trade-offs and challenges of governance approaches
- Develop sector specific theories, frameworks and analytical tools
- Critically examine blue growth and blue economy strategies
- Expand topical and geographic diversity of research

Focal Arenas for Students

For students, this would include important areas of inquiry, growing sectors for employment and forthcoming sustainability challenges in need of young talent and innovation across governance, technology and economy in the twenty-first century.

- Be curious and open-minded about the problems and solutions ahead
- Think critically about the past, present and future of ocean governance
- Explore diverse career paths and opportunities to engage
- Build skill sets for engaging with the future ocean economy
- Develop capacities for cooperation and boundary spanning

Focal Arenas for Policy-Makers

Policy makers face substantial challenges, nonetheless, there are opportunities. This book suggests that seeking diverse and locally relevant perspectives on policy issues during development and implementation will be essential in making higher level policies effective in local arenas. Considering multi-level integration and connectivity with other policy-adjacent and policy-involved organizations can help foster transparency and help avoid the funnelling of singularly focused perspectives into practice and mitigate power imbalances in policy processes.

- Ensure blue growth and blue economy policies account for risks, justice and equity
- Enable cross-sectoral cooperation in policy development, implementation and administration

- Ensure policies and financing align with SDG 14
- Coordinate vertical and horizontal integration of governance
- Enable transparent and participatory governance processes at multiple levels
- Create enabling conditions for innovations and sustainability initiatives
- Enable fair access to financing, loans and subsidy programs
- Support proactive climate mitigation and adaptation strategies
- Invest in research, knowledge and technology transfer programs
- Support transparent monitoring, data collection and evaluation strategies
- Prioritize inclusion and participation of smallholders in governance
- Ensure environmental protections against pollution, degradation and overuse
- Ensure equitable and just social protections

In closing, the future of the oceans dictates the necessity of an ocean governance framework which encompasses our ability to understand the ocean and engage with other actors, but also increase support to and fulfil already made commitments for this framework to ensure ocean sustainability, inclusiveness and justice. We encourage continued engagement, critique and reflection within and beyond the contents of this book in order to jointly continue shaping ocean governance in the years and decades to come.